West & Wood's Introduction to Foodservice

West & Wood's Introduction to Foodservice

EIGHTH EDITION

June Payne-Palacio
Pepperdine University

Monica Theis
University of Wisconsin–Madison

Merrill,
an imprint of Prentice Hall
Upper Saddle River, New Jersey   Columbus, Ohio

Library of Congress Cataloging-in-Publication Data
West and Wood's introduction to foodservice / [edited by] June Payne-Palacio, Monica Theis.—8th ed.

p. cm.

Rev. ed. of: West's and Wood's introduction to foodservice. 8th ed. 1994.

Includes bibliographical references and index.

ISBN 0-13-495425-4

1. Food service management. I. Payne-Palacio, June. II. Theis, Monica. III. West's and Wood's introduction to foodservice.

TX911.3.M27W44 1997

647.95'068—dc20 96-27613

CIP

Editor: Kevin M. Davis
Production Editor: Mary Harlan
Design Coordinator: Karrie M. Converse
Text Designer: Ed Horcharik
Cover Designer: Brian Deep
Cover photo: © Bill Bachmann/Picturesque
Production Manager: Pamela D. Bennett
Director of Marketing: Kevin Flanagan
Advertising/Marketing Coordinator: Julie Shough
Electronic Text Management: Marilyn Wilson Phelps, Matthew Williams, Karen L. Bretz, Tracey Ward
Illustrations: Barry Bell Graphics

This book was set in Garamond by Prentice Hall and was printed and bound by Quebecor Printing/Book Press. The cover was printed by Phoenix Color Corp.

Printed in the United States of America

10 9 8 7 6 5 4 3 2 1

ISBN: 0-13-495425-4

Prentice-Hall International (UK) Limited, *London*
Prentice-Hall of Australia Pty. Limited, *Sydney*
Prentice-Hall of Canada, Inc., *Toronto*
Prentice-Hall Hispanoamericana, S. A., *Mexico*
Prentice-Hall of India Private Limited, *New Delhi*
Prentice-Hall of Japan, Inc., *Tokyo*
Simon & Schuster Asia Pte. Ltd., *Singapore*
Editora Prentice-Hall do Brasil, Ltda., *Rio de Janeiro*

Preface

Since the first edition of *West and Wood's Introduction to Foodservice* (then titled *Foodservice in Institutions*) was published in 1938, the authors have been committed to presenting the basic principles of foodservice management, which can be applied to *all* types of foodservice organizations. This new eighth edition is no exception in giving comprehensive coverage of all aspects of foodservice management.

Earlier editions, however, reflected the distinct difference that existed between commercial, or profit-seeking, organizations and noncommercial, or institutional, not-for-profit foodservice operations. Special emphasis was given to institutional foodservices: schools and colleges, hospitals and health care facilities, and in-plant or industrial foodservices.

In recent years, a philosophical change has taken place—first gradually, then dramatically—in the management of many not-for-profit institutional foodservices. With rising health care costs of recent years, for example, hospitals have become more financially competitive in order to succeed and remain in business. Today, most foodservices are striving for some margin of profit, making less of a distinction between the two types of foodservice. In response to these changes the title of this book was changed with the seventh edition to *Introduction to Foodservice*.

The new title better reflects the major thrust of the text and, it is hoped, gives a good indication of the purpose, scope, and level of the subject matter. While the focus is on basic principles, this edition also reflects the impact of current social, economic, technological and political factors on foodservice operations. Examples and illustrations reflect both noncommercial and commercial applications.

New to This Edition

- Chapter 3, "Food Safety," is a new chapter on what we believe to be one of the most important issues facing foodservice today. The scope of foodborne disease and its relationship to microbiology are discussed. The chapter emphasizes the responsibility of the food manager in assuring safe food and offers specific, practical guidelines on how to design a facility-specific HACCP system.
- Chapter 9, "Facilities Management," is a new chapter that addresses energy and water conservation and provides guidance on solid waste management.

- Chapter 12, "Designing and Managing the Organization," has been revised to include current theory and practical applications of quality management including Total Quality Management and Performance Improvement.
- An Instructor's Manual, of particular interest to professors and teachers, has been newly developed and designed to assist the instructor in lesson planning and student evaluation.

Throughout this edition the material has been updated and revised to reflect current trends and practices. For example, branding and the branded concept as a marketing strategy are thoroughly discussed in Chapter 17. Many new photographs and illustrations are included to help visually interpret the subject matter. In addition, some chapter titles now incorporate new terminology to better reflect the subject matter.

Features

The pedagogical features included in this edition will help students, instructors, and other users maximize the value of this text. Of particular interest are the following:

- Each chapter concludes with a Summary and References, and many have a special section, "For More Information," that helps the reader locate additional sources of information on a particular topic.
- Review Questions at the end of each chapter help pinpoint the important concepts of the subject matter and serve as a study review and test for the reader, ensuring that the more important information is learned.
- Appendixes on Cooking Methods and on Foodservice Equipment serve as ready references for the reader who seeks additional, detailed information. Teachers of foodservice management courses should find these additions helpful in guiding students' learning.
- A Glossary at the end of the book defines and more clearly explains some of the terms unique to our field of study. In the text, these terms are shown in boldface type.

It is expected that users of this text will also supplement their reading with current journals, trade magazines, and research reports, as well as attend seminars and exhibits at conventions and trade shows to keep themselves up-to-date.

Organization of This Edition

Although it is likely no two teachers would organize the subject matter of this course in exactly the same way, we believe that the information in this text is presented in a logical sequence. First, the technical operation of a foodservice is discussed so that students will understand what is to be managed, and then the presentation of management techniques follows.

Introduction to Foodservice is divided into four major parts. Part 1, "Foodservice Organizations," provides an overview. Chapter 1 gives a chronological review of the history of foodservice organizations, while Chapter 2 describes types of current foodservice operations.

Part 2, "Quantity Food Production and Service," includes Chapters 3 through 7. This part begins with a new chapter on food safety and is followed by a function-by-function description of a foodservice operation. These functions include menu planning, purchasing, production, and service. Each chapter includes factors that influence the management of that operational function.

Part 3, "Physical Facilities," is a four-chapter unit that focuses on the maintenance and design of the operational facilities. The unit begins with a chapter on sanitation and cleaning. Guidance is also offered on how to plan, design, and maintain a foodservice operation.

Part 4, "Organization and Administration," provides the reader with the basic knowledge to manage the operational functions of a foodservice. Chapter 12 covers the design and management of organizations. It is followed by a comprehensive chapter on human resource management. The unit concludes with chapters on professional qualities such as administrative leadership and skills including work improvement, financial management, and marketing.

Courses for Which This Text Is Suitable

Although different schools and universities may use this material in a sequence different from that presented here, the subject matter itself is appropriate for courses that include the following (with these or similar titles):

- Introduction to Foodservice Management
- Quantity Food Production
- Purchasing for Foodservices (both food and equipment)
- Organization and Management of Foodservices
- Facility Design and Equipment Arrangement
- Financial Management of Foodservices
- Food Protection and Safety
- Menu Planning for Foodservices
- Foodservice Marketing and Merchandising

It is our hope that this newly revised edition of a classic text continues to meet the needs, as it has in the past, of the present generation of students who are preparing to become administrative dietetics professionals or foodservice managers. We hope too that teachers will find *Introduction to Foodservice*, Eighth Edition, a helpful guide and that foodservice managers in practice will use it as a ready-reference in their work.

ACKNOWLEDGMENTS

Many people have assisted with the preparation of *Introduction to Foodservice*. Without their help, our task would have been impossible.

The current authors are deeply appreciative of the excellent work of the original authors, Bessie Brooks West and LeVelle Wood, in providing a text that has been so widely accepted in the United States and abroad for almost 60 years. The text has been recognized for its authenticity and accuracy, a standard that we have strived to maintain in the new edition. Mrs. West assisted with revisions through the fifth edition, before she passed away in 1984 at the age of 93. Miss Wood was active in all revisions through the sixth edition. Now 93, she lives in a retirement home in Portland, Oregon. Grace Shugart and Virginia Harger retired as co-authors following publication of the seventh edition. Ms. Shugart passed away in 1995. Ms. Harger is enjoying an active retirement in Oregon. Without the pioneering work and writing of these women, this revision would not be a reality.

We are grateful to the peer reviewers, who challenged our thinking and made excellent suggestions for changes or additions to the first drafts of the manuscript. Their comments were honest and open, and many of their ideas have been incorporated into the text. We believe that their input has made the text even more meaningful to our readers. We thank these reviewers for their contributions:

- Evelina W. Cross, Louisiana State University
- Judy Kay Flohr, University of Massachusetts–Amherst
- Sandy Kapoor, California State Polytechnic University
- Jerrold K. Leong, Oklahoma State University
- Elizabeth M. Lieux, University of Delaware
- Louise M. Mullan, University of Minnesota
- Amy Peterson, Florida State University
- Dorothy Pond-Smith, Washington State University

The following individuals helped in special ways in the preparation of this edition. To each of them we are greatly indebted and give our thanks.

Dr. Steve Ingham, Professor, Department of Food Science, University of Wisconsin–Madison, for reviewing and contributing to the new chapter on food safety.

Julie Vincent, Director of Foodservice, Wisconsin Memorial Union, for reviewing the chapter on purchasing.

Secretarial staff of the Department of Food Science, University of Wisconsin–Madison, for their work on the manuscript.

We would like to express special thanks to our editorial staff at Prentice Hall—in particular Kevin Davis, Senior Editor, for his understanding of our subject matter and help in putting it in perspective, for his patience in working with us, and for giving encouragement for the completion of this revision. We are also grateful to Mary Harlan, Production Editor; Lynn Metzger, Editorial Assistant; and Lorretta Palagi, copy editor.

Finally, we wish to acknowledge the support and encouragement of our families—in particular, Monica's husband, Craig Schiestl, and daughter, Emma, and June's husband, Moki Palacio—as well as special friends who have endured the countless hours we have devoted to this work. We appreciate the committed efforts of all of these people.

June Payne-Palacio
Monica Theis

Brief Contents

Contents

1

Foodservice Organizations

1

The Foodservice Industry

OUTLINE

Early History of Foodservice Organizations
- Religious Orders
- Royal and Noble Households

Development of Present-Day Foodservices
- Restaurants
- Colleges and Universities
- School Foodservice
- Clubs and Other Social Organizations
- Hospitals
- Nursing Homes and Other Health Care Centers
- Retirement Residences and Adult Communities
- Industrial and Business Foodservice
- Transportation Companies

Trends in Foodservice

Summary

Perhaps no other industry is as pervasive and touches the lives of so many Americans on a daily basis as the foodservice industry. Those employed in the industry—from farmers, processors, manufacturers, distributors, suppliers, and truckers to those who work in office, plant, and school cafeterias, hotels, hospitals, correctional facilities, and military and in-flight foodservice—can be very proud of the invaluable service they provide to the citizens of the country.

The statistics underscore the size and scope of the industry. Ranked number one among retail employers, foodservice directly employs more than 9 million people. This number is expected to reach 12.4 million by 2005. Annual sales top $275 billion, which represents 5% of the gross national product. Foodservice expenditures account for 42% of every food dollar. Foodservice is the largest employer of ethnic groups, minorities, women, workers with disabilities, and entry-level workers. The millions of jobs provided and created by the industry, the training and teaching of responsibility and skills, the opportunities provided to develop self-esteem and for promotion to management and ownership combine to make this industry an exciting, rewarding, and dynamic career choice.

Today, the foodservice industry is defined in its broadest sense to mean all establishments where food is regularly served outside the home. Such establishments include formal restaurants, hotel or motel and department store dining rooms, coffee shops, family restaurants, specialty and ethnic restaurants, and fast-food outlets. Foodservices that are operated in schools, colleges, and universities; in hospitals, nursing homes, and other health care settings; in recreational facilities; in transportation companies; in the military; in correctional facilities; in office buildings and plants; in convenience stores, supermarkets, and service delis; and in community centers and retirement residences are also included.

The history and development of organizations within the foodservice industry presented in this chapter are intended to give the reader a perspective of, and an appreciation for, foodservices as they are today. The background information should be of special interest to those who already are or are preparing to become managers of foodservice operations. The trends that are shown provide some basis for anticipating the future, and these trends should alert managers to the demands that new developments and changes in this field may bring, so that they can prepare to meet them. As George Santayana so wisely said, "Those who cannot remember the past are condemned to repeat it." History can not only provide people with an opportunity to learn from past mistakes, but can also show which of the seeds that were sown blossomed into successes and why.

Information in the succeeding chapters is basic to the successful operation of all types of foodservices, whatever their philosophies and objectives. All are concerned with providing good food to meet the specific needs of groups of people served outside the home. In Part 1 the stage is set by providing the reader with a picture of the history of the foodservice industry, its current status and trends, followed by a discussion of the various types of foodservice production systems in use today. The technical aspects of managing a foodservice are presented in Part 2 beginning with the critically important topic of food safety. The focus of Part 3 is management of the physical facilities including sanitation, cleaning, energy conservation, solid waste management, environmental safety, design and layout, and equipment and furnishings. In the final section, the organization and management of foodservice operations are discussed. Although the authors believe that the sequence is a logical one in which to study foodservice, each chapter is designed to stand alone and therefore chapters may be read in any order desired.

KEY CONCEPTS

1. Religious orders, royal households, colleges, and inns were among the earliest organizations to practice quantity food production.

2. Advances in the fields of microbiology, physics, and industrial engineering led to improvements in the way food was produced.
3. Innovative and visionary pioneers of the commercial foodservice sector introduced many new concepts that continue to enjoy widespread use today.
4. Several pieces of key legislation have affected school foodservice programs in the past and continue to do so today.
5. Economic conditions and lifestyle changes have led to a desire for convenience, value, and freshness in all food purchased.
6. To provide customer satisfaction and to run a financially sound operation, a foodservice manager must possess an awareness of current trends.

EARLY HISTORY OF FOODSERVICE ORGANIZATIONS

Foodservice organizations in operation in the United States today have become an accepted way of life, and we tend to regard them as relatively recent innovations. However, they have their roots in the habits and customs that characterize our civilization and predate the Middle Ages. Certain phases of foodservice operations reached a well-organized form as early as feudal times in countries that have exerted the most influence on the development of American food habits and customs. These countries are England, France, Germany, and Sweden. In each of those countries, partaking of food was a social event in which the entire family, and often guests, shared. There was no withdrawal for eating that characterizes the customs of certain peoples, no religious beliefs barring participation in eating meals with others. The economic level of the country and the type of food eaten also fostered the need to serve food to groups. Instead of the few grains of parched wheat or corn, the bowl of rice, or a bit of raw fruit that satisfied some races, these people ate meat or a variety of other protein foods from various sources. Because meat and other protein foods could not be transported without danger of spoilage, immediate food production in well-established kitchens and with good supervision was required. These countries, then, have contributed to the evolution of the foodservice industry.

Some of the types of foodservices that existed long ago are contrasted in this chapter with their present-day counterparts. Although in medieval times religious orders, royal households, colleges, and inns were the most prevalent types of organizations in which quantity food production was the rule, other types of foodservice organizations and their development are also considered. These include clubs and other social organizations, schools, the military, correctional facilities, hospitals and health care facilities, employee feeding in industrial plants or offices, retirement homes and residences for other groups of people, restaurants, and transportation companies.

Religious Orders

Religious orders and royal households were among the earliest practitioners of quantity food production, and although these foodservices were far different from those we know today, each has made a contribution to the way in which present-day foodservice is practiced.

Abbeys that dotted the countryside, particularly in England, served not only the numerous brethren of the order, but also thousands of pilgrims who flocked there to worship. The space provided for food preparation indicates the scope of their foodservice operations. At Canterbury Abbey, a favorite site of innumerable pilgrimages, the kitchen measures 45 feet wide.

Records show that the food preparation carried out by the abbey brethren reached a much higher standard than food served in the inns at that time. The vows the brothers took did not diminish their appreciation for good food. Food was grown on the abbey's grounds, and lay contributions were provided liberally for the institution's table. The strong sense of stewardship in the abbeys led to the establishment of a detailed accounting system. These records showed that a specified per capita per diem food allowance was in effect, thus creating an effective early-day cost accounting system.

Royal and Noble Households

The royal household, with its hundreds of retainers, and the households of nobles, often numbering as many as 150 to 250 persons, also necessitated an efficient foodservice. The differing degrees of rank resulted in different food allowances within these groups. In providing for the various needs, strict cost accounting was necessary and here, perhaps, marks the beginning of the present-day scientific foodservice cost accounting. The cost record most often cited is the *Northumberland Household Book*. For this household of more than 140 persons, 10 different daily breakfasts were recorded, the best for the earl and his lady, the poorest for the lowest workman or scullion. A similar range is presumed for the other meals.

The kitchens in these medieval households would appall the present-day foodservice manager in their disregard for sanitary standards in food storage, preparation, and handling. A clutter of supplies that overflowed from inadequate table and shelf space to the wooden plank floors and was handled by children and nosed by dogs commonly composed the background for the preparation of elaborate creations for the table. Since labor was cheap and readily available, a large staff of workers was employed to prepare the food. Rank was evident in the division of labor. The head cook might wear a gold chain over handsome clothing and present his culinary creations to his employer in person. The pastry cook and the meat cook did not rate as high, but were esteemed for their contributions. The average scullion often had scarcely a rag to wear and received broken bread and the privilege of sleeping on the hearth through the chilly winter nights as his wage.

As time passed, discovery of the causes of food spoilage led to improved practices in food storage and in food preparation in these noble households. Advances in the understanding of the laws of physics resulted in the replacement of open hearths with iron stoves and many refinements to the kitchen equipment. A more convenient equipment arrangement led to a reduction in the number of workers required, helping to relieve disorder and confusion. Employees' dress changed to show some regard for the tasks they performed by becoming more practical.

The United States has no equivalent of these royal households. However, the White House, as the president's residence, is one site of official entertainment in this country. Whereas the feasts of the royal households in the past were of "formidable proportions," observers have noted that the White House kitchen staff could prepare anything from "an egg to an ox." The White House kitchen of 100 years ago is shown in Figure 1.1. An early-day state diplomatic dinner at the White House included seven courses, each with many accompaniments. The present trend is toward simplification of menu patterns even for more formal occasions, which is in harmony with current American food consumption habits. A photograph of the 1975 White House kitchen is shown in Figure 1.2, and the White House State Dining Room is shown in Figure 1.3. Both of these rooms are designed for efficient use of space. Equipment in the kitchen has been chosen and placed to maximize productivity.

Notable among the other early foodservices were inns, taverns, and the hostels, which were established by European colleges. From these, the foodservices in schools and colleges, hospitals and other health care facilities, and retirement communities, as well as the wide variety of commercial and industrial foodservices as we know them today, have evolved.

Figure 1.1 1892 White House kitchen.
Reproduced from the Collections, Prints and Photograph Division, Library of Congress.

Figure 1.2 1975 White House kitchen.
Reproduced from the Collections, Prints and Photograph Division, Library of Congress.

DEVELOPMENT OF PRESENT-DAY FOODSERVICES

Restaurants

Historically, the evolution of public eating places was stimulated by people's desire to travel, for both spiritual enrichment and commercial gain. Religious pilgrimages played an important role in establishing the inns of France and England. Merchants traveling from country to country to buy or sell their wares also created the need for places to stop for food and rest. These early inns and taverns, providing for the needs of travelers, were perhaps the forerunners of our present restaurants. However, many of them were primitive and poorly organized and administered. The literature of the time describes unsanitary conditions under which food was prepared and served, monotonous menus, and poor service.

Stagecoach travel in colonial America, tedious at best, created the need for inns where travelers could rest and eat. These inns were much like those in England, but proprietors gave more attention to pleasing the guests. Many inns were family enterprises located in somewhat remote areas. The food offered was the same as that for the family: plain, hearty, and ample. In urban centers during the early decades of the nineteenth century, hotels and inns presented extremely extensive and elaborate menus to attract guests to their facilities. Since revenues from these meals frequently did not cover preparation costs, the bar and lodging income was needed to make up the deficit.

Introduction of the European hotel plan, which separated the charges for room and board, and later the à la carte foodservice were steps toward a rational foodservice in hotels. Through these measures, much of the waste that characterized their foodservice was abolished, leading to the possibility that hotel and inn foodservice could be self-supporting. This led proprietors to establish foodservices separate and distinct from lodging facilities.

The origin of the restaurant concept, however, has been traced to the cook shops of France. They were licensed to prepare *ragoûts*, or stews, to be eaten on the premises or taken to inns or homes for consumption. The shops had *écriteau*, or menus, posted on the wall or by the door to whet the interest of the passerby. The story goes that one Boulanger, a bouillon maker, added a meat dish with a sauce to his menu, contending that this was not a *ragout* and, therefore, did not violate the rights of the *traiteurs*, or restaurant-keepers. In the legal battle that followed, the French lawmakers sustained his point, and his new business was legalized as a restaurant. The word *restaurant* comes from the French verb *restaurer*, which means "to restore" or "to refresh." It is said that the earliest restaurants had this Latin inscription over their doorway: *Venite ad me qui stomacho laoratis et ego restaurabo vos*—Come to me all whose stomachs cry out in anguish, and I shall restore you!

The cafeteria was a further step in the simplification of restaurant foodservices. This style of self-service came into being during the gold rush days of 1849 when the "forty-niners" demanded speedy service. Regarded as an American innovation, its popularity extended throughout the United States. Today, commercial cafeterias still represent an important part of the foodservice industry.

Figure 1.3 White House State Dining Room set for formal service.

Another innovative foodservice was the automat, first opened in Philadelphia in 1902 by Horn and Hardart. Patterned after a "waiterless" restaurant in Berlin, it combined features of a cafeteria with those of vending. Individual food items were displayed in coin-operated window cases from which customers made their selections. This "nickel-in-a-slot" eatery provided good food and high standards of sanitation for nearly 50 years, drawing customers from every walk of life. For many people, it became a haven, especially during the Great Depression years beginning with the stock market crash in 1929, the years of the automat's greatest success. After World War II, the automat's popularity declined as a more affluent society sought greater sophistication in dining. Competition with other types of foodservice became intense.

The passage of the Volstead Act, the Eighteenth Amendment to the Constitution, which prohibited the manufacture, sale, and distribution of alcoholic beverages in the United States, had a major and lasting impact on commercial foodservice. With the loss of alcohol in the menu mix, everyone began to get serious about the food served. Concerned restaurateurs gathered in Kansas City, Missouri, and founded the National Restaurant Association (NRA). Many landmark establishments went bankrupt while, at the same time, a new breed of operation was spawned—the speakeasy. Two of the most famous of the "speaks," the Coconut Grove in Los Angeles and New York's "21" club, became known not only for the bootleg liquor served, but for the quality of food as well.

As mass quantities of automobiles hit the roads, what is considered to be one of America's first drive-in restaurants, the Pig Stand, was opened by J. G. Kirby, a candy and tobacco wholesaler, on the Dallas-Fort Worth Highway in 1921. Service at the barbecue-theme Pig Stand was provided by waitresses who jumped up on the protruding running boards of the automobiles—hence they became known as carhops. The same year Billy Ingram and Walter Anderson started their White Castle operation with a $700 investment. They sold bite-sized hamburgers for 5¢ each. Ingram was a pioneer of many fast-food concepts still in use today such as strict product consistency, unit cleanliness, coupon discounts, heat-resistant cartons for carryout orders, and folding paper napkins.

During the 1920s and 1930s restaurants evolved from being luxuries to necessities. Perhaps no one took better advantage of the growing popularity of automobile transportation than Howard Dearing Johnson of Wollaston, Massachusetts. In 1925 Johnson took a bankrupt pharmacy in Quincy, Massachusetts, and converted it into a soda fountain serving a trio of ice cream flavors he had developed. After expanding his menu to include quick service items such as hamburgers and hot dogs, Johnson set his sights on opening more units. Without capital to do this, he decided to franchise. By 1940 he had 100 franchises and 28 ice cream flavors.

At approximately the same time that Johnson was watching traffic on the highway, a 26-year-old Mormon from Utah was watching pedestrian traffic in Washington, D.C., on a hot July day. J. Willard Marriott saw that the thirsty masses had no place to go for a cold drink. With a $3000 investment he and his future wife, Alice, opened a nine-seat A & W root beer stand that grossed $16,000 the first year. This was the beginning of the Marriott Corporation, currently a multibillion dollar foodservice and lodging empire.

Other milestones during this time were the development of the soft ice cream machine in 1934 by Thomas Carvel, a former compressor mechanic and test driver for Studebaker, and the subsequent opening of the first Dairy Queen in Joliet, Illinois, in 1940.

The repeal of Prohibition in 1933 helped to boost fine dining restaurants and deluxe supper clubs featuring live entertainment. Theme restaurants with fun, but outrageous, gimmicks often thrived. Trader Vic's, Romanoff's, Chasen's, El Morocco, Lawry's Prime Rib, and The Pump Room, among others, became the haunts of the rich and famous.

Meanwhile, at the other end of the dining spectrum, in July 1941 a former bakery delivery man in Los Angeles secured a hot dog cart with $15 cash and a $311 loan against his Plymouth. Carl N. Karcher made $14.75 on his first day in business. The hot dog cart evolved into a drive-in barbecue joint and then a quick service operation featuring hamburgers and chicken sandwiches. Some 50 years later the Carl's Jr. chain would ring up $640 million in sales and number 640 units. Carl Karcher contributed air conditioning, carpeting, piped-in music, automatic charbroilers, salad bars, nutritional guides, and all-you-can-drink beverage bars to the fast-food concept.

While this was going on, the face of fast food was being changed forever just 50 miles east of Los Angeles in the then sleepy little town of San Bernardino. Brothers Mo and Dick McDonald had opened a 600-square-foot facility that violated a basic rule of restaurant design by exposing the entire kitchen to the public. The 25-item menu generated $200,000 in annual sales. Twenty carhops were needed to service the 125-car parking lot. But, faced with increasing competition and the constant turnover of carhops, the brothers made a dramatic decision to eliminate the carhops, close the restaurant, convert to walk-up windows, and lower the hamburger price from 30¢ to 15¢. After a few months of adjustment, annual sales jumped to $300,000. By 1961 the McDonalds had sold 500 million hamburgers and they sold the company to Ray Kroc for $2.7 million. Today McDonald's has about 14,000 units spread over 60 countries with systemwide sales of more than $22 billion.

Spinoffs from the McDonalds' concept included Taco Bell, Burger King, and Kentucky Fried Chicken (now KFC), each with similar success stories. In the 1950s coffee shops began to proliferate, particularly in Southern California. Tiny Nayler's, Ships, Denny's, and the International House of Pancakes (now IHOP) had their beginnings during this time. In New England in 1950 an industrial caterer named William Rosenburg opened a doughnut shop featuring 52 varieties of doughnuts and Dunkin' Donuts was born. In the late 1950s, pizza moved from being served in mom-and-pop, family-run eateries to the fast-food arena. Pizza Hut opened in 1958 and was followed within a few years by Domino's and Little Caesar's.

Innovative marketing concepts were introduced in the 1960s in new chains such as T.G.I. Friday's, Arby's, Subway, Steak and Ale, Victoria Station, Cork 'n' Cleaver, Black Angus, Red Lobster, and Wendy's. The 1970s were marked by the conflict in Vietnam, the rising popularity of ethnic foods, television shows featuring cooking instruction, women entering foodservice management and back-of-the-house executive positions, some interest in health foods and vegetarianism, and the beginnings of California cuisine.

The 1980s have been both good and bad for the restaurant industry. On the positive side, progress has been made with environmental and solid waste proposals and health and nutrition mandates. On the other side, poor economic conditions have led to unbridled expansion, overleveraged buyouts, employee buyouts, a rash of Chapter 11 filings, systemwide restructurings, downsizing, and job layoffs. General Mills opened its Olive Garden chain in 1982 and China Coast in 1990. PepsiCo Inc. acquired Taco Bell, Pizza Hut, and Kentucky Fried Chicken to make it an industry powerhouse. Marriott became the country's largest contract foodservice company when it acquired Gladieux Corporation and Saga Corporation.

Casualties of the 1980s included Sambo's, Flakey Jake's, D-Lites of America, and Popeye's. Some believe that government regulation passed during this decade has been the most harmful since Prohibition. Meal deductibility was reduced to 80% from 100%, the FICA tax-on-tips mandate was instituted, and the Americans with Disabilities Act and the Family Leave Bill went into effect.

Value wars, environmental concerns over packaging waste, and the public's increasing interest in nutrition are all issues that have been faced by the industry in recent past. Operators have responded in various ways to each challenge by offering low-price loss leaders, reducing packaging, and offering lower fat and healthier alternatives on their menus.

Perhaps the post-World War II baby boom generation and the resulting population bulge have influenced the growth of the foodservice industry as much as any other factor in recent years. As this generation raised on "fast food" matures, it continues to seek more sophisticated fast-food dining. Many foodservice trends that seem to be new at the time are in reality, as Woodman (1984) said, "one more repeat in a cyclic phenomenon, wrapped up in a new language and viewed by a new generation."

This cycle is dramatically demonstrated in the uniforms worn by waiters, waitresses, and counter attendants throughout the years. The classic waiter's uniform of the 1900s, consisting of a white shirt, a black bow tie, a jacket, and pants (Figure 1.4) is being seen more often in the fine dining establishments of the 1990s as a unisex garment with interchangeable elements (Figure 1.5). The soda jerks' and waitresses' uniforms of the 1940s and 1950s (Figure 1.6) have given way to livelier, more durable, unisex uniforms as seen in today's fast-food operations.

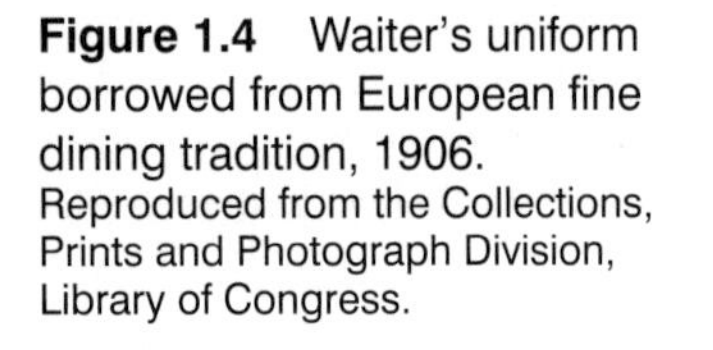

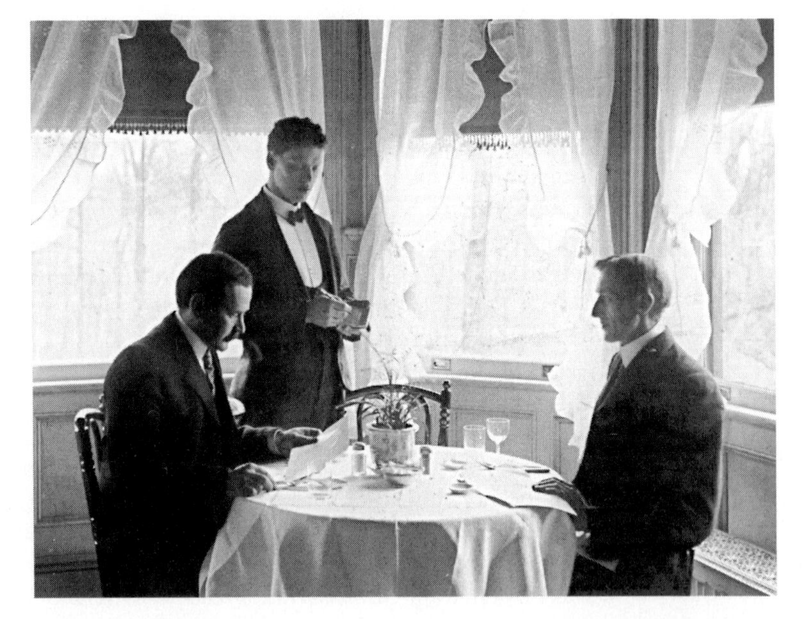

Figure 1.4 Waiter's uniform borrowed from European fine dining tradition, 1906.
Reproduced from the Collections, Prints and Photograph Division, Library of Congress.

Figure 1.5 Modern-day unisex uniform.
Courtesy of Angelica Uniform Group, St. Louis.

Although most would agree that uniform design is not as important as location, food quality, or service, many are finding that what employees wear can have a positive effect on the bottom line. An example of the impact was seen at KFC units in Detroit where employees wear African-American apparel, including kente cloth and khoufi (small, brimless) hats. These changes, in conjunction with new ethnic menu items, have boosted sales from 15 to 30%. Other operators have resisted a rigid dress code. At the California Pizza Kitchens, employees are given guidelines on how to put outfits together, and Angelica Kitchen in Manhattan's bohemian East Village has a strict "no uniform" policy.

The types of restaurant operations management range from the small, independent owners or managers to the complex franchises and chain operators to executives of large corporations where mergers and acquisitions abound. Foodservice managers in today's complex restaurant field must be highly motivated and knowledgeable in order to compete and survive.

Figure 1.6 Uniforms of the 1940s and 1950s worn by soda jerks and waitresses at the "Palace of Sweets."
Reproduced from the Collections, Prints and Photograph Division, Library of Congress.

Colleges and Universities

Providing meals and rooms for college and university students has been the custom for many years. However, responsibility for the kinds of services offered and the administration of these living situations has changed considerably during the years.

From the twelfth century through the medieval ages at European colleges and universities, hostels were the accepted arrangement for student living. On the continent, students managed these hostels. At Oxford, England, however, hostels were endowed to provide board and lodging for students unable to pay these costs for themselves. At least to some degree, the university managed these endowed hostels—a policy that continues today.

Colonial colleges in the United States provided residence halls with dining rooms for all students. Administrators, generally clergymen, were responsible for their operation. They dispatched their duties prayerfully and thriftily—not always with student approval! Later, with an interest in and, hence, a shift toward German educational procedures, which did not include housing as a school responsibility, some colleges lost interest in student living situations. As a result, sororities and fraternities without faculty supervision assumed the feeding and housing of large groups of students. In many cases, this also led to the problem of providing adequate diets for all students.

The twentieth century has witnessed many changes in college and university foodservices in the United States. A shift occurred from the laissez-faire policy of early-day administrators to a very strict one in the late nineteenth century. Until World War II, colleges provided separate dining halls for men and women. Not only did students have their dietary needs satisfied, but they were also trained in the "social graces." Seated table service with students serving in rotation as hostess or host and as waiter or waitress was the accepted procedure in many residence hall dining rooms. Although this service may still be found in some colleges and universities today, it is the exception rather than the rule.

Gradually, with the influx of GI students into American schools of higher education after World War II, the more formal seated service and leisurely dining gave way to the speedy informality of cafeteria service. This service style makes it possible to meet student demands for greater menu variety and to cater to the food preferences of various ethnic groups that make up the student body. Also, with coeducational residences and dining halls now commonly found on the college campus, the dietary requirements of both men and women in the same dining hall can be met by cafeteria or self-service. Student food habits have also changed as a result of today's concern for physical fitness and weight control. Foodservice managers have attempted to comply with this need through suitable menu selections. For example, salad bars, pasta bars, potato bars, and vegetarian bars are standard in most campus dining halls.

Other trends in residence hall dining are longer hours of service, fewer restrictions on the number of services allowed, and greater flexibility in board plans, including a "pay as you eat" plan, rather than having to pay a set rate in advance.

In addition to residence hall dining, a diversity of other campus foodservices is a familiar pattern today. Student union buildings have, for example, set up creative and innovative units catering to students' changing food interests and demands. Commercial fast-food companies have been a major competitor for student patronage in many college towns. In some, therefore, the university has contracted with these companies to set up and operate one of their food units on campus.

Management of college and university foodservices is usually under the direction of well-qualified foodservice managers employed directly by the university or by a contract foodservice company. Both plans are found in colleges today.

The use of college and university dining facilities as laboratories for foodservice management classes is a common practice. Undoubtedly, this has helped establish high requirements for foodservice directors on such campuses. Directing the students' laboratory experiences and the work of numerous part-time student employees presents unique situations not common in other types of foodservice organizations.

School Foodservice

The history of school foodservice is inevitably a part of the larger story of the rapid development of public education. As reforms stemming from the Industrial Revolution began to free society from the supposed necessity of child labor, unemployment among children of school age increased and public concern with education soon became evident. To encourage school attendance, parents and civic-minded townspeople in some European countries banded together to provide low-cost school lunches. It is reported that canteens for schoolchildren were established in France in 1849 and that, in 1865, French novelist Victor Hugo started school feeding in England by providing children from a nearby school with hot lunches at his home in exile. At some time between these dates, school foodservice began in the United States. The Children's Aid Society of New York City opened an industrial school in

1853, in an effort to persuade children from the slums to seek "instruction in industry and mental training," and offered food to all who came.

Beginning in the early 1900s, growing knowledge of nutrition placed emphasis on the importance of wise food selection and the need for nourishing school lunches at little or no cost to the students. Under the notable leadership of Ellen H. Richards, the Boston school committee passed an order that "only such food as was approved by them should be sold in the city schoolhouses." Figure 1.7 shows an early-day school lunchroom kitchen. At about this time, the other men and women likewise concerned with child welfare sponsored similar developments in several urban centers, and the two decades that followed brought significant developments in the school lunchroom movement throughout the country. Although the program was implemented most rapidly in large cities where sustaining organizations and concerned leaders were prevalent, the cause of rural children was also championed by such groups as the Extension Service and PTA councils. Improving the nutrition of schoolchildren through adequate foodservices in schools was the goal.

World Wars I and II again brought into focus the need for improved nutrition among young people because so many were rejected from the military for reasons related to faulty nutrition. Concern was expressed for the future health of the nation

Figure 1.7 Preparation for a school lunch in an early-day institutional kitchen in the United States.

Reprinted by permission of *School Food Service Journal*. The American School Food Service Association, Denver, Colorado, 1986.

if such a trend continued. As a direct result, the first federal legislation designed to assist and direct school foodservices was enacted in 1933. This provided loans enabling communities to pay labor costs for the preparation and service of lunches in schools. In 1935, additional assistance came when the federal government was authorized to donate surplus farm commodities to schools. With these aids, a noon meal became a common part of school activities.

The *major* legislation governing the school lunch program, however, was the National School Lunch Act of 1946 (Public Law 79-396). Through this act, funds were appropriated "as may be necessary to safeguard the health and well-being of the nation's children, and to encourage the domestic consumption of nutritious agricultural commodities and other food by assisting the states through grants in aid and other means, in providing an adequate supply of funds and other facilities for the establishment, maintenance, operation and expansion of non-profit school lunch programs." States were required to supplement federal funding as set forth in Section 4 of the act, and lunches served by participating schools were obligated to meet the nutritional requirements prescribed. Although the National School Lunch Act allowed Type A, B, and C meals, the Type A lunch is the only one now served under the federal school foodservice program and is referred to as "the school meal pattern" (see Chapter 4 for specifics).

Other legislation and amendments have changed the funding policy and circumstances for offering free or reduced-price meals to students, revamped the commodity distribution plan, or provided for supplemental feeding programs. The Child Nutrition Act of 1966 authorized the School Breakfast Program and the Special Milk Program to further help alleviate inadequate nutrition. The Omnibus Reconciliation Act of 1980 reduced the reimbursement rate to schools for the first time and changed the income eligibility standard for students who could receive free or reduced-price meals. Further adjustments were made in 1981 and 1982 to help achieve reductions in federal spending. School foodservice managers work creatively to adjust to these changes while maintaining an attractive meal program to meet nutritional guidelines and appeal to students.

By the late 1980s, U.S. public school enrollment had dropped dramatically, but spending continued to climb. School foodservice directors had to switch from operating subsidized departments to self-supporting ones. Some of the changes that were made in an effort to make the switch include implementing centralized food production; raising prices for paying students; attracting more paying students to the program to offset the free and reduced-price lunches that are served; offering more high-profit fast-food-style à la carte items; and reaching out to service community programs, such as Meals-on-Wheels, senior citizen and day care centers, and community "soup" kitchens.

School foodservice is most effective when dietitians, school administrators, food managers, and allied groups, such as the PTA, recognize its value in the child's mental and physical development. Then they can work together to make the foodservice not just a "feeding program," but rather a nutrition program for all students as part of their learning experience.

The type of organization and management found in school foodservices varies as much as the size and location of schools involved. Small independent schools may have simple on-premise food preparation and service supervised by a cook or manager with one or two employees and/or part-time student helpers. Large city school systems often use a centralized production kitchen and deliver meals for service to individual schools in the system. Centralized management with unit supervisors characterizes this system.

Clubs and Other Social Organizations

Clubs formed by people with a common interest have existed for decades. Member dues and assessments in some clubs provide funds to establish a "clubhouse" or building that usually includes foodservice facilities. These clubs were especially prominent among the wealthy society class in England in the 1800s. As early as 1850, noted chef Alexis Soyer worked closely with the Reform Club of London to provide a sanitary and efficient foodservice setup that would utilize the then fairly recent innovation of stoves, water baths, and refrigeration (Figure 1.8).

Today, city clubs, athletic clubs, faculty clubs, country clubs, and others attempt to rival the settings of better hotels with a similar standard for foodservice. Few homes are spacious enough or adequately staffed to serve meals to 25 or more guests. Clubs provide these facilities and also cater to such functions as receptions and banquets, in addition to providing regular meal service for their members. Sometimes the foodservice helps subsidize the cost of other services such as swimming pools or recreational rooms offered by the club. The number of meals served may be irregular and the stability of income somewhat uncertain, making club foodservice management a real challenge.

Hospitals

The evolution of foodservices in hospitals is as interesting as that of colleges and schools. The first hospital established in the United States was Philadelphia General in 1751. Meals in early-day hospitals were simple to the point of monotony and no attempt was made to provide any special foods or therapeutic diets. Menus in an eighteenth-century American hospital, for example, included mush and molasses for

Figure 1.8 One of the early plans for an institutional kitchen made by Alexis Soyer for the Reform Club of London about 1850.

breakfast on Monday, Wednesday, and Friday, varied by molasses and mush for supper on Monday, Wednesday, Thursday, and Saturday. Oxtail soup and black bread appeared on occasion.

Accounts of the Pennsylvania Hospital in the year 1804 stated that milk, butter, pork, and soap were produced on the hospital grounds for consumption in the hospital. Also, cows, calves, and pigs were sold for income. Salary for a husband and wife serving as steward and matron was $350 for nine months of service.

Dietetics, as a hospital service, had its beginning at the time of the Crimean War (1854–1856). In 1855, Florence Nightingale, whom dietitians, as well as nurses, revere and honor as the pioneer of their profession, established a diet kitchen to provide clean, nourishing food for the ill and wounded soldiers in Scutari (now Uskudar), Turkey. Until then, foods of questionable quality were poorly cooked in unsanitary conditions and served at irregular intervals.

Alexis Soyer (who, as noted earlier, was a chef who had worked with the Reform Club of London) contributed greatly to Nightingale's efforts when he offered to serve gratuitously as manager of the barracks hospital kitchen. Soyer's plan for operating it was as efficient as modern-day practice.

Changes during the next 100 years in hospital foodservice included the introduction of centralized tray service and mechanical dishwashing, establishment of a separate kitchen for special diet preparation and later elimination of such kitchens, and the advent of frozen foods and their use in food preparation. Also, pay cafeterias for staff and employees and separate dining areas for these two groups were introduced during this period. Employing qualified dietitians to administer dietary departments and "therapeutic" dietitians for "special diet" supervision became the usual practice.

Today, hospitals comprise a large group of institutions operated and funded by various governing bodies (e.g., federal, state, county, and city governments) or by religious organizations, or they are privately owned. Regardless of the source of support, the main objective of hospital foodservice is to improve patient health and restore patients to normal activity and a state of well-being. A secondary objective for most hospitals is to provide foodservice for staff, employees, and guests in order to maintain happy, well-nourished personnel and good public relations.

The past 30 years have brought more innovations and changes for greater operating efficiency and increased revenues to hospital dietary departments. Some of the innovations that have been implemented include increasing nonpatient, in-house cafeteria volume; marketing catering services; use of professional chefs to improve menus; contracting professional and food production services to smaller operations; and creating new services, such as diet workshops for the public and take-home employee and patient meals. Vended foods, used to supplement regular meal service, provide round-the-clock food availability. Many hospitals provide a dining room for ambulatory and wheelchair patients in recognition of the therapeutic value of mealtime social contact with others.

Technological advances have influenced foodservice systems, with the introduction of new methods of food production, holding, distribution, and service; computer use for many routine tasks; and use of robots in some hospital foodservices.

Federal policy has influenced trends in hospital procedures and budgeting, as is true with school foodservices. A major source of a hospital's income is Medicare and

Medicaid reimbursement payments for certain patients' hospital and nursing home expenses. In March 1983, Congress approved Medicare-mandated rates for reimbursement of Diagnosis Related Groups (DRGs) for illnesses and medical service for hospital inpatient services. This represents a strict set of cost controls for health care and has created financial incentives for hospitals to contain or reduce health care costs.

Ensuring quality care and service has been the focal point in hospitals and health care facilities in recent years. The Joint Commission on Accreditation of Healthcare Organizations has established standards to enforce quality assurance programs for accrediting its member facilities. One section is devoted to dietetic services to ensure that these services meet the nutritional needs of patients. Compared to other segments of the foodservice industry, hospital foodservice today is unique and complex, requiring a well-qualified dietetic staff to coordinate activities that ensure optimum nutritional care.

Nursing Homes and Other Health Care Centers

Nursing homes and other health care facilities have come into prominence only relatively recently so their history is short compared with that of hospitals. However, the increase in demand for nursing home care has been phenomenal in the last 50 years due to several factors, among them population growth, an increasing number of elderly persons in the population, urban living with condensed family housing units, increased incomes, and greater availability of health care insurance benefits.

The history of nursing homes tends to follow federal legislation enacted to help provide funds to care for individuals in such facilities. The modern nursing home era and the development of the nursing home as a distinct type of health care institution began in 1935 when federal funds to pay for nursing home care for the elderly were made available to the states by the passage of the Social Security Act.

In 1951, the Kerr-Mills bill made federal matching funds for nursing home care available to states that were establishing satisfactory licensing and inspection programs. An amendment to the Hill-Burton program in 1954 provided financial aid to construct facilities for skilled nursing care in order to meet specified requirements. Many large nursing homes were built, and older smaller homes were forced to close or to modernize and expand to meet the standards of the new program.

Medicare and Medicaid legislation (1965) and its amendments (1967) established minimum standards and staffing and inspection requirements that had to be met in order for nursing homes to receive funds for their patients. The Federal Conditions of Participation Regulations (1974) spell out, in detail, standards that skilled nursing facilities must meet.

General guidelines for all nursing homes include licensure and meeting the standards of specific government programs in order to be certified, and one or more levels of care, such as nursing care and related medical services, personal care, and residential care.

Nursing homes provide different categories of nursing care and are classified as skilled, intermediate, and other.

Skilled Nursing Facilities Skilled nursing facilities offer 24-hour nursing care and provide other professional services as prescribed by the physician. Emphasis is on rehabilitation, and the facility is eligible for both Medicare and Medicaid reimbursement.

Intermediate Care Facilities Intermediate care facilities provide basic medical, nursing, and social services in addition to room and board. Patients need some assistance but not intensive care. These facilities are eligible for the Medicaid program only.

Other Facilities In contrast to skilled nursing and intermediate care facilities, which are certified under specific government programs, many nursing homes do not participate in these programs. Therefore, they are not classified as such, even though they may provide the same quality of care and categories of services. A facility may provide more than one category of care. The patients may be of any age. Many facilities specialize in child care or care of persons with mental illness or mental retardation.

Food plays a major part in the lives of most nursing home patients. For many, it may be the one thing they have to look forward to each day. The quality and amount of food offered as well as the care and supervision given to foodservice are important to any nursing home's success and effectiveness. Dining rooms are provided for patients who are ambulatory and able to participate in group dining. Others are served in their own rooms. Many require assistance with eating.

The 1974 Federal Conditions of Participation Regulations, mentioned previously, stated these general "conditions" related to *dietary* services:

> Condition of participation dietetic services: The skilled nursing facility provides a hygienic dietetic service that meets the daily nutritional needs of patients, ensures that special dietary needs are met, and provides palatable and attractive meals. A facility that has a contract with an outside food management company may be found to be in compliance with this condition provided the facility and/or company meets the standards listed herein.

The services of registered dietitians (RDs) are required to ensure that dietary service regulations are adequately met and administered. Part-time or consultant RDs may be employed by small nursing homes. Full-time RDs are needed in the larger, skilled nursing homes.

Other Health Care Centers Corrective institutions and homes for specialized groups such as persons with physical or mental disabilities, orphaned or runaway children, or abused and battered women, and short-term residences for parolees, drug-abuse victims, or other groups during a rehabilitation period come under this category. These organizations usually serve low-income groups and may be supported by the state, county, private charities, or fraternal orders. Residents may or may not qualify for government assistance.

Trained foodservice supervisors are important to the success of these institutions in rehabilitating their residents to good health or for maintaining a healthy body. A well-balanced, nutritious diet that is acceptable to the group being served and that is within the constraints of the budget should be provided.

Day Care Centers Day care centers for children of working parents, the underprivileged, and, more recently, for elderly persons who otherwise would be isolated and lonely offer an excellent opportunity for the foodservice to provide at least one nutritious meal a day in an attractive, friendly atmosphere. Facilities for special groups, such as camps for diabetics, combine social and recreational activities and proper nutrition.

Senior Citizen Centers Many senior citizen centers employ trained supervisory personnel with leadership abilities to guide and assist volunteers who plan and carry out programs of value to the members. Primary among these is a limited meal service, usually a noon meal five days a week. In addition, a dietitian may give talks and advice on meal planning and good nutrition. Eating the noon meal together offers participants an opportunity to socialize. Often the meals served to senior citizen groups are subsidized in part by church, fraternal, or philanthropic organizations within the community. Free meals to those unable to pay is a common practice.

Meals-on-Wheels, a program for delivering meals (generally once a day, five days a week) to the homes of persons unable to go to a center or to prepare their own food, is usually funded through local charities. A large number of volunteer workers are involved in the program to help make it possible. Food for such meals may be prepared by contract with a senior citizen center, a school, a college, or a restaurant.

Health care of persons who are ill, disabled, and needy in the United States, whether at home, in hospitals, or in other health care facilities, attracts unprecedented attention from legislative groups, patients, administrators, and the general public. Recent government cuts in federal spending and the uncertain state of the American economy, together with the high cost of health care, have tended to slow the growth of and census in nursing homes and hospitals. Also, some of the funds available for long-term care have been diverted to home care or to newly organized, cost-containing alternate care delivery entities. Among the latter are health maintenance organizations (HMOs), which emphasize preventive care; immediate care centers for urgent situations; surgical centers for surgical procedures not requiring hospitalization; birthing stations; cardiac rehabilitation centers; and similar specialized units. Foodservice is a necessary part of such facilities. These developments have occurred during a short time and more changes are certain. On the part of all concerned, constant alertness to any pending or new legislation that may affect our present methods of health care is required.

Retirement Residences and Adult Communities

During the past few decades, many persons reaching retirement age expressed interest in an easier type of living than the total upkeep required by their own homes. The result was the development of group living and dining facilities for retirees. These provide comfortable, congenial, independent living with minimal responsibility and offer freedom from coping with a large house, its taxes, and maintenance problems, and also little, if any, food preparation.

Many types of residences or adult communities are available at various price levels. Some are high-rise apartment-style buildings; others are on the cottage plan. Either type may have individual kitchenettes, central dining facilities, or both. Figure 10.3, Chapter 10, shows an attractive retirement home dining room. Other variations include purchase or rental units with full meal service, one meal a day, or no meals included in the monthly charge. Usually housekeeping, recreational and social activities, and medical and emergency nursing care units are provided. Access to public transportation is a factor, but adequate health care is considered the most important provision.

Foodservice for this older clientele should be directed by a well-qualified dietitian or foodservice manager who is knowledgeable about their dietary needs and food preferences and who can help contribute to the group's good health and relaxed, happy, gracious living.

Industrial and Business Foodservice

Providing food for employees at their workplace has been necessary since early times when labor was forced, or hired, to work in the fields or on the monuments of antiquity, such as the pyramids and the great walls of the world. The apprentices and journeymen in the guilds and manor houses of the Middle Ages had to be fed by their owners, as did the slaves of ancient Greek and Roman households.

The Industrial Revolution brought great changes to social and economic systems. The plight of child laborers, in particular, resulted in legislation in England that forced managers to provide meal periods for the young workers.

Robert Owen, a Scottish mill owner near Glasgow during the early nineteenth century, is considered the "father of industrial catering." He so improved working conditions for his employees that his mill became a model throughout the industrial world. Among other things, it contained a large kitchen and eating room for employees and their families. Prices for meals were nominal, and so began the philosophy of subsidizing meal service for employees. In the United States in the 1800s, many employers provided free or below-cost meals to their employees, a practice that continues to some extent today.

The importance of industrial feeding was not fully realized, however, until the World War II period. Most of the workers in plants at that time were women who demanded facilities for obtaining a hot meal while at work. Because new plants were often in isolated locations and the competition for workers was intense, plant managers complied with worker demands and provided meal service facilities. Many plants have continued this service as an indispensable part of their operations, either under plant management or on a contract/concessionaire basis. Serving units within an industrial facility may include a central cafeteria, vending machines and snack bars throughout the plant, and mobile carts to carry simple menu items to workers at their stations. Some also include a table-service dining room for executives and guests.

A newer segment of employee feeding, for office building workers, has mushroomed since the 1970s. Enterprising contract foodservice caterers or companies have found this a ready market in larger cities. Many office workers enjoy the time-saving convenience of having meals delivered to them at their desks, eliminating the need to find an eating place during their brief meal period. This type of employee feeding appears to be increasing in popularity.

Today, industrial and business foodservice managers face challenges on many fronts. Their audiences are no longer quite as captive, and the employers are no longer as benevolent as they were in the 1800s. No longer subsidized, in-plant foodservice must offer a variety of menu selections, at a reasonable price, that are served in pleasant surroundings in order to compete successfully with the myriad of neighborhood fast-food outlets, mobile catering companies, sidewalk vendors, and "white tablecloth" restaurants.

Transportation Companies

In the 1940s, noncommercial foodservice, a large but quiet segment of American foodservice, began to branch into some nontraditional areas such as airline catering. J. Willard Marriott, who by this time had renamed his A & W restaurants "Hot Shoppes," began to supply box lunches to passengers on Eastern, American, and Capital airlines leaving Washington, D.C.'s old Hoover Airport. Full meals were supplied later on and delivered in special insulated carriers and placed on board by a custom-designed truck with a loading device attached to the roof.

Today a major segment of the foodservice industry—titled "food on the move"—is provided by airlines, trains, and cruise ships. For all three, food and foodservice make up a marketing tool used to "sell" travel on a particular line. Both airlines and railroads face the unique problem of planning foods that will "go the distance"; the logistics of food they serve is a challenge. Management must test the market to determine who the travelers are and what they want or like to eat. Keeping up with changing fads and fluctuations in food preferences is an important aspect of meal service on trains and planes.

Most cruise ships offer such a variety of foods, styles of service, and presentations that their passengers can find their preferences well met. Usually, there is more good food available than any person can consume. Food storage and preparation facilities on these ships are adequate to make them self-sufficient for the length of the cruise or ocean voyage.

Career opportunities in foodservice management with transportation companies offer many possibilities for the future, especially for persons who have good business backgrounds in addition to their training in foodservice.

TRENDS IN FOODSERVICE

Foodservice operators who stay on top of emerging trends have a better chance of attracting and satisfying customers and thus boosting sales and beating the competition. However, predicting trends is not always an easy task. Fads in foodservice are common. In contrast to a trend, which grows and matures, a fad is a fleeting interest. Fads are usually fun innovations that add interest and excitement, whereas trends are fueled by such present conditions as the state of the economy and changes in lifestyles.

Macro trends include an increase in the number of chain restaurant outlets (*points of access*), the increase in the use of technology, expansion of menus, and *family value marketing.*

Among the various segments of the market, the following trends are emerging:

- Corrections foodservice is expanding rapidly as prison populations increase and the use of the *cook/chill* process continues to grow.
- The fine dining restaurant business is down, but interest in cafes and bistros is increasing.
- Recreational facility foodservices are expanding with *upscaled menus.*
- School foodservice faces budget battles and legislative changes with an increase in the use of brand-name foods (*branding*) and the development of a business mentality.
- Hospital foodservice is employing *benchmarking* statistics to justify costs, introducing "grab-and-go" food in the staff cafeteria, espresso bars, limited patient menus, restaurant-style menus, comfort foods, and *satelliting* (selling food to other facilities).
- Foodservice in the lodging sector is incorporating mini-marts, ethnic fare, simpler foods, healthier selections, and buffets.
- College foodservice will see more self-service, grab-and-go options, extended hours of operation, authentic vegetarian dishes, and full-flavored ethnic choices.
- Military foodservice faces base closings but also better food quality, consistency, and pricing with more branding, catering to inactives, high-energy nightclubs, and kiosks and mini-units.
- Foodservice in nursing homes will serve to sicker and younger patients with more convenience products, more liberal diets, and a room service option.

Food trends include specialty coffees; entrée salads; high flavor condiments; spicy food with Mexican pegged as the next biggest ethnic cuisine after Italian; specialty desserts; *comfort foods* such as meat loaf, roast chicken, mashed potatoes, and fruit cobbler; pasta; and beef. One research group found that the top five growth foods between 1968 and 1992 were pasta, chips, turkey sandwiches, cereal, and soft drinks. Convenience and value were found to be the driving force behind consumers' choices. A diminishing of the importance of nutrition in food selection, with the percentage of consumers very concerned about nutrition dropping to its lowest point since 1985, has led to a stabilizing of the consumption of foods such as beef, eggs, and ice cream.

The *display cooking* trend in upscale restaurants is finding its way into institutional foodservices. Kitchenless, storage-free designs where all food is displayed and prepared in full view appeals to all the senses as customers see, hear, smell, and taste food go from raw to cooked.

In the noncommercial sector, whether to operate the foodservice with *in-house management* or to contract it to a *contract foodservice* company continues to be an important and difficult decision to make. After years of cutting the bottom line to control food and labor costs, contractors are shifting their focus to improving promotions, services, and price-value perceptions. Those who choose to stay with or return to in-house management cite the opportunity to increase revenue, improve quality and control, and stamp the operation with a unique signature as the reasons.

SUMMARY

This brief history of the development of the foodservice industry should give readers an appreciation of the industry as it is today. The present status of foodservice is impressive. NRA statistics indicate dramatic growth in the foodservice industry: a sales volume of $313 billion was predicted for 1996.

Individuals in the United States today eat 45% of their meals away from home, and one of four meals eaten at home was prepared outside the home. Foodservice is the number one employer among all retail business, with some nine million persons employed, two-thirds of whom are women and one-fourth of whom are teenagers.

The large patronage given restaurant enterprises greatly exceeds any casual estimate. Statistics on the number of persons in the military services, in hospitals, in correctional facilities, and in other federal, state, or municipal institutions show the enormous scope of necessary foodservices. Consider also the thousands of students whose nutrition the schools, colleges, and universities accept as a major responsibility. Industry, too, recognizes the importance of feeding its millions.

All U.S. citizens have direct personal contact with foodservices in institutions at some time during their lives. By paying taxes, they support federal, state, county, and municipal institutions that serve food to their residents. All persons probably have had meals at school, some in college dining halls, and others at their workplaces. Some people enjoy dining at their clubs, while others may have meals "en route" as they travel. At some time, a large percentage of U.S. citizens may require the services of a hospital or nursing home; still others may choose retirement home living for later years. Whatever our involvement, we expect the food to be of high quality, prepared to conserve its nutritive value, and served in the best condition and manner possible.

Essentials in attaining this goal are knowledge of food safety, menu planning, food purchasing, preparation, delivery service systems, successful personnel direction including delegation and supervision, wise planning and clear-cut organization, a good system of financial control, and efficient facility design and equipment arrangement. These are the responsibilities of the foodservice director and are discussed in the following chapters.

REVIEW QUESTIONS

1. Where did institutional foodservice get its start?
2. How have present-day lifestyles made an impact on commercial and institutional foodservices?
3. What are the trends in college and university foodservice? In hospital foodservice?
4. What major legislation established the National School Lunch Program?
5. Where did foodservice cost accounting get its start?
6. What sciences led to the improvements in methods used in institutional foodservice?
7. Who is considered the pioneer of dietetics and why?
8. What are the two main objectives of hospital foodservice today?

SELECTED REFERENCES

Allen, R. L.: Corporate contract feeders aim to beef up revenue. *Nation's Rest. News.* 1994; 28(23):4.

Barber, M. I.: *History of the American Dietetic Association.* Philadelphia: J. B. Lippincott, 1959.

Bartlett, M.: Industry surges ahead: Restaurants and institutions: '95 industry forecast: Industry overview. *Restaurants and Institutions.* 1995; 105(1):44.

Batty, J.: Uniforms through time. *Natl. Rest. Assoc.* 1992; 12(3):24.

Chaudry, R.: Companies to watch. *Restaurants and Institutions.* 1994; 104(1):119.

Deckard, L.: Bringing it all back home: In-house food and beverage service. *Amusement Bus.* 1994; 106(21):19.

Doherty, J. C.: Saluting 75 years: National Restaurant Association's 75th anniversary. *Nation's Rest. News.* 1994; 20(special issue):5.

Editorial. *J. Am. Diet. Assoc.* 1934; 9:104.

Feingold, S. N.: The futurist quiz. *The Futurist.* 1992; 26(2):56.

Hersch, V.: You are what you wear: Uniforms of restaurant personnel. *Rest. Bus. Mag.* 1993; 92(7):184.

Lang, J.: Money-making megatrends. *Voice of Foodservice Distr.* 1994; 30(15):48.

Muller, C. C., and Woods, R. H.: An expanded restaurant typology. *Cornell Hotel Rest. Admin. Quart.* 1994; 35(3):27.

NRA projects growth in 1992. *Food Mgmt.* 1992; 27(3):49.

Ryan, N. R.: Under the silver dome: Midyear trends run a wild and wacky gamut, from cold coffee to "hot pants," from daring dietitians to database dining. *Restaurants and Institutions.* 1994; 104(20):78.

Schuster, K., and Staff, F. M.: Crossing the line. *Food Mgmt.* 1988; 23(1):94.

Social Security Administration: Skilled nursing facilities: Standards for certification and participation in Medicare and Medicaid programs. *Fed. Reg.* 1974; 39:2241.

Study weighs influence of nutrition, convenience; study of consumer eating habits conducted by NPD Group Inc. for National Pasta Association. *Milling & Baking News.* 1993; 72(12):1.

Trends: School foodservice in the year 2000 and beyond. In *National Food Service Management Institute Conference Proceedings.* National Food Service Management Institute, 1992.

Woodman, J.: Twenty years of "400" translates into light years of change for foodservice. *Restaurants and Institutions.* 1984; 94(15):98.

2

Types of Foodservice Operations

OUTLINE

The foodservice industry is complex, fast growing, and ever changing. Many factors affect its growth and status, including socioeconomic conditions, demographic shifts, and the changing food habits and desires of the American people. Being alert to these changes will help foodservice managers adapt their operations to meet the demands of the times. This background information, together with a classification of the many types of foodservice markets, is provided in this chapter.

The systems approach to management is introduced. This concept is based on the idea that complex organizations are made up of interdependent parts (subsystems) that interact in ways to achieve common goals. Application of the systems concept is made to foodservice organizations.

Managers face decisions about how to organize foodservice departments for the efficient procurement, production, distribution, and service of their food and meals. Many options are available based on the type of food purchased, where the food is prepared in relation to where it is served, the time span between preparation and service, and the amount and kind of personnel and equipment required.

Foodservices with similar characteristics are grouped as particular types of production or operating systems. Each of the four types of foodservice operating systems found in the United States today is described with its identifying features, advantages, and disadvantages. The typical foodservice organizations that use each type are also identified. This description should provide a basis for managers to decide on the type of operation suitable for a particular situation.

KEY CONCEPTS

1. Socioeconomic trends and demographic changes affect the foodservice industry. Being able to predict these changes is an important attribute for a foodservice manager to possess.
2. The foodservice industry is vast and complex. The wide range of establishments in the industry may be classified into three major categories: commercial, institutional, and military. Each of these three may then be further categorized by specific type of operation.
3. The mission of a foodservice organization is the foundation on which all decisions should be made.
4. A system is a set of interdependent parts that work together to achieve a common goal. A foodservice organization is a system.
5. Systems theory evolved from earlier management theories such as scientific management, the human relations movement, and operations research.
6. The four major types of foodservice systems in operation today are conventional, ready-prepared, commissary, and assembly/serve. These classifications are based on differences in location of preparation, amount of holding time and method of holding cooked food, the purchase form of the food, and labor and equipment required.

SCOPE AND STATUS OF FOODSERVICES

As described in Chapter 1, foodservice in the United States today is a complex and fast-changing industry, one that has expanded rapidly in the last half-century. It ranks as the number one retail employer with more than nine million workers. A conservative estimate is that 45% of meals consumed are planned, prepared, and served outside the home in a variety of establishments.

Factors Affecting Growth

The growth in patronage of foodservices may be attributed in part to socioeconomic trends and other demographic changes. For example, the *changing status of women* has had an influence on the workforce. In 1970, approximately 43% of women over 16 years of age were working, and in 1993, 59% of women in that age group were in the workforce. Today, two-thirds of the industry's employees are women and 7 out of

10 supervisors in food preparation and service occupations are women. This may indicate a dual income for some families and may contribute toward the creation of a more affluent society. More people can afford to dine out; more women are lunchtime customers.

Another factor influencing the foodservice industry is the increasing number of *single-person households* and the potential for people living alone to eat out. They tend to spend a larger portion of their food budget on meals away from home than do family groups. Also, the population growth in the United States seems to be slowing. If this trend continues, there will be fewer young people and an increasing number of older persons in our society. The *average age of the U.S. population*, now nearly 34 years, will continue to increase as the number of babies born remains low and the life span of adults continues to lengthen. These facts seem to indicate a need for more retirement and health care facilities, an older target market for restaurants, and a change in the age groups in the labor market. Recent *changes in the American workplace* have also had far-reaching effects on the foodservice market. With the shift toward high technology and the computer industry, there are more office-bound jobs and white collar workers. In-plant feeding is down and contract foodservice in business offices is increasing. The shortened work week of recent years has added leisure time and promoted the recreational foodservice segment of the industry.

The awakened *interest in the health and well-being* of people and concern about improving the nutritional status of individuals have also had an impact on foodservice. In fact, much research is being conducted and reported by the media concerning the impact of nutrition on health. People are becoming generally more knowledgeable about nutrition and food safety. As a result, most types of foodservices, from schools and colleges to airline and commercial operations, are offering healthier choices in their menus.

All of these factors have helped shape the foodservice industry into what it is today. Managers must always be alert to societal trends and have the ability to adjust their operations to the changing situation in order to be competitive and successful in this market.

CLASSIFICATION OF FOODSERVICES

The foodservice industry is broad in scope and encompasses a wide range of establishments. *Restaurant USA* classifies them into three major groups:

- Commercial,
- Institutional—business, educational, governmental or institutional organizations that operate their own foodservice, and
- Military.

Further, these three groups can be divided into more specific categories according to type of operation:

Commercial Foodservices

- Eating places
    - Full-service restaurants
    - Limited service (fast-food) restaurants
    - Commercial cafeterias
    - Social caterers
    - Ice cream, frozen-custard, yogurt stands
- Food contractors
    - Manufacturing and industrial plants
    - Commercial and office buildings
    - Hospitals and nursing homes
    - Colleges and universities
    - Primary and secondary schools
    - In-transit foodservice (airlines)
    - Recreation and sports centers
- Lodging places
    - Hotel restaurants
    - Motor-hotel restaurants
    - Motel restaurants
    - Retail-host restaurants (store and gas station foodservices)
    - Recreation and sports (includes movies, bowling alleys, recreation and sports centers)
    - Mobile caterers
    - Vending and nonstore retailers

Institutional Foodservice

- Employee foodservice
- Public and parochial elementary and secondary schools
- Colleges and universities
- Transportation
- Hospitals
- Nursing homes and homes for the aged, blind, orphans, and the physically and mentally challenged
- Clubs, sporting and recreational camps
- Community centers

Military Foodservice

- Officers' and NCO clubs ("open mess")
- Military exchanges

Each of the establishments in these classifications has its own objectives, goals, and types of organization and management. Although they may seem widely divergent, each is concerned with providing a foodservice to some segment of the public. There is a commonality among them that can be identified for the purpose of grouping them into specific types of foodservice operations.

FOODSERVICE OPERATIONS

The Nature of Foodservice Management

All organizations have a *mission* that evolves from their reason for existence. A written *mission statement* is rapidly becoming a common document for guiding organizational decision making. To achieve this mission effectively, the organization must then develop specific targets or *objectives*. For example, a foodservice organization's mission might be to satisfy customers by serving high-quality, nutritious food at reasonable prices while achieving a desired profit for the organization. The objectives in this case might be such *benchmarks* as percent of customers marking satisfied and above on a rating scale, increase in total sales and/or number of customers, number of "regular" customers, and net profit. It is the responsibility of management to achieve the organization's objectives.

A generic definition of management is that it is the effective and efficient integration and coordination of resources to achieve the desired objectives of the organization. *Managerial effectiveness* may be measured by how well the organization achieves its objectives over time. *Efficiency,* in contrast to effectiveness, is a measure of short-term objectives. If a foodservice was paying $1 for a head of lettuce and using an entire head for an individual salad, we would surmise that a lot of lettuce was being wasted. This is a comparison of input of lettuce to output of one salad—an inefficient use of resources, a short-term measure. The effectiveness measure would be to produce a high-quality, nutritious salad at a reasonable price in order to satisfy potential customers and return a profit to the organization.

Of prime importance to any organization in this increasingly competitive world is how well it is able to adapt, reach its objectives, and serve its mission. Viewing the organization as a system is essential in this endeavor, as is choosing the correct production system for the particular needs or characteristics of the operation.

The Systems Concept and Approach

Before discussing foodservice organizations as "systems," this section briefly reviews the systems *concept* and systems *approach* and how systems theory has evolved from other theories of management. This review will establish a common basis of understanding and make application of the systems concept to foodservice an easy transition.

The word *system* is used freely and in many different contexts. We read and speak of the solar system, defense system, transportation system, school system, and even of the human body as a system. A system has been defined in many ways and with so many different words that it may seem confusing. However, this commonality is found among systems: A system is a set of interdependent parts that work together to achieve a common goal. The interrelated parts are known as subsystems, each dependent on the others for achieving its goals. For example, a train cannot achieve its goal of transporting passengers from one destination to another if the wheels are off the track even though all parts of the train are in good working order. All elements must be coordinated to function together for success.

Organizations are systems. This concept has evolved gradually from earlier theories of management. Traditional views in prominence in the late nineteenth and early twentieth centuries included the scientific management theory, which puts emphasis on efficient work performance. Workers were trained to perform a task in what was perceived to be the *one* best way. If all performed efficiently, the goals could be reached.

In the late 1920s, research conducted by Elton Mayo and his associates at the Hawthorne Plant of the Western Electric Company led to the findings that social and psychological factors were critically important determinants of worker satisfaction and productivity. Thus, the human relations movement in industry began.

After World War II, quantitative methods began to be employed for the purposes of decision making. The application of computer technology and mathematical models was collectively called operations research or management science.

Each of these theories contributed to the development of systems theory as we view it today: The organization is an entity composed or made up of interdependent parts—the subsystems. Each subsystem contributes to the whole and receives something from the whole while working to achieve common goals. Management's role is considered a "systematic endeavor," one that recognizes the needs of all of the parts. Decisions are made in light of the overall effect of management on the organization as a whole and its objectives. This type of leadership is the systems approach, that is, an acceptance of the systems theory of management and the use of it as a style of managing. The recognition that a change made in one part of the system has an impact on all parts of the system is an example of the use of the systems approach. Three areas of common usage of this approach are:

- **Systems philosophy,** which is described as a way of thinking about phenomena in terms of wholes, including parts, components, or subsystems, with emphasis on their interrelationships,
- **Subsystems analysis,** as a method for problem solving or decision making, and
- **Systems management,** as the application of the systems theory to managing organizational systems or subsystems.

Money, materials, time, equipment, utilities, facilities, and personnel, together with the necessary information, are the *inputs* into the system. The work that is performed, known as *operations*, transforms the raw material into the finished product or services. These are the *outputs*. The outputs provide the information on how the operations worked or failed, or how they should be changed or modified. This information is known as *feedback* and provides management with data for decision making.

An organization is also an open system that interacts regularly with external forces in its surrounding environment. These forces include various regulatory agencies, customers, competitors, and suppliers. Practices within the organization are affected by these external forces and, conversely, the organization has an effect on the forces in its environment.

Various diagrams can be used to illustrate an organization as a system with its inputs, the subsystems that perform the operations, and the outputs, together with their interactions with the environment. One that is clear, simple, and easily adaptable to specific organizations is shown in Figure 2.1.

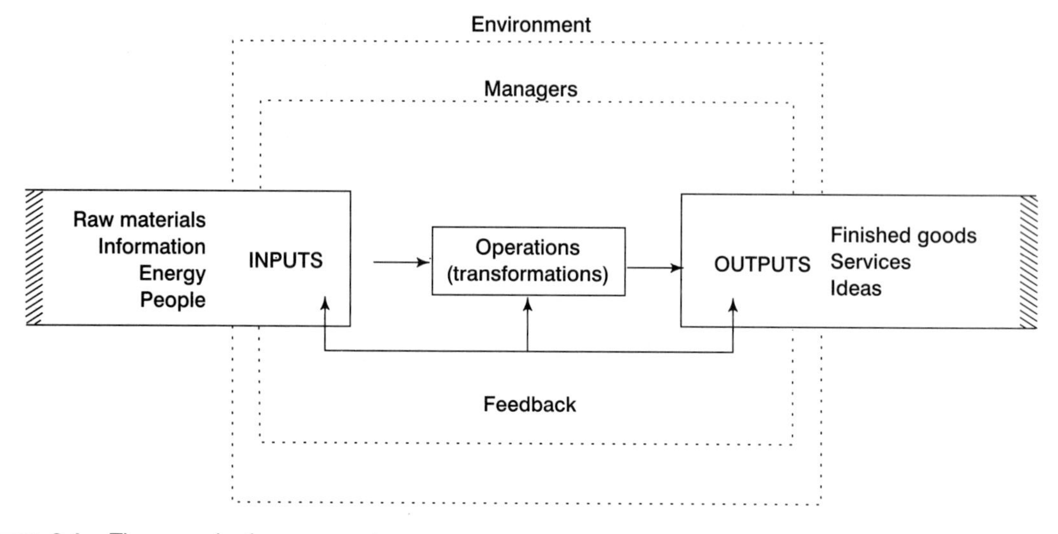

Figure 2.1 The organization as a system.

The inputs in a foodservice system are food products, supplies, personnel labor and skills, money, utilities, information, time, facilities, space, and equipment. Transformation of these inputs into outputs takes place in the functional subsystems shown in Figure 2.2. The resulting outputs are ready-to-serve foods, clientele and personnel satisfaction, and financial accountability. Ideas generated from the results of operations are the feedback for use in improving the operation as necessary. All parts of the system are linked by management functions, such as planning, organizing, and staffing, which are discussed in Part 4. To accomplish unification of the system, managers use various linking processes, such as communication and decision making. Surrounding the system are environmental factors, such as regulatory agencies, the economy, social and cultural aspects, and the various constituents of the operation, such as customers and suppliers.

Types of Foodservice Systems

Those foodservices that operate in a similar manner, or with common elements, give the basis for grouping them into specific types of systems. Four major types of foodservice systems are in operation in the United States today. The systems differ in *where* the food is *prepared* in relation to where it is served, the *time span* between preparation and service, the *forms of foods* purchased, *methods of holding* prepared foods, and the amount and kind of *labor and equipment required*. These four types of foodservice systems are conventional, ready-prepared (cook/chill or cook/freeze), commissary (central production kitchen), and assembly/serve.

Conventional As the name implies, the **conventional system** has been used traditionally throughout the years. Menu items are prepared in a kitchen in the same facility where the meals are served and held a short time, either hot or cold, until serving time. In earlier years, all preparation as well as cooking took place on the premises and foods were prepared from basic ingredients. Kitchens included a butcher shop, a bakery, and vegetable preparation units.

Over the years, a modified conventional system has evolved because of labor shortages, high labor costs, and the availability of new forms of food. To reduce time and labor costs, foodservice managers began to purchase some foods with "built-in" labor. Butcher shops, in which meats were cut from prime cuts, and bake shops are gone from most "conventional" kitchens today. Meats are now purchased ready to cook or portion controlled; bread and many bakery items are purchased from a commercial bakery or prepared from mixes; and produce is available in prepeeled, cut, frozen, or canned forms, all of which reduce the amount of production and labor required on the premises. Foods with varying degrees of processing are now used in conventional foodservice systems.

This system is most effective in situations and locales where the labor supply is adequate and of relatively low cost, where sources of food supplies, especially raw foods, are readily available, and when adequate space is allocated for foodservice equipment and activities.

Typical users of the conventional system are smaller foodservice operations such as independent restaurants, schools, colleges, hospital and health care facilities, homes for specialized groups, and in-plant employee feeding.

Advantages The conventional system has many advantages. Quality control is considered of primary importance. Through the menus, recipes, and quality of ingredients selected by the manager, the foodservice achieves its individuality and standard of quality desired. It is not dependent on the availability and variety of frozen entrées and other menu items commercially prepared. This system is more adaptable to the regional, ethnic, and individual preferences of its customers than is possible with other systems. From an economic standpoint, greater flexibility is possible in making menu changes to take advantage of good market buys and seasonal fluctuations. Also, less freezer storage space is required than with the other systems and distribution costs are minimal, both of which save on energy use and costs.

Disadvantages The conventional system produces an uneven, somewhat stressful workday caused by meal period demands. Because the menu differs each day, the workloads vary, making it difficult for workers to achieve high productivity. Skilled workers may be assigned tasks that could be completed by nonskilled employees just to fill their time between meal periods. When three meals a day are served, two

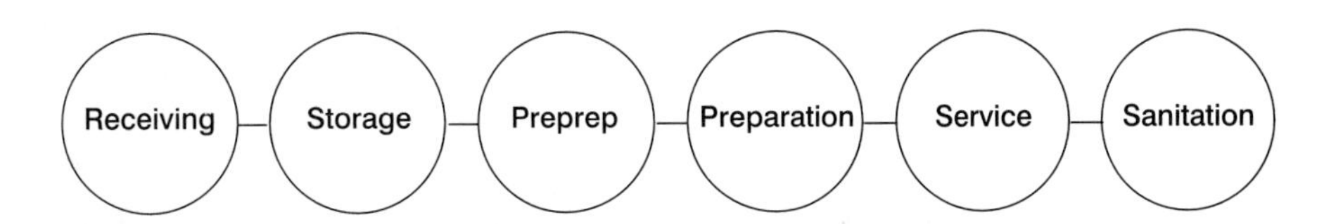

Figure 2.2 The functional subsystems of a traditional foodservice operation.

shifts of employees are required to cover the 12- to 15-hour or more workday. Scheduling workers may be difficult with overlapping shifts.

Ready-Prepared (Cook/Chill or Cook/Freeze) In the ready-prepared system, foods are prepared on the premises, then chilled or frozen and stored for use at some later time. Thus, foods are "ready," prepared well in advance of the time needed. This is the distinct feature of ready-prepared foodservice systems—the separation between time of preparation and service. In addition, the food is not for immediate use as in the conventional system. Unlike the commissary system, foods *are* prepared on site; however, the place of preparation is not the place of service.

The **cook/chill method** can be accomplished in a variety of ways, but basically the food is prepared and cooked by conventional or other methods, then its temperature is brought down to 37°F in 90 minutes or less, and refrigerated for use at a later time. In one variation, prepared food is either preplated or put into bulk containers, such as hotel pans, chilled in a blast chiller, stored in a refrigerator for up to five days, rethermalized, and served. In another method, food items are prepared in kettles, pumped into special air- and water-tight plastic packages (Figure 2.3) that hold

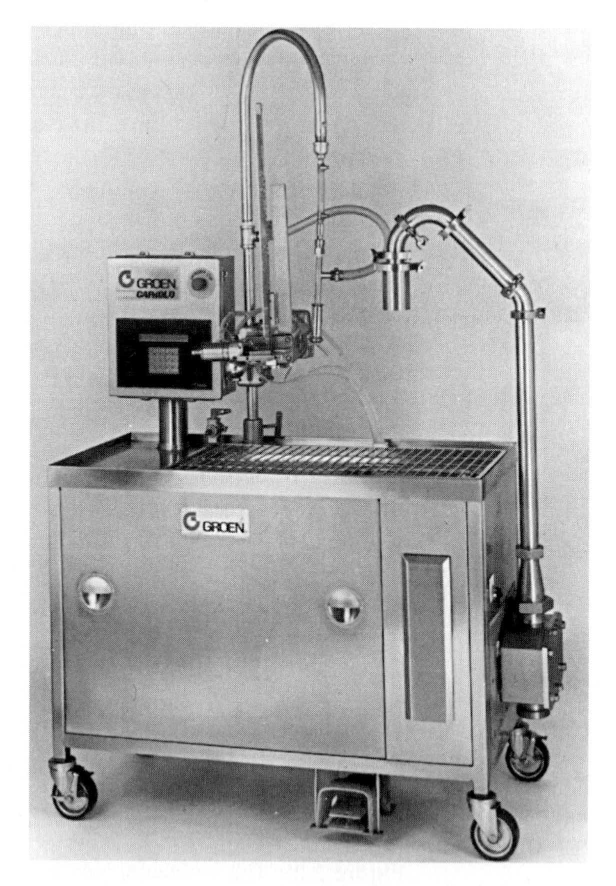

Figure 2.3 This packaging station is designed to pump products from the kettle into flexible plastic casings or pouches or into vacuum packages for cook-tank processing.
Courtesy of Groen, Elk Grove Village, Illinois.

Figure 2.4 A tumble chiller/cook tank that tumbles flexible casings in an ice-water bath or slow cooks vacuum-packaged products in circulating hot water.
Courtesy of Groen, Elk Grove Village, Illinois.

1.5 to 3 gallons, given an ice-water bath in a tumble chiller (Figure 2.4), and stored in the refrigerator. Food items prepared by this method may be held for up to 45 days. Meat is prepared in this method by putting it in a large tank that automatically cooks the meat and then chills it in ice water as soon as the cooking cycle is over. Meats can then be refrigerated for up to 60 days.

In the **cook/freeze method,** a blast freezer or cryogenic freezing system must be available to freeze foods quickly and thus prevent cell damage. Foods for freezing may be preplated, but more often they are stored in bulk, which requires less freezer storage space.

Note that the ready-prepared entrées and vegetables undergo two heating periods: first, when foods are prepared and, second, after storage to reheat them for service to the consumer. Ready-prepared systems were developed to offset the critical shortage and high cost of skilled foodservice employees. Also, it was seen as a way of evening out the workload from day to day and during each day because only certain menu items are prepared on any given day to build up an inventory for future use.

Advantages The advantages of the ready-prepared system are related to reducing the "peaks and valleys" of workloads that may be found in the conventional system. Production scheduling to build up the menu item inventory can be on a 40-hour week, 8-hour day, without early morning and late evening shifts. Employee turnover is decreased and recruitment of new employees is enhanced by offering staff a more normal work week and reasonable hours.

Other advantages are reductions in production labor costs, improved quality and quantity control by decreasing job stress related to production deadlines, and improved nutrient retention by decreasing time food is held within the serving temperature range. There can be more balanced use of equipment when preparation is spread over eight hours, rather than at mealtime only.

Management has close control over menu selections, the quality of ingredients, and portion size and quantity. This is not always true in other systems, especially with the assembly/serve system. Menu variety is potentially greater with this system, because many items can be prepared and stored for future use.

One advantage that the ready-prepared system has over the commissary system is the lack of worry about delivery from the central production kitchen. When foods are prepared and stored on the premises, menu items are available on call and no waiting is involved.

Disadvantages One disadvantage is the need for large cold storage and/or freezer units, which take space and add to energy costs. Depending on the method, a blast chiller or blast freezer is required, and they are expensive to purchase and operate. Control for food safety is especially essential with the cook/freeze method. Longree (1980) warns that "the production of precooked frozen foods must not ever be handled in a haphazard fashion; unless the freezing operation can be a continuous, streamlined, bacteriologically controlled, short-time process, the bacteriological hazards could be formidable." (See Chapter 3 for more information of food safety.)

Because frozen foods are prone to structural and textural changes, extensive modifications in the recipe and ingredients are usually necessary to offset cell damage and to assure high-quality products.

Appropriate and adequate equipment for rethermalizing foods prior to service is essential and can be costly. Microwave and/or convection ovens are the equipment usually used in service units located near the consumers.

Although ready-prepared systems have been used primarily by large-volume institutions and centralized commissary chain setups, such as health care units, employee feeding facilities, airlines, and correctional institutions, lower volume applications have begun to appear. Schools, supermarkets, fast-food companies, and large restaurants are now utilizing this technology.

Commissary (Central Production Kitchen) The **commissary foodservice system** is described as a large, central production kitchen with centralized food purchasing and delivery of prepared foods to service (satellite) units located in separate, remote areas for final preparation and service. This system was made possible by the development of large, sophisticated equipment for preparing and cooking large quantities of food from the raw, unprocessed state. Foodservice organizations with many serving units, sometimes widely separated as in a large city school system, sought ways to consolidate operations and reduce costs. The commissary system is the result.

Prepared foods may be stored frozen, chilled, or hot-held. Menu items may be distributed in any one of several forms: bulk hot, bulk cold, or frozen for reheating and portioning at the satellite serving units; or preportioned and preplated for service and chilled or frozen before delivery.

Typical users of this system are airline caterers, large city school systems, and franchised or chain restaurant organizations that provide food for their various outlets and vending companies.

Advantages The commissary foodservice system can realize cost savings due to large-volume purchasing and reduced duplication of the labor and equipment that would be required if each serving unit prepared its own food. Some facilities where food is served may not have adequate space for a production kitchen, or the space can be better utilized for some other purpose. Quality control may be more effective and consistent with only one unit to supervise.

Disadvantages Food safety and distribution of prepared foods may be concerns. There are many critical points in mass food production where contamination could occur. Employment of a food microbiologist or someone knowledgeable about safe techniques in mass food handling with specialized equipment is highly desirable yet often costly.

Food must be loaded and transported in such a manner that it is maintained at the correct temperature for safety and is of good quality and appearance when received for service. This requires specialized equipment and trucks for delivery. Poor weather conditions, delivery truck breakdowns, or other such catastrophes can result in food arriving late, causing irritating delays in meal service.

Another disadvantage is the high cost of purchase, maintenance, and repair of the sophisticated and specialized equipment needed for this type of production and distribution.

Assembly/Serve The **assembly/serve system** requires no on-site food production. This has led to the use of the term "kitchenless kitchen." Fully prepared foods are purchased and require only storage, final assembling, heating, and serving. Assembly/serve systems evolved with the development of a variety of high-quality frozen entrées and other food products that have appeared on the market in recent years. Also, foodservice managers confronted with high labor costs and few skilled employees turned to this system to relieve the labor situation. Often with this system, "single-use" disposable tableware is used, thus eliminating the need for a dishwashing unit.

With the availability of frozen entrées with a starch that are low in fat and sodium, some hospitals have begun to purchase these retail-size commercially prepared frozen entrées for their patient foodservice. They are then "popped out" onto the service plate and rethermalized with IQF (individually quick frozen) vegetables and served. These pop-out food items have resulted in the system being characterized as "pick, pack, pop, and pitch!" In addition to the regular production line items, some companies are willing to produce items according to individual purchaser's recipes and specifications. In addition to frozen foods, assembly/serve systems are also beginning to use *sous vide*. Sous vide is a method of food production in which foods are precooked and vacuum packed. Rethermalization is accomplished by boiling the food in the vacuum packages in which they are stored.

The primary users of the assembly/serve system are hospitals, yet some health care institutions and restaurants also use it. Although foodservices of all classifications can use prepared entrée items, few have adopted them exclusively. Hotels and restaurants that employ unionized chefs can be prohibited from using frozen entrées.

Advantages The foremost advantage of the assembly/serve system is the built-in labor savings. Fewer personnel are required and they do not have to be highly skilled or experienced. Procurement costs are lower due to better portion control, less waste, reductions in purchasing time, and less pilferage. Equipment and space requirements are minimal as are operating costs for gas, electricity, and water.

Disadvantages The availability in some markets of a good selection of desired menu items or those that have regional appeal is limited. However, more and better quality frozen entrées are becoming available. The higher cost of these prepared foods, however, may not be offset by the labor savings realized. Managers must carefully weigh the overall cost of this system.

Another disadvantage may be the quality of available prepared products and customer acceptability. The proportion of protein food (meat, fish, seafood, etc.) to sauce or gravy in some menu items may not be adequate to meet the nutritional requirements of the clientele. For example, two ounces of protein are required in the school meal pattern in school foodservice programs. Many frozen entrées may contain much less than that. Evaluation of products under consideration for use in the assembly/serve system is essential.

A manager considering a change from another system to the assembly/serve system should carefully evaluate the change in amount and kind of equipment needed. It may be excessively high in cost and in energy consumption to operate the duplicate pieces of heating equipment. Additional freezer space required for storage of the inventory of frozen entrées may not be available or may be too costly to install. Recycling or disposal of the large quantities of packaging materials and single-use tableware, if used, must be part of the total concern.

Each type of foodservice system has proved successful in providing acceptable quality food in specific organizations with the conditions described for each. However, foodservice managers attempting to decide on one system over another should undertake an extensive investigation and study before making any decisions. Among the factors to consider are cost comparisons, availability of foods in all forms, quality, and nutritional value of fully prepared items, customer needs and acceptability, equipment and space requirements, energy use as estimated by the amount and kinds of equipment needed for each system, and availability and cost of labor.

A summary of the major characteristics of each system is given in Table 2.1. A flowchart of the step-by-step processes of the four foodservice systems is shown in Figure 2.5.

SUMMARY

Today's foodservice managers should view their organization as a system composed of various elements or subsystems that are united by a common goal and that are

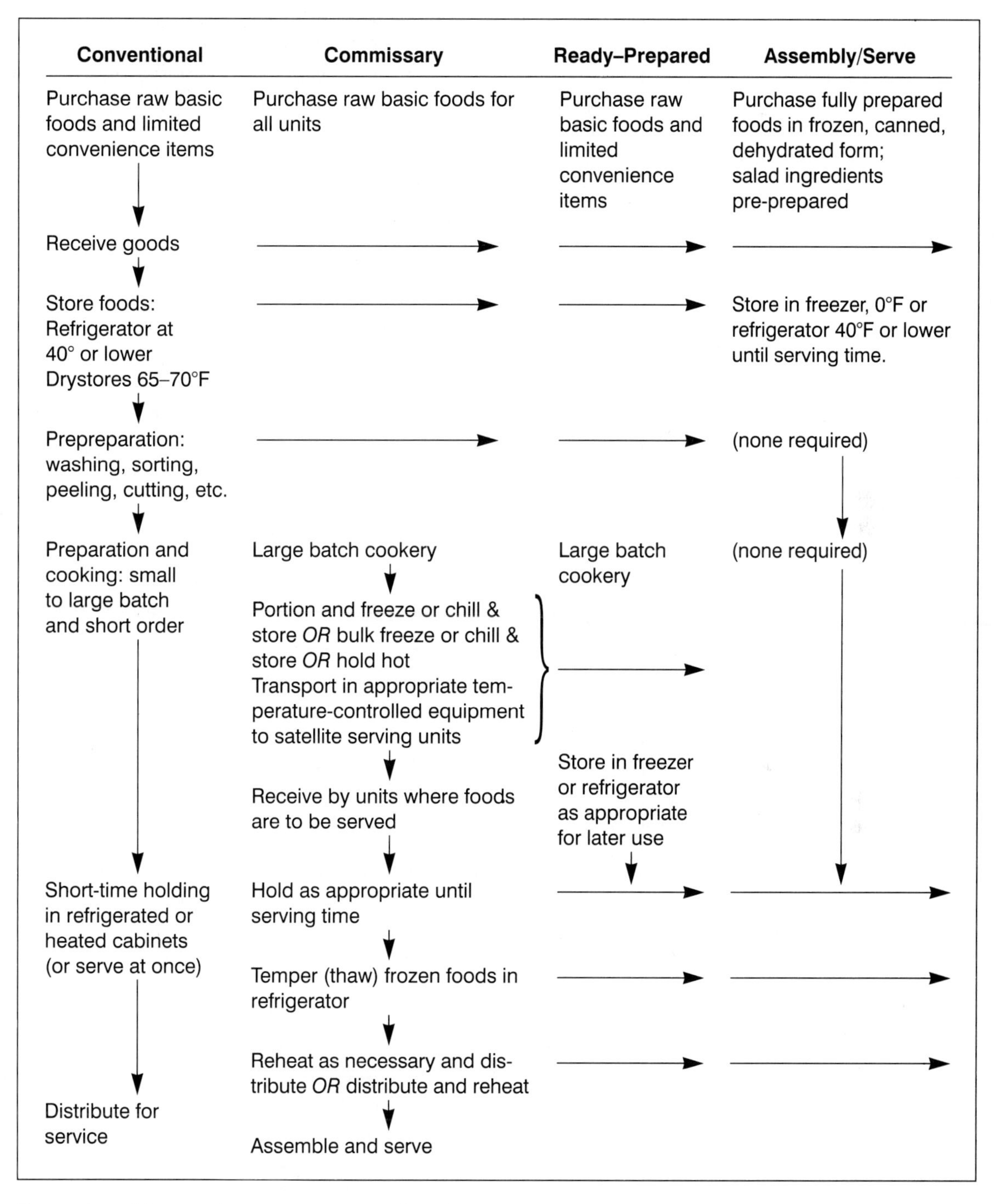

Figure 2.5 The step-by-step processes of the four foodservice systems.

Table 2.1 Summary of the characteristics of the four types of foodservice systems.

	Conventional	Ready Prepared Cook/Chill	Ready Prepared Cook/Freeze	Commissary	Assembly/Serve
Location of food preparation kitchen in relation to where served:	On premises where food is served	On premises where food is served		Central production kitchen in building separate from service units. Food transported to satellite serving units.	Off premises (commercially prepared foods are purchased)
Form of food purchased:	Raw; some convenience foods	Raw; some convenience foods		Primarily raw ingredients	All convenience and pre-pared foods—frozen, canned, dehydrated, or prepeeled fresh
Food procurement:	Purchase for its own unit	Purchase for its own unit		Centralized purchasing for all service units	Purchase for own use
Time span between preparation and service, and method of holding:	Food prepared for immediate service (may be held hot, or chilled for a few hours)	Food prepared and cooked then chilled and held for 1–3 days, or 45–60 days depending on the system.	Food prepared and fast frozen; held for later use up to 3–4 months	Food prepared and may be (a) distributed to satellite units for immediate service, (b) chilled and either pre-plated or put into bulk, (c) chilled and frozen and stored for later use either pre-plated or in bulk	No on-premises preparation. Foods purchased pre-prepared are stored and ready for reheating and service at any time needed

Amount and kind of equipment required:	All pre-preparation, cooking and serving equipment needed. Both skilled and unskilled employees needed	All pre-preparation, and cooking and serving equipment. One or more blast-chillers—large amounts of refrigerated storage space, or cook tank, water bath, and pumping equipment	A "blast" or cryogenic freezer—large amounts of freezer holding space.	Large, sophisticated equipment for pre-preparation and cooking. Some robots may be used—can be reprogrammed for various tasks. Suitable containers for packaging and delivery; trucks to deliver prepared foods to service units; reheating equipment if foods frozen or chilled	Equipment for reheating as steamers, steam jacketed kettles, convection or microwave ovens. Equipment for setting up and serving. Reheating equipment as convection or microwave ovens and kettles for immersion heating
Labor needs:	Skilled cooks and preparation workers as well as less skilled for pre-preparation and serving	Fewer highly skilled cooks needed compared with conventional because of "production line" type of work and only one or two items prepared per day; workers needed to reheat foods, operate that equipment, and assemble and serve meals		Highly trained in technological aspects of food production in mass quantities. Food microbiologists to assure food safety. Employees must be able to operate highly specialized equipment used for food production.	No skilled cooks or other pre-preparation employees needed. Workers for assembling salads and desserts, etc. Workers for reheating and serving foods must be able to operate equipment.
Typical foodservices using this:	Independent restaurants and cafeterias; hospitals and health care facilities; homes for specialized groups; in-plant foodservices; colleges and universities; schools	Large hospitals, some large colleges and universities		Airlines; chain restaurants; large school districts; commercial caterers and vending companies	Hospitals and nursing homes. Some commercial foodservices and colleges

interdependent and interact so that the processes or functions involved produce outcomes to meet stated objectives.

An overall definition of foodservice systems is well stated by Livingston and Chang (1979): A food service system is an integrated program in which the procurement, storage, preparation and service of foods and beverages, and the equipment, methods (and personnel) required to accomplish these objectives are fully coordinated for minimum labor, optimum customer satisfaction, quality, and cost control.

The arrangement of subsystems, procurement, food preparation, delivery and service, and sanitation into varying ways is the basis for grouping foodservices into types of systems, each with common elements and procedures. Four major types of foodservice systems found in the United States are conventional, ready-prepared, commissary, and assembly/serve. An evaluation of the merits of each system based on its characteristics, advantages, and disadvantages should be made before any one is adopted for use in a specific foodservice organization. The North Central Regional research group (1977) has stated the rationale for each of the four systems, which should be helpful in this decision making:

> *Rationale for Conventional Foodservice Systems.* Traditionally, effective foodservice administrators with conventional foodservice systems have utilized a skilled labor force for food production 13–14 hours per day. Given adequate food production equipment and available skilled labor, foods may be procured with limited amounts of processing. However, with constantly rising labor costs within the foodservice industry, the current trend in conventional foodservice systems is to procure more extensively processed foods.
>
> *Rationale for Commissary Foodservice Systems.* The commissary foodservice principles have been adopted in systems where service areas are remote from, yet accessible to, the production center. This concept can be applied to reduce the duplication of production labor and equipment which occurs if production centers are located at each foodservice site. Space requirements at the service sites are minimized because limited production equipment is required. By centralizing food procurement and production, the economies of volume purchasing may be realized. Commissary foodservice concepts are employed to meet various operational objectives related to effect use of resources.
>
> *Rationale for Ready-Prepared Foodservice Systems.* Mass producing and freezing food may reduce labor expenditures by more effective use of labor in selected situations. Peak demands for labor may be removed because production is designed to meet future rather than daily needs. Furthermore, fewer skilled employees can be trained to heat and serve menu items, thus reducing the number of highly skilled workers required by the system. Food procurement in volume may decrease food costs for the system. A foodservice system based on ready-prepared products is contraindicated if additional expenditures for storage facilities, equipment, and food inventory cannot be absorbed by the organization.
>
> *Rationale for Assembly/Serve Foodservice Systems.* Assuming a lack of skilled food production employees, and an available supply of highly processed, quality food products, an assembly/serve foodservice operation may achieve operational objectives to provide client satisfaction. Managerial decisions to adopt this form of foodservice system should consider the availability of these resources to the foodservice operation.

Recent research studies on foodservice systems in relation to time and temperature effects on food quality have been summarized and reported in another North Central Research bulletin. These microbiological safety, nutrient retention, and sensory quality studies provide specific data useful to persons deciding on a system to install or to those contemplating a change in systems. Further investigations are needed to advance understanding of the interrelationships among food products, resources, processes, and management in foodservices and so improve food quality in foodservice establishments.

The vast and ever-changing foodservice industry continues to be shaped by socioeconomic changes, demographic shifts, and the varying food habits and desires of the American people. Foodservice managers must keep abreast of these conditions and adapt their operations to the changing times in order to be competitive and successful.

REVIEW QUESTIONS

1. What are the socioeconomic conditions and demographic changes that have influenced the foodservice industry, and how have they had an impact on foodservices?
2. Define the systems concept.
3. Compare and contrast the systems theory to the scientific management theory.
4. Diagram a foodservice organization as a system.
5. Compare and contrast the four major types of foodservice systems described.
6. Which foodservice system(s) should be considered in each of the following situations?:
   a. high labor cost in the area
   b. very low equipment budget
   c. close quality control is desired
   d. high food cost in the area

SELECTED REFERENCES

Berkman installs 1st "kitchenless" kitchen. *FoodService Director.* 1991; 4(8):1.

Employment and Earnings. Washington, D.C.: U.S. Department of Labor, January 1995.

1994 Foodservice industry forecast: Restaurants reengineer. *Restaurants USA*. 1993; 13(11):15.

Foodservice Systems: Product Flow and Microbial Quality and Safety of Foods. North Central Regional Research Publication No. 245, Research Bulletin 1018. Columbia, Mo.: University of Missouri Agriculture Experiment Station, 1977.

Hitt, M. A., Middlemist, R. D., and Mathis, R. L.: *Management Concepts and Effective Practice,* 3rd ed. Saint Paul, Minn.: West Publishing Company, 1989.

Livingston, G. E., and Chang, C. M., eds.: *Food Service Systems: Analysis, Design and Implementation*. New York: Academic Press, 1979.

Longree, K.: *Quantity Food Sanitation,* 3rd ed. New York: John Wiley and Sons, 1980.

Rising food and labor costs top "worry list." *FoodService Director.* 1991; 4(8):71.

Schuster, K., and Staff, F. M.: Crossing the line. *Food Mgmt.* 1988; 23(1):94.

Soap box. *FoodService Director.* 1992; 5(1):34.

Statistical Abstract of the United States, 14th ed. Washington, D.C.: U.S. Department of Commerce, September 1994.

Ward, B.: Cold facts on quick chillers. *Foodservice Equipment & Supplies*. 1991; 44(9):65.

2

Quantity Food Production and Service

3

Food Safety

OUTLINE

Foodborne illness continues to be a major cause of illness, distress, and preventable death in the United States. Each year thousands of people become ill from contaminated food and it is estimated that 9000 people die each year as a result of unsafe food. Knowledge of and a commitment to food safety by the food manager and the entire foodservice staff are critical to the prevention of foodborne illness. The purpose of this chapter is to review foodborne illness and basic food microbiology as a basis for designing a food safety program. This basic information is followed by specific issues related to food safety in a foodservice operation. These issues include personal hygiene, food safety regulations, and food safety programs. The Hazard Analysis and Critical Control Points (HACCP) program is presented in detail and a model HACCP plan for foodservice is offered as a guideline for implementing this food safety program. *Note:* Due to the seriousness of foodborne illness, food safety is presented here as a chapter in itself. Cleaning and sanitation are covered later in Chapter 8.

KEY CONCEPTS

1. Foodborne illness is a major cause of illness, distress, and death in the United States.
2. The incidence of foodborne illness appears to be increasing.
3. Populations susceptible to foodborne illness include children, the elderly, pregnant women, and the immunocompromised.
4. Costs of foodborne illness are estimated to be in the billions of dollars.
5. The primary organisms that cause foodborne illness are the salmonellae strains, *Campylobacter jejuni*, pathogenic *Escherichia coli*, and *Vibrio parahaemolyticus*.
6. Emerging pathogens are those organisms that have been increasingly identified as causing foodborne illness.
7. Implementing standards for good personal hygiene is critical for infection control.
8. Employees need to be trained in proper food handling techniques to prevent microbial growth and cross-contamination.
9. An official system of government laws, regulations, codes, and standards has been designed to protect the public from unsafe food.
10. The Hazard Analysis and Critical Control Points program is a food safety program that identifies and controls hazards during the entire food production process from receiving to service.

FOODBORNE ILLNESS

Foodborne illness or disease is defined by the World Health Organization (WHO) as "a disease of an infectious or toxic nature caused by, or thought to be caused by, the consumption of food or water." A more inclusive statement defines foodborne illness as any illness that results from something that has been eaten. This definition then includes microbial, as well as physical and chemical causes of foodborne illness. Microbial and chemical causes of foodborne illness are discussed in detail later in this chapter. Physical contamination refers to the presence of any particle that is typically not a component of the food item. Examples include metal and glass.

It is important for the reader to distinguish between food that is unsafe for human consumption and that which is spoiled. Any food that is not fit for human consumption is spoiled. A spoiled food however is not necessarily unsafe in the sense that it would result in foodborne illness. For example, a badly bruised apple may be spoiled and undesirable for consumption but would not cause foodborne illness if eaten. A number of criteria are used to assess whether a food is fit to eat. These criteria include the following:

- State of maturity, development, or ripeness;
- Freedom from contamination such as pollution, insects, rodents, chemicals;
- Freedom from microorganisms and parasites that cause foodborne illness;
- Freedom from contamination that can occur during handling (receiving, production, and service); and
- Freedom from decomposition and/or contamination that can occur during storage (freezer-burn, excess ripening, or fresh produce contaminated with drippings from thawing meat).

Incidence of Foodborne Illness

The **Centers for Disease Control and Prevention** (CDC) in Atlanta, Georgia, an agency of the U.S. Public Health Service (USPHS), is the clearinghouse for receiving reports of foodborne illness outbreaks. These food poisoning incidents are published each week in the *Morbidity and Mortality Weekly Report* (see the section titled "For More Information" at the end of this chapter). From 1973 to 1987, the CDC recorded 2841 outbreaks of foodborne illness representing 124,994 individual cases. An **outbreak** is defined as an incident of foodborne illness that involves two or more people who ate a common food, which has been confirmed through laboratory analysis as the source of the outbreak. Exceptions to this are botulism and chemical poisoning, in which a single case is considered an outbreak. A **case** is simply an individual who experiences illness after eating an incriminated food. The number of cases per outbreak can vary widely (Table 3.1).

Recorded cases are thought to represent only a fraction of actual cases. Several food disease experts have estimated the actual incidence of foodborne illness in the United States to range from 6.5 million to 99 million annually. The underreporting of actual incidences is likely due to the difficulty associated with recognizing foodborne illness. The common symptoms of vomiting, cramps, and diarrhea are thought to be dismissed as "the flu" and therefore not reported.

Of current concern is the growing realization that the actual incidence of foodborne illness is on the increase. For example, the CDC reports that the number of egg-related outbreaks of foodborne illness from the salmonellae strains increased nationally by 90% from 1985 to 1993.

Table 3.1 Number and percentage of foodborne disease outbreaks and cases of known etiology, by etiologic agent, 1973–87.

	Outbreaks		Cases	
Etiologic agent	**No.**	**%**	**No.**	**%**
Bacterial				
Bacillus cereus	58	2	1123	1
Brucella	4	<1	43	<1
Campylobacter	53	2	1547	1
Clostridium botulinum	231	8	494	<1
Clostridium perfringens	190	7	12234	10
Escherichia coli	10	<1	1187	1
Salmonella	790	28	55864	45
Shigella	104	4	14399	12
Staphylococcus aureus	367	13	17248	14
Streptococcus, group A	12	<1	1917	2
Streptococcus, other	7	<1	248	<1
Vibrio cholerae	6	<1	916	1
Vibrio cholerae, non–01	2	<1	11	<1
Vibrio parahaemolyticus	23	1	535	<1
Yersinia enterocolitica	5	<1	767	1
Other bacterial	7	<1	373	<1
Total bacterial	1869	66	108906	87
Viral				
Hepatitis A	110	4	3133	3
Norwalk virus	15	1	6474	5
Other viral	10	<1	1023	1
Total viral	135	5	10630	9
Parasitic				
Giardia	5	<1	131	<1
Trichinella spiralis	128	5	843	1
Other parasitic	7	<1	30	<1
Total parasitic	140	5	1004	1
Chemical				
Ciguatoxin	234	8	1052	1
Heavy metals	46	2	753	1
Monosodium glutamate	18	1	58	<1
Mushroom poisoning	61	2	169	<1
Paralytic shellfish poisoning	21	1	160	<1
Histamine (scombroid) fish poisoning	202	7	1216	1
Other chemical	115	4	1046	1
Total chemical	697	25	4454	4
Total	**2841**	**100**	**124994**	**100**

Source: From Bean, N. H., and Griffin, P. M.: Foodborne disease and outbreaks in the United States, 1973–1987: Pathogens, vehicles, and trends. *J. Food Protection.* Vol. 53, No. 9, pp. 804–817, September 1990.

A number of factors have been attributed to this increase in outbreaks of foodborne illness, including the following:

- The increase in "at-risk" populations including the elderly, children, and people with compromised immune systems such as individuals with AIDS,
- More meals prepared and eaten outside of the home,
- Changes in food preparation and handling, and
- Changes in the microorganisms that cause foodborne illness.

Costs Associated with Outbreaks of Foodborne Illness

It is difficult to account for the total and true costs of foodborne illness, but the economic loss associated with foodborne disease outbreaks can be much more startling than most foodservice directors realize. Todd (1989) reports that in the United States the preliminary cost estimates are $1000 per reported case. Medical care, lost business, and lawsuits against the foodservice contribute the most to the cost, but loss of income for victims and infected food handlers is also considerable. The social costs of pain and suffering are impossible to measure. Other estimates place the total cost of foodborne illness in the United States at $4.8 to $23 billion per year.

The CDC has estimated that commercial and noncommercial foodservice establishments account for 79% of reported outbreaks of foodborne illness. This is due in part to the fact that a single error in a high-volume operation results in a high number of cases. It is essential then that the food manager understand basic food microbiology as it applies to building a food safety program.

The understanding and application of microbiological principles to the operation of a foodservice department cannot be overemphasized. Too many establishments receive poor ratings by health department inspectors because of lack of cleanliness or improper food handling procedures. Reports of outbreaks of foodborne illness due to contaminated food or workers continue to occur. Such outbreaks have devastating effects on the reputation of an establishment and possibly on its success or failure.

Foodservice managers should have an understanding of the microbiological aspects involved in such outbreaks and be able to implement procedures to prevent their occurrence.

BASIC FOOD MICROBIOLOGY

Certain microscopic organisms, such as bacteria, are able to invade the human body and cause illness and sometimes death. Because contaminated food and workers are major sources for transmitting the organisms to people, it is essential that foodservice managers understand the conditions involved and insist that employees follow correct procedures to prevent such illness from occurring.

Bacteria whose forms are recognizable under the microscope are *cocci*, round in shape; *bacilli*, rod shaped; and *spirilli*, corkscrew-like. The conditions necessary for bacterial growth are nutrients; moisture; favorable temperatures, pH, and atmosphere;

and time. Most bacteria grow best in low-acid food; a few in acid food. Some grow best if sugar is present in the food, others if proteins are present. Some need oxygen for growth, and others thrive in its absence. The temperature most favorable to growth of pathogenic bacteria is body temperature (about 98°F); temperatures below 41°F inhibit their growth markedly, and temperatures above 140°F for a period of time are lethal to many varieties of organisms. The time required for growth and multiplication depends on the other environmental conditions present and the type of food in question.

Any food can be a vehicle for foodborne illness, but some are more hazardous than others. The U.S. Food and Drug Administration (FDA) 1995 Food Code defines **potentially hazardous foods** as

(a) food that is natural or synthetic and is in a form capable of supporting:
  (i) The rapid and progressive growth of infectious or toxigenic microorganisms;
  (ii) The growth and toxin production of *Clostridium botulinum;*
  (iii) In shell eggs, the growth of *Salmonella enteritidis.*

(b) an animal food (a food of animal origin) that is raw or heat-treated; a food of plant origin that is heat-treated or consists of raw seed sprouts; cut melons; and garlic and oil mixtures.

One means by which bacteria are transmitted from person to person is *direct contact.* Direct contact, either by carriers or infected persons who harbor the disease-causing bacteria and convey them to other individuals, accounts for a large part of the spread of communicable disease. According to the USPHS, there are some 62 different communicable diseases, each caused by a specific kind of organism. A *carrier* is a person who, without symptoms of communicable disease, harbors and gives off from his or her body the specific bacteria of a disease, usually without being aware of it. An *infected* person is one in whose body the specific bacteria of a disease are lodged and have produced symptoms of illness. Thus, others are aware of the possible danger of contamination.

Consumers can become infected by ingesting water, milk, or other food products that have been contaminated with the fecal material of an infected person or animal. Still another path of infection is drinking raw milk drawn from cows with infected udders. A now rare source of infection is eating meat from hogs infected with the parasitic organism *Trichinella spiralis.*

An infectious disorder of the respiratory system such as a common cold can be spread by a droplet of spray from a cough or sneeze. An *indirect route* of infection spread through respiratory discharges is the used handkerchief, or the contaminated hand, and the subsequent handling of food or plates and cups in serving a patron.

The modes of transmission or locomotion for pathogens are diagrammed in Figure 3.1. Note that human wastes, particularly fecal material, are especially hazardous. An individual who has used the toilet is certain to have contaminated hands. If careful and thorough hand-washing is ignored, the worker's hands can be a dangerous "tool" in the kitchen.

Common foodborne diseases are classified as either **foodborne infections** or **foodborne intoxications.** A foodborne infection is an illness caused by the ingestion of harmful living organisms usually found on the food consumed. In a food intoxication the bacteria grow on the food prior to its ingestion and produce **toxins** or poisonous substances that cause the illness. In this case, the microorganisms

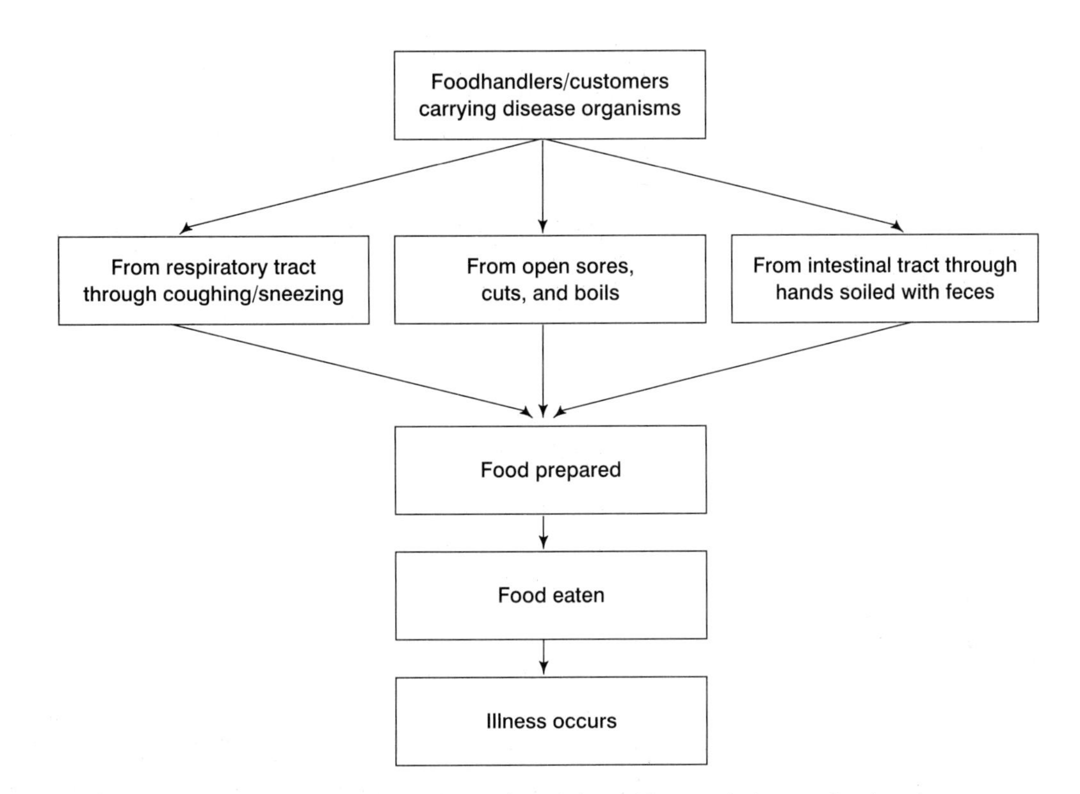

Figure 3.1 Transmission of a foodborne illness from infected human beings to food and back to other human beings.
Source: From *Applied Foodservice Sanitation,* 4th ed. Copyright © 1992 by the Educational Foundation of the National Restaurant Association.

themselves are not harmful, but the toxins they produce are. Illness occurs more quickly from intoxication than from infection. In both cases cramps, nausea, diarrhea, and sometimes vomiting can occur.

Microorganisms Causing Foodborne Infections

Primary organisms that cause foodborne infections are salmonellae, *Campylobacter jejuni,* pathogenic *Escherichia coli,* and *Vibrio parahaemolyticus.* The first two are especially prevalent and require special care in food handling.

Salmonellae infections account for numerous cases of gastrointestinal disorders. The causative organism may be any of the 2300 known serotypes of the genus *Salmonella,* of which the three most common are *S. enteritidis, S. typhimurium,* and *S. heidelberg* (Cliver, 1990). The intestinal tract of animals and humans is the principal reservoir of salmonellae. It is excreted in the feces and can contaminate food and water if personal hygiene and proper food handling are not practiced. Primary food sources of salmonellae infections include poultry, poultry products, beef, and pork. Proper cooking will kill salmonellae, but cross-contamination of cooked foods with raw foods must be avoided.

Salmonella enteritidis infections accounted for 25% of all reported cases of salmonellosis in 1985. Infections have been a particular concern in the Northeast portion of the United States where there has been a fivefold increase since 1975. This increase has been associated with raw and undercooked commercial Grade A shell eggs. Many states have adopted various guidelines regarding the preparation and service of eggs. The food manager is encouraged to contact the local health department for regional guidelines (see the discussion of egg handling techniques later in this chapter).

Another organism, long known as an animal pathogen, has only recently been recognized for its importance and prevalence in human gastroenteritis. It is *Campylobacter jejuni* and is transmitted by the consumption of raw milk, contaminated water, and undercooked chicken, beef, pork, and raw clams. Poultry has an especially high carriage rate of this contamination. Cross-contamination from knives or cutting boards may be another source of contamination. Evidence of the seriousness of *C. jejuni* is that it is now isolated from human diarrheal stools more frequently in the United States than are salmonella and *Shigella* spp. *Campylobacter jejuni* is readily killed by heat and is highly sensitive to chlorine-containing sanitizers.

Various types of pathogenic *Escherichia coli* have been responsible for a number of foodborne illness outbreaks in this country. Imported soft cheese was the cause of the first identified outbreak, which occurred in 1971. The most serious and highly publicized outbreak of *E. coli* occurred in 1993 when four children died after eating undercooked hamburgers. *Escherichia coli* is found in the feces of humans and other animals, and so may contaminate soil, water, and food plants. Illnesses caused by *E. coli* range from "traveler's diarrhea" to the life-threatening hemolytic uremic syndrome. The organism is easily killed by heat, but foods that have been heat-processed can be contaminated after heating, and improper refrigeration and temperature control can result in the organism increasing in numbers. In addition to cheese, both raw and processed shellfish, raw ground beef, and inadequately cooked ground beef have caused some outbreaks of gastrointestinal illness due to the presence of pathogenic *E. coli*.

Vibrio parahaemolyticus has been the cause of the majority of foodborne illness in Japan, but only in recent years has it been recognized as a cause in other countries. Because this microorganism lives in saltwater, contamination of fish and shellfish can occur, especially in those from warm waters. Most outbreaks have resulted from eating raw or undercooked seafood or cooked seafood that has come into contact with raw product or contaminated containers. Examples of problems foods include raw oysters and sushi. At present, *V. parahaemolyticus* gastroenteritis has not been considered a serious health problem in the United States. But if seafood, especially crustaceans, is mishandled during storage, cooking, and holding, the organism will multiply to levels that, if ingested, cause mild abdominal pain, diarrhea, and often nausea and vomiting.

Microorganisms Causing Food Intoxication

Organisms that cause food intoxication include *Staphylococcus aureus, Clostridium botulinum,* and *Clostridium perfringens.*

Staphylococcal food intoxication, the most frequent type of food intoxication, results from toxin production by *Staphylococcus aureus* in high-protein menu items such as cooked meats, eggs, and milk, as well as cream pie. This organism is commonly found on the human skin and is abundantly present in pimples and suppurating wounds. In addition, many people without skin problems carry the organism, with the nose being the most common site. Major points of control of this food hazard include (1) training employees to avoid hand contact with the face, and wash hands after such contact occurs, (2) excluding from work any employee with pimples, pus pockets, or suppurating wounds so that gross contamination of food may be avoided, and (3) refrigerating foods that could support *S. aureus* growth and toxin production. The toxin produced by *S. aureus* is not destroyed by cooking so refrigerated temperature control is critical.

General estimates are that a majority of the cases of food intoxication are due to staphylococcal organisms. The illness is not fatal, but causes severe diarrhea and nausea to its victims for several hours. Usually the symptoms are evident within 0.5 to 6 hours after the ingestion of the infected food, which may have shown no visible indications of the contamination at the time of consumption.

Clostridium botulinum is a spore-forming organism that causes a far more serious food intoxication known as botulism. The organism can grow and produce toxin in various low-acid foods under anaerobic conditions. The toxin is highly poisonous and often fatal. Commonly, food that supports the growth of *C. botulinum* shows some noticeable changes from normal. Thus, food that appears abnormal should not be "taste tested" because just a small amount of ingested toxin can cause illness. The toxin can be destroyed by boiling vigorously for 20 minutes. Fortunately, botulism is an uncommon occurrence in foodservice operations. Commercially processed foods are usually fully sterilized at high temperature under pressure. However, the hazard is great enough to stimulate constant watchfulness of food condition and quality.

Clostridium perfringens, which is an anaerobic, spore-forming bacterium, is often placed in the group of organisms causing food intoxication. Although the toxin may be present in food, it is believed to be most often produced in the intestinal tract, and thus the illness is usually an infection. The incubation period varies between 8 and 20 hours, after which illness occurs. Symptoms are relatively mild and the duration of the illness is usually about one day.

Clostridium perfringens is found widely distributed in soil, water, dust, sewage, and manure, and is also found in the intestinal tracts of humans and healthy animals. Many foods purchased by foodservices, especially meats, are probably contaminated with this organism. Also, foodservice workers may carry this organism into the kitchen on their hands. Extreme care must be taken to keep hands clean and equipment clean and sanitized, especially meat slicers. Meats to be sliced should never be left to be "cut as needed" over a long serving period, and slicers must be thoroughly cleaned after each use. One food handling error often associated with illness caused by *C. perfringens* is slow cooling of large batches of high-protein foods such as roasts, stews, and gravy. These foods should be cooled rapidly to 40°F or colder (see the section, Proper Food Handling, later in this chapter for proper chilling techniques). *Clostridium perfringens* grows rapidly, faster than almost any of the other bacteria discussed here. Gravy, a frequent offender, should be held above 140°F or below 41°F. Table 3.2 summarizes the most common foodborne diseases.

Some agents have only recently been identified as important causes of foodborne illness. These agents are referred to as **emerging pathogens** and cause diseases

Table 3.2 Major foodborne diseases.

Disease	Salmonellosis Infection	Staphylococcal Intoxication	Clostridium Perfringens Infection/Intoxication	Bacillus Cereus Intoxication	Botulism Intoxication
Foods implicated	Poultry, poultry salads, meat and meat products, milk, shell eggs, egg custards and sauces, other protein foods	Reheated foods, ham and other cooked meats, dairy products, custards, potato salad, cream-filled pastries, and other protein foods	Meat that has been cooked at low temperature for long period of time or cooled slowly	Rice and rice dishes, custards, seasonings, dry food mixes, puddings, cereal products, sauces, vegetable dishes, meat loaf	Improperly processed canned low acid foods, non-acidified garlic-in-oil products, grilled onions, stews, meat/poultry
Incubation period	6 to 72 hours	½ to 6 hours	8 to 22 hours	½ to 16 hours	12 to 36 hours
Symptoms	Abdominal pain, headache, nausea, vomiting, fever, diarrhea	Nausea, vomiting, diarrhea, dehydration	Abdominal pain, diarrhea	Nausea and vomiting, diarrhea, abdominal cramps	Vertigo, visual disturbances, inability to swallow, respiratory paralysis
Duration	2 to 3 days	24 to 48 hours	24 hours	6 to 24 hours	Several days to a year
Precautionary measures	Avoid cross-contamination, refrigerate food, cool cooked meats and meat products properly, avoid fecal contamination from food handlers by practicing good personal hygiene	Avoid contamination from bare hands, exclude sick food handlers from food preparation and serving, practice sanitary habits, proper heating and refrigeration of food	Use careful time and temperature control in cooling and reheating cooked meat dishes and products	Use careful time and temperature control and quick chilling methods to cool foods, hold hot foods above 140°F (60°), reheat leftovers to 165°F (74°C)	Do not use home-canned products, use careful time and temperature control for sous-vide items and all large, bulky foods, keep sous-vide packages refrigerated, purchase only acidified garlic-in-oil, cook onions only on request

Source: Adapted from *Applied Foodservice Sanitation,* 4th ed., page 35. Copyright © 1992 by the Educational Foundation of the National Restaurant Association.

such as listeriosis, hemorrhagic colitis, and Norwalk virus infection. Table 3.3 summarizes some emerging pathogens.

Chemical Causes of Foodborne Illness

Microbial causes of foodborne illness are common but disease and illness can also be caused by chemical contaminants in food. This type of foodborne illness results from eating food to which toxic chemicals have been added accidentally. Reports from the CDC for 1987 showed that 29% of all foodborne outbreaks were of chemical etiology.

Chemical poisoning may result from contamination of food with foodservice chemicals such as cleaning and sanitizing compounds, excessive use of additives and preservatives, or contamination of food with toxic metals. The foodservice manager is responsible for implementing the necessary precautions to ensure that food is protected from these hazards. Minimum precautions include proper labeling and storage of all chemicals and frequent inservice training for employees on the hazards and proper use of chemicals.

THE ROLE OF THE FOOD MANAGER

Food managers, especially those responsible for providing food to "at-risk" populations, have an important responsibility in the prevention of foodborne illness. Food managers must instill a sense of urgency about and educate foodservice employees on the realities of foodborne disease. To do this, the foodservice manager must be well educated on food safety and related topics of microbiology, epidemiology, and food science. Then, to effect change, the foodservice manager must take a proactive role to design a food safety plan that accomplishes the following objectives. A foodservice manager should be able to:

- Design, implement, and maintain an effective food safety program such as HACCP;
- Train, motivate, and supervise foodservice employees to ensure that the food safety program is maintained;
- Stay current with government regulations, codes, and standards related to food safety; and
- Commit to lifelong learning about food safety, emphasizing science-based information.

The following sections of this chapter provide the base for achieving these objectives.

EMPLOYEE HEALTH AND PERSONAL HYGIENE

The provision of safe food begins during the hiring process. As discussed earlier in this chapter, many cases of foodborne illness can be linked directly to lack of attention to personal hygiene, cleanliness, and food handling procedures. The CDC issues a list of infectious and communicable diseases that are often transmitted through food contaminated by infected food handlers. The pathogens that can cause diseases after an infected person handles that food include:

Table 3.3 Emerging pathogens increasingly identified as causing foodborne illness.

Disease	**Campylobacteriosis Infection**	**E. Coli 0157:H7 Infection**	**Listeriosis Infection**	**Norwalk Virus Infection**
Pathogen	Campylobacter jejuni	Escherichia coli 0157:H7	Listeria monocytogenes	Norwalk and Norwalk-like viral agent
Foods implicated	Raw vegetables, unpasteurized milk and dairy products, poultry, pork, beef, and lamb	Raw and undercooked beef and other red meats, imported cheeses, unpasteurized milk, raw finfish, cream pies, mashed potatoes, and other prepared foods	Unpasteurized milk and cheese, vegetables, poultry and meats, seafood, and prepared, chilled, ready-to-eat foods	Raw vegetables, prepared salads, raw shellfish, and water contaminated from human feces
Incubation	3 to 5 days	12 to 72 hours	1 day to 3 weeks	24 to 48 hours
Duration of illness	1 to 4 days	1 to 3 days	Depends on treatment—high mortality in those with compromised immune system	24 to 48 hours
Symptoms	Diarrhea, fever, nausea, abdominal pain, headache	Bloody diarrhea; severe abdominal pain; nausea, vomiting, and occasionally fever, hemolytic uremic syndrome	Nausea, vomiting, headache, fever, chills, backache, meningitis, spontaneous abortion	Nausea, vomiting, diarrhea, abdominal pain, headache, and low-grade fever
Precautionary measures	Avoid cross-contamination, cook foods thoroughly	Cook beef and red meats thoroughly, avoid cross-contamination, use safe food and water supplies, avoid fecal contamination from food handlers by practicing good personal hygiene	Use pasteurized milk and milk products, cook food to proper temperatures, avoid cross-contamination	Use safe food and water supplies, avoid fecal contamination from food handlers by practicing good personal hygiene, thoroughly cook foods

Source: Adapted from *Applied Foodservice Sanitation,* 4th ed., page 48. Copyright © 1992 by the Educational Foundation of the National Restaurant Association.

- Hepatitis A,
- Norwalk and Norwalk-like viruses,
- *Salmonella typhi,*
- *Shigella* species,
- *Staphylococcus aureus,* and
- *Streptococcus pyogenes.*

There is no scientific evidence that the human immunodeficiency virus (HIV) or acquired immunodeficiency syndrome (AIDS) can be transmitted through food.

As mentioned, the manager can implement preventive measures at the hiring stage to minimize the **risk** of food contamination and mishandling. This is accomplished through health screening and careful training of foodservice employees. Individuals being considered for positions that involve food handling should undergo a health examination before being hired and at routine intervals thereafter. The exam should include a tuberculin test, and many operations, especially those in health care organizations, require screening for hepatitis A. Many state and local regulatory agencies require specific health tests prior to hiring. The manager should consult the local health department for specific requirements.

The successful hiring process should be followed by a thorough orientation and training on the standards of personal hygiene established for the foodservice operation. Personal hygiene is simply the application of principles for maintaining health and personal cleanliness. Policies should be designed, implemented, and monitored that cover proper attire, personal hygiene habits, and employee illness. The specific methods designed to fulfill the intent of these policies are frequently referred to as **infection control** procedures. Minimally, policy on infection control should address the areas outlines in the following sections.

Proper Attire

- Employees should wear clean, washable clothing. Uniforms are recommended but, if not feasible, clean aprons are essential.
- Effective hair restraints must be worn to cover head and facial hair. Commonly used restraints include nets, bonnets, and caps. The purpose of hair restraints is to prevent hair from falling into the food and to discourage the food handler from touching his or her hair.
- Jewelry is discouraged because bacteria can lodge in settings and contaminate food.

Proper Hand Washing

The single most important practice in preventing the spread of foodborne illness is proper hand washing. Foodservice employees should wash their hands using the procedure illustrated in Figure 3.2. This technique is referred to as the double hand-washing technique and is recommended by the FDA under the following circumstances:

- After using a rest room, contacting body fluids and discharges, or handling waste containing fecal matter, body fluids, or other bodily discharges (for exam-

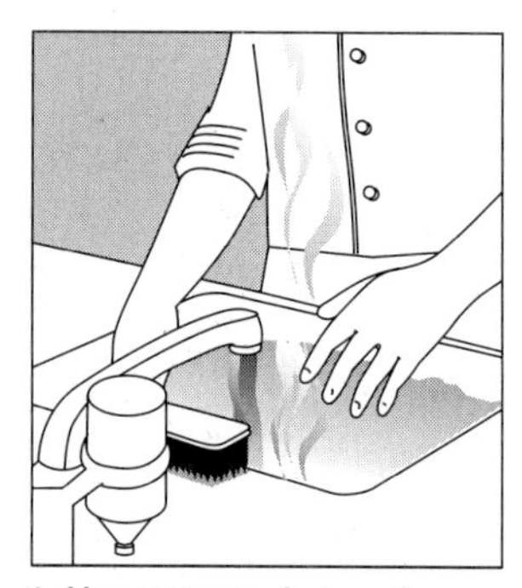

1. Use water as hot as the hands can comfortably stand.

2. Moisten hands, soap thoroughly, and lather to elbow.

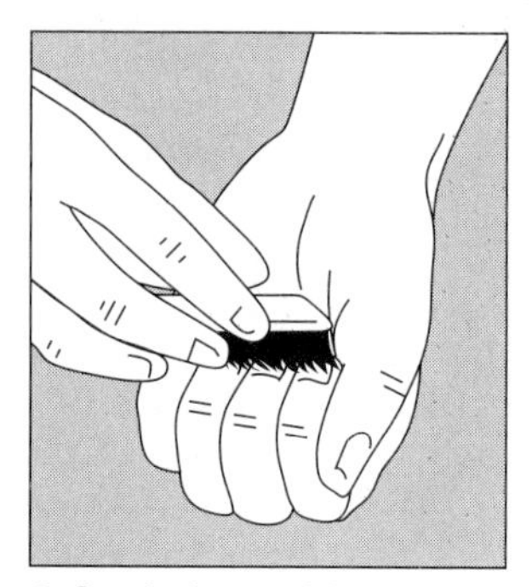

3. Scrub thoroughly, using brush for nails. Rinse.

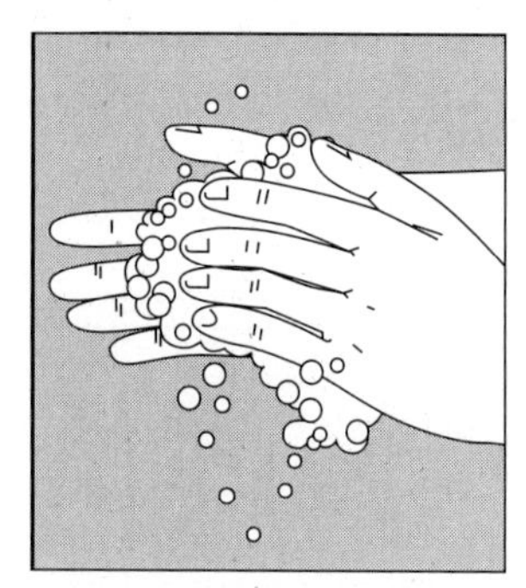

4. Resoap and lather, using friction for 20 seconds.

5. Rinse thoroughly under running water.

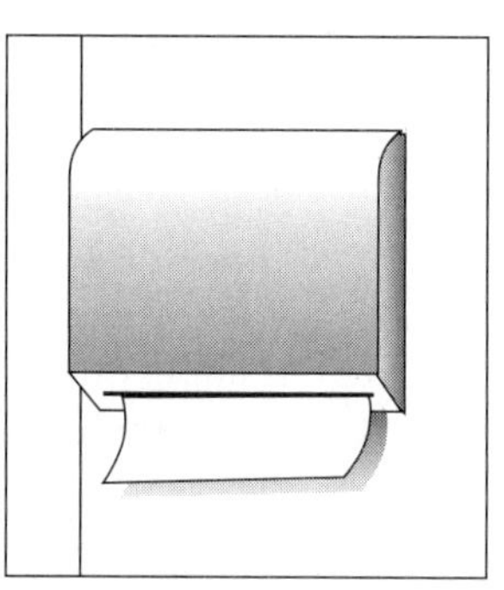

6. Dry hands, using single-service towels or hot-air dryer.

Figure 3.2 Proper hand-washing techniques.
Source: Adapted from *Applied Foodservice Sanitation,* 4th ed., Copyright © 1992 by the Educational Foundation of the National Restaurant Association.

ple, personal care attendants in day care centers and nursing homes may be responsible for changing diapers and serving food);

- Before beginning work or before returning to work following a break;
- After coughing, sneezing, or using a handkerchief or disposable tissue;
- After smoking, using tobacco, eating, or drinking;
- After handling soiled equipment or utensils;
- Immediately before food preparation such as working with food, clean equipment, utensils, and supplies; and
- When switching from working with raw to cooked food.

Other Personal Hygiene Habits

- Foodservice personnel should keep their fingernails trimmed and maintained.
- Hands should be kept away from face, hair, and mouth.
- Disposable gloves should be encouraged for direct food contact and are required by law in some areas of the country. Employees need to understand

that gloves can become contaminated, thus gloves need to be changed frequently to prevent cross-contamination.
- Smoking should be permitted in designated areas only and away from food preparation and service areas.
- Only authorized personnel should be allowed in the production areas.

Cuts, Abrasions, and Employee Illness

- All cuts and abrasions such as burns and boils should be covered with a waterproof bandage.
- Cuts on hands should be covered with a waterproof bandage and a watertight disposable glove.
- Employees with symptoms of vomiting, diarrhea, fever, respiratory infection, or sore throat should not handle food.
- Any employee suspected of having a **communicable disease** as listed by the CDC should be referred to employee health or their personal physician for clearance before returning to work.

FLOW OF FOOD THROUGH THE FOODSERVICE OPERATION

Gaining a basic knowledge of food microbiology and applying it to personal hygiene practices are preliminary steps to designing an effective food safety program for the foodservice operation. A well-designed food safety program will address the entire foodservice operation. It is therefore essential that the manager understand how food moves through the operation.

The movement of food through a foodservice operation is referred to as the **flow of food.** It begins at the point where a decision is made to include a food item on the menu and ends with the final service to the customer. The functions basic to food flow in any operation include receiving, storage, preparation, holding, service, cooling leftovers, and reheating (or rethermalization). Figures 3.3 through 3.6 illustrate how these functions relate to one another in the various types of foodservice systems and how food flows through each type of system. The foodservice manager must be able to identify potential hazards at each step in the food flow and design a food safety program that will prevent the potential hazards from being realized. Part of the program design will include procedures for safe and proper food handling at each stage of the food preparation process. HACCP, a food safety program specifically designed to monitor food safety during the entire production process, is presented later in this chapter.

PROPER FOOD HANDLING

Hiring healthy employees and providing thorough, ongoing training in personal hygiene are important aspects of food safety but by no means a guarantee against

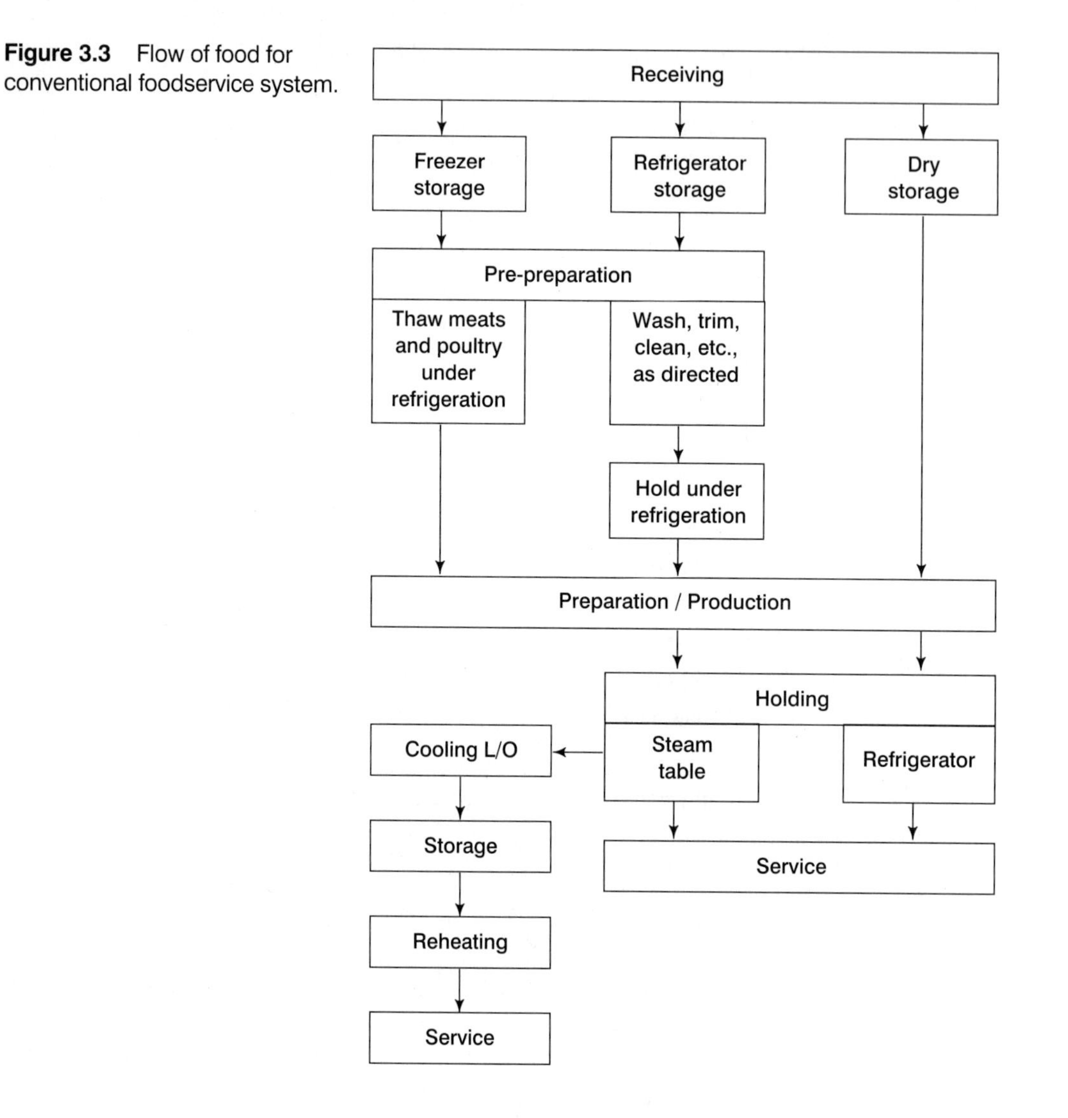

Figure 3.3 Flow of food for conventional foodservice system.

outbreaks of foodborne illness. Proper food handling techniques must be used to avoid conditions suitable for microbial growth and cross-contamination. *Cross-contamination* is the transfer of harmful microorganisms from one item of food to another via a nonfood surface such as human hands, equipment, or utensils. It may also refer to a direct transfer from a raw to a cooked food product.

Precautions for Safe Food Production

Proper food handling throughout the purchasing, storage, production, and service of food is critical in safeguarding the food against contamination. Legal safeguards are provided by federal, state, and local regulatory agencies, which are responsible for setting and enforcing standards for raw and processed foods (see Chapter 5). Minimum standards for sanitation in foodservice establishments are monitored by city and state agencies, but managers are responsible for the maintenance of sanitation standards in their respective foodservices.

Numerous factors can contribute to the outbreak of foodborne illness, but errors in food handling are often implicated in outbreaks of foodborne illness. The National Sanitation Foundation International list the following as frequently cited factors in outbreaks of foodborne illness:

- Failure to cool food properly,

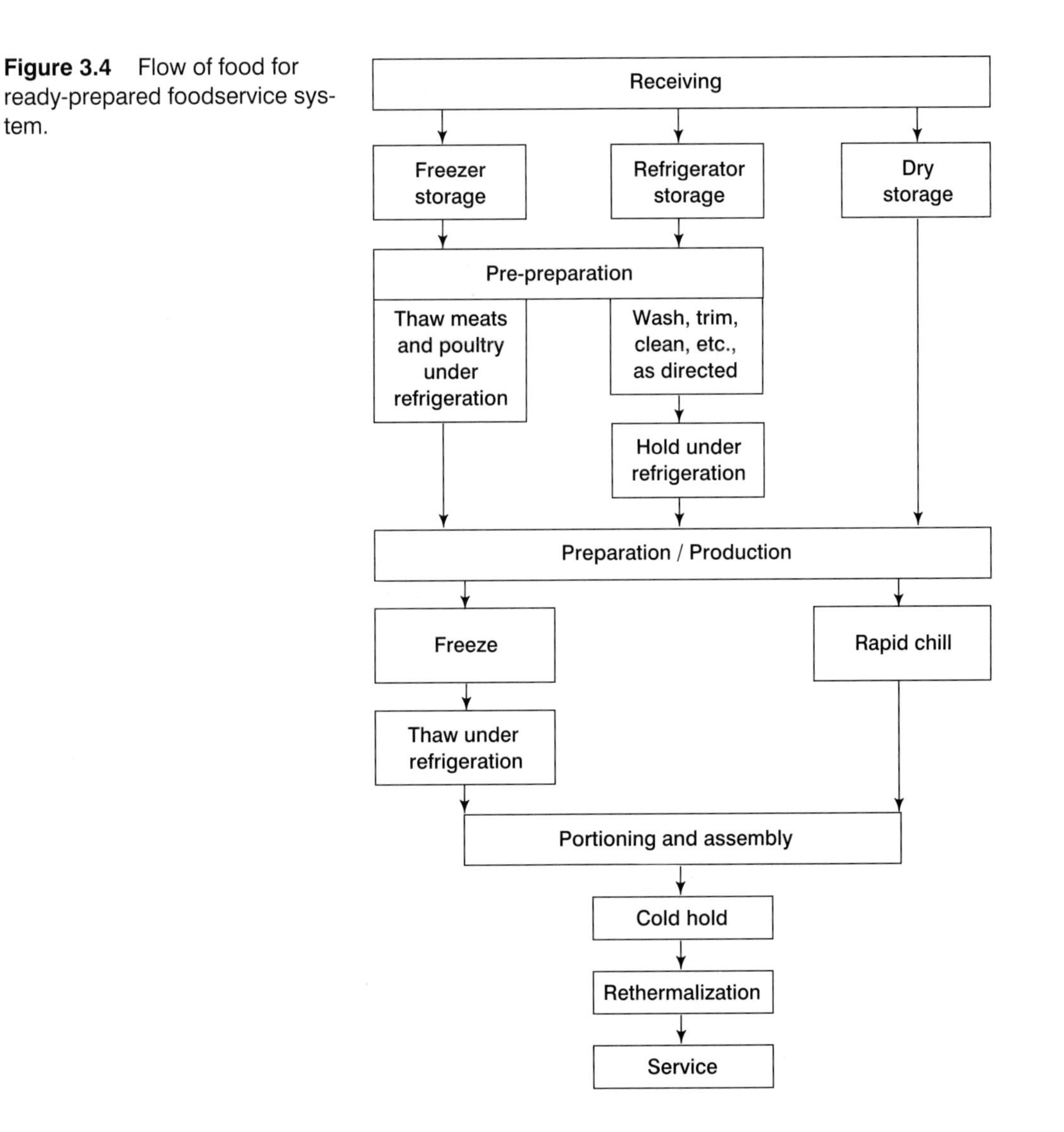

Figure 3.4 Flow of food for ready-prepared foodservice system.

Figure 3.5 Flow of food for commissary foodservice system.

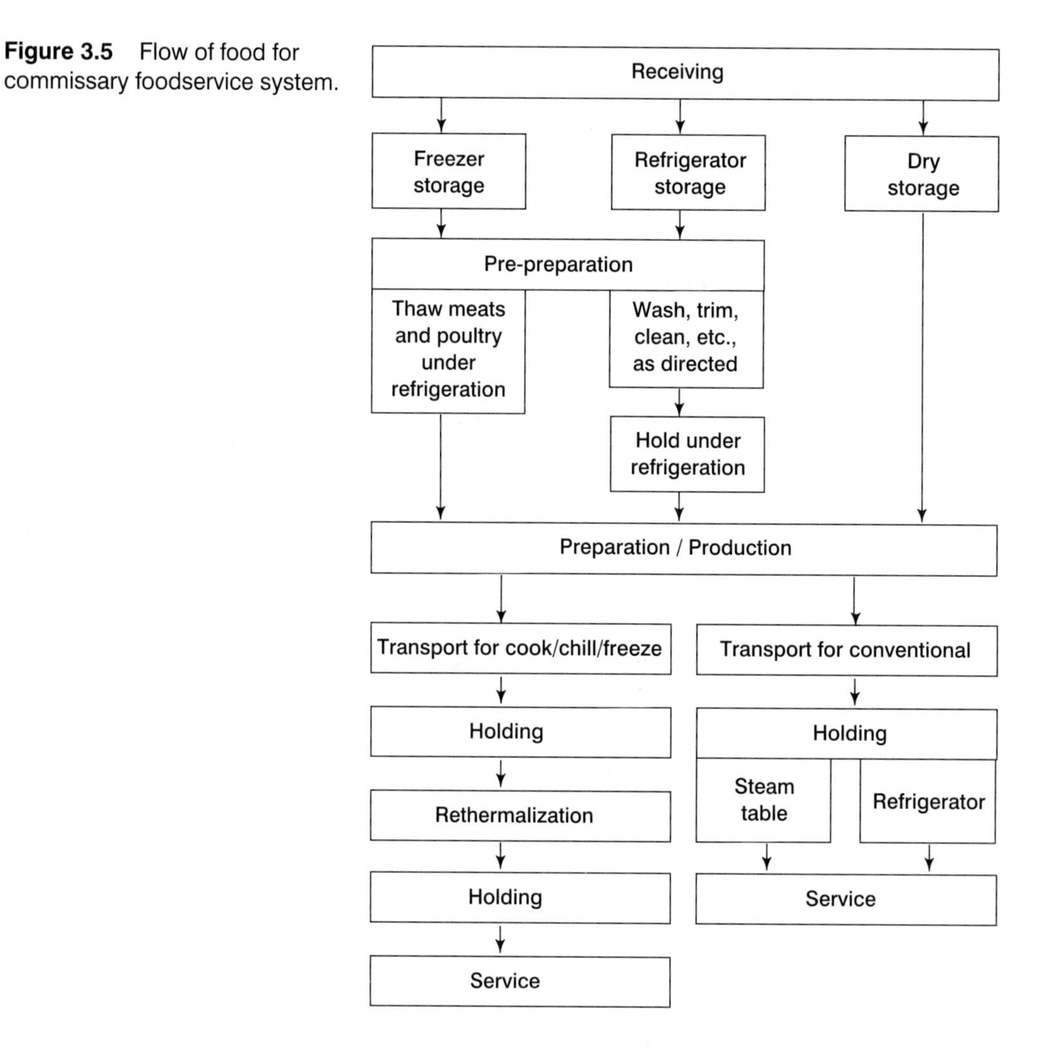

- Failure to thoroughly heat or cook food,
- Infected employees who practice poor personal hygiene at home and at the workplace,
- Foods prepared a day or more before they are served,
- Raw, contaminated ingredients incorporated into foods that receive no further cooking,
- Foods allowed to remain at bacteria-incubation temperatures, and
- Cross-contamination of cooked foods with raw food, or by employees who mishandle foods, or through improperly cleaned equipment.

These errors can be avoided through thorough, ongoing training. Employees should understand time-temperature relationships and practice proper food handling techniques.

Time-Temperature Relationships

Temperature has long been recognized as a particularly important factor in the control of harmful organisms. Time is an equally important factor in minimizing microbial growth during food storage, production, holding, transportation, and service. An important rule in food protection then is the time-temperature principle, which is based on three tenets regarding the handling of potentially hazardous foods:

1. Food items must be rapidly cooled to 41°F or less.
2. Cold food should be held at an internal temperature of 41°F or less.
3. Hot foods should be held at 140°F or higher.

According to the 1995 Food Code (see the section on Food Regulations and Standards later in this chapter), the temperature range of 41 to 140°F is referred to as the **danger zone** because disease-causing bacteria are capable of rapid multiplication in this temperature range. Figure 3.7 is a temperature guide for food safety and highlights the danger zone. The period of time that food is allowed to remain in this critical temperature zone largely determines the rate and extent of bacterial growth. Most food handling techniques are designed to keep food items, especially potentially hazardous foods, out of this temperature range. Various stages of food preparation require that foods be in the danger zone at various times. For example, cooked meat will be at room temperature while it is being sliced and again while it is being used to make sandwiches. The National Sanitation Foundation International recommends that the total time in the danger zone should be limited to 4 hours for any given food product.

The food manager must be aware of time-temperature relationships throughout the entire food production process. This concept is explained fully later in this chapter. It is imperative that the internal temperature of potentially hazardous food be kept *below* 41°F or *above* 140°F to ensure safety. This means that the temperature of

Figure 3.6 Flow of food for assembly/service system.

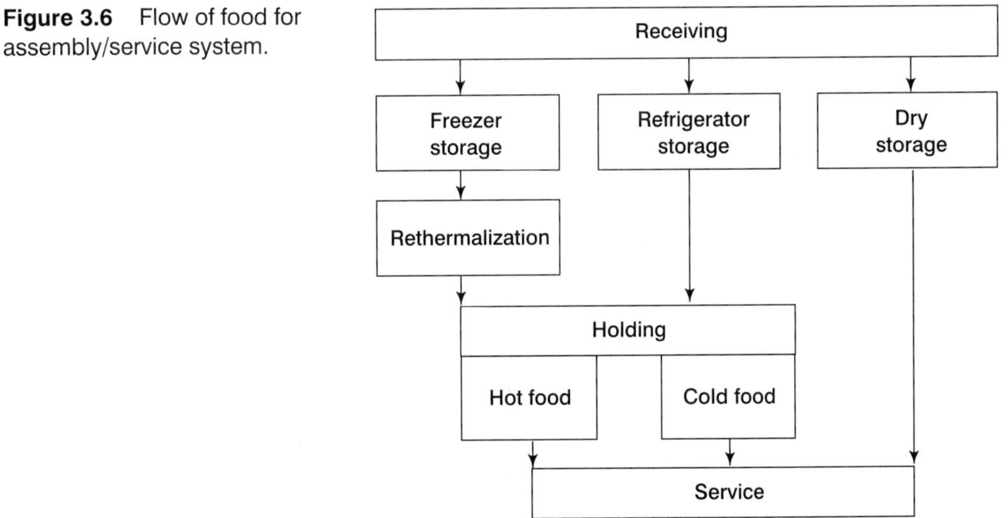

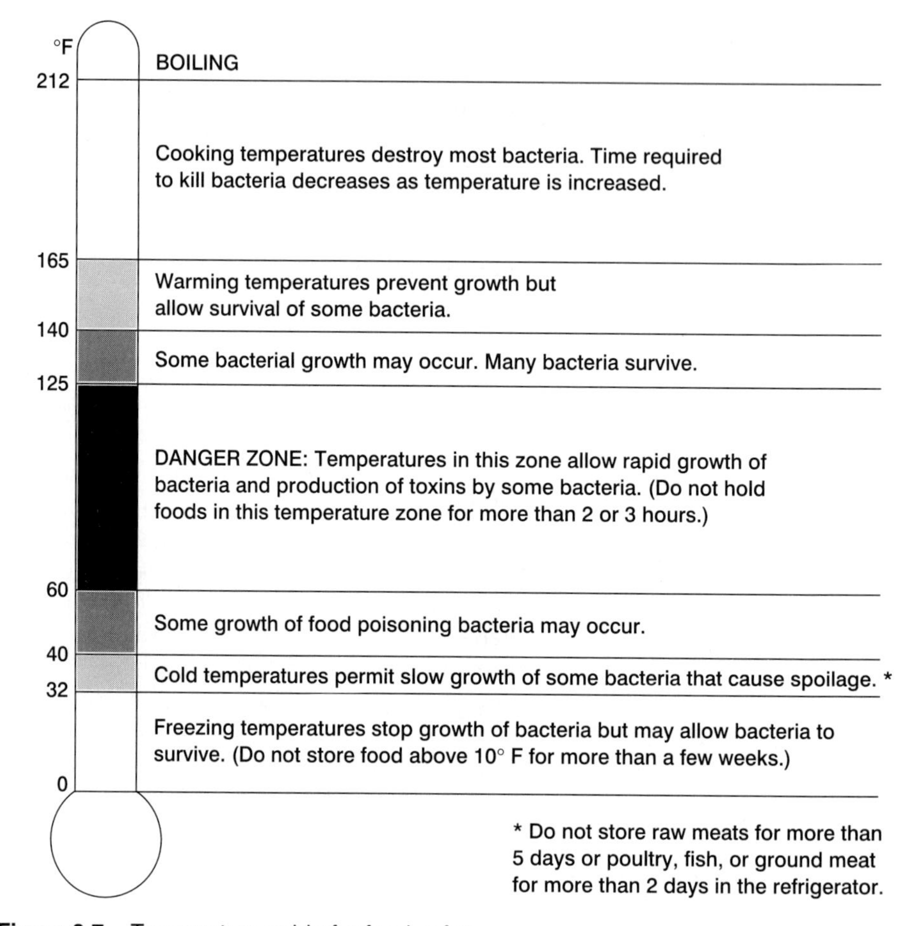

Figure 3.7 Temperature guide for food safety.
Source: From Keeping Food Safe, *Home and Garden Bulletin,* No. 162. Washington, D.C.: U.S. Department of Agriculture.

the refrigerator should be colder or the holding equipment hotter to maintain the proper internal temperature in the food. Temperature controls on walk-in and other refrigerators should be in good working order and checked and documented daily to make certain that temperatures are maintained below 41°F as appropriate for the specific foods stored in them. Figure 3.8 is an example of a temperature documentation form for refrigerator units. Proper cooling methods are illustrated in Figure 3.9.

Thermometers

Well-maintained thermometers are essential to ensure that food temperatures are properly monitored. Thermometers should be used for checking incoming deliveries of frozen and refrigerated foods and for monitoring internal temperatures during all phases of production and service. Thermometers should be metal stemmed, numerically scaled from 0 to 220°F, and accurate to ±2°F. Other features include easy-to-read numbers and a stem or probe of at least five inches. A thermometer with a calibration nut is recommended so that the scale can be easily adjusted for accuracy. Thermometers should be cleaned and sanitized after each use. Thermometers that have been approved by the National Sanitation Foundation International are recommended.

POTENTIAL HAZARDS IN FOOD PRODUCTION

Foods that are particularly hazardous include meat, poultry, fish, and eggs. These products are frequently contaminated with foodborne pathogens, which can spread

Heartland Country Village
Refrigerator/Freezer Temperatures
Month of:____________________

Day	Walk-in Freezer		Walk-in Refrigerator		Cook's Holding Refrigerator	
	AM Temp	PM Temp	AM Temp	PM Temp	AM Temp	PM Temp
1						
2						
3						
4						
5						
6						
7						
8						
9						
10						
11						
12						
13						

Figure 3.8 Temperature documentation chart.

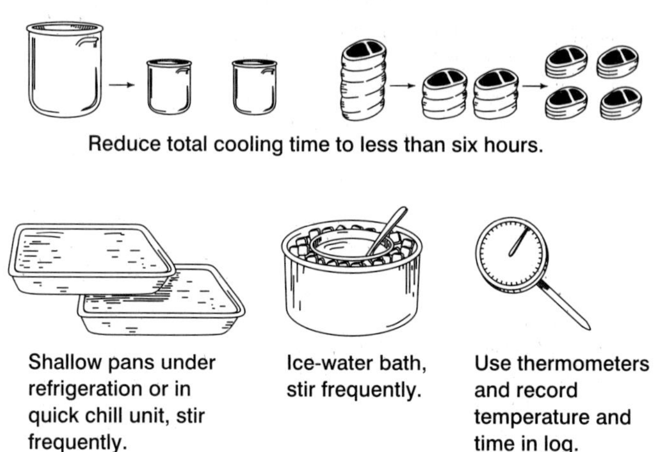
Reduce Food Mass
Reduce total cooling time to less than six hours.
Shallow pans under refrigeration or in quick chill unit, stir frequently.
Ice-water bath, stir frequently.
Use thermometers and record temperature and time in log.

Figure 3.9 Safe procedures for cooling hot foods.
Source: From *Applied Foodservice Sanitation,* 4th ed. Copyright © 1992 by the Educational Foundation of the National Restaurant Association.

to surfaces of equipment, to the hands of workers, and to other foods. If frozen turkeys are to be cooked whole, they should be thawed completely in the refrigerator before being cooked, and if cooked the day before, they should not be refrigerated without first reducing their bulk. The practice of cooking, chilling, and then reheating beef roasts is also potentially hazardous because reheating may not produce a temperature high enough (165°F) to destroy any bacteria that may have survived in the meat.

Food requiring preliminary preparation, which may include cooking prior to the final preparation, should be refrigerated following the preliminary steps. This includes items such as sandwich and salad mixtures; sliced, chopped, cut, and boned poultry and meats; ground, mixed, and shaped cooked meats; cream pie fillings and puddings; and sliced ham and similar items.

Cream pie fillings, puddings, and other products made with eggs, if not served hot, should be refrigerated as soon as possible after cooking. Attempting to cool at room temperature to save refrigeration is a practice to be discouraged. Masses of hot food cool slowly, even in large walk-in refrigerators. To be cooled quickly, the food should be poured into shallow containers to a depth of no more than two inches and refrigerated, as shown in Figure 3.9. Other suggestions for cooling large amounts of food quickly include stirring the food and placing the pan of food into an ice bath or vat of cold running water. The FDA 1995 Food Code recommends that potentially hazardous cooked foods be cooled from 140 to 70°F within 2 hours, and from 70 to 41°F or below within 4 hours.

As mentioned earlier, the incidence of foodborne illness resulting from salmonella contamination is increasing. This problem has been associated with raw or undercooked Grade A shell eggs. In the past, contamination was thought to be due to dirty or cracked shells. The more recent outbreaks however suggest that *Salmonella enteritidis* is, in some cases, transmitted directly from the hen to the *inside* of the egg. This means that more stringent guidelines must be implemented to handle eggs safely.

The following are general egg handling recommendations:

- Purchase Grade A or better eggs from a reliable source.
- Check eggs on delivery to ensure that they have been kept refrigerated during transport.
- Keep eggs refrigerated, pulling eggs from such storage only as needed; never store eggs at room temperature.
- Raw eggs should not be used as an ingredient in the preparation of food that will not be thoroughly cooked.
- Rotate eggs in inventory using the first-in/first-out (FIFO) method.
- Use only clean, crack-free eggs.
- Thoroughly wash hands before and after handling eggs and make sure equipment is clean and sanitized.
- Avoid pooling large quantities of eggs; cook eggs in small batches; no more than three quarts per batch.
- Never combine eggs that have been held on a steam table with a fresh batch of eggs.

Mishandling of food by cooks and other production workers also constitutes a hazard. Cooked ingredients in potato salad, for instance, can be contaminated by food handlers during peeling, slicing, chopping, or mixing operations. **Cross-contamination** by a worker or equipment that has been in contact with raw meat or poultry, and then with the cooked product, is to be avoided. Table 3.4 is a detailed summary of basic food handling techniques for safeguarding the procurement, preparation, and service of food. The foodservice manager should check federal, state, and local regulations for specific standards. Once employees are trained in the principles and techniques outlined in Table 3.4, frequent follow-up supervision on the job is necessary to ensure that workers actually understand and observe the procedures taught. Only then will the desired high standards for food handling be maintained from day to day.

FOOD SAFETY REGULATIONS AND STANDARDS

There are basically two systems of guidance and control that function to protect the public from unsafe food. First there is the official system of government laws, **regulations, codes,** and **standards** specifically designed for the food industry. Many of these agencies are discussed in detail in Chapter 5. Although much of food regulation is triggered at the federal level, local jurisdictions such as state, county, and city agencies assume regulatory enforcement responsibilities once food enters a foodservice establishment.

The second system is made up of controls and standards the industry unofficially imposes on itself. These controls and standards come from trade associations, such as the National Restaurant Association and the National Sanitation Foundation International, and from professional organizations such as the Joint Commission on Accreditation of Healthcare Organizations.

It is the responsibility of the manager to:

- Be aware of all federal, state, and local regulations that apply to his or her operation.
- Accurately interpret those regulations.

Table 3.4 Food handling techniques and safeguards against contamination.

Procurement	Food Preparation/Production	Service
Purchase meat, poultry, dairy products and shellfish from officially inspected and approved sources.	Use separate cutting boards for meat, poultry, fish, raw fruits and vegetables, and raw and cooked foods.	
Purchase canned goods from approved sources. Do not use home-canned foods. Reject or discard canned foods with swollen, leaking, or bulging lids or those with lids that are rusted or pitted.	Thoroughly clean cutting boards and work areas after each use.	
Store foods at proper temperatures.	Hold cold foods at 41°F or less; hold hot foods at 140°F or higher.	
Cover, label and date all foods.	Thaw frozen food under refrigeration.	
Specify that frozen products be maintained at 0°F or lower during delivery.	Do not use eggs with cracked shells; thoroughly cook eggs.	
Do not store cleaning supplies and other chemicals in the same area as food.	Wash fresh fruits and vegetables.	
	Refrigerate cooked foods immediately using proper cooling techniques.	Use tongs or gloves to handle ice and individual food items such as rolls.
	Use clean utensils in food preparation and service.	Protect salad bar and buffet table items with transparent shields (sneeze guards).
	Use a clean spoon each time for taste testing.	Discard cracked or chipped service ware.
	Clean and sanitize knives in slicers immediately after use with raw or cooked meats, fish, or poultry.	Avoid handling rims or glasses and other food contact surfaces.
	Check internal temperatures during cooking to assure proper end-point time and temperature have been met.	Check hot-holding temperatures to assure they hold at no less than 140°F during the service time.

- Identify an appropriate standard for compliance with each regulation.
- Design, implement, and monitor policies, procedures, and programs to ensure that operations are in compliance with regulations.

Regulations and interpretations may vary from one jurisdiction to another. For example, one state may have more restrictive standards on serving temperatures as compared to another state. It is recommended that the manager contact an appropriate local agency, such as the department of health, to become familiar with local regulations and interpretations that apply to a specific type of foodservice operation.

Definitions

To effectively design, implement, and monitor policies, procedures, and programs of food safety, the manager must understand the various terms that apply to government mandates. A **regulation** is a written government control that has the power or force of law. A **code** is a collection of regulations, usually pertinent to a specific type of organization. For example HSS-132 Wisconsin Administrative Code is a collection of regulations that apply to Medicare-funded, long-term-care facilities in Wisconsin. The following is an example of a food safety regulation from this code:

HSS-132.63 (5)(e) *Temperature.* Food shall be served at proper temperatures.

Note the vagueness of this statement. A logical question on the part of a food manager is "What are proper serving temperatures?" The manager has two options in dealing with this question. One is to contact the local regulatory agency and request an official interpretation. If one exists, the agency will provide the interpretation in writing. Figure 3.10, for example, is the official rule interpretation for the regulation on proper temperature in Wisconsin.

The second option is to refer to industry standards. A standard is a measure used to define and evaluate compliance with a regulation. For example, the standard for proper serving temperatures in Wisconsin is at least 120°F for hot food and no more than 50°F for cold food items. Many of these standards are documented in model food codes. The most recently released and widely used model code is the FDA's 1995 Food Code.

The 1995 Food Code

The 1995 Food Code was developed by the FDA with the cooperation of the U.S. Department of Agriculture (USDA). The code is neither law nor regulation but is provided for guidance and consideration for adoption by jurisdictions that have regulatory responsibility for food service, retail, and vending operations. According to the FDA, the code provides the latest and best scientifically based advice about preventing foodborne illness. Highlights include the importance of time, temperature control, and safe hand-washing. A most important and useful feature of this code is the framework it provides for designing a food safety program. The code promotes the Hazard Analysis and Critical Control Points program, discussed in the next section, as the best available system for assurance of food safety.

HAZARD ANALYSIS AND CRITICAL CONTROL POINT

HACCP is an acronym that stands for **Hazard Analysis and Critical Control Point.** It is a proactive process of consecutive actions to ensure food safety to the

State of Wisconsin \ DEPARTMENT OF HEALTH AND SOCIAL SERVICES

DIVISION OF HEALTH
MAIL ADDRESS: P. O. BOX 309
MADISON, WISCONSIN 53701

RULE INTERPRETATION

IN REPLY PLEASE REFER TO:

Date: Ocotber 29, 1982

To: See below*

Wis. Administrative Code HSS 132
Current Edition 8-1-82
Code Reference 132.63(5)(e)

From: Louis E. Remily LER
Acting Director
Bureau of Quality Compliance

Subject: SERVING FOOD AT PROPER TEMPERATURES

Question:

What are "proper" serving temperatures?

Answer:

Hot foods are to be served at a temperature generally accepted by residents or not less than 120°F. Temperatures less than 120°F. would not be considered palatable. In addition, if the food is placed in front of the resident at above 120°F., it would have been maintained at a temperature above the optimum level to prevent bacterial growth prior to reaching the resident.

Cold foods are to be served at a temperature generally accepted by residents or at a temperature of not more than 50°F. If served at that temperature, foods would have to have been maintained at 40-45°F. prior to reaching the resident, thus preventing bacterial growth.

Past Edition H-32 Reference H-32.17(7)(a) Effective Date 11/1/74

Request No. 055

This rule interpretation was issued as a result of a question on the above current edition. It applies to all previous and subsequent editions in which the test of the rule remains substantially unchanged.

cc - BQC Sections
- BQC Districts
- Nursing Home Associations

Figure 3.10 Rule interpretation of proper temperatures.

highest degree through the identification and control of any point or procedure in a specific food system, from receiving through service, where loss of control may result in an unacceptable health risk. HACCP differs from traditional endpoint food safety programs in that it is preventive in nature and focuses on the entire process of food preparation and service. In this sense it is a self-inspection process sometimes described as a self-control safety assurance program. HACCP plans are designed to prevent the occurrence of potential food safety problems.

HACCP is not new; the concept originated more than 40 years ago. The Pillsbury Company is frequently credited for pioneering the application of HACCP to the food processing industry when, in 1971, they worked in cooperation with the National Aeronautic and Space Administration (NASA) to create food for the U.S. space program that approached 100% assurance against contamination by bacterial and viral pathogens, toxins, and **chemical hazards** or **physical hazards** that could cause illness or injury to the astronauts.

HACCP has been used extensively in the food processing industry for many years. Since the mid-1980s HACCP has been evaluated as an appropriate means of monitoring food safety in all segments of the food industry including foodservice operations. On March 20, 1992, the National Advisory Committee on Microbiological Criteria for Foods (NACMCF) adopted a revised document on HACCP that included seven principles that provide guidance on the development of an effective HACCP plan. The following are core concepts of HACCP as defined by NACMCF:

- Emphasizes the concept of prevention and universal application.
- Incorporates a decision tree for use in identifying critical control points (see Figure 3.11).

Figure 3.11 Critical control point (CCP) decision tree (apply at each step in food preparation that has an identified hazard).

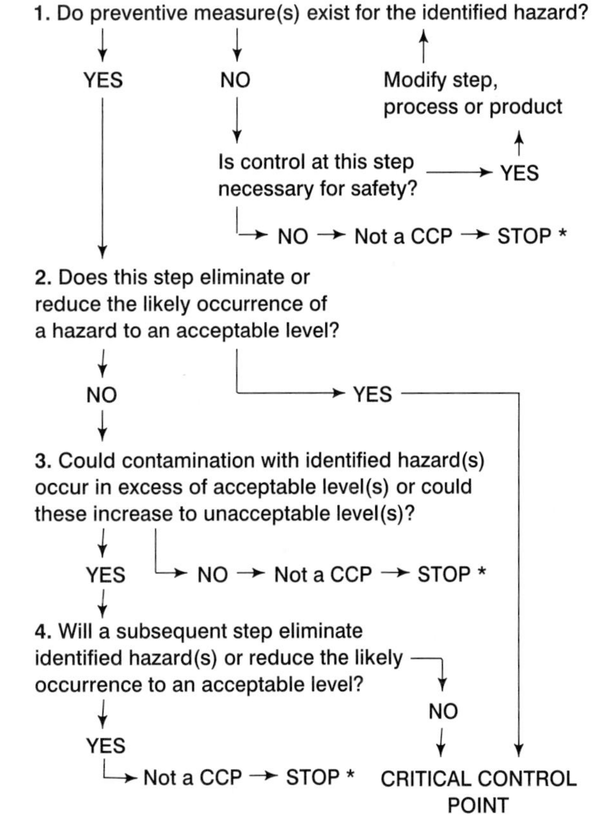
1. Do preventive measure(s) exist for the identified hazard?
YES
NO
Modify step, process or product
Is control at this step necessary for safety?
YES
NO → Not a CCP → STOP *
2. Does this step eliminate or reduce the likely occurrence of a hazard to an acceptable level?
NO
YES
3. Could contamination with identified hazard(s) occur in excess of acceptable level(s) or could these increase to unacceptable level(s)?
YES
NO → Not a CCP → STOP *
4. Will a subsequent step eliminate identified hazard(s) or reduce the likely occurrence to an acceptable level?
NO
YES
Not a CCP → STOP *
CRITICAL CONTROL POINT
* Proceed to next step in the described process.

- Provides a detailed explanation of the application of HACCP.

Unique to HACCP is that, in order to be effective, by definition it must be a documented system that delineates the formal procedures for complying with the seven principles. HACCP continues to evolve, especially for the foodservice segment of the food industry. Further refinements will evolve as new food products and systems are developed and as hazards and their control measures are more clearly understood.

Several issues have been raised specific to the foodservice segment as perceived barriers to the effective implementation of HACCP. These issues include:

- Lack of resources including time and personnel,
- Complexity of foodservice operations,
- High turnover of personnel, and
- Burden of required documentation procedures.

Barriers are inherent to any new concept or procedure. It is also important to note that HACCP does not replace programs for personal hygiene or cleaning and sanitation. And, finally, it is not a panacea; it does not address every conceivable, plausible hazard in a foodservice operation. The professional manager, however, accepts barriers and limitations as part of the challenge of implementing a system in the spirit that advantages far outweigh the perceived disadvantages.

HACCP is the best food safety system available to foodservice operators at this time. The primary benefit of HACCP is that it emphasizes prevention of hazards to the food at all stages of the processing continuum. Another advantage is that it clearly identifies the food establishment as the final party responsible for ensuring the safety of the food and handling procedures. HACCP is a rational, scientific approach and monitors both current and past conditions under which food is processed.

Because of its many advantages it is anticipated that HACCP will soon be mandated by the federal government as a means to ensure effective, efficient food safety in all segments of the food industry. The seafood industry is the first processing segment to be required to implement HACCP, and this mandate is likely to be followed by the meat and poultry industries. Since the early 1990s the foodservice industry has been under increasing pressure to adopt the principles of HACCP and there is abundant evidence that HACCP regulations for the foodservice industry are coming. For example, on August 4, 1994, the FDA released an Advanced Notice of Proposed Rules seeking comment on extending mandatory HACCP to all segments of the food industry including foodservice operations.

Some state regulatory agencies have already adopted the HACCP principles for use in survey processes. The Joint Commission on Accreditation of Healthcare Organizations (JCAHO) has some of the guidelines in place and it is anticipated that future JCAHO standards will fully adopt the HACCP concept.

The Seven Principles of HACCP

The seven principles of the HACCP program are as follows:

1. *Identify hazards and assess their severity and risks:* A **hazard,** as defined in the 1995 Food Code, means a biological, chemical, or physical property that may cause an unacceptable consumer health risk. An example of a **biological hazard** would be the presence of *Salmonella* bacteria on raw chicken as it enters the foodservice operation. The best means to evaluate hazards is to draw a diagram of the flow of food and then analyze each specific step.

2. *Identify the **critical control points** (CCP) in food preparation:* A CCP is any point or procedure in a specific food system where loss of control may result in an unacceptable health risk. A critical control point for raw chicken would be the final cooking step because this is the last opportunity to eliminate or reduce the *Salmonella* to a safe level.

3. *Establish **critical limits** for preventive measures associated with each identified CCP:* For example, time and endpoint cooking temperatures should be established for cooking procedures.

4. *Establish procedures to monitor CCPs:* Examples of these procedures may include visual evaluation and time-temperature measurements.

5. *Establish the corrective action to be taken when monitoring shows that a critical limit has been exceeded:* For example, the receiving procedures should indicate that frozen products with evidence of thawing be rejected.

6. *Establish effective record-keeping systems that document the HACCP system:* Traditional records such as receiving records, temperature charts, and recipes can serve as the basis for documentation.

7. *Establish procedures to verify that the system is working:* This may be as simple as reviewing records on a timely, routine basis or as complex as conducting microbiological tests.

These guidelines were designed for the food processing industry and may seem complicated, if not overwhelming, as applied to foodservice operations. For example, initial HACCP guidelines for the food processing industry treat each food product as a separate HACCP plan. If literally applied to foodservice, this would imply that each menu item be treated as a HACCP plan and a **flowchart** similar to the one in Figure 3.12 would need to be designed for *each* menu item. This may simply not be realistic for foodservice operations especially those of high volume and hundreds of menu items.

The model presented here is one example of how HACCP might be adapted and applied from receiving to point of service (POS) in a small facility (see Figure 3.13). The intent is for each phase of this model to be supported with sound policies on food handling that include critical limits rather than starting at receiving for each menu item. Documentation requirements are achieved through existing records including receiving records, storage temperature charts, standardized recipes, and service records (see, for example, the time-temperature documentation sheet shown in Figure 3.14).

Figure 3.13 represents the flow of food from the time the ingredients are received to the point of service. Receiving, storage, and preparation are seen as individual HACCP plans because identified hazards, CCPs, critical limits, and monitoring procedures are similar for all ingredients regardless of the recipes in which they are used (see, for example, the HACCP plan for receiving shown in Figure 3.15). Each recipe then is also an individual HACCP plan (see the sample recipe of Figure 3.16). Each recipe form includes identified hazards, CCPs, and critical limits (time and temperatures). For facilities with a great number of recipes, it is recommended that the initial HACCP plan focus on recipes that include potentially hazardous ingredients such as raw eggs, poultry, meat, and milk products. Some large facilities are experimenting with computerized HACCP programs. Figure 3.17 is an example of a food flowchart that was generated by a computer.

Critical Control	Hazard	Standards	Corrective Action
		Receiving	
Receiving beef	Contamination and spoilage	Accept beef at 45°F (7.2°C) or lower; verify with thermometer	Reject delivery
		Packaging intact	Reject delivery
		No off odor or stickiness,	Reject delivery
Receiving vegetables	Contamination and spoilage	Packaging intact	Reject delivery
		No cross-contamination from other foods on the	Reject delivery
		No signs of insect or rodent activity	Reject delivery
		Storage	
Storing raw beef	Cross-contamination of other foods	Store on lower shelf	Move to lower shelf away from other foods
		Label, date, and use FIFO rotation	Use first; discard if maximum time is exceeded
	Bacterial growth and spoilage	Beef temperature must remain below 45°F (7.2°C)	Discard if time and temperature abused
Storing vegetables	Cross-contamination from raw potentially hazardous foods	Label, date, and use FIFO rotation	Discard product held past rotation date
		Keep above raw potentially hazardous foods	Discard contaminated, damaged, or spoiled products

Figure 3.12 The flow of food through the operation.

Critical Control	Hazard	Standards	Corrective Action If Standard Not Met
		Preparation	
Trimming and cubing beef	Contamination, cross-contamination, and bacteria increase	Wash hands	Wash hands
		Clean and sanitize utensils	Wash, rinse, and sanitize utensils and cutting board
		Pull and cube one roast at a time, then refrigerate	Return excess amount to refrigerator
Washing and cutting vegetables	Contamination and cross-contamination	Wash hands	Wash hands
		Use clean and sanitized cutting board, knives, utensils	Wash, rinse, and sanitize utensils and cutting board
		Wash vegetables in clean and sanitized vegetable sink	Clean and sanitize vegetable sink before washing vegetables
		Cooking	
Cooking stew	Bacterial survival	Cook **all** ingredients to minimum internal temperature of 165°F (73.9°C)	Continue cooking to 165°F (73.9°C)
		Verify final temperature with a thermometer	Continue cooking to 165°F (73.9°C)
	Physical contamination during cooking	Keep covered, stir often	Cover
	Contamination by herbs and spices	Add spices early in cooking procedure	Continue cooking at least 1/2 hour after spices are added
		Measure all spices, flavor enhancers and additives, and read labels carefully	
	Contamination of utensils	Use clean and sanitized utensils	Wash, rinse, and sanitize all utensils before use
	Contamination from cook's hands or mouth	Use proper tasting procedures	Discard product
		Holding and Service	
Hot holding and serving	Contamination, bacterial growth	Use clean and sanitary equipment to transfer and hold product	Wash, rinse, and sanitize equipment before transferring food product to it
		Hold stew above 140°F (60°C) in preheated holding unit, stir to maintain even temperature	Return to stove and reheat to 165°F (73.9°C)
		Keep covered	Cover
		Clean and sanitize serving equipment and utensils	Wash, rinse, and sanitize serving utensils and equipment

Figure 3.12 *Continued*

Critical Control	Hazard	Standards	Corrective Action If Standard Not Met
		Cooling	
Cooling for storage	Bacterial survival and growth	Cool rapidly in ice water bath and/or shallow pans (<4" deep)	Move to shallow pans
		Cool rapidly from 140°F (60°C) to 45°F (7.2°C) in 4 hours or less	Discard, or reheat to 165°F (73.9°C) and re-cool one time only
		Verify final temperature with a thermometer; record temperatures and times before product reaches 45°F (7.2°C) or less	If temperature is not reached in less than 4 hours, discard; or reheat product to 165°F (73.9°C) and re-cool one time only
	Cross-contamination	Place on top shelf	Move to top shelf
		Cover immediately after cooling	Cover
		Use clean and sanitized pans	Wash, rinse, and sanitize pans before filling them with product
		Do not stack pans	Separate pans by shelves
	Bacterial growth in time or after prolonged storage time	Label with date and time	Label with date and time or discard
		Reheating	
Reheat for service	Survival of bacterial contaminants	Heat rapidly on stove top or in oven to 165°F (73.9°C)	Reheat to 165°F (73.9°C) within 2 hours
		Maintain temperature at 140°F (60°C) or above; verify temperature with a thermometer	Transfer to preheated hot holding unit to maintain 140°F (60°C) or above
		Do not mix new product into old product	Discard product
		Do not reheat or serve leftovers more than once	Discard product if any remains after being reheated

Figure 3.12 *Continued*
Source: From *Applied Foodservice Sanitation,* 4th ed. Copyright © 1992 by the Educational Foundation of the National Restaurant Association.

Figure 3.13 HACCP flowchart for conventional foodservice system.
Source: *From rule interpretation of "proper serving temperatures," Wisconsin Administrative Code, October 29, 1982.

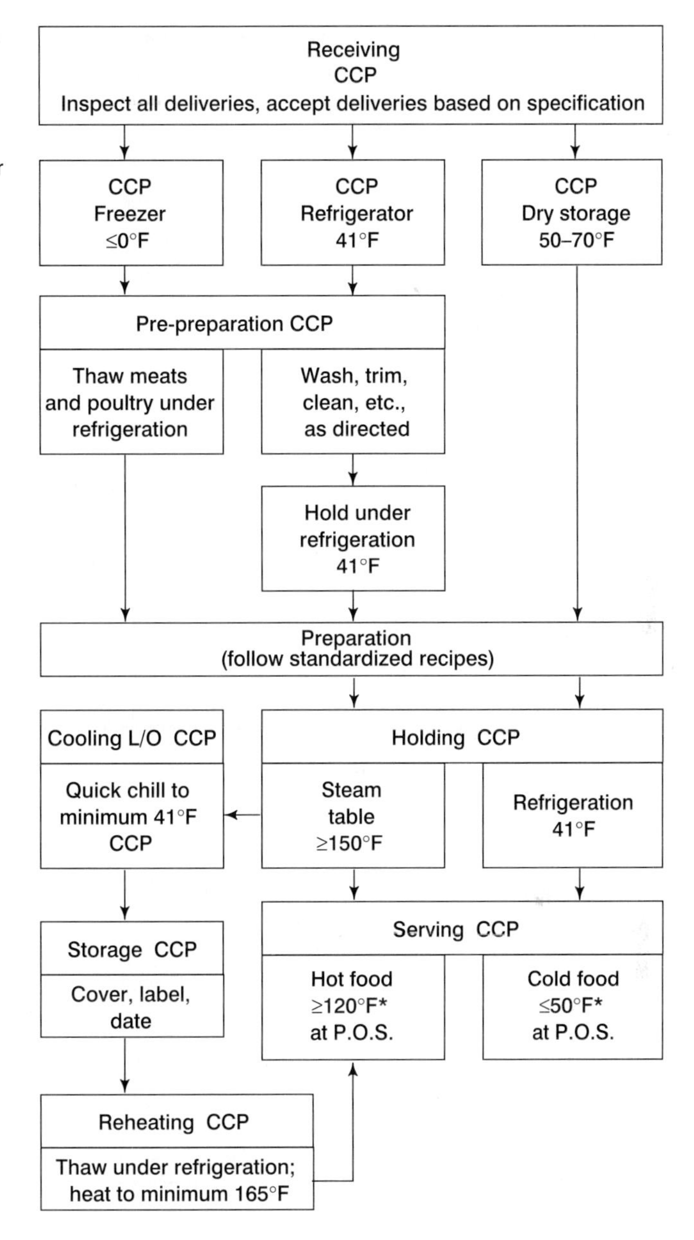

Heartland Country Village
HACCP Plan
Cooking Temperature
Date: ________________

Menu Item	Cook Time Start	Cook Time Stop	Final Temp	Comments	Cook Initial

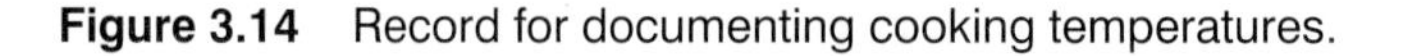

Figure 3.14 Record for documenting cooking temperatures.

SUMMARY

Thousands of people become ill each year as a result of consuming a food that was microbially, chemically, or physically contaminated. A single error in food handling in a foodservice operation can cause a major outbreak. It is the responsibility of the food manager to have the necessary knowledge base and an understanding of food handling principles to design, implement, and monitor a successful food safety program.

REVIEW QUESTIONS

1. How is a foodborne illness defined?
2. What is the difference between a case and an outbreak?

Heartland Country Village
HACCP Plan

Process Step	Hazards	Preventive Measures and Critical Limits	Monitoring Process	Corrective Action	Records	Verification
Receiving-CCP	Bacterial contamination, physical contamination	1. Frozen foods must be received at a product temperature of no higher than 0°F. 2. All refrigerated product, including fresh meat, produce, dairy, and eggs, must be received at a product temperature of no more than 40°F. 3. No off odor. 4. Packing intact.	All deliveries will be checked against specifications immediately upon arrival. Check temperatures of refrigerated items and conduct visual analysis for physical damage (bulging cans, open containers, etc.).	Reject all product that does not meet standards established by specification.	Standard receiving records.	Supervisor to review receiving records on weekly basis.

Figure 3.15 HACCP plan for receiving.

Heartland Country Village

Recipe Title *Scrambled Eggs*

Yield Information			
Portions	*50*	Cooking Temperature	*350° F (conventional oven)*
Pan Size	*4" – 1/2 pan*	Cooking Time	*1 hour*
Number of Pans	*1*	Portion Size	*1/4 c*
		Portion Utensil	*#16 scoop*

Ingredient	Amount	Procedures
Vegetable spray		– Spray pan with vegetable spray; set aside.
Eggs	5 dozen	– Remove eggs from refrigerator, check shells for cracks and soil; discard cracked eggs, remove soil. – Break clean eggs into mixer bowl. – Beat slightly on medium speed, using wire attachment.
1% Milk Salt Pepper	4 cups 1 Tbsp. ¼ tsp.	– Add milk, salt and pepper. Beat until well blended (3 to 5 minutes). – Pour mixture into prepared pan. – (CCP) Bake for 1 hour at 350°F to minimal internal temperature of 165°F and until product is firm in center (do not overbake). – Transfer to steamtable just prior to service. **Critical Control Point** Measure internal temperature of scrambled eggs. If internal temperature of scrambled eggs is less than 165°F, continue to bake until internal temperature is at least 165°F.

Figure 3.16 Sample recipe including HACCP.

3. What is the role of the food manager in food safety?
4. Identify at least three "at-risk" populations as defined by the U.S. Public Health Service.
5. What conditions contribute to the growth of bacteria?
6. Define *intoxication* and *infection* as related to foodborne disease.
7. Cite examples of chemical, physical, and biological hazards.
8. What is cross-contamination?
9. Describe how *Salmonella* and *Staphylococcus aureus* contamination can occur. How can they be controlled?
10. What is HACCP and how can it be applied in the foodservice setting?
11. What is the purpose of a flowchart?
12. Describe time-temperature principles in relation to food safety.
13. Describe at least three appropriate cooling methods.
14. Compare and contrast government regulations, standards, and codes.

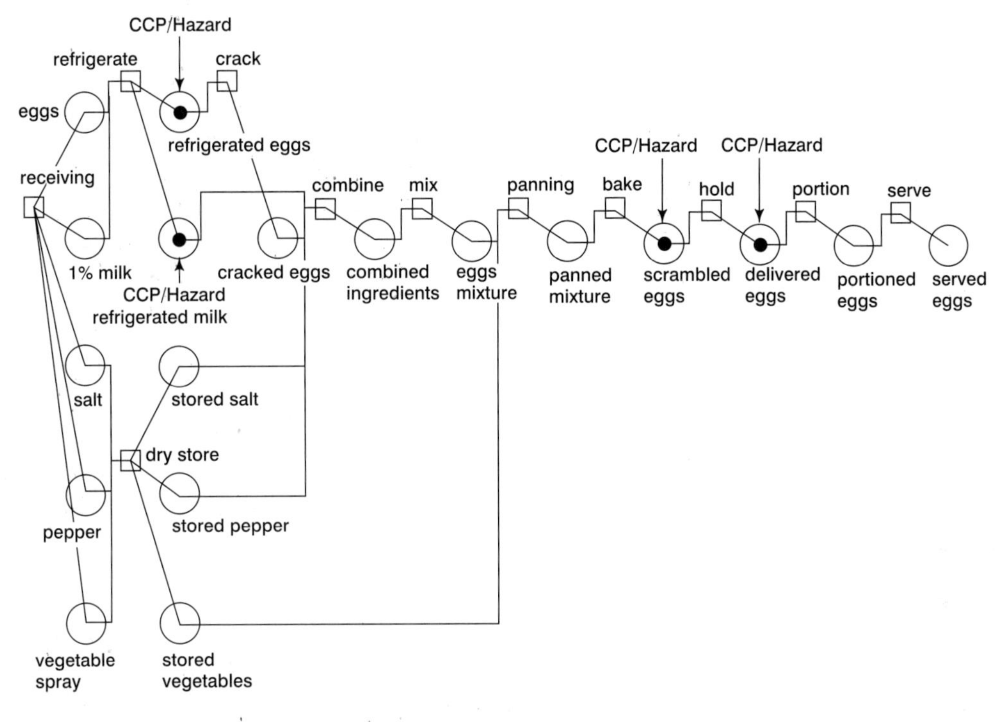

Figure 3.17 Computerized flow of food for scrambled egg recipe.

SELECTED REFERENCES

Applied Foodservice Sanitation, 4th ed. Chicago, Ill.: National Restaurant Association Educational Foundation, 1992.

Bean, N. H., and Griffin, P. M. Foodborne disease and outbreaks in the United States, 1973–1987. Pathogens, vehicles and trends. *J. Food Prot.* 1990; 53(9):804.

Bean, N. H., Griffin, P. M., Goulding, J. S., and Ivey, C. B. Foodborne disease outbreaks, 5-year summary, 1983–1987: Pathogens, vehicles, and trends. *J. Food Prot.* 1990; 53:711–728.

Beasley, M. A. Implementing HACCP standards. *Food Mgmt.* 1995; 30(1):40.

Bryan, F. L. Applications of HACCP to ready-to-eat chilled foods. *Food Technol.* July 1990, pp. 70–77.

Bryan, F. L. Hazard analysis and critical control point (HACCP) concept. *J. Food Prot.* July 1990, pp. 416–418.

Cliver, D. O.: *Foodborne Diseases.* San Diego, Calif.: Academic Press, 1990.

Cody, M. M., and Keith, M. *Food Safety for Professionals: A Reference and Study Guide.* Chicago, Ill.: The American Dietetic Association, 1991.

Education Foundation of the National Restaurant Association. (1993). HACCP reference book. Chicago, Ill.: National Restaurant Association Educational Foundation.

Food Code—1995, Recommendations of the U.S. Department of Health and Human Services, U.S. Public Health Service. Washington, D.C.: U.S. Food and Drug Administration, 1996.

LaVella, B., and Bostic, J. L. *HACCP for Food Service.* LaVella Food Specialists, 332 Halcyon, St. Louis, MO 63122, 1994.

National Advisory Committee on Microbiological Criteria for Foods. *Hazard analysis and critical control point system. Int. J. Food Microbiol.* 1992; 16:1–23.

Norton, C. Preparing your operation for the HACCP process. *Food Mgmt.* May 1992, p. 64.

Todd, E. C. D. Preliminary estimates of costs of foodborne disease in the United States. *J. Food Prot.* 1989; 52:595–601.

Todd, E. Foodborne illness. Epidemiology of foodborne illness: North America. *Lancet* 1990; 336:788–790.

FOR MORE INFORMATION

The American Dietetic Association (ADA)
National Center for Nutrition and Dietetics (NCND)
216 W. Jackson Boulevard, Suite 800
Chicago, IL 60606-6995
Phone: (312)899-0040 or (800)877-1600

Centers for Disease Control and Prevention (CDC)
1600 Clifton Road, NE
Atlanta, GA 30333
Phone: (404)639-3311

Council for Agricultural Science and Technology (CAST)
137 Lynn Ave
Ames, IA 50010-7120
Phone: (515)292-2125

USDA's Food Safety Inspection Service (FSIS)
USDA Meat and Poultry Hotline
1400 Independence Avenue, SW
Room 2925-South
Washington, DC 20250
Phone: (800)535-4555

The National Restaurant Association
Educational Foundation
250 S. Wacker Drive, Suite 1400
Chicago, IL 60606

United States Department of Agriculture (USDA)
14th Street and Independence Avenue, SW
Washington, DC 20250

United States Environmental Protection Agency (EPA)
401 M Street, SW
Washington, DC 20460

United States Food and Drug Administration (FDA)
5600 Fishers Lane
Rockville, MD 20857
Phone: (301)443-1544

United States Public Health Service (USPHS)
200 Independence Avenue, SW
Washington, DC 20201

4

The Menu: The Focal Point of the Foodservice Operation

OUTLINE

The **menu** is indeed the focal point of the foodservice operation and, to varying degrees, influences all aspects of foodservice. The menu serves as a catalyst that drives all operational functions: purchasing, production, assembly, distribution, service, and sanitation. The menu is also a control tool that determines resource acquisition and utilization in the form of labor, equipment, and facilities, and in this sense it has a profound impact on the budget.

A menu is a detailed list of food items that may be ordered (as in a restaurant) or served (as in a hospital, school, or corrections facility). The menu should reflect a particular operation's mission and, therefore, it will vary greatly from one organization to the next. In the commercial setting, the menu is designed to attract customers and generate sales, whereas noncommercial operations prepare menus to meet the needs and wants of a known population. Regardless of the type of foodservice organization for which menus are being considered, careful planning, implementation, and evaluation are essential to the success of meeting customer needs and preferences within the budgetary constraints of the organization.

The purpose of this chapter is to review the many factors that influence menu procedures, and then to describe the menu process of planning, writing, and evaluating. The chapter begins with a comprehensive review of the many factors that influence the menu planning process. The most important factor is the customer profile. This review is followed by specific guidelines on how to write menus and includes step-by-step procedures and a timetable to ensure that the menu process is completed in a timely fashion. The menu writing section is followed by a discussion on menu design and layout. The chapter concludes with strategies for menu evaluation from the customer's perspective.

It is important for the reader to understand that the following guidelines serve as an excellent basis for menu planning. It is equally important to appreciate the need to be flexible and creative to ensure that the planned menus meet the needs of the customer and reflect the philosophy of the foodservice organization.

KEY CONCEPTS

1. Numerous customer and facility factors must be considered during the menu planning process.
2. Menu planning is a process of development, implementation, and evaluation.
3. Numerous types of menus are available from which to choose; the final choice must be appropriate for the type of facility and customers served.
4. Most health care facilities must comply with federal and state menu planning regulations.
5. Written menus for modified diets are essential for most health care facilities.
6. Written menus must comply with federal and state truth-in-menu and nutritional labeling legislation.
7. The written menu can serve as a marketing tool for the foodservice operation.
8. Menu pricing strategies are an important aspect of financial management.

9. Customer satisfaction with menus can be measured by means of a number of observation and survey techniques.

MENU PLANNING

The primary goal of a foodservice operation is to serve food that is acceptable to its clientele. It is important to be familiar with the target market: who the consumers are, their characteristics, and their food preferences. They may be located in a variety of settings, such as schools, hospitals, restaurants, or extended care facilities. The menus for each type of organization and its patrons are different.

Certain management factors must also be considered. Allowing adequate time to complete the menu planning process is essential to smooth implementation of the menu. The menu planner should aim for the best menu and service, with optimal use of personnel and equipment. Often, a compromise must be made between what a foodservice would like to offer and what it is capable of producing and serving. The following subsections provide a more detailed description of menu planning considerations.

Organizational Mission and Goals

The planned menu must be appropriate for the foodservice and consistent with its organizational mission and goals. Whether the major goal is to provide nutritionally adequate meals at a reasonable cost, as in school foodservice, or whether profit is the major goal, the menus must reflect the organization's stated purpose as defined in the mission statement. Whatever the specific goals, most foodservices are offering food choices that reflect the current emphasis on nutrition and that meet the quality and service expectations of the customer.

The Customer

The menu planner should carefully study the population to be served regardless of whether menus are being planned for a commercial or noncommercial operation. Data on demographics, sociocultural influences, and eating habits will generate a composite profile of the customer.

Demographics. The term *demographics* refers to the statistics of populations. Specific indicators include but are not limited to age, gender, health status, and level of education. Economic information such as personal income may also be included in this definition. Trends in this information are important to the menu planner because eating habits vary among population groups.

It is well known, for example, that the American population is getting older. Persons 65 years or older currently represent approximately 13% of the U.S. population. This number is expected to increase to 30% by the year 2030. The eating habits and preferences of this population are very different compared to those of younger populations. "Baby Boomers" are now middle aged and eat many of their meals outside the home. They are also the parents of "Generation X" and may have a significant influence over the eating habits of these young people.

Along with demographic information, the geographic distribution of populations may be of interest to the menu planner. Midwest states, including the Dakotas, Iowa, Nebraska, Kansas, and Missouri, for example, have a high percentage of individuals over the age of 65. This population segment is expected to grow in Southwestern states including Arizona and New Mexico. These and other population shifts will have an impact on the sociocultural makeup of customers.

Sociocultural Influences. The term *sociocultural* refers to the combining of the social and cultural factors of a population. These factors include marital status, lifestyle, ethnic background, values, and religious practices. These issues have a greater impact on menu planning than ever before given the increase in the cultural diversity of the U.S. population during the past several years. For example, Hispanic and Asian populations have increased dramatically in several areas of the United States.

Food plays an important role in our social lives. The wise menu planner becomes knowledgeable about social influences and respects the personal preferences of the customer.

Closely related to sociocultural influences are psychological needs. To many, food offers comfort and emotional satisfaction. As indicated in Chapter 1 there appears to be a trend toward "comfort foods" such as homestyle soups and stews, and meat and potato type meals.

Nutritional Requirements. Meeting the nutritional requirements of individuals in the group to be served is important in meal planning. This is especially true in foodservices that are responsible for providing all of a group's daily nutritive requirements, such as in long-term care facilities or prisons. These menus should fulfill the most recent recommended dietary allowances (RDAs) defined by the National Research Council. The RDAs specify nutrient needs for various age groups by sex. Schools participating in the National School Lunch Program are required to meet certain nutrition requirements, as are many hospitals and government-sponsored nutrition programs (see Figures 4.1 and 4.2). Before planning menus for a government-funded operation, the menu planner should contact an appropriate government agency to verify regulatory requirements for menu composition. For example, you might contact a state health department to acquire regulatory information applicable to health care facilities, such as hospitals, long-term care facilities, or child care operations.

The general public is increasingly concerned about health and more interested in nutrition than ever before. In response to this changing attitude, restaurants and other foodservices are featuring foods that comply with today's definition of healthful eating. Many people are observing the dietary guidelines issued by the United States Department of Agriculture (USDA) and the United States Department of Health and Human Services. The guidelines were revised in 1995 and are presented in Figure 4.3.

School Lunch Patterns		**Minimum Quantities**		**Recommended Quantities**
Components		**Grades K–3** ages 5–8 (Group III)	**Grades 4–12** age 9 & over (Group IV)	**Grades 7–12** age 12 & over (Group V)
Meat or meat alternate	A serving of one of the following or a combination to give an equivalent quantity:			
	Lean meat, poultry, or fish (edible portion as served)	1½ oz	2 oz	3 oz
	Cheese	1½ oz	2 oz	3 oz
	Large egg(s)	¾	1	1½
	Cooked dry beans or peas	⅜ cup	½ cup	¾ cup
	Peanut butter	3 Tbsp	4 Tbsp	6 Tbsp
	Nuts/seeds	¾ oz = 50%	1 oz = 50%	1½ oz = 50%
Vegetable and/or fruit	Two or more servings of vegetable or fruit or both to total	½ cup	¾ cup	¾ cup
Bread or bread alternate	Servings of bread or bread alternate	8 per week	8 per week	10 per week
Milk	A serving of fluid milk	½ pint (8 fl oz)	½ pint (8 fl oz)	½ pint (8 fl oz)

Figure 4.1 National School Lunch patterns.
Source: Supplement to 1984 Food Buying Guide for Child Nutrition Programs, PA-1331, U.S. Department of Agriculture.

The food guide pyramid (Figure 4.4) is a graphic depiction of the dietary guidelines and was developed to offer a visual outline of what healthy Americans should eat each day. The pyramid is designed to emphasize a variety of food choices and moderation of intake for some food groups and nutrients such as fat and sodium.

Food Consumption, Trends, Habits, and Preferences. As stated earlier, the clientele of a foodservice operation is generally composed of individuals from different cultural, ethnic, and economic backgrounds, most of whom have definite food

School Breakfast Patterns		Minimum Required Serving Size		
Food Components/Items		**Ages 1 and 2**	**Ages 3, 4, 5**	**Grades K–12**
Milk (fluid)	As a beverage, on cereal, or both:	½ cup	¾ cup	½ pint
Juice/fruit/ vegetable*	Fruit and/or vegetable OR full-strength fruit juice or vegetable juice	¼ cup	½ cup	½ cup
Select one serving from each of the following components/items or two servings from one component/item.				
Bread/bread alternates**	One of the following or an equivalent combination:			
	Cereal (whole-grain or enriched or fortified)	¼ cup vol or ⅓ oz wt	⅓ cup vol or ½ oz wt	¾ cup vol or 1 oz wt
	Bread (whole-grain or enriched)	½ slice	½ slice	1 slice
	Biscuit, roll, muffin, corn-bread, etc. (whole-grain or enriched meal or flour)	½ serving	½ serving	1 serving
Meat or meat alternate	One of the following or an equivalent combination:			
	Lean meat, poultry, or fish	½ oz	½ oz	1 oz
	Cheese	½ oz	½ oz	1 oz
	Egg (large)	½	½	½
	Peanut butter or other nut or seed butters	1 Tbsp	1 Tbsp	2 Tbsp
	Nuts and/or seeds*** (as listed in program guidance)	½ oz	½ oz	1 oz
	Cooked dry beans and peas	2 Tbsp	2 Tbsp	4 Tbsp

* Recommended daily: Citrus juice or fruit or a fruit or vegetable that is a good source of vitamin C (See *Menu Planning Guide for School Food Service,* PA-1260).

** For serving sizes of breads and bread alternates, see *Food Buying Guide for Child Nutrition Programs,* PA-1331 (1984).

*** No more than one ounce of nuts and/or seeds may be served in any one meal.

Figure 4.2 National School Breakfast patterns.
Source: Supplement to 1984 Food Buying Guide for Child Nutrition Programs, PA-1331, U.S. Department of Agriculture.

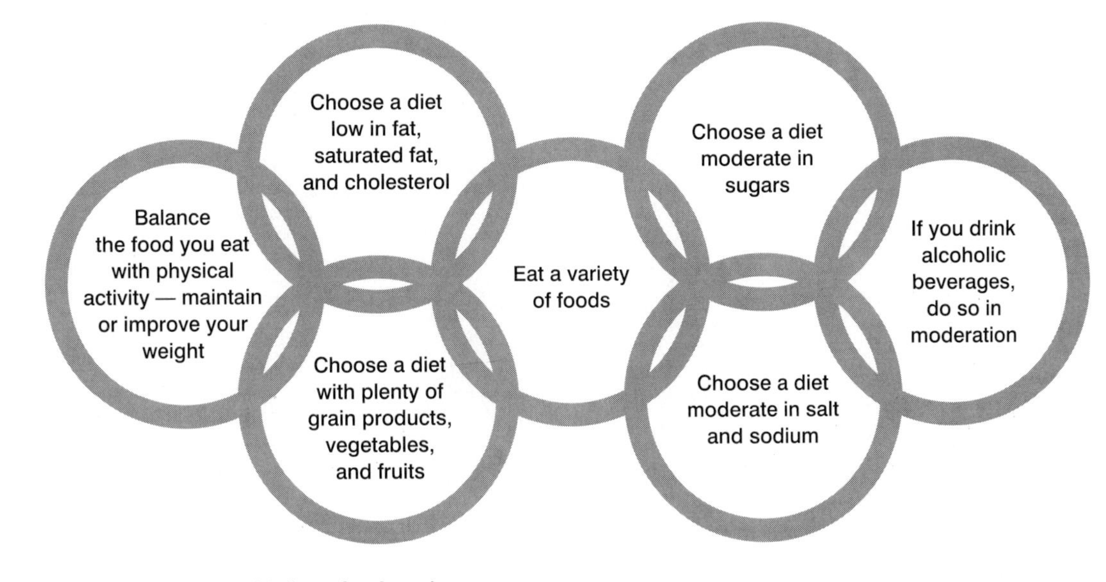

Figure 4.3 Dietary guidelines for Americans.
Source: U.S. Department of Agriculture/U.S. Department of Health and Human Services, 1995.

likes and dislikes. The menu planner must keep this in mind when selecting foods to satisfy this diverse group.

Food habits are based on many influences, the most direct being the attitude of the parents in the home about food and eating. The family's ethnic and cultural backgrounds and its economic level combine to determine the foods served and enjoyed. These habits are passed down from generation to generation, and when several different cultural or ethnic backgrounds are represented in one group for which a menu is to be made, the task of planning presents a challenge.

In today's mobile society, however, people are becoming more knowledgeable about ethnic and regional foods. Interest in Mexican, Oriental, Italian, and other international foods is evident from the number of specialty restaurants. Many health care facilities, schools, colleges, and similar foodservices include these foods in their menus to add variety and to contribute to the cultural education of their clientele. The menu planner should be aware of local and regional food customs and religious restrictions. For example, a menu planner should be well aware of kosher and Muslim dietary restrictions.

In addition, the traditional three-meals-a-day pattern, with the entire family eating together, has changed. People are eating fewer meals at home. They are eating more frequently and at less regular hours. To accommodate this change in eating habits, a more flexible meal schedule is evident in most institutional foodservices, and continuous service is available in many restaurants. In spite of the desire for fast-paced services, some experts predict that with the variety of prepared foods available today, many people will eat frozen and other easily prepared foods at home and will seek a restaurant with interesting decor and atmosphere when they dine away from home. The person planning menus for any type of foodservice should monitor such trends.

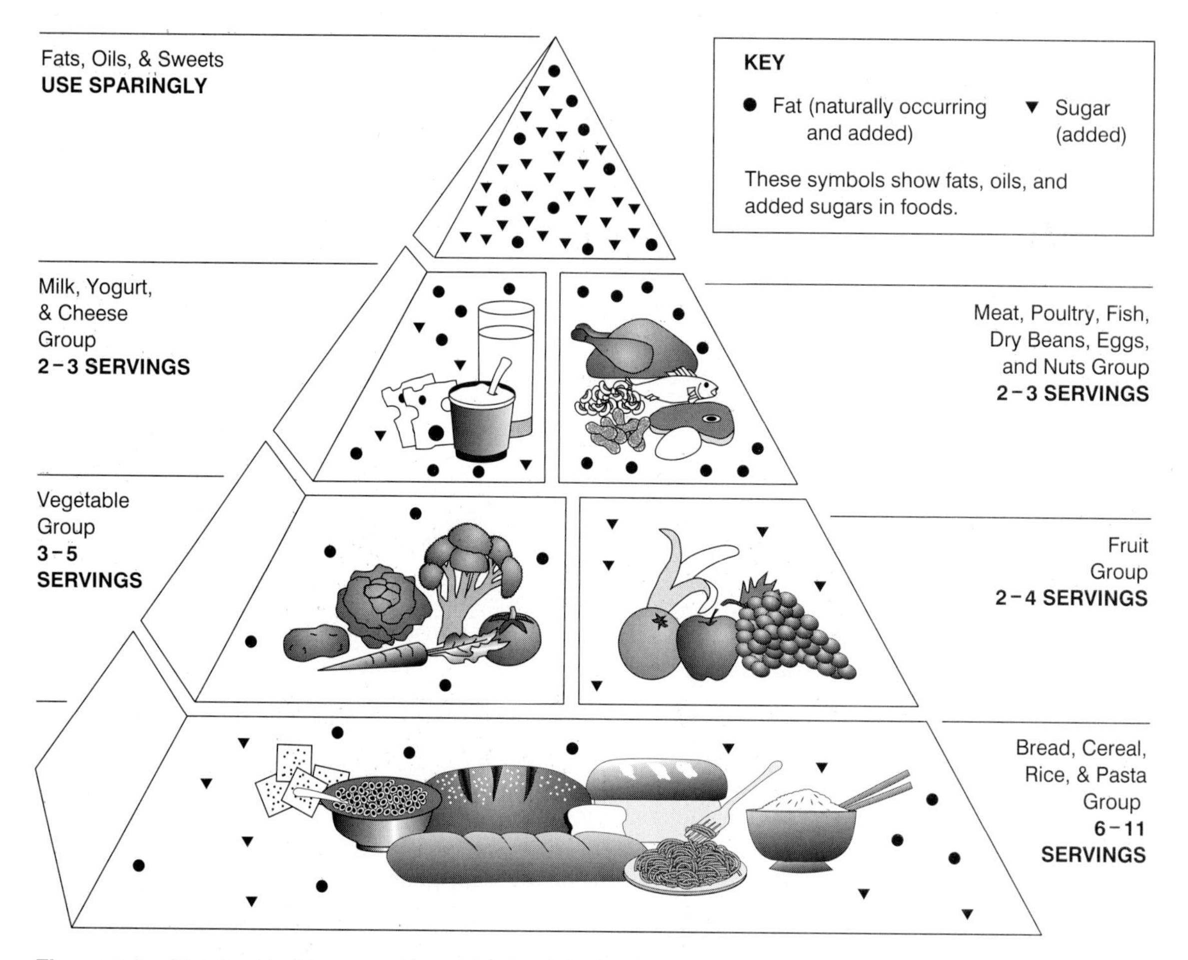

Figure 4.4 The food guide pyramid: a guide to daily food choices.
Source: U.S. Department of Agriculture/U.S. Department of Health and Human Services, 1992.

Budgetary Guidelines

Before any menu is planned, the amount of money that can be spent on food must be known. In commercial operations, the amount to be budgeted is based on projected income from the sale of food. This income must generate adequate revenue to cover the cost of the raw food, labor, and operating expenses and must allow for reasonable profit. Management determines these financial objectives through strategic menu pricing. Table 4.1 is a summary of three common menu pricing methods.

In a school, health care facility, or other nonprofit organization, a raw-food-cost allowance per person, per meal, or per day may be determined. This does not mean that the cost of every item must fall below the budgeted figure. The planner looks at the total weekly or monthly food cost. This cumulative average can be made readily available by a computer-assisted system. By balancing more costly items with less

Table 4.1 Summary of three common menu pricing methods.

Method	Concept	Formula	Example	Advant.	Disadvantages	Comments
Factor (also referred to as fixed factor and markup)	The raw food cost is multiplied by a predetermined factor that takes into account labor, supplies and any projected profit margin.	Selling price = Food cost × Factor Pricing factor = 100% ÷ Desired Food Cost Objective (stated as a percentage of the selling price).	Desired Food Cost % = 40 100 ÷ 40 = 2.5 If menu item food cost is $.90 then: Selling Price = .90 × 2.5 = $2.25	Simple; easy to use.	Results in prices too low when applied to labor-intensive, low food cost menu items and vice versa. Can create large price deviation among menu items in a specific menu category.	Institutions and subsidized operations generally use a factor of 2.5 while high profit operations may use a factor in the range of 3.7 to 4.0 for a food cost percent of 25% to 27%.
Prime Cost	The sum of food and labor costs is divided by a predetermined pricing factor that accounts for food and labor through separate percentages.	Selling price = Prime Cost × Pricing Factor Prime Cost = Food Cost + Direct Labor Cost Pricing factor = 100% ÷ (Food Cost % + Labor Cost %)	Food cost % = 40 Labor cost % = 38 Food cost for roast beef = $.90 Labor cost for roast beef = $.24 Pricing factor = 100 ÷ (40+38) = 1.28 Selling price = ($.90+$.24) × 1.28 = $1.46	More accurately reflects the direct labor cost on a per item basis. Reduces price deviations among menu items in a specific menu category	Can be very time consuming to collect labor costs on a per menu item basis.	A modified version of this method uses a labor factor that reflects the labor intensity (low, medium, high) of each menu item. Labor factors are developed by and specific to individual operations.
Cost Plus Profit (also referred to as recovery plus profit and actual cost)	Prices are calculated for menu items based on the entire actual cost to produce the item. Based on per portion costs for food, direct labor, operating expenses, fixed expenses, and profit.	Selling Price = Food Cost ($) + Direct Labor ($) ÷ 100 – Operating Expenses (% of Sales) + Fixed Expenses (% of Sales) + Profit (%)	Roast Beef: If: Food cost = $.90 Labor cost = $.24 Op. exp. = 10% Fixed exp. = 7% Profit = 5% selling price = ($.90 + $.24) ÷ (100 – 22) = $1.46	Expresses cost in detail	Very time consuming to calculate and track all costs per month per menu item.	

expensive foods, a more interesting variety can be offered and the budget can still be maintained. For example, the raw food cost of rib roast or beef tenderloin steak may be offset by low-cost poultry or ground beef items. On selective menus that offer a choice of two or more entrées, a well-liked lower cost item should be offered with a more expensive food. For example, tacos or burritos are popular and relatively low-cost entrées that could be offered as alternatives to more costly baked pork chops. Costs, then, may determine the choices, but it is important to remember that variety in the menu may be enhanced by balancing the use of high-cost and low-cost items.

Production and Service Capabilities

Equipment and Physical Facilities. The menu plan for the day must be one that can be produced in the available work space and with the available equipment. Care should be taken to distribute the workload evenly for ovens, ranges, mixers, and other large pieces of equipment. The ovens are especially vulnerable to overuse. The inclusion of too many foods at one meal that require oven use may cause an overload or complicate production schedules. For example, it may not be possible to bake Swiss steak, potatoes, and a vegetable casserole at the same meal. If equipment must be shared among production units, the menu should not include items that will cause conflict. For example, unless separate ovens are available to the bakery unit, it may not be possible to bake hot breads if other menu items must be baked just prior to serving. Equipment usage errors can be alleviated by involving the production staff in the menu planning process.

The planner should be aware of restrictions on equipment and space, and be familiar with the methods of preparation, equipment capacity, and the pans or other utensils needed before choosing the menu items. Refrigerator and freezer space must also be considered. A chilled dessert, molded gelatin salad, and individual cold plates may be difficult to refrigerate if all are planned for the same day's menu.

The amount of china, glassware, or silverware available may influence the serving of certain menu items at the same meal. For example, fruit cobbler and a creamed vegetable may both require sauce dishes. Foods such as parfaits or shrimp cocktail should not be placed on the menu unless there is appropriate glassware for serving. For operations that use tray service, tray design must also be taken into consideration before food items are included on the menu.

Personnel. The availability and skill of employees are factors to consider when determining the variety and complexity of a menu. Understanding the relationship between menu and personnel helps the planner to develop menus that can be prepared with the available staff. Work schedules must be considered for all days because some foods require advance preparation, whereas others are prepared just prior to service. Menu items should be planned that enable employees' workloads to be spread evenly throughout the day and that do not result in too much last-minute preparation.

Availability of Food

The location of the market or other sources of food may have a limiting effect on the menu. This is especially true of fresh foods. A knowledge of fruits and vegetables and their seasons enables a planner to include them on the menu while they are at their peak of quality and at an affordable price. Regardless of whether the menu planner is responsible for purchasing the food, he or she should keep abreast of new items on the market and be alert to foods that would add interest to the menu or that would improve the variety and quality of menu items offered.

Style of Service

The style of service influences food item selection and the number of choices on the menu. Some foods are more adaptable to seated service than to cafeteria service; for example, cherries jubilee requires tableside preparation. Baked Alaska would be difficult to manage in cafeteria service because of last-minute preparation and the need for it to be served immediately after preparation. Tray design may limit the number or form of foods offered. For example, a layered cake with whipped topping may not work if a covered, insulated tray is being used.

The distance between the point of preparation and the point of distribution should also be considered, along with the elapsed time between the completion of preparation and service. If the food is prepared in a central kitchen and sent to service areas in remote locations, the menu should not include foods, such as soufflés, that change during transportation. Foods transported in bulk to a service unit must be of a type that will hold up well and still be appetizing when served.

Types of Menus

After careful analysis of the numerous factors that must be considered during the menu planning process, the menu planner needs to decide what type of menu will be used. There are many types of menus from which to choose. The decision is primarily influenced by the type of foodservice operation and the needs of the customer to be served. All types of menus are defined, at least in part, by the degree of choice offered.

Extent of Selection. A **selective menu** includes two or more choices in some or all menu categories. The exact number of options will vary with different types of foodservices. The menu mix, or the selection of food items to be offered in each category, must be carefully planned to meet the needs of the customer and to ensure even workloads and balanced utilization of equipment.

A *full-selective menu* offers at least two choices in every category. The advantage of this approach is that it allows maximum choice to the customer. The primary disadvantage of full-selective menus is the obvious demand on operational resources. Ingredients and food products must be available in inventory to meet menu demand and the production staff must have the skills and flexibility to respond to the variety of choices. In response to these demands and as a result of shorter hospital stays for patients, many health care facilities are implementing limited or semiselective menus.

A limited or **semiselective menu** allows one or more selections in some of the menu categories. For example, a long-term care facility may offer two entrée and two dessert selections at lunch and dinner, but only one choice in the vegetable and salad categories. Restaurants, on the other hand may offer a choice of entrée accompanied by standard side dishes.

A **nonselective menu** or preselective menu offers no choices. Organizations using the nonselective menu usually have a list of alternatives to offer in the event that a customer does not want any of the menu items offered. These are frequently referred

Table 4.2 Sample menus for the various types of selective menus.

Full Selective	Limited Selective	Nonselective
Appetizers	*Appetizers*	
Chilled tomato juice	Chilled tomato juice	
Cream of mushroom soup		
Entrées	*Entrées*	*Entrée*
Roast beef with gravy	Roast beef with gravy	Roast beef with gravy
Grilled tuna steak with dill sauce	Chicken salad on croissant with relishes	
Chicken salad on croissant		
Fresh fruit and cottage cheese plate with muffin		
Vegetables	*Vegetables*	*Vegetables*
Mashed potatoes with gravy	Mashed potatoes with gravy	Mashed potatoes with gravy
Boiled red potatoes	Steamed broccoli spears	Steamed broccoli spears
Steamed broccoli spears	Fresh vegetable plate	
Creamed carrots		
Salads	*Salads*	*Salads*
Garden salad with French dressing	Garden salad with French dressing	Garden salad with French dressing
Mandarin orange gelatin salad		
Desserts	*Desserts*	*Desserts*
Pecan pie with whipped topping	Pecan pie with whipped topping	Pecan pie with whipped topping
German chocolate cake with coconut icing	Butter brickle ice cream	
Butter brickle ice cream	Fresh fruit	
Fresh fruit		
Breads	*Breads*	*Breads*
Dinner roll	Dinner rolls	Dinner roll
White bread	Whole wheat bread	
Whole wheat bread		
Bread sticks		
Beverages	*Beverages*	*Beverages*
Coffee 2% milk	Coffee	Coffee
Tea Skim milk	Tea	2% milk
Hot cocoa Chocolate milk	2% milk	

to as "write-ins" in the health care industry because they are hand written directly onto the patient menu. Table 4.2 illustrates the different types of selective menus.

Menus may be *static,* or *set,* which means that the same menus are used each day. This type of menu is found in restaurants and other foodservices where the clientele changes daily or where there are enough items listed on the menu to offer sufficient variety. Many hospitals are now experimenting with nonselective and restaurant-style menus due to shorter patient stays. For example, mothers on maternity wards may be discharged within 24 hours after delivery. Some flexibility may be built into the menu by changing an item or two daily or offering daily specials. The static menu may be quite limited in choice, as in most fast-food restaurants. Changes in these menus are made only after careful development of a new product and extensive market research and testing.

A **single-use menu** is one in which the menu is planned for a certain day and is not repeated again in exactly the same form, although some food combinations may be used again in future meals. This type of menu is often used for special functions, holidays, or catering events.

The **cycle menu** is a carefully planned set of menus that is rotated at definite intervals—the length of cycle depending on the type of foodservice. With the current 3-day average patient stay, hospitals could have a 7-day cycle, but may prefer 8- or 12-day cycles to avoid repetition of foods on the same day each week. A three-week cycle, however, would accommodate a wider variety of foods and might result in greater patient satisfaction. An extended care facility normally has a relatively long cycle because of the length of stay of the residents.

Cycle menus have several advantages. After the initial planning has been completed, time is freed for the planner to review and revise the menus to meet changing needs such as holidays, vacations, changes in personnel, or availability of a food item. Repetition of the same menu aids in standardizing preparation procedures and in efficient use of equipment. Forecasting and purchasing are simplified and, with repeated use of the menus and needed adjustments, employee workloads can be balanced and distributed fairly.

Cycle menus do, however, have some potential disadvantages. They may become monotonous if the cycle is too short or if the same food is offered on the same day each week. The cycle menu may not include well-liked foods often enough, or it may include unpopular items too frequently. The cycle menu may not provide for foods that come into the market at varying times of the year, but many foodservices solve this problem by developing summer, fall, winter, and spring cycles; others note the seasonal alternatives on the menu. If these disadvantages can be resolved and the menu properly developed to meet the needs of a particular foodservice system, the cycle menu can become an effective management tool.

Whatever the length of the cycle, the menus must be carefully planned and evaluated after each use. A cycle menu should be flexible enough to handle emergencies, to utilize leftover food, and to accommodate new ideas and seasonal variations.

Menus may also be categorized by the method of pricing. In the **à la carte menu,** food items are priced separately. This type of menu allows the patron to select only the food wanted. The **table d'hôte menu** offers a complete meal at a fixed price, usually with a choice of some items. The **du jour menu** refers to the menu of the day. It must be planned and written daily.

Menu Patterns

The **menu pattern** is an outline of the menu item categories offered at each meal and the extent of selections within each category. The term *meal plan* is frequently used interchangeably with the term *menu pattern*. For the purposes of this text a *meal plan* refers to the number of meal opportunities offered over a specified period of time, usually 24 hours. For example, a small cafe may offer only breakfast and lunch; a day care may offer two snacks and lunch; and a long-term care facility may offer breakfast, lunch, dinner, and an hour of sleep (HS) snack.

For years, the traditional schedule has been three meals a day, including breakfast, lunch, and dinner, served within a certain time span. In some cases, the larger meal has been served at noon, resulting in a pattern of breakfast, dinner, and supper. In foodservice, the trend is moving away from this traditionally structured plan because of the desire of many patrons for fast food and instant service and because many people prefer to eat at different times of the day.

To accommodate this trend, cafeterias are offering snack-type foods as part of their regular menu and lengthening their service hours. Some hospitals offer light evening meals as an alternative to the traditional dinner menu. Many foodservices with multiple-choice menus have essentially the same menu for the noon and evening meals. The following is an example of a three-meal plan with corresponding meal patterns:

Breakfast

Fruit or juice
Cereal, hot or cold
Eggs and breakfast meats
Toast or hot bread
Choice of beverages

Lunch

Soup (optional)
Entrée or sandwich
Salad or vegetable
Bread with margarine or butter
Fruit or light dessert
Choice of beverages

Dinner

Soup (optional)
Entrée (meat, fish, poultry, or vegetarian)
Two vegetables (one may be potato or pasta)
Salad
Bread with margarine or butter
Dessert
Choice of beverages

Menus for most institutions that serve three meals a day are based on this pattern. The number of choices offered varies with the type of foodservice; with the type of service or method of delivery of the food; and with the personnel, equipment, and money available.

Food Characteristics and Combinations

When planning menus, one must visualize how the food will look on the plate, tray, or cafeteria counter and sense the combinations of foods presented. This is referred to as *presentation* and is based on the sensory and aesthetic appeal of food. Consider how the flavors will combine and whether there is contrast in texture, shape, and consistency; in other words, an overview of the final menu as it is served.

Color gives eye appeal and helps to merchandise the food. At least one or two colorful foods should be included on each menu. A green vegetable adds color to an otherwise colorless combination of broiled fish and creamed potatoes, or buttered parsley potatoes and glazed carrots also improve the appearance of a plate. Other green vegetables, tomatoes, and beets also add color, as do garnishes of fruit, watercress, or radishes. It is just as important to have pleasing color combinations on the cafeteria counter as on the individual plate.

Texture refers to the structure of foods and can best be detected by the mouth. Crisp, soft, smooth, and chewy are adjectives describing food texture. A variety of textures should be included in a meal. A crisp vegetable salad accompanying a chicken and rice casserole, along with fruit cobbler or other fruit dessert, would offer more contrast in texture than would a gelatin salad and chocolate pudding.

Consistency is the way foods adhere together—their degree of firmness, density, or viscosity—and may be described as firm, thin, thick, or gelatinous. Serving two creamed foods on the plate would be unattractive. A menu including baked ham with cherry sauce, scalloped potatoes, and creamed peas would be unappetizing because the foods would intermingle on the plate.

Shape of food plays a big part in eye appeal, and interest can be created through variety in the form in which foods are presented. One way to add interest to the menu is to vary the way in which vegetables are cut; for example, carrots can be cut into julienne strips or circles, cubed, or shredded; green beans can be served whole, cut, or French cut. Dicing and cutting machines provide an easy method for obtaining different forms and sizes. Variation in height of food as presented on a plate also contributes to eye appeal for the customer.

Flavor combinations are important in menu planning. In addition to the basic flavors of sweet, sour, bitter, and salty, vegetables may be thought of as strong and mild flavored, and chili or other foods may be thought of as spicy or highly seasoned. A variety of flavors in the meal is more enjoyable than duplication of any one flavor. Foods with the same basic flavors, such as spaghetti with tomato sauce and sliced tomato salad, should be avoided in the same meal.

Certain food combinations complement each other, such as turkey and cranberries, roast beef and horseradish sauce, or pork and applesauce. The planner should avoid exclusive use of stereotyped combinations, however, and explore other accompaniments to make menus more interesting. Red currant jelly instead of mint with lamb is an example.

Variety in preparation should be considered in menu planning. For example, a meal of baked chicken, baked potatoes, and baked squash obviously relies on only one preparation technique. Variety may be introduced by marinating or stir-frying foods in addition to the traditional fried, broiled, baked, braised, or steamed methods. Foods can be varied further by serving them creamed, buttered, or scalloped, or by adding a variety of sauces.

If the menu pattern provides entrée choices, the selection should include at least one meat and one vegetarian entrée, along with poultry, fish, and meat extenders to complete the number of items required. Fewer luncheon entrées may be offered than for the dinner meal, depending on the type of foodservice.

MENU WRITING

An inherent liking for good food, a lack of prejudice, a flair for planning based on creativity and imagination, and the ability to merchandise food attractively are traits that aid the menu planner. If one individual is responsible for menu planning, it is helpful to have input from purchasing, production, and service personnel. Many foodservices assign this responsibility to a committee rather than to one individual, a practice that is especially appropriate for a multiple-unit foodservice. Input from the public through marketing research, food preference studies, test marketing, and participation on food or menu committees can be of assistance. The planner should be alert to new products and to trends in consumer preferences and also be aware of menu items that are offered successfully by the competition, whether it is a nonprofit or commercial situation. Menu planning should be ongoing, current, and flexible enough to respond to changing conditions.

Information on consumer preference is important in menu planning. The relative popularity of a food may serve as a guide to the frequency with which a given item appears on the menu. A file of previous menus with comments concerning the reactions of guests, the difficulty or ease of preparation, and the cost helps prevent repetition and indicates combinations found satisfactory and profitable.

Timetable for Planning, Development, and Implementation

How far in advance of actual production and service should menus be planned? The answer depends greatly on the type of menu used, the extent of selections offered, and the size and complexity of the foodservice system. For example, a single-use menu for a holiday meal in a restaurant may require as little as a week of planning time, assuming the recipes are tested and standardized. A selective, cycle menu with several selections and never-before-tried items for a large hospital can take several months of advanced planning to ensure proper implementation (see Table 4.3).

Steps in Menu Development

A planning worksheet on which to record menu items is shown in Figure 4.5. For a selective menu offering certain items daily, time is saved by having the names of

Table 4.3 Example of a timetable for menu planning, development, and implementation.

Time Period	Recommended Activities
6 months in advance	• Gather menu planning information: recipes, customer, survey summaries, etc. • Review budget • Organize menu planning team
5 months in advance	• Meet with menu planning team • Establish goals of menu plan • Assign responsibility
4 months in advance	• Begin writing master menu • Have menu planning team evaluate menu • Review availability of equipment and china needed to produce menu • Begin recipe testing and standardization
3 months equipment in advance	• Allow production staff to review master menu for utilization and work load • Have purchasing agent review new menu items and write specifications
2 months in advance	• Develop production sheets and update forecasting methods • Design menus to be distributed to customers • Send first draft to printers
1 month before implementation date	• Proofread printed menus • Return to printers for final revisions • Conduct final implementation procedures

these foods printed on the worksheet. A suggested step-by-step procedure for planning menus follows.

1. *Entrées:* Plan the meats and other entrées for the entire period or cycle. Because entrées are generally the most expensive items on the menu, costs can be controlled to a great extent through careful planning at this point by balancing the frequency of high-cost versus low-cost entrées offered.

2. *Soups and sandwiches:* If a soup and sandwich combination is to be an entrée choice, it should be planned with the other entrées. In a cafeteria, a variety of sandwiches may be offered, and these may not change from day to day. If more than one soup is included, one should be a cream or hearty soup and one a stock soup.

3. *Vegetables:* Decide on the vegetables appropriate to serve with the entrées. Potatoes, rice, or pasta may be included as one choice. On a selective menu, pair a less popular vegetable with the one that is well accepted.

4. *Salads:* Select salads that are compatible with the entrées and vegetables. If a protein-type salad, such as chicken, tuna, or deviled egg, is planned as an entrée choice, it should be coordinated with the other entrée selections. If only one salad is

Menus

Week of ______________________

	Monday	Tuesday	Wednesday	Thursday
Breakfast				
Fruit	1.			
Fruit juice	1.			
	2.			
Cereal	1.			
	2. Assorted dry	Assorted dry	Assorted dry	Assorted dry
Entrée	1.			
Bread	1. Toast	Toast	Toast	Toast
	2.			
Beverages	C.T.M.	C.T.M.	C.T.M.	C.T.M
Lunch				
Soup	1.			
Entrées	1.			
	2.			
Vegetable	1.			
Bread	1. Assorted	Assorted	Assorted	Assorted
Salads	1. Salad bar	Salad bar	Salad bar	Salad bar
	2.			
Desserts	1.			
	2.			
Beverages	1. C.T.M.	C.T.M.	C.T.M.	C.T.M.
	2.			
Dinner				
Soup	1.			
Entrées	1.			
	2.			
Potato or pasta	1.			
Vegetables	1.			
	2.			
Salads	1. Salad bar	Salad bar	Salad bar	Salad bar
	2.			
Desserts	1.			
	2.			
	3.			
Beverages	1. C.T.M.	C.T.M.	C.T.M.	C.T.M.

Figure 4.5 Suggested worksheet for menu planning.

to be offered, choose one that complements or is a contrast in texture to the other menu items.

On a selective menu, include a green salad plus fruit, vegetable, and gelatin salads to complete the desired number. Certain salads, such as tossed salad, cottage cheese, deviled egg, or cabbage slaw, can be offered daily. A salad bar that includes a variety of salads and relishes may also be a standard feature.

5. *Desserts:* If no choice is to be offered, plan a light dessert with a hearty meal and a richer dessert when the rest of the meal is not too filling. On a selective menu, the number of choices may be limited to two or three plus a daily offering of fruit, ice cream or sherbet, and yogurt. For a commercial cafeteria, the dessert selection may be quite extensive and include a two-crust pie, a creme pie, cake or cookies, pudding, fruit, ice cream or sherbet, and gelatin dessert.

6. *Garnishes:* To maximize plate appearance, it is recommended that a planned garnish be considered for each meal. The garnishes should be part of the master menu or a separate cycle. The planned garnishes eliminate last-minute decision making and allow adequate time to ensure that proper ingredients are available to assemble garnishes for each meal. Menu planning books and trade publications are excellent resources for garnish ideas.

7. *Breads:* Vary the kinds of breads offered or provide a choice of white or whole-grain bread and a hot bread. Many foodservices use homemade breads as one of their specialties. Vary the shape and ingredients of bread selections to maximize variety.

8. *Breakfast items:* Certain breakfast foods are standard and generally include fruit juices, hot and cold cereals, and toast. It is customary to offer eggs in some form and to introduce variety through the addition of other entrées, hot breads, and fresh fruits.

9. *Beverages:* A choice of beverages that includes coffee, tea, whole milk, and low-fat milk is offered in most foodservices. Decaffeinated coffee and tea are generally provided, and soft drinks and a variety of juices also may be included. Some hospitals now offer wine selections to their patients when approved by the attending physician.

Menu Evaluation

Menu evaluation is an important part of menu planning and should be an ongoing process. The menu as planned should be reviewed prior to its use and again after it has been served. A foodservice manager can best evaluate menus by looking at the entire planned menu and responding to the following questions. The use of a checklist helps to make certain that all factors of good menu planning have been met. (Figures 4.6 and 4.7 are examples of menu evaluation tools.)

Checklist for Menu Evaluation

1. Does the menu meet nutritional guidelines and organizational objectives?
2. Are the in-season foods that are offered available and within an acceptable price range?
3. Do foods on each menu offer contrasts of color? texture? flavor? consistency? shape or form? type of preparation? temperature?

Menu Evaluation Form

Cycle _______________ Dates _______________ Evaluator _______________

Place a check mark on days when a problem is noted for any characteristic. Comment on the problem.

CHARACTERISTICS	DAYS							COMMENTS
	S	M	T	W	T	F	S	
Menu Pattern—Nutritional Adequacy Each meal is consistent with the menu pattern. All food components specified met the nutritional needs of the clientele.								
Color and Eye Appeal A variety of colors is used in each meal. Color combinations do not clash. Colorless or one-color meals are avoided. Attractive garnishes are used.								
Texture and Consistency A contrast of soft, creamy, crisp, chewy, and firm-textured foods is included in each meal, as much as possible, for clientele served.								
Flavor Combinations Foods with compatible, varied flavors are offered. Two or more foods with strong flavors are avoided in the same meal. For example, onions, broccoli, turnips, cabbage, or cauliflower; tomato juice and tomato-base casserole; and macaroni and cheese and pineapple-cheese salad, are not served together.								
Sizes and Shapes Pleasing contrasts of food sizes and shapes appear in each meal. Many chopped or mixed items are avoided in the same meal. For example, cubed meat, diced potatoes, mixed vegetables, and fruit cocktail are not served together.								

Figure 4.6 Menu evaluation form.

Walnut Grove Health Care Center
General and Modified Diets

Cycle 1 Day Wednesday
Dietitian ____________

	General Menu	Portion	Modified Diets				
			Mechan	Puree	2 gm Na^+	1500 ADA	1200 ADA
B	Orange Jc.	1/2c	√	√	√	√	√
	Scr. Egg	1/4c (#16)	√	√	SF	√	√
	WW Toast	2 sl.	√	Hot Cereal	√	√	1
	Margarine	2 pats	√	–	√	√	1
	Jelly	2 pkt.	√	–	√	diet	diet
	Milk–2%	8 oz	√	Whole	√	√	skim
L	Baked Chic.	3 oz	ground	puree	√	2 oz	2 oz
	Mashed pota.	1/2c (#8)	√	√	√	√	√
	Gravy	2 T (1 oz)	√	√	SF	FF	FF
	Broccoli	1/2c	√	puree	√	√	√
	Orange Garnish	1 slice	√	–	√	√	√
	WW Roll	1	√	–	√	√	–
	Margarine	1 pat	√	–	√	√	–
	S.B. Shortcake	1	√	puree	√	1/2c Berries	1/2c Berries
	Milk 2%	8 oz	√	whole	√	√	skim
D	Veg. Soup	3/4c (6 oz)	√	puree	SF	√	√
	Crackers	3	√	puree	SF	–	–
	Ham Sand.	1	w/ground meat	puree	Beef	1/2	1/2
	Sweet Pickle	2	–	–	√	√	√
	Mixed Melon	1 cup	√	puree	√	√	√
	Milk 2%	8 oz	√	whole	√	√	skim
HS	Gr. Crax	1	√	√	√	√	√
	Milk 2%	4 oz	√	√	√	√	skim

Figure 4.7 Modified diet extension form.

4. Can these foods be prepared with the personnel and equipment available?
5. Are the workloads balanced for personnel and equipment?
6. Is any one food item or flavor repeated too frequently during this menu period?
7. Are the meals made attractive with suitable garnishes and accompaniments?
8. Do the combinations make a pleasing whole, and will they be acceptable to the clientele?

Writing Menus for Modified Diets

In many foodservice operations, especially those affiliated with health care, the foodservice department is responsible for ensuring that physician-ordered diets are provided accurately. A qualified dietetics professional, such as a registered dietitian or dietetic technician, works with the foodservice manager to implement these special menus. Modified menu extensions are an excellent management tool for monitoring this responsibility. The modified menu extensions are generated from the master menu and a diet manual that defines the modified diets for a particular facility. Many dietetic associations and hospitals have written diet manuals that are available for sale. It is important to select a diet manual that best represents the diets needed in a given situation. For example, a manual developed for a hospital may not be the best choice for a long-term care facility.

Once diets are defined, the foodservice administrator should meet with a dietitian knowledgeable in modified diets and develop the menu extensions (Figure 4.7). A menu extension should be planned for each day. The extended portion of the menu illustrates how the modified diet, as defined in the manual, can be adapted from the master.

Extended menus have several advantages. These menus serve as a tool for menu analysis to ensure that modified diets are prepared and served according to physicians' written diet orders. The extensions also serve as a reference for the foodservice employees so they can be certain that diets are prepared and served accurately. Finally, the extensions are a useful purchasing tool, clearly identifying the need for special dietary foods (i.e., low-sodium items).

Menus as Documents. Printed master menus for both general and modified diets are excellent documents for department evaluation and budget planning functions. Any changes made from the master should be noted on the master menu for future evaluation. Master menus should be signed and dated by the person responsible for menu content.

THE PRINTED MENU

As indicated at the beginning of this chapter, the menu is an itemized list of foods served at a meal. From it, a working menu and production schedules evolve. The term also refers to the medium on which the menu is printed, which presents the food selection to the restaurant customer, the hospital patient, or other clientele.

The menu may also be posted on a menu board, as is the custom in most cafeterias and fast-food restaurants.

Menu Design and Format

A menu card must be designed and worded to appeal to the guest, to stimulate sales, and often to influence clientele to select items that the foodservice wants to sell. The menu card should be of a size that can be easily handled. It should also be spotlessly clean, simple in format with appropriate print size and type, and have ample margin space. The menu should be highly legible and interesting in color and design, harmonizing with the decor of the foodservice. The printed menu is a form of merchandising and an important marketing tool. It should not be thought of as a price sheet alone, but as a selling and public relations device.

Descriptive Wording. Menu items are usually listed in the sequence in which they are served and should present an accurate word picture of the foods available so that the patron can properly visualize the menu items. It is disappointing for the customer to imagine one thing and be served something entirely different.

Truth-in-Menu Legislation. Giving misleading names to menu items is unfair to the customer and is illegal where truth-in-menu legislation has been enacted. In general, these laws require that the menu accurately describe the foods to be served. If baked Idaho potatoes are listed on the menu, they must indeed be Idaho potatoes. The same is true when listing Maine lobster, or the point of origin for other foods. "Fresh" foods listed on the menu must be fresh, not frozen or canned. If the word "homemade" is used on the menu, it means that the food was made on the premises. If a menu lists a grade such as U.S. Choice beef and indicates portion size, the meat must be of that grade and size.

Descriptive words do enhance the menu and, if accurate, may influence the customers' selections. Here are some examples of descriptive wording: sliced, red tomatoes on Bibb lettuce, fresh spinach salad with bacon-mayonnaise dressing, old-fashioned beef stew with fresh vegetables, chilled melon wedge, and warm peach cobbler with whipped topping. The menu should not include recipe names that are unknown to the customer or that do not indicate the contents. Even where truth-in-menu legislation is not in effect, accuracy in menu wording helps to ensure customer satisfaction.

Menu Marketing

The manner in which food choices are presented to potential customers can have a significant impact on sales. It is estimated that two-thirds of menu choices made are influenced by the menu itself. Menu boards are a common means of communicating with the customer.

Menu Boards and Signage. Menu boards and signage can have a significant impact on food sales in commercial operations. The purpose of a menu board is to describe the food and beverage items available for purchase. Menu boards are sometimes referred to as "silent" sales representatives that encourage potential customers to make a choice and at the same time contribute to financial objectives through sales. Menu boards are designed to attract attention and come in a variety of sizes, shapes, and colors. Some are illuminated and many can be custom designed. Figure 4.8 shows examples of menu boards.

CUSTOMER SATISFACTION

The ultimate test of a successfully planned menu is to determine the degree of satisfaction on the part of the customer. Although highly subjective, satisfaction can be

Figure 4.8 Examples of menu boards.

Figure 4.8 *Continued.*
Courtesy of Main Street Menu Systems, Brookfield, Wisconsin.

measured and evaluated over time using a number of techniques including surveys, comment cards, frequency ratings, and sales data.

Satisfaction Surveys

Satisfaction surveys can be done formally via written surveys and comment cards (Figure 4.9). Many health care facilities contract with outside agencies for survey activities to ensure that the procedure is not biased. Statistics and trends reported by these agencies can be used in a concept referred to as *benchmarking,* where satisfaction levels determined for one facility can be compared to those of other similar facilities. Other, less formal survey techniques include observations of actual eating habits in dining rooms and plate waste in the warewashing area. Children's reaction to food is frequently measured using a facial hedonic scale as illustrated in Figure 4.10.

Frequency Ratings or Popularity Indexes

Frequency ratings and popularity indexes are established via formal or informal surveys in which customers are asked to rate or rank menu items according to preference. This technique is commonly used in schools and long-term care facilities where advisory groups representing the students and residents participate in menu planning.

Sales Data

In commercial operations sales data are the primary means by which satisfaction is measured. Modern cash registers can track and evaluate the contribution that each menu item makes to the financial objectives of foodservice operation.

What Do You Think About Our Food and Nutrition Services?

In an effort to provide you with appetizing and nourishing meals and further improve our nutrition services, we ask you to take a few moments to fill out this questionnaire. It will be collected from you today. If you have any questions regarding your diet, please ask to see a dietitian or call extension 624. Our staff is anxious to serve you in the best possible manner during your hospital stay.

Thank you.

Name ____________________

Room ____________________

Diet __________ Level __________

Days Since Admission ____________________

					How can we improve?
1. The foods I receive meet my personal needs and expectations......	❑ Always	❑ Usually	❑ Sometimes	❑ Never	________
2. If I required a dietitian's help to improve food intake or to tailor the diet to meet my therapeutic needs, this was done...........	❑ Very Well	❑ Well	❑ Fairly Well	❑ Poorly	________
3. On the menu I select from, I find the variety of foods to be...........	❑ Very Good	❑ Good	❑ Fair	❑ Poor	________
4. The portion sizes for foods on my tray are adequate......................	❑ Always	❑ Usually	❑ Sometimes	❑ Never	________
5. I receive exactly what I order on my trays....................................	❑ Always	❑ Usually	❑ Sometimes	❑ Never	________
6. Someone is available to answer any dietary questions I have.........	❑ Always	❑ Usually	❑ Sometimes	❑ Never	________
7. Meals are served at times when I feel the most like eating............	❑ Always	❑ Usually	❑ Sometimes	❑ Never	________
8. I have been able to receive between meal snacks when I want.....	❑ Always	❑ Usually	❑ Sometimes	❑ Never	________
9. The appearance of trays and serviceware is neat and attractive.....	❑ Always	❑ Usually	❑ Sometimes	❑ Never	________
10. I find the taste of the foods to be..	❑ Very Good	❑ Good	❑ Fair	❑ Poor	________
11. If I require assistance with meals, it is provided when needed.......	❑ Always	❑ Usually	❑ Sometimes	❑ Never	________
12. If ever I have a food-related problem, someone quickly responds..	❑ Always	❑ Usually	❑ Sometimes	❑ Never	________
13. Hot foods such as soups, entrees, and coffee are hot enough........	❑ Always	❑ Usually	❑ Sometimes	❑ Never	________
14. Cold foods such as juice, milk and desserts are cold enough........	❑ Always	❑ Usually	❑ Sometimes	❑ Never	________
15. The attitude of the personnel serving my tray has been................	❑ Very Good	❑ Good	❑ Fair	❑ Poor	________
16. Someone visited me routinely to see how well I liked my diet......	❑ Daily	❑ Most Days	❑ Some Days	❑ Never	________
17. My overall opinion of quality of Food & Nutrition Service is.........	❑ Very Good	❑ Good	❑ Fair	❑ Poor	________

We would appreciate any further suggestions as to how we can better serve you. Please write additional comments below. You may also list foods you would like to see on our menu or employees from Food & Nutrition Services that have been most memorable.

____

____

____

Figure 4.9 Sample patient satisfaction questionnaire for use in hospitals.
Source: Hospital Food and Nutrition Focus, Aspen Publishing, Frederick, Maryland, March 1995.
Copyright © 1992 Aspen Publishers, Inc.

SUMMARY

The menu is the focal point from which many functions and activities in a foodservice organization begin. It determines the foods to be purchased, it is the basis for planning production and employee schedules, and it is an important factor in controlling costs.

In planning foodservice menus, many factors must be considered: (1) the nutritional requirements, food habits, and preferences of the individuals in the group for which menus are being planned; (2) the goals of the organization; (3) the amount of money available; (4) limitations on equipment and physical facilities; (5) the number and skill of employees; and (6) the type of service. The menu must offer a selection of foods that is satisfying to the clientele, but it must be one that can be produced within the constraints of the physical facility and limitations dictated by management policies.

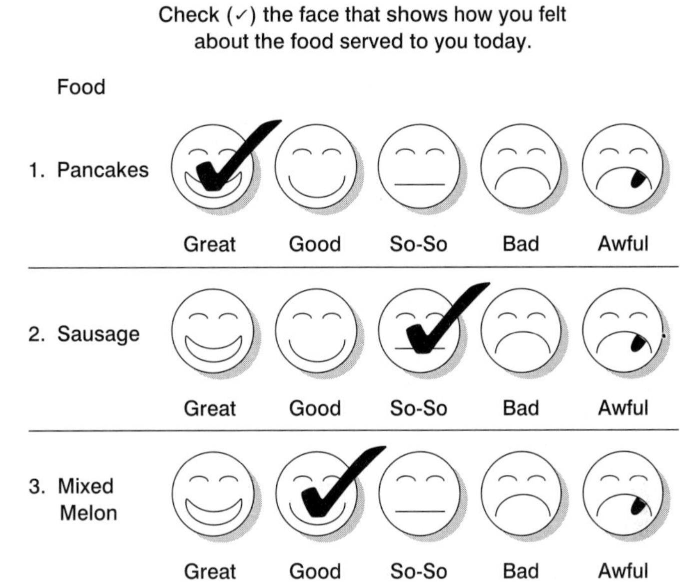

Figure 4.10 Facial hedonic scale used to evaluate child satisfaction.

The menu can take different forms, each written for the needs of a particular type of foodservice. The static or set menu, in which the same menu items are offered each day, is found mainly in commercial foodservices. A selective menu offers two or more choices in each menu category and is widely used in various types of foodservices. A nonselective menu offers no choice, but in schools and health care facilities where this type of menu is often used, choices in some categories may be limited. A cycle menu is a carefully planned set of menus that are rotated at definite intervals. The single-use menu is planned for a certain day and is not repeated in the same form.

Systematic planning procedures that include continuous evaluation of the menus as served should be followed. The menu planner should keep abreast of new products on the market and should be alert to the preferences of the clientele and the need for changes in the menu. Innovation is a key word in today's menu planning. New menu ideas and marketing techniques must be developed if the foodservice is to satisfy a clientele that is becoming increasingly sophisticated about food.

REVIEW QUESTIONS

1. What are the preliminary planning factors to consider in a menu planning process?
2. How might these factors vary between menus for a hospital and those for an extended-care facility?
3. What are the various types of menus from which to choose? What might be an appropriate type of menu for a school lunch program?

4. Plan a one-week, limited, select menu for a psychiatric facility; age range is 18 to 85 years. What factors did you consider in planning this menu?
5. Obtain a menu from a local organization (i.e., hospital, restaurant, school, or nursing home). Evaluate the menu on food characteristics and combinations. What changes would you recommend?

SELECTED REFERENCES

Briley M. E., Roberts-Gray, C., and Simpson, D.: Identification of factors that influence the menu at child care centers: A grounded theory approach. J. Am. Diet. Assoc. 1994; 94:276–281.

Connelly, D. J.: The silent salesperson. School Food Service J. 1993; pp. 50–51.

Dubé, L., Trudeau, E., and Bélanger, M. C.: Determining the complexity of patient satisfaction with foodservices. J. Am. Diet. Assoc. 1994; 94:394–398, 401.

Kittler, P. G., and Sucher, K.: *Food and Culture in America.* New York: Van Nostrand Reinhold, 1989.

Kotschevar, L. H.: *Management by Menu,* 2nd ed. Dubuque, Iowa: William C. Brown Publishers, 1987.

Measuring patient satisfaction. Hospital Food and Nutrition Focus. 1995; 11(7):7–8.

National Research Council: *Recommended Dietary Allowances,* 10th ed. Washington, D.C.: National Academy of Sciences, 1989.

Seaberg, A. G.: *Menu Design; Merchandising and Marketing,* 4th ed. New York: Van Nostrand Reinhold, 1991.

Shugart, G., and Molt, M.: "Menu planning," part 3 in *Food for Fifty,* 9th ed. Englewood Cliffs, N.J.: Merrill/Prentice Hall, 1993.

U.S. Department of Agriculture/U.S. Department of Health and Human Services: *Nutrition and Your Health: Dietary Guidelines for Americans.* Washington, D.C.: Office of Governmental and Public Affairs, 1990.

FOR MORE INFORMATION

The National Restaurant Association
Association Headquarters
1200 Seventeenth Street NW
Washington, DC 20036-3097

The National Restaurant Association
Educational Foundation
250 S. Wacker Drive, Suite 1400
Chicago, IL 60606

5

Purchasing, Receiving, and Storage

OUTLINE

Purchasing is an essential function in the foodservice system and is the first step in the production and service of quality food. Although the procurement process for an institutional or commercial foodservice involves food, supplies, and equipment, major emphasis in this chapter is given to the buying of food.

Today's market offers a large variety of products from which selections must be made in order to meet the needs of a particular foodservice. Whether the buying decisions are made by the manager, a chef, a purchasing agent, or other qualified personnel, they must be based on quality standards, the economic structure of the organization, and a thorough knowledge of the marketing system.

Procedures for the selection of vendors, determination of food needs, and writing of specifications are discussed, as are the methods of purchasing. The emphasis in this chapter is on information that will assist the buyer in making purchasing decisions.

An important part of the purchasing process is the receiving, storage, and issuing of food and supplies. Sound receiving procedures are of particular importance to ensure that delivered products meet predetermined standards for quality. A discussion of specific receiving practices is included in this chapter. Requirements for storage facilities and records needed for inventory control are also included.

KEY CONCEPTS

1. Purchasing is a market-driven function and is influenced by market regulations.
2. The market is regulated by federal, state, and local agencies.
3. There are formal and informal methods of purchasing. The prime vending and just-in-time purchasing methods are growing in popularity and acceptance.
4. Food and supply purchases should be based on written specifications.
5. An effective buyer is knowledgeable about markets, purchasing methods and procedures, and regulations and upholds high standards of professional ethics.
6. Par stock and mini-max are inventory stock methods that trigger reorder points.
7. Vendors should be carefully evaluated to ensure that they can meet the necessary purchasing and delivery requirements.
8. Receiving and storage procedures are important for cost and quality control.
9. The physical and perpetual methods are commonly used for monitoring inventory.
10. A number of key documents are important for recording the purchasing process.

WHAT IS PURCHASING?

Purchasing is the process of getting the right product into a facility and to the consumer at the right time and in a form that meets preestablished standards for quantity, quality, and price. Other related terms include *procurement, buying,* or *shopping.* Regardless of the term used, purchasing as a process involves several key functions:

- Identification of a need to be filled by an outside source,
- Development of specifications or detailed descriptions of desired products,
- Determination of quantity needed,
- Market research regarding product availability,
- Negotiations with sellers and order placement,
- Receiving of items and transfer into proper storage, and
- Issuing of items.

The purchasing process begins with a well-planned menu and a thorough understanding of the market and current market conditions.

THE MARKET

The market is the medium through which commodities are moved from the producer to the consumer—an exchange of ownership takes place. Markets are classified as primary, secondary, and local. A market is classified depending on the type of food sold, the location of the market, and the marketing channels involved.

In the food industry, a **primary market** is the basic source of supply. Examples include roasted coffee from New York City, meats from Chicago stockyards, and fruits and vegetables from the agricultural areas of California. Primary markets influence the sales and distribution of their goods through various activities, such as setting prices and quality standards. Raw agricultural products are delivered to primary markets where they are processed or prepared for distribution to secondary markets.

The **secondary market** is the physical, functional unit of the market system where products are accepted from the primary markets and distributed to local buyers. Secondary markets greatly influence activities in primary markets and have gradually assumed many of the functions traditionally performed in the primary markets. For example, secondary markets now receive direct shipments of fresh fruit and vegetables.

The third type of market available to the food buyer is the **local market.** Farmers' markets may offer savings on seasonal products, and the local supermarkets may be a good source of supply for the restaurant or small foodservice operator who only has money and storage space to buy what is currently needed.

Food is distributed through marketing channels with the aid of numerous marketing agents often referred to as middlemen. For example, wholesalers in secondary markets purchase products in large quantities from primary markets and redistribute in smaller quantities to local buyers. Full-service wholesalers handle all items typically required by their customers, whereas specialty wholesalers restrict their sales to only three or four items within a particular line, such as fresh and frozen meats, fresh produce, or canned foods and groceries. Food also may be distributed through brokers and commissioned agents (see Figure 5.1).

A broker is a sales representative for a manufacturer or a group of manufacturers, but is in business for him- or herself. Food brokers do not buy or own any merchandise; they find buyers for the manufacturers' products. A commissioned agent is similar to a broker except that the commissioned agent takes ownership of the merchandise from the manufacturer and then seeks interested buyers.

The marketing system involves the growing, harvesting, transporting, processing, packaging, storing, selling, financing, and supplying of market information for the many foods and food products available. Each process and transfer of ownership adds to the cost of the end product so that the final consumer cost is far in excess of the amount paid to the original producer of the commodity. The tendency today is to bypass as many individual marketing agents as possible and, thus, reduce the marketing cost contributed by these agents' charges. For example, large-volume buyers may be able to purchase directly from the food processor, thus saving the middleman's charges for handling, storing, and selling the product.

The market is dynamic and ever changing and the food buyer must be alert to trends and conditions that affect it. Government policy, economic trends, and adverse weather conditions are but a few of the factors that demand the attention of the buyer. Exchange of information between seller and buyer is an important function of the market and is made possible through various channels, such as trade association newsletters, local and federal market reports, and the press. Other sources of market information are technical and trade association meetings and magazines, research reports, talks with sales representatives, and visits to the produce markets and wholesale distributors.

Adverse growing conditions can affect food prices, as can unusual consumer demands and seasonal variations. Some foods are relatively stable in price and follow general economic conditions. Others are more perishable and have greater price fluctuations during the year. Most fresh fruits and vegetables are considered best at

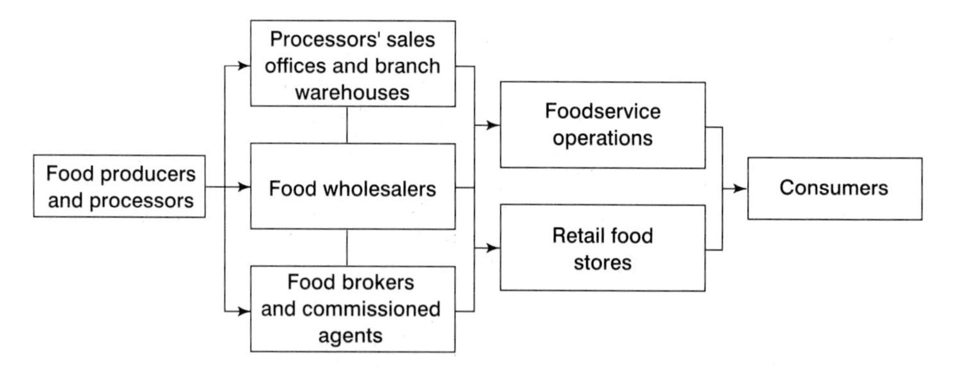

Figure 5.1 Structure of the food distribution system.
Source: Reprinted with permission from *Foodservice Manual for Health Care Institutions*. 1994 edition, published by American Hospital Publishing, Inc., Chicago, Illinois. Copyright © 1994.

the height of the production season, particularly those grown within a given market area. However, treatment of fresh foods and improved transportation, refrigeration, and storage facilities have greatly reduced the so-called "seasonability" of foods. Stocks of processed foods may be high or depleted at times, which will affect both price and availability.

Food Safety and Inspection Programs

The safety and wholesomeness of our food supply are ensured through government safety and inspection programs. Quality is defined and ensured through grading services, which are not to be confused with the inspection programs. Grading is discussed later in this chapter.

Government safety and inspection programs are used to evaluate foods for signs of disease, bacteria, chemicals, infestation, filth, or any other factor that may render the food item unfit for human consumption. All foods shipped in interstate commerce must meet the requirements of one or more federal laws and regulations. Foods sold intrastate must meet state and local regulations that are *at least equal to* the federal requirements.

The major responsibility for ensuring safe, wholesome food lies with the U.S. Department of Agriculture (USDA) and the Food and Drug Administration (FDA). Numerous other government departments and agencies are also involved in specific aspects of food safety regulation. The following is a summary, by both government department and enforcement agency, of food safety programs in the United States.

U.S. Department of Agriculture. Within the USDA, the Agricultural Marketing Service is responsible for enforcing the Meat Inspection Act, Poultry Products Inspection Act, and Egg Products Inspection Act. Inspection of commodities for wholesomeness is *mandatory* for meats, poultry, and other processed foods. An official stamp affixed to the product indicates that it is a high-quality product and processed under sanitary conditions. Figure 5.2 illustrates examples of federal food inspection stamps.

Food and Drug Administration. The FDA is an enforcement agency within the Department of Health and Human Services. The FDA is responsible for the enforcement of the federal Food, Drug, and Cosmetic Act, the Fair Packaging and Labeling Act, and the Nutritional Labeling and Education Act, and it covers the production, manufacture, and distribution of all food involved in interstate commerce except meat, poultry, and eggs.

Under the Food, Drug, and Cosmetic Act no food may enter interstate commerce that is deemed adulterated or misbranded. A food is adulterated under the following conditions:

- It contains substances that are injurious to health.
- Any part of it is filthy or decomposed.
- It has been prepared or held under unsanitary conditions.
- It contains portions of diseased animals.

Figure 5.2 Federal inspection stamps for meats, poultry, seafood, and eggs.

Federal Inspection Stamp for Meat

Federal Inspection Stamp for Fish

USDA Poultry Inspection Mark

USDA Egg Products Inspection Mark

A food is **misbranded** if the label does not include the information mandated by law or if it gives misleading information.

The FDA also determines and enforces standards of identity, quality, and fill. **Standards of identity** define what a food product must contain to be called by a certain name. For example, percent butterfat is defined for the different types of fluid milk (i.e., whole, 2% low fat, skim). **Standards of quality** apply primarily to canned fruits and vegetables, describing the ingredients that go into those products. These standards limit and define the number and kinds of defects permitted. They do not provide a basis for comparing foods as do grades, but they do establish minimum quality requirements. **Standards of fill** regulate the quantity of food in the container. They tell the packer how full the container must be to avoid deceiving the consumer. All of these standards are mandatory for foods in interstate commerce and may be used voluntarily for others.

The FDA is responsible for enforcing federal labeling requirements. Such requirements were first made mandatory with the passage of the Pure Food and Drug Act of 1906. Since then, several laws have been passed by Congress to define these labeling requirements in more detail. Table 5.1 provides an historical review of this legislation.

National Marine and Fisheries Service. A voluntary inspection system for fish, fish products, and grade standards for some products is controlled by the National Marine and Fisheries Service, an agency of the Department of Commerce. If the product carries a U.S. grade designation, the purchaser is ensured of continuous in-plant inspection during processing by federal inspectors. An ungraded product may or may not have been inspected during processing.

Table 5.1 Federal labeling requirements.

Legislation	Pure Food and Drug Act, 1906	Food and Drug Cosmetic Act (FDCA), 1938	Fair Packaging and Labeling Act (FPLA), 1966	Nutrition Labeling Regulations, 1973	Nutritional Labeling and Education Act (NLEA), 1990
Intent	Protects the public. Defines misbranding and adulteration. Prohibits foods that *are* injurious to a person's health.	Establishes standards of identity. States specific labeling requirements. Prohibits foods that *may be* injurious to a person's health.	Provides consumer with accurate information for value comparison. Prevents use of unfair or deceptive methods of packaging or labeling of consumer products.	Educates the consumer about the nutrient content of foods.	Provides extensive nutrient information of packaged foods. Improves nutritional content of packaged foods.
Key requirements	Authorized food processing plant inspections to assure sanitary conditions.	Name and address of manufacturer or distributor. Name of the food. Quantity of content. Statement of ingredients listed by common or usual name in decreasing predominance.	Same information as the FDCA. Defines a food label in terms of format and information. Name of food/net quantity must appear on "principal display panel." Net content must be in legible type and in distinct color contrast. Defines type size and location.	*Voluntary* except for foods fortified with vitamins, minerals, or protein, or in situations where a nutritional claim is made.	*Mandatory* for all packaged food. Includes provisions for the nutritional labeling of the 20 most common produce and seafood items. Serving sizes are stated in household units. Regulates nutrient content and health claims including those made by restaurants on signs and placards.

U.S. Public Health Service. The U.S. Public Health Service (PHS) is concerned primarily with control of infections and contagious disease but is also responsible for the safety of some foods. This agency is responsible for the inspection of some shellfish, and they advise state and local governments on sanitation standards for production, processing, and distribution of milk. The PHS standard for Grade A fresh milk is a standard of wholesomeness, which means that it has met state or local requirements that equal or exceed federal requirements.

Environmental Protection Agency. The Environmental Protection Agency (EPA) regulates pesticides. Responsibilities include setting tolerance levels for pesticide residues in foods, establishing the safety of new pesticides, and providing educational materials on the safe use of pesticides. The EPA also determines quality standards for water.

Department of the Treasury. The Bureau of Alcohol, Tobacco, and Firearms (BATF) in the Department of the Treasury is responsible for monitoring the production, distribution, and labeling of alcoholic beverages. This includes all alcoholic beverages except those that contain less than 7% alcohol, which are monitored by the FDA.

THE BUYER

Food and supplies for a foodservice organization may be purchased by an individual, by a purchasing department, or through a cooperative arrangement with other institutions, depending on the size and ownership of the organization and its procurement policies. In a small operation, the buying may be done by the manager as part of his or her responsibilities.

Whatever the arrangement, it is the responsibility of the foodservice department or the individual functional unit of a foodservice to communicate its needs to the buyer to assure delivery of the needed amount of food and supplies at an appropriate time and of the desired quality. This requires cooperation on the part of the buyer, as well as the foodservice personnel, and a willingness to honor the quality standards set by the foodservice.

Purchasing the amount and quality of food required for the foodservice within the limitations imposed by the budget and financial policies of the organization requires knowledge of internal and external factors. Internal factors include the customers, the menus, recipes, labor availability and skills, equipment, storage facilities, and quantities of foods needed. External factors include the marketing system, food standards and quality, what is available on the market, and purchasing methods.

The buyer represents the institution in negotiations with market representatives and should have some knowledge and understanding of legal requirements, especially as they relate to orders and contracts. There should be a clear understanding as to the extent of the buyer's decision-making authority and of the institutional policies within which the buyer must operate.

Ethics in Purchasing

Buying demands integrity, maturity, bargaining skills, and maintenance of a high standard of ethics. Acting as an agent for the institution, the buyer is entrusted with making decisions concerning quality, prices, and amounts to purchase, and cannot afford to compromise either money or position. Buyers may be subjected to bribes and other kinds of inducements to influence a buying decision. *Collusion* refers to a secret arrangement or understanding between buyer and seller for fraudulent purposes. The most common example of this comes in the form of "kickbacks" where the buyer accepts something of value (money or merchandise) from the vendor. Violations of professionalism in purchasing should be clearly identified in the organization's policy on ethics (frequently referred to as a Code of Ethics). No gifts or other favors should be accepted that might obligate the buyer.

A buyer must be able to deal successfully with sales representatives, purveyors, and other marketing agents. Courtesy and fair treatment contribute to establishing a satisfactory working relationship with these agents, who can be valuable sources of information on new products and the availability of foods on the market. If orders are placed through a salesperson, scheduling regular office appointments or a specific time for telephoning saves time for both the buyer and seller.

Products should be evaluated objectively and buying decisions made on the basis of quality, price, and service. Information received in confidence from one company should not be used to obtain an unfair advantage in competitive negotiations.

Types of Purchasing

Foodservice operations work under different types of purchasing arrangements depending on a number of factors including organizational size, ownership, and geographic location. Centralized and group purchasing are common types of purchasing and are described on the following pages. It is important for the reader to realize that many single-unit operations conduct their purchasing functions as part of the departmental operations. For example, a chef in a single-unit restaurant may have full responsibility for purchasing, independent of a purchasing department or group contract.

A number of independent or single-unit operations are discovering and taking advantage of substantial cost savings by purchasing through warehouse clubs. Warehouse clubs, sometimes referred to as "alternative format" grocery stores, offer bulk food items at rock-bottom prices to dues-paying members. Operating costs for handling and distribution are kept to a minimum and savings are passed along to customers. Product is made available in large, nondescript warehouses where items are kept in their original containers and simply stacked on pallets. Customer service is generally not available and members assume the responsibility for pickup and delivery.

Centralized Purchasing. Centralized buying, in which a purchasing department is responsible for obtaining needed supplies and equipment for all units in the organization, is used in many universities, schools, multiple-unit restaurants, and hospitals. By relieving the individual units of the responsibility of interviewing sales representatives, negotiating contracts, and placing orders, this system has proven to be cost effective and time saving for the foodservice, especially if the system also has central storage. Where centralized purchasing is used, the authority to buy some product, such as fresh produce or other perishables, may be delegated to the foodservice, or in multiple-unit organizations to the individual units.

One disadvantage of centralized purchasing is that excessive paperwork can result. In addition, friction can develop between the purchasing department and the foodservice unit if there is not a clear understanding of decision-making authority, especially on quality standards. The possibility for friction exists in all large-scale purchasing unless the limits of authority are well defined and the lines of communication are clearly defined and kept open.

Group or Cooperative Purchasing. It is beneficial for buyers to increase volume and/or lower service requirements to improve leverage with suppliers and, thus, buy at lower prices. The efforts to increase volume have led some foodservice directors to consolidate their buying power with that of other organizations in cooperative purchasing arrangements, sometimes referred to as purchasing consortiums. For example, several hospitals in a metropolitan area may combine their purchases to obtain lower prices and possibly more favorable service arrangements; or in smaller communities, two or more dissimilar foodservices, such as a school, hospital, and nursing home, may join in a group purchasing agreement. Central warehousing may be part of a group purchasing plan, but if the volume is large enough, vendors may agree to deliver merchandise directly to the individual units.

Group buying differs from central purchasing in that members of the group are independent organizations and are not under the same management. In central purchasing, the members are usually units of a larger system, such as schools in a city- or county-wide school system. Obviously, the main advantage of group buying is the price advantage gained by increased volume, which in turn may attract more prospective vendors. Other advantages of group purchasing for the foodservice managers include the following:

- Freedom from having to meet with sales representatives,
- Time savings through streamlined paperwork and administration of the purchasing function, and
- An annual review of specifications.

One more abstract advantage of group purchasing is the potential opportunity for "value-added" services. For example, the group purchasing agreement may include continuing education opportunities for members.

The buyer is selected by the group and usually maintains an office independent of the participating organizations. The purchasing service generally is supported by a fee paid by each institution that is based on the percentage of its orders. To be effective, all members of the group must commit their time and the majority of their purchase orders to the group's efforts. Participating agencies must agree on common specifications and establish a bid schedule. Food preferences may differ from system to system so members must occasionally be willing to compromise their requirements for the benefit of the group.

THE VENDOR

The selection of suppliers is one of the most important decisions that must be made in a purchasing program. Management and the buyer should work together to establish quality standards for food and supplies to be purchased, and to conduct a market search for reliable vendors that are able to furnish the desired products.

A buyer new to a foodservice operation can locate vendors through numerous resources, including these:

- The phone book,
- Other foodservice operators in the area,
- Trade journals and publications, and
- Local trade shows.

A key responsibility of the buyer in initiating a professional working relationship with a vendor is to evaluate carefully the vendor's scope of products and services. To accomplish this, the buyer must make the necessary inquiries into the vendor's product line, available services, and reliability in meeting contract specifications.

The buyer should carefully evaluate the product line of the potential vendor to assess availability of needed products and to ensure that the products meet the quality standards of the organization. Details of delivery schedules, payment policies, and contingency plans for situations in which standards are not met should be known by the buyer. For example, a buyer would want to know the vendor's policy regarding corrected billings when a damaged or spoiled product is delivered. Policies on emergency deliveries should also be reviewed. Some vendors have policies on how to deliver food supplies in the event of a disaster such as a tornado, snowstorm, or earthquake. It is very important to review discount policies on early payments, rebates, and volume purchasing discounts. Visits to the vendor's local headquarters and talks with other buyers can supply much of this information.

Many vendors provide support equipment or service programs with the purchase of certain food products. For example, a vendor may offer to give coffee-making equipment at no extra cost with the purchase of its coffee products or a popcorn popper with the purchase of the vendor's popcorn line. It is important for the buyer to inquire about the technical assistance provided by the vendor for service and maintenance of these programs. Necessary information would include service availability in the event of equipment failure and scheduling of routine maintenance on the equipment.

The location and size of the foodservice are important factors in the selection of a supplier. If the operation is located in or near a large metropolitan area, there may be several suppliers that could meet quantity and quality needs and whose delivery schedules are satisfactory for the foodservice.

For an operation in a small or remote location, part or all of its supplies may be purchased locally. In this situation, the buyer should be sure that the vendors carry adequate stocks and are able to replenish products quickly. If there are not enough local suppliers to offer competitive prices, the buyer may prefer to purchase only certain products locally, such as dairy products, bakery items, and fresh produce, and place less frequent orders for canned foods and groceries with a larger wholesaler that will break, or split, cases of food or supplies into quantities more appropriate for the foodservice operation.

In some situations, large-volume operators may be able to purchase canned foods or other nonperishable items directly from the processor and work out satisfactory arrangements for delivery of their products. Since the quantity of foods purchased would be large, the amount and kind of storage space and the financial resources can be determining factors in whether direct buying is possible.

METHODS OF PURCHASING

The two principal methods of buying are *informal* or *open-market buying* and *formal competitive bid buying.* Both may be used at various times for different commodities. Variations of these methods or alternative buying arrangements may be preferred by some foodservices or may be used during uncertain market conditions.

Purchasing is a management function and, as such, the foodservice administrator will have policies and procedures to guide him or her in setting up a course of action. The methods of buying that are selected depend on these institutional policies, on the size of the organization, the amount of money available, location of the vendors, and the frequency of deliveries.

Informal or Open-Market Buying

Informal purchasing is probably the most commonly used method of buying, especially in smaller foodservice operations. The system involves ordering needed food supplies from a selected list of dealers based on daily, weekly, or monthly quotations. Prices are based on a set of specifications furnished to interested vendors. The buyer may request daily prices for fresh fruits and vegetables but may use a monthly quotation list for grocery items. The order is placed after consideration of price in relation to quality, delivery, and other services offered.

Contact between the buyer and vendor is made by fax, computer, telephone, a visit to the market, or through sales representatives who call on the buyer. The use of price quotation and order forms on which to record the prices submitted by each vendor (Figure 5.3) is an aid to the buyer. If the quotations are provided via the telephone, the prices should be recorded. For large orders of canned goods or groceries, or where the time lapse between quotations and ordering is not important, requests for written quotations can be made by mail, as shown in Figure 5.4.

Considering new vendors from time to time and visiting the market when possible enable the buyer to examine what is available from other vendors and to note the current prices. When using informal purchasing, the buyer and vendor must agree on quantities and prices before delivery. Only vendors who give reliable service and competitive prices should be considered for open-market buying.

Name of Food Service Fruit and Vegetable Quotation and Order Sheet							
For Use on ________			Delivery Date ________				
				Price Quotes			
				Vendors			
Specs	Amount Needed	Amount on Hand	Amount to Order	A	B	C	D
Fruits:							
Vegetables:							

Figure 5.3 A suggested form for recording telephone price quotations for fresh produce. Combines a listing of total needs, the inventory on hand, and the resulting quantities to order. When an order is placed, the vendor's price quotation is circled.

Formal Competitive Bid Buying

In formal competitive bid buying, written specifications and quantities needed are submitted to vendors with an invitation for them to quote prices, within a stated time, for the items listed. The request for bids can be quite formal and advertised in the newspaper, and copies can be printed and widely distributed, or the request can be less formal with single copies supplied to interested sellers. Bids are opened on a designated date, and the contract generally is awarded to the lowest responsible bidder.

Purchasing agents for local, state, or federal government-controlled institutions are usually required to submit bids to all qualified vendors, especially those bids over

_______________ **University**

Food Stores Department

____
(Date)

To:

INQUIRY NO.

Quote on this sheet your net price f.o.b. for the items specified below. We reserve the right to accept or reject all or part of this proposal. Quotations received until 4:00 P.M. _______________

Important: Read instructions on reverse side before preparing bid.

Quantity	Unit	REQUEST FOR QUOTATIONS —This is NOT an order	Price Unit	Total Price
		Return—TWO COPIES—To: Food Stores Department		

We quote you f.o.b. _______________

____
Sign Firm Name Here

Delivery can be made { immediately / _______ days.

_________ Per _________
Date

Cash Discount: _______________

Figure 5.4 Suggested form for requesting price quotations by mail.

a certain dollar limit. Buyers for private organizations, however, may select the companies whom they wish to invite to bid and the buyer may include on the list only those vendors whose performance and reliability are known. The procedure for competitive bid buying is discussed in more detail later in this chapter in the section on "Purchasing Procedures."

Advantages and Disadvantages. Bid buying is usually required by government procurement systems and is found to be advantageous by large foodservices or multiple-unit organizations. The formal bid, if written clearly, minimizes the possibility of misunderstandings occurring with regard to quality, price, and delivery. The bid system is satisfactory for canned goods, frozen products, staples, and other nonperishable foods. Food that is purchased by standing order, such as milk and bread, is also appropriate for this type of buying, but it may not be practical for perishable items because of the day-to-day fluctuation in market prices.

There are two main disadvantages to formal competitive bidding. The system is time consuming, and the planning and requests for bids must be made well in advance so that the buyer has time to distribute the bid forms and the suppliers have time to check availability of supplies and determine a fair price. Although this type of buying was designed to ensure honesty, it does lend itself to manipulation when large amounts of money are involved, especially if the buyers and the purchasing department are open to political pressure.

Competitive Bidding Variations. Many variations and techniques are found in formal competitive bidding, depending on the type of institution, financial resources of both vendor and buyer, and storage facilities of the foodservice and delivery capabilities of the vendor. Bids can be written for a supply of merchandise over a period of time at prices that fluctuate with the market; for example, a six-month supply of flour may be required, with 500 pounds delivered each month, at a price compatible with current market conditions.

In a firm fixed price (FFP) contract, the price is not subject to adjustment during the period of the contract, which places maximum risk on the vendor and is used when definite specifications are available and when fair and reasonable prices can be established at the outset. A buyer may request bids for a month's supply of dairy products, to be delivered daily as needed. Another variation involves the purchase of a specific quantity of merchandise, such as a year's supply of canned goods, but because of inadequate storage the foodservice may wish to withdraw portions of the order as needed throughout the duration of the contract.

Many different forms are used in the written bidding system and the terms may differ in various parts of the country, but all of them basically are invitations to bid with the conditions of the bid clearly specified. Attached to the invitation is a listing of the merchandise needed, specifications and quantities involved, and any conditions related to supply and fluctuations in the market. (An example of a bid request is given later in this chapter in Figure 5.11.)

Other Methods of Purchasing

Cost-Plus Purchasing. In *cost-plus purchasing,* a buyer agrees to buy certain items from a purveyor for an agreed-on period of time based on a fixed markup over the vendor's cost. The time period may vary and could be open for bid among different vendors. Such a plan is most effective with large-volume buying.

The vendor's cost generally is based on the cost of material to the buyer plus any costs incurred in changes to packaging, fabrication of products, loss of required trim, or shrinkage from aging. The markup, which must cover overhead, cost of billing or deliveries, or other expenses that are borne by the vendor, may vary with the type of food being purchased. When negotiating a cost-plus purchasing agreement, a clear understanding should be reached of what is included in the cost and what is considered part of the vendor's markup. Some way of verifying the vendor's cost should also be part of the agreement.

Prime Vending. *Prime vending* is a method of purchasing that has gained popularity and acceptance among restaurant and noncommercial buyers during the past several years. The method involves a formal agreement (secured through a bid or informally) with a single vendor to supply the majority of product needs. Needs are generally specified in percent of total need by category. Categories may include fresh meat and poultry, frozen, dairy, dry, produce, beverages, and nonfood categories such as disposables, supplies, equipment, and chemicals. The percents range from 60 to 80 with the higher levels being most common. The agreement is based on a commitment to purchase the specified amount for a specified period of time.

The primary advantages of this method are reduced prices, which are realized through high volume and time savings. The time savings results from not having to fulfill administrative and accounting requirements for numerous vendors. Additional advantages include the development of a strong, professional partnership with the vendor and the potential for value-added services such as computer software for submitting and tracking orders.

The buyer must be alert to potential problems with prime vendor contracts. For example, prices may increase over time, therefore, procedures for periodically auditing prices should be clearly defined as part of the agreement.

Blanket Purchase Agreement. The *blanket purchase agreement* (BPA) is sometimes used when a wide variety of items are purchased from local suppliers, but the exact items, quantities, and delivery requirements are not known in advance and may vary. Vendors agree to furnish—on a "charge account" basis—such supplies as may be ordered during a stated period of time. BPAs should be established with more than one vendor so that delivery orders can be placed with the firm offering the best price. Use of more than one vendor also allows the buyer to identify a "price creep," which can occur when only one vendor is involved.

Just-in-Time Purchasing. *Just-in-time purchasing* or JIT is yet another variation of purchasing. It is in fact a production planning strategy where product is purchased in the exact quantities required for a specific production run and delivered "just-in-time" to meet the production demand. The goal is to have as little product in inventory for as little time as possible in an effort to maximize cash flow. Some products such as milk, bread, and possibly fresh meat can go directly into production and avoid inventory costs altogether. Other benefits include better space management and fresher product. This method has an impact on all functional units, the most obvious being production. This arrangement must be carefully planned and orchestrated to ensure that shortages do not occur.

PRODUCT SELECTION

Market Forms of Foods

Deciding on the form in which food is to be purchased is a major decision that requires careful study. Costs involved in the purchase and use of fresh or natural forms of food versus partially prepared or ready-to-eat foods and the acceptability of such items by the consumer are major factors to consider. Several options may be available to prepare the same menu item. Fruit pies, for example, can be made from scratch, or by using partially prepared ingredients such as ready-to-bake crusts and ready-to-pour fillings. Other options include ready-to-bake pies and fully baked, ready-to-serve pies.

Due to a lack of space, equipment, or personnel, a foodservice manager may wish to consider purchasing partially or fully processed convenience items. Before making this decision, the cost, quality, and acceptability to patrons of the purchased prepared food should be compared with the same menu item made on the premises. Table 5.2 lists the factors to consider in a make-or-buy decision. If the decision is to

Table 5.2 Make-or-buy decisions.

Factor	Considerations
Quality	Evaluate whether quality standard, as defined by and for the organization, can be achieved.
Equipment	Assess availability, capacity, and batch turnover time to assure that product demand can be met.
Labor	Evaluate availability, current skills, and training needs.
Time	Evaluate product setup, production, and service time based on forecasted demand for the product.
Inventory	Gauge needed storage and holding space.
Total cost	Conduct complete cost analysis of all resources expended to make or buy product.
	Use cost as decision basis after other factors have been carefully analyzed.

buy the prepared product, the manager and the buyer must establish quality standards for these foods.

For foodservices preferring the preparation of product in their own kitchens, there are alternatives in purchasing that can save preparation time. The market offers a variety of processed ingredients from which to choose. Dehydrated chopped onion, precut melons, shredded cheese, frozen lemon juice, cooked chicken and turkey, and various baking, soup, sauce, and pudding mixes are examples. Casserole-type entrées, which often require time-consuming processes, may be made by combining several convenience items, such as frozen diced chicken, sauce made from a mix or from canned soups, and pasta or vegetables.

Choosing fresh, frozen, or canned foods depends on the amount of labor available for preparation, comparative portion costs, and acceptability by patrons. The high cost of labor has caused many foodservices to limit the use of fresh fruits and vegetables except for salads or during times of plentiful supply when costs are lower. There may be times when a menu change must be made because of the price differential among fresh, frozen, and canned food items.

Keeping in mind the quality standard established for the finished product, the manager must find the right combination of available foods in a form that will keep preparation to a minimum yet yield a product of the desired quality. Figure 5.5 illustrates a make-or-buy calculation.

Food Quality

Before food can be purchased, the quality of foods most appropriate to the foodservice operation and their use on the menu must be decided. The top grade may not always be necessary for all purposes. Foods sold under the lower grades are wholesome and have essentially the same nutritional value, but they differ mainly in appearance and, to a lesser degree, in flavor.

Foods that have been downgraded because of lack of uniformity in size or that have broken or irregular pieces can be used in soups, casseroles, fruit gelatin, or fruit cobblers. Also, more than one style or pack in some food items may be needed. Unsweetened or pie pack canned peaches can be satisfactory for making pies, but peaches in heavy syrup would be preferable for serving in a dish as a dessert.

Quality Standards. Quality may refer to wholesomeness, cleanliness, or freedom from undesirable substances. It may denote a degree of perfection in shape, uniformity of size, or freedom from blemishes. It may also describe the extent of desirable characteristics such as color, flavor, aroma, texture, tenderness, and maturity. Assessment of quality may be denoted by grade, brand, or condition.

Grades. Grades are market classifications of quality. They reflect the relationship of quality to the standard established for the product, and they indicate the degree of variation from that standard. Grades have been established by the USDA for most agricultural products, but their use is voluntary.

Scenario: Is it more cost effective to purchase precut or whole head lettuce for use on a salad bar, assuming quality is comparable?

	Head Lettuce	Pre-cut
Information Needed:		
Price/case	$10.90	$8.20
Amount/case (in pounds) as purchased (AP)	36# AP	20# AP
Yield	76%	100%
Labor time per case for cleaning and chopping	.317 hours (19 min.)	0 minutes
Labor costs/hour	$8.00	$8.00

Calculations for head lettuce

1. Edible portion (E.P.) in pounds = 36# × .76 = 27.36#
2. Labor costs per case = $8 × .317 = $2.546
3. Labor cost per usable pound = $2.546 ÷ 27.36 = $.09
4. Food cost per usable pound = $10.90 ÷ 27.36 = $.40
5. Total cost per usable pound = $.09 + $.40 = $.49

Calculations for pre-cut lettuce

1. Edible portion (E.P.) in pounds = 20# per case
2. Labor costs per case = 0
3. Food costs per usable pound = $8.20 ÷ 20 = $.41
4. Total cost per usable pound = $.41

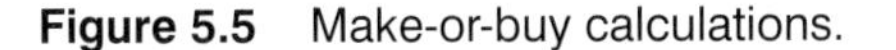

Figure 5.5 Make-or-buy calculations.

Grading and Acceptance Services. The USDA Agricultural Marketing Service, in cooperation with state agencies, offers official grading or inspection for quality of meat and meat products, fresh and processed fruits and vegetables, poultry and eggs, and manufactured dairy products. Grading is based on U.S. grade standards developed by the USDA for these products.

Included in the grading and inspection programs is a USDA *Acceptance Service* available to institutional food buyers on request. This service provides verification of the quality specified in a purchase contract. The product is examined at the processing or packing plant or at the supplier's warehouse by an official of the Agricultural Marketing Service or a cooperating state agency. If the product meets the specifications as stated in the contract, the grader stamps it with an official stamp and issues a certificate indicating compliance. If the purchases are to be certified, this provision should be specified in contracts with vendors. The inspection fee is then the responsibility of the supplier.

USDA grades are based on scoring factors, with the total score determining the grade. The grades vary with different categories of food as noted in the following list:

- *Meats:* U.S. Prime, U.S. Choice, U.S. Select, and U.S. Standard. Quality grades are assigned according to marbling, maturity of the animal, and color, firmness, and texture of the muscle. Yield grades of 1, 2, 3, 4, or 5 are used for beef and lamb to indicate the proportion of usable meat to fat and bone, with a rating of 1 having the lowest fat content. Veal and pork are not graded separately for yield and quality. (See "For More Information" at end of the chapter for guidance or specifications.)
- *Poultry:* Consumer grades are U.S. Grades A, B, and C, based on conformation, fleshing, fat covering, and freedom from defects. Grades often used in institutional purchasing are U.S. Procurement Grades 1 and 2. The procurement grades place more emphasis on meat yield than on appearance.
- *Eggs:* U.S. Grades AA, A, and B. Quality in shell eggs is based on exterior factors (cleanliness, soundness, shape of shell, and texture) and interior factors (condition of the yolk and white and the size of the air cell, as determined by candling). Shell eggs are classified according to size as Extra Large, Large, Medium, and Small.
- *Cheddar cheese:* U.S. Grades AA, A, B, and C. Scores are on the basis of flavor, aroma, body and texture, finish and appearance, and color.
- *Butter:* U.S. Grades AA (93 score), A (92 score), B (90 score), and C (89 score), based on flavor, body and texture, color, and salt.
- *Fresh produce:* U.S. Fancy, U.S. Extra No. 1, U.S. No. 1, U.S. Combination, and U.S. No. 2. Fresh fruits and vegetables are graded according to the qualities deemed desirable for the individual type of commodity, but may include uniformity of size, cleanliness, color, or lack of damage or defects. Grades are designated by name or by number. Because of the wide variation in quality and the perishable nature of fresh fruits and vegetables, visual inspection may be as important as grade; or a buyer might specify that the condition of the product at the time of delivery should equal the grade requested.
- *Canned fruits and vegetables:* U.S. Grade A (or Fancy), U.S. Grade B (or Choice for fruits and Extra Standard for vegetables), U.S. Grade C (or Standard), and U.S. Grade D (or Substandard). The factors for canned fruits and vegetables include color, uniformity of size, absence of defects, character, flavor, consistency, finish, size, symmetry, clearness of liquor, maturity, texture, wholeness, and cut. In addition to these factors, general requirements must be met, such as fill of container, drained weight, and syrup density. The grading factors vary with individual canned fruits and vegetables, but the scoring range is the same. Figure 5.6 shows suggested standards for canned foods. A summary follows of the total scoring ranges for grades of canned fruits and vegetables:

Grades	**Scores**
A (Fancy)	90–100
B (Choice, Extra Standard)	75–89
C (Standard)	60–74
Substandard	0–59

- *Frozen fruits and vegetables:* U.S. grade standards are available for many frozen fruits and vegetables but not standards of identity, quality, or fill of container. Fruit can be packed with sugar in varying proportions such as four or five

Standards for Canned Foods

Fruits

Grade	*Quality of Fruit*	*Syrup*
U.S. Grade A or Fancy	Excellent quality, high color, ripe, firm, free from blemishes, uniform in size, and very symmetrical.	Heavy, about 55%. May vary from 40 to 70%, depending on acidity of fruit.
U.S. Grade B or Choice or Extra-Standard	Fine quality, high color, ripe, firm, free from serious blemishes, uniform in size, and symmetrical.	About 40%. Usually contains 10 to 15% less sugar than Fancy grade.
U.S. Grade C or Standard	Good quality, reasonably good color, reasonably free from blemishes, reasonably uniform in size, color, and degree of ripeness, and reasonably symmetrical.	About 25%. Contains 10 to 15% less sugar than Choice grades.
Substandard	Lower than the minimum grade for Standard.	Often water-packed. If packed in syrup, it is not over 10%.

Vegetables

Grade	*Quality of Vegetable*
U.S. Grade A or Fancy	Best flavored, most tender and succulent, uniform in size, shape, color, tenderness; represents choice of crop.
U.S. Grade B or Extra-Standard (sometimes called Choice)	Flavor fine; tender and succulent; may be slightly more mature, more firm in texture, and sometimes less uniform than Fancy grade.
U.S. Grade C or Standard	Flavor less delicate; more firm in texture, often less uniform in size, shape, color; more mature.
Substandard	Lower than the minimum grade for Standard.

Figure 5.6 Suggested standards for USDA grades of canned fruits and vegetables.

parts of fruit to one part of sugar by weight or without sugar. Fruits or vegetables can be individually quick frozen or frozen in solid blocks.

Designation of U.S. grades and marking in the form of a shield are permitted only on foods officially graded under the supervision of the Agricultural Marketing Service of the USDA. Figure 5.7 shows examples of grade stamps.

Brands. Brands are assigned by private organizations. Producers, processors, or distributors attempt to establish a commodity as a standard product and to develop demand specifically for their own brands. The reliability of these trade names depends

Figure 5.7 Federal grade stamps for meat, poultry, and eggs.

on the reliability of the company. Brand names may represent products that are higher or lower in quality than the corresponding government grade. However, some brand-name products are not consistent in quality. Private companies may set up their own grading system, but such ranking may show variation from season to season.

Some knowledge of brand names is essential in today's marketing system. USDA grades are used for most fresh meats and for fresh fruits and vegetables, but very few canners use them, preferring instead to develop their own brands. If USDA grades are specified and bidders submit prices on brand products, the buyer should be familiar with USDA grades and scores in order to evaluate the products. The buyer may wish to request samples or, if the order is large enough to justify it, request a USDA grading certificate.

PURCHASING PROCEDURES

The complexity of the purchasing system depends on the size and type of an organization, whether the buying is centralized or decentralized, and established management policies. Procedures should be as simple as possible, with record keeping and paperwork limited to those essential for control and communication.

Good purchasing practices include the use of appropriate buying methods, a systematic ordering schedule, maintenance of an adequate flow of goods to meet production requirements, and a systematic receiving procedure and inventory control.

The process of purchasing, using the informal and formal methods of buying, is shown in Figure 5.8 and discussed in the following sections.

Identifying Needs

Quantities of food needed for production of the planned menus are identified from the menus and from recipes used to prepare them. Added to this are staples and other supplies needed in the various departments or production and service areas.

Inventory Stock Level. A system of communicating needs from the production areas and the storeroom to the buyer is essential. Establishment of a minimum and maximum stock level provides a means of alerting the buyer to needs. The minimum level is the point, established for each item, below which the inventory should not fall. This amount depends on the usage and time required for ordering and delivery. If canned fruits and vegetables, for example, are purchased every three months through the formal bidding procedure, the time lapse would be longer than for fresh produce that is ordered daily or weekly through informal buying.

The *minimum stock level,* then, includes a safety factor for replenishing the stock. The *maximum inventory level* is equal to the safety stock plus the estimated usage, which is determined by past experience and forecasts. From this information, a reorder point is established. Figure 5.9 illustrates the mini-max and par stock systems for establishing reorder points.

Another factor to be considered in the amount to reorder is the quantity that is most feasible economically. For example, if 5 cases of a food are needed to bring the stock to the desired level, but a price advantage can be gained by buying 10 cases, the buyer may consider purchasing the larger quantity. The buyer is encouraged to weigh carefully the true economic advantages of these price incentives. Stored food ties up cash, and unused food that spoils is literally money down the drain.

Quantity to Buy. The amount of food and supplies purchased at one time and the frequency of ordering depend on the amount of money on hand, the method of buying, the frequency of deliveries, and storage space. With adequate and suitable storage, the purchase of staples may vary from a two- to six-month supply, with perishables being ordered weekly and/or daily.

Meat, poultry, fish, fresh fruits and vegetables, and other perishable foods may be purchased for immediate use on the day's menu or more likely are calculated for

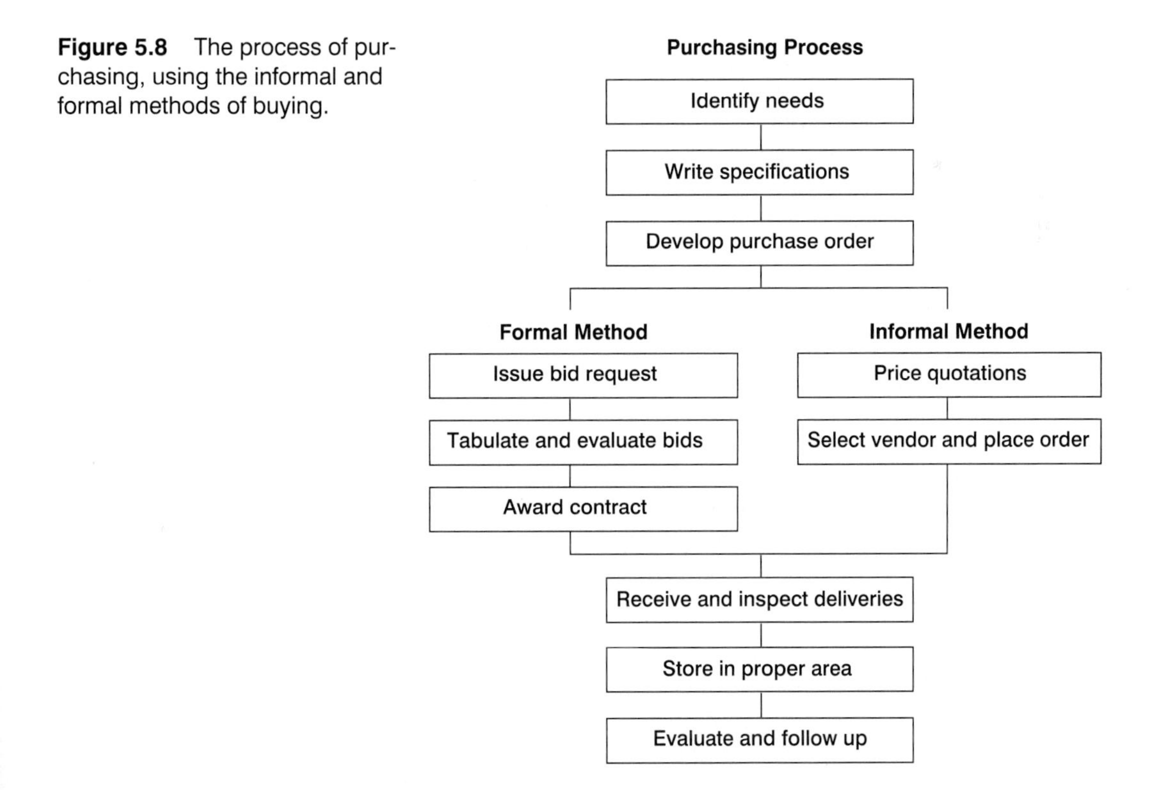

Figure 5.8 The process of purchasing, using the informal and formal methods of buying.

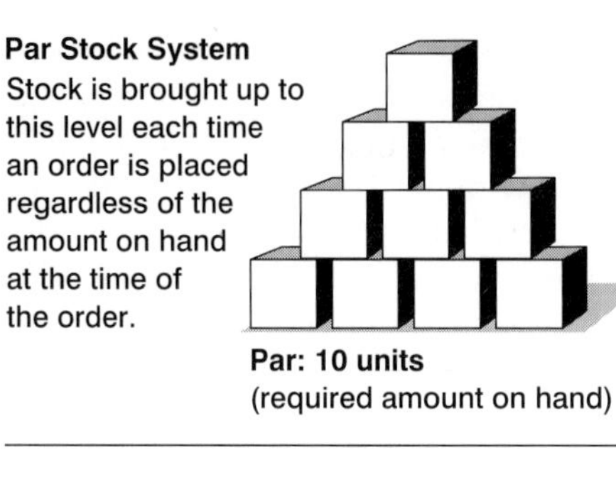

Par Stock System
Stock is brought up to this level each time an order is placed regardless of the amount on hand at the time of the order.
Par: 10 units
(required amount on hand)

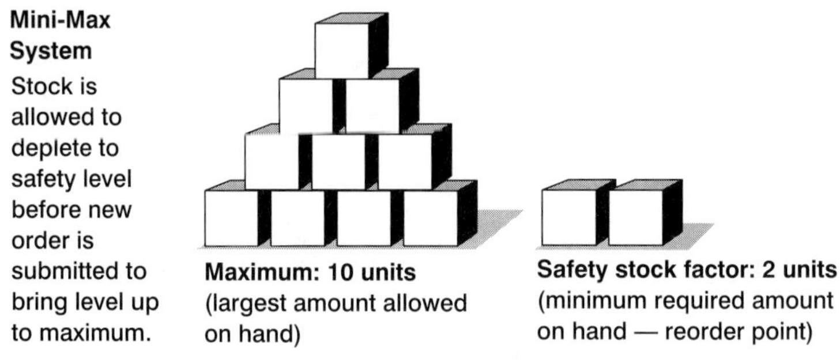

Mini-Max System
Stock is allowed to deplete to safety level before new order is submitted to bring level up to maximum.
Maximum: 10 units
(largest amount allowed on hand)

Figure 5.9 Comparison of par stock and mini-max inventory systems with reorder points.

two or more days, depending on delivery schedules, storage facilities, and preparation requirements. Quantities are based on the portion size and projected number of servings needed, taking into consideration the preparation and cooking losses. If the recipes are stored in a computer, it is a simple task to calculate the amount needed for the desired number of servings. Recipe calculations are provided in Chapter 6.

Some products, such as milk and bread, are delivered daily or several times a week, and the orders are based on the amount needed to keep the inventory up to a desired level. The price can be determined by a contract to furnish certain items as needed for a period of a month or longer. A stock level of butter and margarine, cheese, eggs, lettuce, celery, onions, and certain other fruits and vegetables can be established and maintained, whereas other produce is ordered as needed from the menu. Figure 5.10 is a suggested form for recording supplies on hand and the amounts to order.

Canned foods and groceries are generally purchased less often than are perishable foods, the frequency depending on storage space and money available. Assuming that adequate storage is available, a year's supply of canned goods can be purchased at one time if bought on competitive bid or if growing conditions indicate a possible shortage or a rise in prices. In some cases, an arrangement can be made for the supplier to store the food and deliver it as needed. A projection is made of the quantity that will be needed for the designated period, based on past purchases. That amount less the inventory is the quantity to purchase.

Specifications

Many foodservice operations have a set of specifications that have been developed as different food and supply items have been purchased. If specifications do not exist for the items being purchased, they should be written before bids or quotations are sought.

A **specification** is a detailed description of a product, stated in terms that are clearly understood by both buyer and seller. Specifications should be brief and concise but should contain enough information so that there can be no misunderstanding. Certain information is included in all specifications for food products:

Daily Purchase Order

Date ____________

ON HAND ORDER

Dairy:

____ gal whole milk ____
____ cs ½ pt whole ____
____ cs ½ pt choc ____
____ cs ½ pt B. milk ____
____ cs ½ pt skim ____
____ lb cot. cheese ____
____ Ice cream ____

Bread:

____ White bread ____
____ Wheat bread ____
____ Rye ____
____ Sandwich white ____
____ Sandwich wheat ____
____ Sandwich rye ____
____ Crumbs ____

Sweet Rolls:

____ Raisin bread ____
____ Cinnamon ____
____ Butterscotch ____
____ Raised donuts ____
____ Bismark ____
____ Twist ____
____ Pecan strip ____
____ Stick donuts ____
____ Jelly donuts ____

Potato Chips:

____
____
____
____
____
____

ON HAND ORDER

Meats:

____
____
____
____
____
____
____

Fish:

____
____
____

Poultry:

____ Chicken ____
____ Turkey ____
____ Eggs ____

Fresh Vegetables:

____ Cabbage ____
____ Carrots ____
____ Cauliflower ____
____ Celery ____
____ Celery cabbage ____
____ Cucumbers ____
____ Egg plant ____
____ Head lettuce ____
____ Leaf lettuce ____
____ Onions ____
____ Parsley ____
____ Peppers ____
____ Potatoes ____
____ Spinach ____
____ Squash ____
____ Tomatoes ____

ON HAND ORDER

Frozen Vegetables:

____ Asparagus ____
____ Green beans ____
____ Lima beans ____
____ Broccoli ____
____ Brussels sprouts ____
____ Cauliflower ____
____ Peas ____

Fresh Fruits:

____ Apples ____
____ Bananas ____
____ Berries ____
____ Cantaloupe ____
____ Grapefruit ____
____ Grapes ____
____ Lemons ____
____ Oranges ____
____ Peaches ____
____ Pineapple ____
____ Plums ____
____ Watermelon ____

Frozen Fruits and Juices:

____ Apples ____
____ Cherries ____
____ G. fruit sections ____
____ Lemon juice ____
____ Orange juice ____
____ Peaches ____
____ Rhubarb ____
____ Strawberries ____

Miscellaneous:

____
____

Figure 5.10 Daily purchase order form.

- *Name of the product:* The common, or trade, name of an item generally is simple, but the name of some products may vary in different parts of the country.
- *Federal grade or brand:* As already noted, the USDA has established federal grades for most agricultural products, but many packers or food processors have developed their own brands or trade names for canned, frozen, or other processed foods. If a bidder submits a quotation on a brand name product in lieu of a federal grade, buyers may request verification of quality by the USDA Acceptance Service; see "Grading and Acceptance Services" earlier in this chapter under "Food Quality."
- *Unit on which price is quoted:* This refers to the size and type of unit, such as pound, gallon, can, bunch, or other unit in common use.
- *Name and size of container:* Examples of container size include a case holding six No. 10 cans, a 30-pound can of frozen cherries, or a 30-pound lug of tomatoes.
- *Count per container or approximate number per pound:* Examples include 30/35 count canned peach halves per No. 10 can; eight-per-pound frankfurters; size 36 grapefruit, which indicates the number of fruit in a 4/5 bushel box. Sizes 27 through 40 are acceptable for half grapefruit servings. Oranges and apples also are sized according to the number in the box. Apples 80 to 100 are large, 113 to 138 medium, and 150 to 175 small.

Additional information may be included for different categories of food:

- *Fresh fruits and vegetables:* Variety, weight, degree of maturity, geographical location; for example, Jonathan apples, Indian River grapefruit, or bananas turning ripe, pale yellow with green tips. If needed immediately, specify fully ripe, bright yellow flecked with brown, and no green.
- *Canned foods:* Type or style, pack, size, syrup type, drained weight, specific gravity. Examples include cream style corn; whole vertical pack green beans; No. 4 sieve peas; apricot halves in heavy syrup or 21 to 25 degrees brix (syrup density); diced beets, drained weight 72 ounces (per No. 10 can); tomato catsup with total solids content of at least 33%.
- *Frozen foods:* Variety, sugar ration, temperature during delivery and on receipt; for example, sliced strawberries, sugar ratio of 4:1, or delivered frozen, 0°F or less.
- *Meats and meat products:* Age, market class, cut of meat, exact cutting instructions, weight range, fat content, condition on receipt.
- *Dairy products:* Milk fat content, milk solids, bacteria count, temperature during delivery and on receipt.

A well-written specification includes all of the information needed to identify the food item and to ensure that the buyer is getting exactly the quality desired. It should be identifiable with products or grades currently on the market and capable of being checked by label statements or USDA grades. Information for use in writing specifications is available from the USDA and from material published by the industry such as the National Association of Meat Purveyors. See the section titled "For More Information" at the end of this chapter.

Issuing Bid Requests

An invitation to bid provides vendors with an opportunity to submit bids for specific items advertised. Bid requests originate in the office of the purchasing agent or the person authorized to purchase for the foodservice. A bid request includes quantities required and purchase specifications for each item. In addition, the general conditions of acceptance are outlined, including the date and method of delivery, terms or payment, willingness to accept all or part of the bid, discounts, the date of closing bids, and other terms of the negotiations. Figure 5.11 is a sample bid request that includes general requirements and that would be used with Figure 5.12, which would list needed items and specifications.

The bid request may also ask for samples to be tested. This is especially important when large quantities are involved and often is requested when purchasing canned foods. Testing of canned foods is done by "can cutting," which involves opening of the sample cans and evaluating the products according to USDA scoring factors. If samples from more than one company are being tested, the labels on the cans should be covered so the test will be impartial.

Copies of USDA specifications and score sheets are available from the Government Printing Office. An example of a score sheet for canned foods is shown in Figure 5.13. Scores from different samples would be summarized on a form similar to Figure 5.14. Product evaluation forms may be developed for testing of other foods and should include specific qualities to be judged. A panel composed of persons who are involved in foodservice quality control should participate.

A bid schedule outlining the bid periods and delivery frequency should be established and, when possible, planned so that new packs of processed fruits and vegetables, usually available in October and January, can be used. This step is omitted in informal buying. Quotations are requested from two or more vendors, usually by telephone or from price sheets.

Developing Purchase Orders

The procedure for authorizing purchases differs in various foodservices. The process may begin with a purchase request, to be used along with quality specifications, as the basis for a purchase order and bid request to be issued by an authorized purchasing agent. A foodservice director, who is also the buyer, may develop a purchase order that has been compiled from requisitions from the various production and service units or from individual units in a school system or multiple-unit organization. In central purchasing for these operations, the requisitions originating in individual units do not necessarily need to include specifications, since the quality is determined at a central point and is uniform throughout the system.

Regardless of the method used, there should be a clear understanding of who is authorized to issue purchase requests or orders and the vendors should be aware of the name, or names, of authorized purchasing personnel. Authorization to sign for goods received and to requisition supplies from the storage areas should also be understood.

The purchase order specifies the quantity of each item needed for the bid period, quality specifications, and required date of delivery. The order must include the name of

Bid request for Frozen Fruits and Vegetables

Issued By ____ Date ____________________

____ Date to be Delivered __________

Address ____

____

Bid Request for ____

Notice is hereby given that the Board of Education of _______________ County, _______________ State (hereinafter referred to as the Board) requests written and sealed bids on the following items to be submitted to said Board on or before 10:30 a.m. on ___________, 19_____ and on each subsequent bidding period date indicated in the bid specifications. Sealed bids will not be opened until 10:30 A.M. ___________, 19____ if the outside is marked: Do not open until ________________________. Respond on attached bid form.

Contract Period: The bid covers the period from date of award through ___________, 19____, inclusive and vendors receiving awards shall be the sole suppliers to all schools for items for the period.

Samples: Bidders will be required to submit samples of the items bid upon. Samples are to be furnished without cost to ______________________ and are to be sent to ______________________ on ____________ by ____________.

Quality: Successful bidders must furnish United States Department of Agriculture Grade Certificates indicating each fruit and vegetable item to be U.S. Grade _____.

Grade Certificates: A U.S. Grade Certificate shall be submitted for required items prior to delivery. These certificates must cover the specific brand name of items being delivered. The code numbers on the item being delivered shall be the same as the codes listed on the certificate.

Estimated Quantities: Quantities indicated on Bid Proposal Forms are estimated total requirements based on anticipated use. They will provide the basis for determining the total low bid complying with specifications for each group of items and are submitted as information.

Actual Quantities: Quantities on attached Bid Sheet are estimated to cover the period from date of award to ___________, 19_____, inclusive. The School Food Service Department will furnish the successful bidder with actual quantities as are needed. Purchaser guarantees to purchase during contract period only, the actual requirements needed.

Delivery (Equipment): Carrier shall utilize only properly insulated, mechanical or thermostatic temperature control refrigeration equipment. Such equipment must be capable of maintaining temperature to protect the product. All products must be delivered in a hard frozen state, 0°F. or below.

The Board of Public Education reserves the right to reject the use of any equipment by a carrier if it is not in a clean, sanitary condition and suitable for hauling of all goods.

Figure 5.11 Sample bid request.

Each carrier shall furnish a Certificate of Insurance issued by an insurance company showing that the Board of Public Education will be protected from loss or damage to property of third persons or to the carriers' own property, loss or damage to Board of Public Education commodities, and injury or death to third persons or to the carrier's employees. Carrier will assume full common liability for all shipments.

Orders: All orders will be placed directly with awardees by telephone by the individual qualified purchasing official who in some cases may be the individual lunchroom manager. They may order fractional cases. Regular orders should be placed at least seventy-two (72) hours (3 work days) before the delivery time requested; but each emergency order should be filled within two (2) hours after the order is placed. ALL VENDORS MUST SUBMIT SEPARATE DELIVERY TICKETS AND/OR INVOICES FOR NONBID ITEMS.

Deliveries: Deliveries shall be made to the receiving area of individual schools between the hours of 7:00 a.m. and 2:15 p.m. These deliveries must be made in mechanically refrigerated trucks maintaining a temperature below freezing at all times.

Invoices & Statements: Invoices for the purchases of food and miscellaneous supplies made by schools are paid by the central accounting department. In order to facilitate the handling of these invoices, ALL VENDORS MUST ADHERE TO THE FOLLOWING INSTRUCTIONS:

Code number for each school listed on each invoice. (A list of schools with code number is attached.)

All items on delivery tickets MUST be billed according to description of item quoted on bid. Unit prices for all items shall be recorded and invoices shall be accurately extended. SEPARATE DELIVERY TICKETS AND/OR INVOICES SHALL BE MADE FOR ALL NONBID ITEMS.

All vendors must issue delivery tickets and credit memos in QUADRUPLICATE, and all four (4) copies must be signed by qualified purchasing official.

2 copies (original and 1 carbon) left with proper person at time of delivery.

2 copies to be returned to vendor.

The vendor shall forward as per attached list, weekly statements, with one signed delivery ticket attached, directly to the School Food Service Department.

All delivery tickets supporting weekly statements must be in exact agreement with copy of delivery tickets left with manager. If for any reason it is necessary to make a change on the delivery ticket, MAKE AN ADDITIONAL CHARGE OR CREDIT MEMO.

All cancellations or merchandise returns must be recorded by driver on all FOUR COPIES of delivery tickets, or "pick-up tickets."

2 copies left with manager at time of pick-up.

2 copies to be returned to vendor.

Do not mail statement to individual schools.

A monthly statement for each school should be sent to the official responsible for paying bills by the 10th working day or by the 10th calendar day of every month, following date of purchase.

Figure 5.11 *Continued.*
Source: Food Buying Guide for School Food Service, PA-1257, USDA.

Bid Form

Name of School District ____________________ Date __________

Item no.	Quantity	Unit size	Item and specification	Brand quoted	Quantity quoted	Unit price	Total price

Figure 5.12 Bid form to be used with Figure 5.11 to list items and specifications.

the organization, the individual making the request, and the signature of the person officially authorized to sign the order. Purchase order forms can be prenumbered or the number can be added at the time of final approval, but a number, as well as the date of issue, is essential for identification. Figure 5.15 shows a suggested purchase order form.

Tabulating and Evaluating Bids

Bids should be kept sealed and confidential until the designated time for opening. Sealed envelopes containing the bids should be stamped to indicate the date, time, and place of receipt. Bids received after the time and date specified for bid opening must be rejected and returned unopened to the bidder.

The opening and tabulation of bids should be under the control of an appropriate official. When schools and other public institutions are involved, the quotations and contents of bids should be open to the public. The bids and low bids should be carefully examined. In most instances, public purchasing laws specify that the award be made to the lowest responsible bidder. The following points should be considered before accepting bids:

1. Ability and capacity of the bidder to perform the contract and provide the service,
2. Ability of the bidder to provide the service promptly and within the time specified,
3. Integrity and reputation of the bidder,
4. Quality of bidder's performance on previous contracts or services,

5. Bidder's compliance with laws and with specifications relating to contracts or service, and
6. Bidder's financial resources.

Before the bid is awarded to a vendor the buyer may request test samples of the product from each qualified bidder to compare the actual product against the prede-

Score Sheet for Canned Tomatoes

Number, Size, and Kind of Container

Label

Container Mark or Identification		Cans/Glass		
		Cases		
Net Weight (oz)				
Vacuum (in.)				
Drained Weight (oz)				
Factors		*Score Points*		
I. Drained weight	20	(A) 18–20 (B) 15–17 (C) 12–14 (SStd) 0–11		
II. Wholeness	20	(A) 18–20 (B) 15–17 (C) 12–14		
III. Color	30	(A) 27–30 (B) 23–36 (C) 19–22 (SStd) 0–18		
IV. Absence of defects	30	(A) 27–30 (B) 22–26 (C) 17–21 (SStd) 0–16		
Total Score	100			
Normal flavor and odor				

Grade

Figure 5.13 Government score sheet for grading canned tomatoes.

Kind	Code	Label Net Weight	Actual Weight	Sp. Gr. or Drained Weight	Brix	Count	Remarks and Ratings (Defects, Color, etc.)	Price Per Dozen	Price Per Can	Price Per Piece

Figure 5.14 Suggested form for recording data on samples of canned products.

termined specifications. Can cutting, as mentioned earlier, is a formal process for evaluating the actual quality of canned goods against those identified in the bid specification. This process is recommended to ensure that the products meet or exceed the standards as specified. If specifications are not met and the contract award is not awarded to the lowest bidder, a full and complete statement of the reasons should be prepared and filed with other papers relating to the transaction.

Awarding Contracts

The contract should be awarded to the most responsive and responsible bidder with the price most advantageous to the purchaser. Buying on the basis of price alone can result in the delivery of products that are below the expectations of the foodservice. Purchasing should be on the basis of price, quality, and service.

The general conditions of the contract should include services to be rendered, dates and method of deliveries, inspection requests, grade certificates required, procedure for substitutions, and conditions for payment. The following information should also be provided: name and address of the foodservice, a contract number, type of items the contract covers, contract period, date of contract issue, point of delivery, quantities to be purchased, and the signature of an authorized representative of the firm submitting the bid. The terms of sale must be clearly stated in the contract. Table 5.3 is a summary of the various FOB (free on board) terms of sale methods used in formal purchasing. Point of origin is defined as the manufacturer's loading dock.

A contract that is issued represents the legal acceptance of the offer made by the successful bidder, and it is binding. All bidders, both successful and unsuccessful, should be notified of the action. When the contract is made by a purchasing agent, the foodservice should receive a copy of the contract award and specifications.

RECEIVING

Receiving is the point at which foodservice operations inspect the products and take legal ownership and physical possession of the items ordered. The purpose of receiv-

Purchase Order

Name of Institution ______________________________ Date ____________________

______________________________ Purchase Order # __________

Address *(Please refer to above number on all invoices)*

____

Address

____

Requisition No. __________

Department ______________________________ Date Required __________

To ______________________________

____

____

____

Instruction for Completing Order. Prepare in triplicate for the vendor, business office, and the manager.

Shipped to:__________ FOB__________ Via__________ Terms__________

Unit	Total quantity	Specification	Price per unit	Total cost

Approved by______________________________

Figure 5.15 Suggested purchase order form.

Table 5.3 Terms of sale and freight charges.

	Responsibilities			
Terms of Sale	**Pays Freight Charges**	**Bears Freight Charges**	**Owns Product During Transit**	**Files Damage Claims (if necessary)**
FOB origin—freight collect	Buyer	Buyer	Buyer	Buyer
FOB origin—freight prepaid	Seller	Seller	Buyer	Buyer
FOB origin—freight prepaid and charged back	Seller	Buyer	Buyer	Buyer
FOB destination—freight collect	Buyer	Buyer	Seller	Seller
FOB destination—freight prepaid	Seller	Seller	Seller	Seller
FOB destination—freight collect and allowed	Buyer	Seller	Seller	Seller

FOB = Free on Board

ing is to ensure that the food and supplies delivered match established quality and quantity specifications. The receiving process also offers an opportunity to verify price.

A well-designed receiving process is important to cost and quality control and, therefore, warrants careful planning and implementation. Minimally, a good receiving program should include clearly written policies and procedures on each of the following components:

- Coordination with other departments (e.g., production and accounting),
- Training for receiving personnel,
- Facility and equipment needs,
- Written specifications,
- Authority and supervision,
- Scheduled receiving hours,
- Security measures, and
- Documentation procedures.

Consequences of a poorly planned receiving program include the following:

- Short weights,
- Substandard quality,
- Double billing,
- Inflated prices,
- Mislabeled merchandise,
- Inappropriate substitutions,
- Spoiled or damaged merchandise, and
- Pilferage or theft.

In simple terms, a poorly planned and executed receiving process results in financial loss for the operation.

Coordination with Other Departments

The receiving function needs to be coordinated with other departments in the foodservice organization. Production and accounting are two key departments that need a well-defined working relationship with receiving personnel. The production department needs in-house food and supply products to meet the demands of the menu and service centers.

In many organizations, the accounting department is responsible for processing the billing of food and supply purchases. Receiving records must be completed and submitted to accounting in a timely fashion so that payments are made on time. Prompt billing allows the organization to take advantage of early payment discounts and avoid late payment penalties.

Personnel

The responsibility of receiving should be assigned to a specific, competent, well-trained person. However, in reality this job is often not specifically assigned at all, but simply handled by the person working closest to the dock. During training, this individual must learn to appreciate the importance of the receiving function in cost and quality control.

Qualifications of the receiving clerk should include knowledge of food quality standards and awareness of written specifications, the ability to evaluate product quality and recognize unacceptable product, and an understanding of the proper documentation procedures. The receiver's authority must be well defined by policy, which should clarify the scope of the receiver's authority and to whom he or she reports. Even though the receiver is well trained and trustworthy, consistent and routine supervision of the receiving area is recommended to ensure that procedures are followed and that the area is kept secure.

Facilities, Equipment, and Sanitation

A well-planned receiving area should be as close to the delivery docks as possible, with easy access to the storage facilities of the operation. This arrangement helps to minimize traffic through the production area and reinforce good security measures.

The area itself should be large enough to accommodate an entire delivery at one time. If a receiving office is in the area, it should have large glass windows so receiving personnel can easily monitor the activities of the area.

The amount and capacity of receiving equipment depends on the size and frequency of deliveries. Large deliveries may require a forklift for pallet deliveries. A hand truck may be adequate in medium to small operations. Scales, ranging from platform models to countertop designs, are needed to weigh goods as they arrive. A policy should be in place such that the scales are calibrated on a regular basis to ensure accuracy.

Some small equipment is also needed, including thermometers for checking refrigerated food temperatures and various opening devices, such as shortblade knives and crate hammers. Specifications, purchase orders, and documentation records need to be readily available and the receiving office should carry an ample supply of pens, paper, tape, and markers. Filing cabinets to house documents are recommended.

Cleaning and sanitation procedures for the receiving area should be defined by policy. Plans for pest control need to be determined and some cleaning supplies should be readily available to keep the area clean during the daily activities.

Scheduled Hours for Receiving

Hours of receiving should be defined by policy, and vendors should be instructed to deliver within a specific time range. The purpose of defined receiving times is to avoid the busiest production times in the operation and too many deliveries happening at the same time. Thus, many operations instruct vendors to deliver midmorning or midafternoon to avoid high food production and service times.

Security

The receiving components already discussed contribute to the security of the receiving process. A few additional practices, however, can contribute to a secure receiving program. Deliveries should be checked immediately upon arrival. After the receiving personnel is confident that the order meets specifications, he or she can sign the invoice and the delivery should be moved immediately to proper storage. This practice minimizes quality deterioration and opportunity for theft.

Doors to the receiving area should be kept locked. Some facilities keep doors locked at all times and use a doorbell system for the delivery personnel to use when they arrive. Finally, all unauthorized personnel should be kept out of the receiving area. This is particularly difficult in facilities where the area is used for other purposes, such as trash removal. More frequent supervision is recommended in these situations.

The Receiving Process

Once the components of a receiving program are planned and implemented an organization is ready to receive goods. The receiving process involves five key steps:

1. Inspect the delivery and check it against the purchase order.
2. Inspect the delivery against the invoice.
3. Accept order only if all quantities and quality specifications are met.
4. Complete receiving records.
5. Immediately place goods in appropriate storage.

Methods. The two main methods of receiving are the blind method and the invoice receiving method. The *blind method* involves providing an invoice or purchase order, one in which the quantities have been erased or blacked out, to the receiving clerk. The clerk must then quantify each item by weighing, measuring, or counting and recording it on the blind purchase order. The blind document is then compared with the original order. This method offers an unbiased approach by the receiving clerk but is time consuming and, therefore, more labor intensive.

A frequently used and more traditional method is *invoice receiving*. Using this method, the receiving clerk checks the delivered items against the original purchase order and notes any deviations. This method is efficient but requires careful evaluation by the clerk to ensure that the delivery is accurate and quality standards are met.

Tips for Inspecting Deliveries. The following are some additional tips that the receiver should keep in mind when evaluating food and supply deliveries:

- Inspect foods immediately upon arrival for quality and quantity ordered.
- Anticipate arrival and be prepared.
- Check adequacy of storage space.
- Have purchase orders and specifications ready.
- Make certain the receiving personnel are well trained.
- Check temperatures of refrigerated items upon arrival.
- Check frozen items for evidence of thawing or burn.
- Check perishable items first.
- Randomly open cases or crates of product to determine that the container includes entire order.

Numerous resources are available for use in evaluating food quality and several of these are listed at the end of this chapter.

Evaluation and Follow-Up. Evaluation of products should be continued as they are issued for use because some discrepancies may not be detected until the item is in use. When products are found to be defective, some type of adjustment should be made. If the products are usable but do not meet the specifications, the buyer may request a price adjustment. The manager may refuse to accept the shipment or, if some of the food or supplies are used but found to be unsatisfactory, the buyer may arrange to return the remaining merchandise or request some type of compensation. The purchasing agent or other proper official should be notified of deficiencies in quality, service, or delivery.

STORAGE

The flow of material through a foodservice operation begins in the receiving and storage areas. Careful consideration should be given to procedures for receiving and storage, as well as to the construction and physical needs of both areas. In planning, there should be a straight line from the receiving dock to the storeroom and/or refrigerators and, preferably, on the same level as the kitchen. A short distance between receiving and storage reduces the amount of labor required, reduces pilferage, and causes the least amount of deterioration in food products.

The proper storage of food immediately after it has been received and checked is an important factor in the prevention and control of loss or waste. When food is left unguarded in the receiving area or exposed to the elements or extremes of temperature for even a short time, its safekeeping and quality are jeopardized.

Adequate space for dry, refrigerator, and freezer storage should be provided in locations that are convenient to receiving and preparation areas. Temperature and humidity controls and provision for circulation of air are necessary to retain the various quality factors of the stored foods. The length of time foods can be held satisfactorily and without appreciable deterioration depends on the product and its quality when stored as well as the conditions of storage. Suggested maximum temperatures and storage times for a few typical foods are given in Figure 5.16. The condition of stored food and the temperature of the storage units should be checked frequently.

Dry Storage

The main requisites of a food dry-storage area are that it be dry, cool, and properly ventilated. If possible, it should be in a location convenient to the receiving and preparation areas.

Dry storage is intended for nonperishable foods that do not require refrigeration. Paper supplies often are stored with foods, but a separate room should be provided for cleaning supplies, as required in many health codes. The separation of food and cleaning materials that could be toxic prevents a possible error in identification or a mixup in filling requisitions.

Temperature and Ventilation. The storage area should be dry and the temperature not over 70°F. A dark, damp atmosphere is conducive to the growth of certain organisms, such as molds, and to the development of unpleasant odors. Dry staples such as flour, sugar, rice, condiments, and canned foods are more apt to deteriorate in a damp storage area. The storeroom is more easily kept dry if located at or above ground level, although it need not have outside windows unless required by code. All plumbing pipes should be insulated and well protected to prevent condensation and leakage onto food stores. If the storage area does have windows, they should be equipped with security-type sash and screens and painted opaque to protect foods from direct sunlight.

Ventilation is one of the most important factors in dry storage. The use of wall vents, as shown in Figure 5.17, is the most efficient method of obtaining circulation of air, but other methods are possible. The circulation of air around bags and cartons of food is necessary to aid in the removal of moisture, reduction of temperature, and elimination of odors. For this reason, containers of food are often cross-stacked for better air circulation.

Storeroom Arrangement. Foods and supplies should be stored in an orderly and systematic arrangement. A definite place should be assigned to each item with similar products grouped. The containers are dated and usually left in the original package or placed in tightly covered containers if the lots are broken. All items should be stored on racks or shelves instead of directly on the floor or against walls.

Food	Suggested Maximum Temperature (°F)	Recommended Maximum Storage	
Canned products	70	12 months	
Cooked dishes with eggs, meat, milk, fish, poultry	36	Serve day prepared	
Cream filled pastries	36	Serve day prepared	
Dairy products			
Milk (fluid)	40	3 days	In original container, tightly covered
Milk (dried)	70	3 months	In original container
Butter	40	2 weeks	In waxed cartons
Cheese (hard)	40	6 months	Tightly wrapped
Cheese (soft)	40	7 days	In tightly covered container
Ice cream and ices	10	3 months	In original container, covered
Eggs	45	7 days	Unwashed, not in cardboard
Fish (fresh)	36	2 days	Loosely wrapped
Shellfish	36	5 days	In covered container
Frozen products	0 (to −20)		
Fruits and vegetables		1 growing season to another	Original container
Beef, poultry, eggs		6–12 months	Original container
Fresh pork (not ground)		3–6 months	Original container
Lamb and veal		6–9 months	Original container
Sausage, ground meat, fish		1–3 months	Original container
Fruits			
Peaches, plums, berries	50	7 days	Unwashed
Applies, pears, citrus	50 (to 70)	2 weeks	Original container
Leftovers	36	2 days	In covered container
Poultry	36	1–2 days	Loosely wrapped
Meat			
Ground	38	2 days	Loosely wrapped
Fresh meat cuts	38	3–5 days	Loosely wrapped
Liver and variety meats	38	2 days	Loosely wrapped
Cold cuts (sliced)	38	3–5 days	Wrapped in semi-moisture-proof paper
Cured bacon	38	1–4 weeks	May wrap tightly
Ham (tender cured)	38	1–6 weeks	May wrap tightly
Ham (canned)	38	6 weeks	Original container, unopened
Dried beef	38	6 weeks	May wrap tightly
Vegetables			
Leafy	45	7 days	Unwashed
Potatoes, onions, root vegetables	70	7–30 days	Dry in ventilated container or bags

Figure 5.16 Suggested maximum storage temperatures and times.

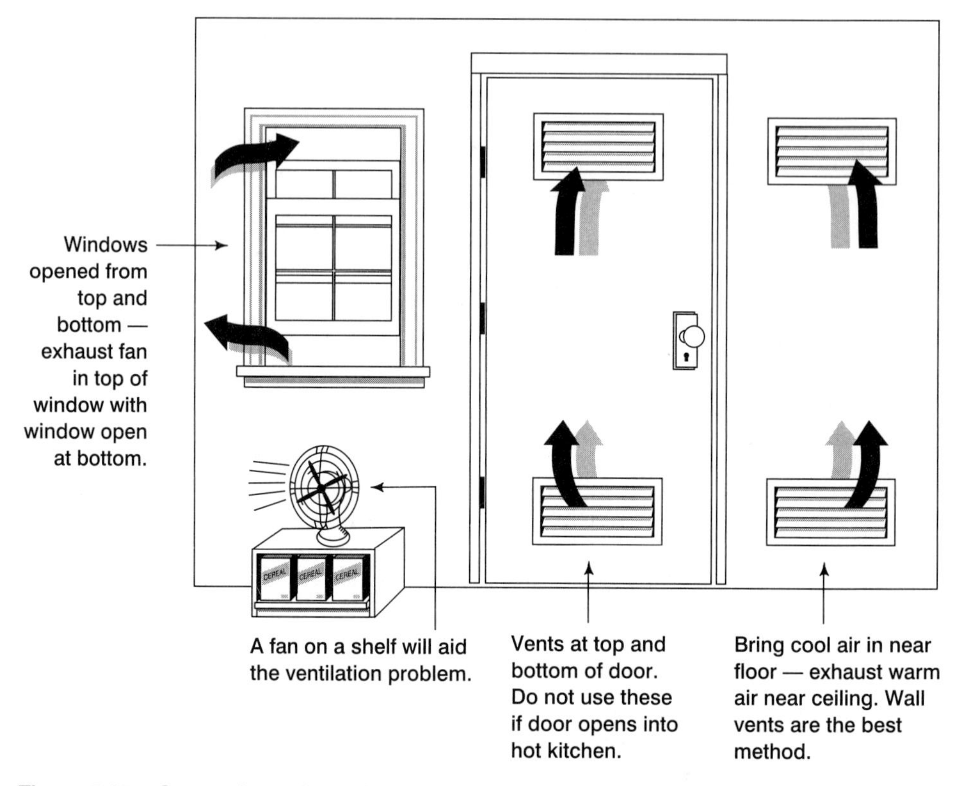
Windows opened from top and bottom — exhaust fan in top of window with window open at bottom.
CEREAL
CEREAL
CEREAL
A fan on a shelf will aid the ventilation problem.
Vents at top and bottom of door. Do not use these if door opens into hot kitchen.
Bring cool air in near floor — exhaust warm air near ceiling. Wall vents are the best method.

Figure 5.17 Suggestions of possible ways to provide circulation of air in a dry storage area.
Courtesy of Ohio Department of Health.

Cases and bags of food can be stacked on slatted floor racks, pallets, or wheeled metal platforms. Hand or power lifts are useful for moving loaded pallets from one location to another, but the aisles between shelves and platforms should be wide enough for the use of such mobile equipment.

Shelving, preferably metal and adjustable, is recommended for canned foods or other items that have been removed from cases. Shelves should be far enough off the floor and away from the wall to permit a free flow of air. Some state regulations may have specific measures. Broken lots of dry foods, such as sugar and flour, should be stored in metal or plastic containers with tightly fitted lids. The items can be arranged according to groups, and foods in each group can be placed on the shelves in alphabetical order—for example, canned fruits would be shelved as follows: apples, apricots, etc. Food should also be stored using the FIFO (first-in/first-out) method. New shipments should be placed in back to ensure use of the oldest stock first. Alphabetical arrangement facilitates counting when the physical inventory is taken and locating items when filling storeroom requisitions. A chart showing the arrangement of supplies is helpful to storeroom personnel. It should be posted near the door or other place where it can be easily seen.

Sanitation. Food in dry storage must be protected from insects and rodents by means of preventive measures, such as the use of proper insecticides and rodenticides, the latter under the direction of persons qualified for this type of work. Many operators contract with a pest control service to provide routine monitoring. Floors in the dry storage area should be slip resistant and easily cleaned. A regular cleaning schedule designed according to the volume of traffic and other activity in this area is vital to the maintenance of clean and orderly storage rooms. No trash should be left on the shelves or floor, and spilled food should be wiped up immediately.

Refrigerated and Freezer Storage

The storage of perishable foods is an important factor in their safety and retention of quality. Fresh and frozen foods should be placed in refrigerated or frozen storage immediately after delivery and kept at these temperatures until ready to use. Recommended holding temperatures for fresh fruits and vegetables are 40°F to 45°F and 32°F to 40°F for meat, poultry, dairy products, and eggs. Frozen products should be stored at 0° to -20°F.

In some foodservices, separate refrigerators are available for fruits and vegetables, dairy products and eggs, and for meats, fish, and poultry. Fruits and vegetables, due to their high moisture content, are susceptible to freezing and, therefore, should be kept at a slightly higher temperature than meats or dairy products. As in dry storage, foods under refrigeration should be rotated so that the oldest is used first. Fruits and vegetables should be checked periodically for ripeness and decaying pieces removed to prevent further spoilage. Some vegetables, like potatoes, onions, and squash, can be kept at temperatures up to 60°F and, in some foodservices, are placed in dry storage. Foods that absorb odors must be stored away from those that give off odors.

In many operations, walk-in refrigerators are used for general and long-term storage, with reach-in units located near workstations for storage of daily perishables and foods in preparation and storage. In a large foodservice, individual refrigerator units can be grouped together for convenience to receiving and preparation areas and for servicing. Separate cooling equipment makes it possible to control and maintain the proper temperature for the food stored in each unit. All refrigeration and freezer units should be provided with thermometers, preferably the recording type. Walk-in refrigerators can have remote thermometers mounted outside the door so that temperatures can be read without opening the door, as shown in Figure 5.18. Temperatures should be checked twice daily and any irregularity reported to the appropriate supervisor. Prompt action can result in saving food as well as money. Employees should be aware of the correct temperatures for the refrigerators and should be encouraged to open the doors as infrequently as possible.

Cleanliness is vital to food safety. Refrigerators should be thoroughly cleaned at least weekly and any spillage wiped up immediately. Hot food should be placed in shallow pans to chill as soon as possible after preparation unless it is to be served immediately. Cooked foods and meat should be covered to reduce evaporation losses and to limit odor absorption and damage from possible overhead leakage or dripping. Cooked meats should be stored above raw meats in the refrigerator to ensure that cooked foods are protected from raw meat drippings. Daily checks on the contents of refrigerators are advisable so that leftover and broken package foods are incorporated into the menu without delay.

Figure 5.18 Some of the built-in features of a walk-in refrigerator include a door that opens at floor level, an easily visible temperature indicator outside the unit, and a tight-fitting, well-hinged door with a lock that releases from the inside.
Courtesy of Arctic Industries, Inc.

Self-contained refrigeration units are used for ice makers, water dispensers, counter sections for display of salads, and storage for individual milk cartons. Each is adjusted to maintain the temperature needed. Freezer storage generally is in walk-in units, which may open from a walk-in refrigerator or the dry storage area. Ice cream and other frozen desserts may be kept in separate freezer cabinets.

The maintenance of refrigeration equipment requires the regular inspection by and services of a competent engineer to keep the equipment in good working order. However, the manager and other employees must be able to detect and report any noticeable irregularities, because a breakdown in the system could result in heavy loss of food and damage to equipment. In most installations, the refrigerator system is divided into several units so that failure in one will not disrupt the operation of the others.

INVENTORY RECORDS AND CONTROL

Accurate records are essential to inventory control and provide a basis for purchasing and for cost analysis. The exact procedure and forms used will vary according to policies of the institution and degree of computerization, but an adequate control system requires that a record be made of all food products and supplies as they are received and stored, and again as they are issued for use in production or other areas of the foodservice.

Receiving

All incoming supplies should be inspected, as explained earlier, and recorded on a receiving record form such as the one shown in Figure 5.19. A journal in which to list the items received, with date of receipt, can also be used as a receiving record. Whatever form is used, the information should be checked with the purchase order, the delivery slip, and the invoice to be sure that the merchandise has been received as ordered and that the price is correct.

Storeroom Issues

Control of goods received cannot be effective unless storerooms are kept locked and authority and control over the merchandise are delegated to one person. Even if the foodservice is too small to justify the employment of a full-time storekeeper, an

Receiving Record Date____________

Quantity	Unit	Description of item	Name of vendor	Inspected and quantity verified by	Unit price	Total cost	Distribution	
							To kitchen	To store room

Figure 5.19 Sample receiving record form.

employee may be made responsible for receiving, putting away, and issuing goods from the storeroom in addition to other assigned duties.

No food or other supplies should be removed from the storeroom without authorization, usually in the form of a written requisition. An exception may be perishable foods that are to be used the same day they are received and are sent directly to the production units. In that case, they are treated as direct issues and are charged to the food cost for that day. All foods that are stored after delivery are considered storeroom purchases and in most operations can be removed only by requisition.

A list of supplies needed for production and service of the day's menu is compiled by the cook or other person responsible for assembling ingredients. If the foodservice uses an ingredient room for weighing and measuring ingredients for all recipes, the personnel in this unit are responsible for requesting supplies. (The ingredient room is discussed in more detail in Chapter 6.) The list of needed supplies is then submitted to the storekeeper, who completes the requisition. The order is filled and delivered to the appropriate department or section. The exact procedure for issuing supplies varies with the size of the operation and whether there is a full-time storekeeper.

Requisitions should be numbered and made out in duplicate or triplicate as the situation requires. Prenumbering of the requisitions makes it possible to trace missing or duplicate requisitions. An example of a storeroom requisition is shown in Figure 5.20. Columns should be included for unit price and total cost unless a com-

Storeroom Requisition

Issue following items to

Date:

________________ Department

Signed:

Item	*Description*	*Quantity Ordered*	*Quantity Received*	*Unit Price*	*Total Cost*	*Authorized Signature*

Figure 5.20 Requisition for storeroom issues.

puter-assisted program is used, in which case the data will be available from the stored information in the computer. An inventory number is needed for each item on the requisition if a computer is used in calculating costs.

The requisition should be signed by a person authorized to request supplies and should be signed or initialed by the individual who fills the order. The requisitioning of food and supplies is an important factor in controlling costs and in preventing loss from pilferage, and it should be practiced in some form even in a small foodservice.

Perpetual Inventory

The perpetual inventory is a running record of the balance on hand for each item in the storeroom. The use of computers has simplified the process of maintaining the perpetual inventory and is used by many foodservices. In organizations where data processing is not used, the inventory information is usually recorded on cards. Standard forms for perpetual inventory are available from suppliers, or the foodservice manager can design forms that will work best for an individual operation. An example is given in Figure 5.21.

The perpetual inventory provides a continuing record of food and supplies purchased, in storage, and used. Items received are recorded on the inventory from the invoices, and the amounts are added to the previous balance on hand. Storeroom issues are recorded from the requisitions and subtracted from the balance. Additional information usually includes the date of purchase, the vendor, the brand purchased, and the price paid.

If minimum and maximum stock levels have been established, as discussed earlier in the section titled "Identifying Needs," these figures should be indicated on the inventory. When computers are used, the reorder point is included in the program. If cards are used, a colored marker or some other method of flagging will alert the storekeeper or manager to reorder the item when the inventory reaches that level.

These inventory records are recommended for all items except perishable foods that are delivered and stored in the production area. A physical inventory taken at the time perishable foods are ordered is more realistic. However, if there is a need for purchasing information on prices or total amounts of these foods used during a certain period of time, a purchase record as illustrated in Figure 5.22 may be used to record the date of purchase, amounts, prices, and vendors. In Figure 5.23, a purchase record and summary of purchases are combined with a perpetual inventory.

Time and strict supervision are required if the perpetual inventory is to be an effective tool, but it is a useful guide for purchasing and serves as a check on irregularities, such as pilferage or displacement of stock. It also provides useful information on fast-moving, slow-moving, or unusable items.

Physical Inventory

An actual count of items in all storage areas should be taken periodically, usually to coincide with an accounting period. In some organizations, an inventory is taken at

SPECIFICATIONS

NO.	*FIRMS*	*NO.*	*FIRMS*
1		4	
2		5	
3		6	

Date	*Firm*	*Brand*	*Size*	*Amt. Rec'd*	*Cost Doz*	*Cost Uni*	*Date*	*Firm*	*Brand*	*Size*	*Amt. Rec'd*	*Cost Doz*	*Cost Unit*

PERPETUAL INVENTORY

Date	*In*	*Out*	*Bal.*	*Date*	*In*	*Out*	*Bal.*	*Date*	*In*	*Out*	*Bal.*	*Date*	*In*	*Out*	*Bal.*	*Date*	*In*	*Out*	*Bal.*

PERPETUAL INVENTORY

Food Classification	Item	Description	Size	Maximum/Minimum

Figure 5.21 Perpetual inventory form.

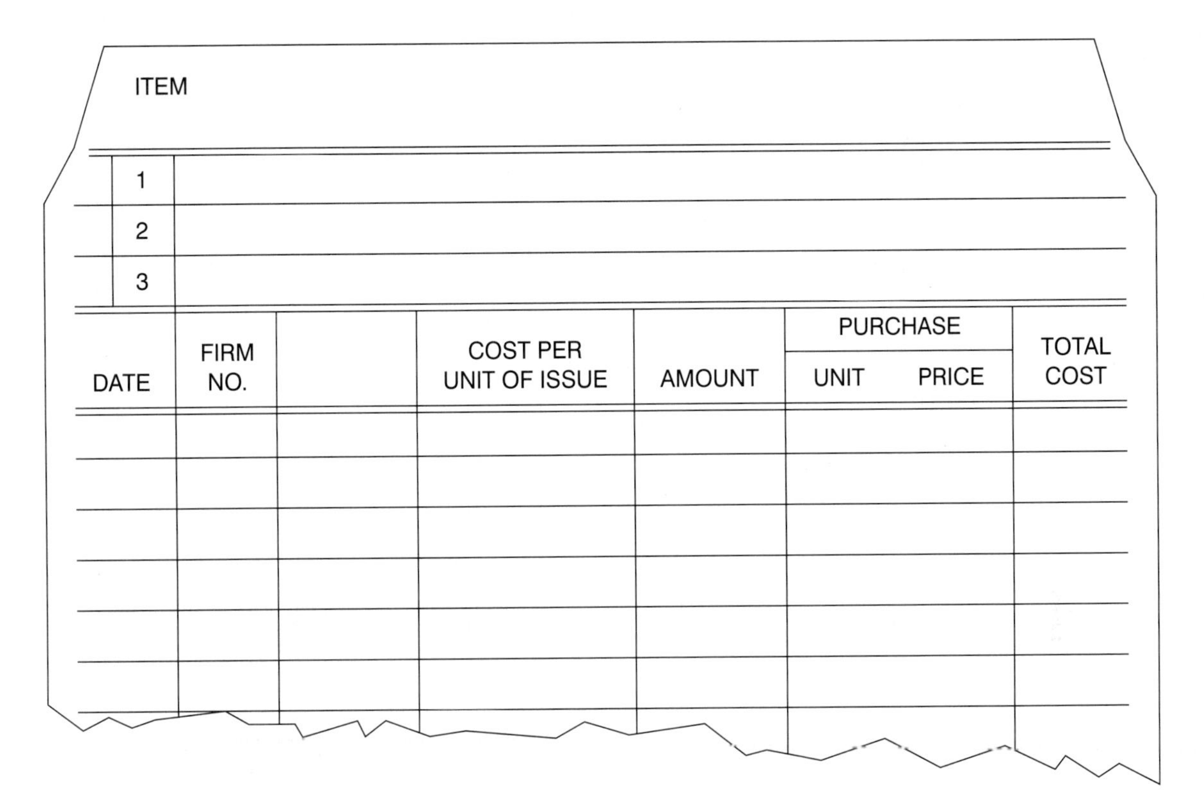

ITEM

1	
2	
3	

DATE	FIRM NO.		COST PER UNIT OF ISSUE	AMOUNT	PURCHASE		TOTAL COST
					UNIT	PRICE	

Figure 5.22 Purchase record form.

the end of each month, in others two or three times a year. The inventory is simplified if two people work together, one in a supervisory position or not directly involved with the storeroom operation. As one person counts the number of each item on hand, the other enters it on the inventory.

The procedure for taking a physical inventory is simplified by developing a printed form on which are listed the items normally carried in stock and their unit sizes, as shown in Figure 5.24. For convenience and efficiency in recording, the items on the inventory form can be classified and then arranged alphabetically within the group or listed in the same order as they are arranged in the storeroom and in the perpetual inventory. Space should be left on the form between each grouping to allow for new items to be added.

After the physical inventory is completed, the value of each item is calculated and the total value of the inventory determined. Inventory figures are used to calculate food costs by adding the total food purchases to the beginning inventory and subtracting the ending inventory. The physical count also serves as a check against perpetual inventory records. Minor differences are expected, but major discrepancies should be investigated. Carelessness in filling requisitions or in record keeping is the most common reason for these errors, which may indicate a need for tighter storeroom controls or more accurate record keeping.

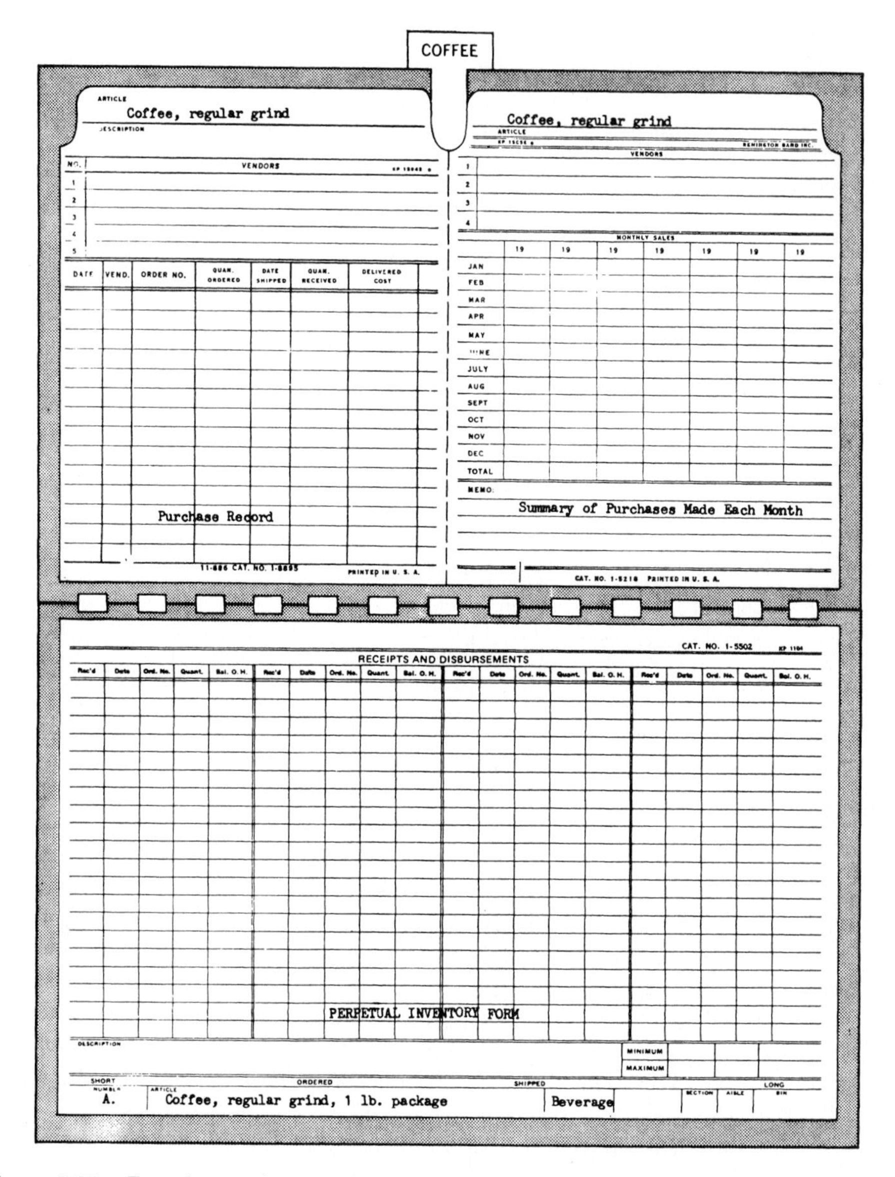

COFFEE

ARTICLE
Coffee, regular grind
DESCRIPTION

NO.	VENDORS
1	
2	
3	
4	
5	

DATE	VEND.	ORDER NO.	QUAN. ORDERED	DATE SHIPPED	QUAN. RECEIVED	DELIVERED COST

Purchase Record

11-886 CAT. NO. 1-8895 PRINTED IN U. S. A.

Coffee, regular grind
ARTICLE
REMINGTON RAND INC.

	VENDORS
1	
2	
3	
4	

MONTHLY SALES

	19	19	19	19	19	19	19
JAN							
FEB							
MAR							
APR							
MAY							
JUNE							
JULY							
AUG							
SEPT							
OCT							
NOV							
DEC							
TOTAL							

MEMO:
Summary of Purchases Made Each Month

CAT. NO. 1-5218 PRINTED IN U. S. A.

CAT. NO. 1-5502

RECEIPTS AND DISBURSEMENTS

Rec'd	Date	Ord. No.	Quant.	Bal. O. H.	Rec'd	Date	Ord. No.	Quant.	Bal. O. H.	Rec'd	Date	Ord. No.	Quant.	Bal. O. H.	Rec'd	Date	Ord. No.	Quant.	Bal. O. H.

PERPETUAL INVENTORY FORM

DESCRIPTION
MINIMUM
MAXIMUM

SHORT ORDERED SHIPPED LONG

NUMBER	ARTICLE			SECTION	AISLE	BIN
A.	Coffee, regular grind, 1 lb. package	Beverage				

Figure 5.23 Form for purchase and summary records combined with perpetual inventory.

Both perpetual and physical inventories should be kept of china, glassware, and silverware. These items should be revalued at least once a year on the basis of physical inventory, although it may be desirable to revalue them at more frequent intervals. An inventory of other kitchen and dining room equipment and furniture normally is taken once a year.

Management of inventory is practiced both to determine quantities to keep on hand and to determine the security methods used to control how the stock influences overall foodservice costs. Each organization should decide on maximum and minimum quantities desirable to maintain in the storeroom. This decision is based on storage facilities and capacities, delivery patterns, and the volume of business. Established standards for quantities desirable to keep on hand aid in purchasing—both as to quantity to order and when to order.

Carrying an inventory is costly and some thought should be given to the total value of such an investment. Carrying costs normally range from 15% to 35% of the inventory value, but can be as high as 50%, and fast-food operations with a minimum of inventory and rapid turnover may be as low as 2% to 4% of the inventory value. This percentage includes capital invested in inventory that prevents the use of money for other purposes; storage costs of space, personnel, materials handling, records, and utilities; deterioration and spoilage; obsolescence of items no longer in use; and insurance on this asset.

The Student Union Food Division **Page 1**

Physical Inventory ________________ 19 ____

Classification	Item	Unit	Quantity	Unit Price	Total Cost
Beverages:					
	Coffee	14 oz pkg			
	Tea, iced	1 gal			
	Tea, individual	100/Box			
Cereals:					
	Assorted individual	50/carton			
	Corn Flakes	100/cs			
	Cream of Wheat	1# 12 oz box			
	Hominy grits	1# 8 oz box			
	Oats, rolled	3# box			
	Ralstons	1# 6 oz box			
	Rice, white	1# box			
Cereal Products and Flour:					
	Cornmeal	Bulk/lb			

TOTAL PAGE 1 ____________

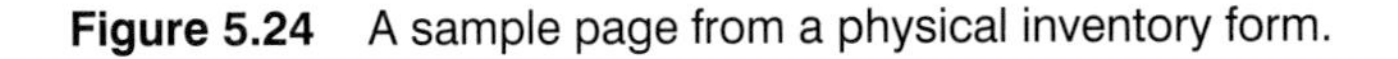

Figure 5.24 A sample page from a physical inventory form.

SUMMARY

Purchasing is an essential function in the operation of a foodservice organization and is vital to maintaining an adequate flow of food and supplies to meet production and service requirements.

Informal or formal methods of buying may be used, sometimes varying for different commodities. The buyer should be knowledgeable about the market and should understand the legal implications of contracts and bid buying. Purchasing may be the responsibility of the foodservice administrator or may be done centrally through a purchasing department. Group buying, in which several organizations combine their purchasing volume, has been successful in many cases.

The safety of food products is protected by several federal agencies, and quality grades have been established for many products. Detailed specifications should be used to ensure the purchase and delivery of products of the desired quality. Decisions must be made by the foodservice on the market form preferred, the quality to buy, and whether to make or buy prepared foods.

Good purchasing procedures include the use of appropriate buying methods, establishment of ordering schedules, and a system of communicating needs from production and service areas to the buyer. Foods and supplies should be received and checked by trained personnel and properly stored at appropriate temperatures. A storeroom control system that includes authorized issuing of supplies and complete inventory records is essential, but the procedures and paperwork should be limited to those necessary for control and communications.

REVIEW QUESTIONS

1. What are the most recent trends and changes in the food market? How have these changes influenced the purchasing function of the foodservice organization?
2. Read a local newspaper for several consecutive days. What events (economic, political, environmental, social) might have an impact on purchasing decisions made in a foodservice department?
3. How does the marketing channel have an impact on the price paid for a food product by the consumer?
4. Which food products are under the regulatory control of one or more government agencies?
5. What qualities should a foodservice buyer possess?
6. What are the advantages and disadvantages of group and centralized purchasing?
7. What are the advantages and disadvantages of the two principal methods of buying? What are some alternative methods of purchasing?
8. Explain the following recommendation: "Level of quality purchased should match intended use."
9. Compare and contrast a perpetual and physical inventory.
10. How does a well-planned receiving program contribute to cost and quality control?

11. What are some potential consequences of a poorly planned and monitored receiving process?

SELECTED REFERENCES

Byers, A. B., Shanklin, C. W., and Hoover, L. C.: *Food Service Manual for Healthcare Institutions,* 1994 edition. Chicago, Ill.: American Hospital Publishing, Inc., 1994.

Khan, M.: *Foodservice Operations.* New York: Van Nostrand Reinhold, 1987.

Kotschevar, L. H., Donnelly, R.: *Quantity Food Purchasing,* 4th ed. New York: Macmillan, 1994.

The Meat Buyers Guide, 7th ed. McLean, Va.: National Association of Meat Purveyors, 1984.

National Restaurant Association, The Education Foundation: *Applied Foodservice Sanitation,* 4th ed. New York: John Wiley & Sons, 1992.

Sneed, J., and Kreese, K. H.: *Understanding Foodservice Financial Management.* Rockville, Md: Aspen Publishers, Inc., 1989.

Spears, M. C.: *Foodservice Organizations; A Managerial and Systems Approach,* 3rd ed. Englewood Cliffs, N.J.: Prentice Hall, 1995.

Warfel, M.C., and Cremer, M.L.: *Purchasing for Food Service Managers,* 2nd ed. Berkeley, Calif.: McCutchin Publishing Corporation, 1990.

FOR MORE INFORMATION

Food Buying Guide for School Food Service
U.S. Government Printing Office
Washington, DC 20402

Food Service Directory and Buyers' Guide for Fresh Products
Produce Marketing Association
700 Barksdale Road
Newark, DE 19711

Grade Standards for Fresh and Processed Fruits and Vegetables
USDA Fruit and Vegetable Division
Washington, DC 20250

Grade Standards for Poultry, Rabbits, and Shell Eggs
USDA Poultry and Dairy Division
Washington, DC 20250

Quality and Yields Grades for Beef, Veal, . . . and Hog Carcasses
USDA Livestock Meat, Grain, and Seed Division
Washington, DC 20250

National Association of Meat Purveyors
Meat Buyer's Guide
8365B Greensboro Drive
McLean, VA 22101

National Dairy Council
6300 N. River Road
Rosemont, IL 60018

National Livestock and Meat Board
36 S. Wabash Avenue
Chicago, IL 60603

6

Production Management

OUTLINE

Production of high-quality food involves a number of interrelated steps, each dependent on the other. The transformation of raw or processed foods into an acceptable finished product ready for service is an essential operation in the overall foodservice system. It requires the purchase of food products of good quality, initial storage and "holding" at optimum temperatures at various points in its production, and, generally, one or more processing procedures under controlled conditions.

Traditionally, these procedures have been carried out in the individual foodservice, and menu items were prepared "from scratch." Today, however, there are alternatives to this conventional system. Foodservice organizations composed of several individual units may centralize all or part of their food production in a commissary or central production kitchen. Preparation in these facilities can range from controlled production of items such as desserts and baked goods; preparation of meats ready for cooking; preparation of fruits and vegetables for salads or for final cooking in the individual foodservice units; or complete preparation and cooking of menu items, packaged in individual or bulk containers, and chilled or frozen for delivery to serving units.

Many foodservices prepare either all or part of the food for immediate service or hot-holding. In some, however, food is cooked then chilled or frozen for later service. Others purchase certain menu items in ready-to-cook or ready-to-serve forms, and most use some type of convenience ingredients or components. These foodservice systems—conventional, commissary (central production kitchen), ready prepared (cook/chill or cook/freeze), and assembly/serve—are discussed in detail in Chapter 2.

Regardless of the system used, production planning and control are vital to the successful production of high-quality food. Recipe standardization, forecasting, and scheduling of production are discussed in this chapter. Various elements of production and quality control are included to illustrate the importance of management such that established standards are consistently met.

KEY CONCEPTS

1. It is the responsibility of the foodservice manager to establish and implement quality standards for food production through application of food production principles and techniques.
2. Numerous processing steps are needed to transform raw food into an acceptable finished product.
3. A standardized recipe is one that has been evaluated, tested, and adjusted for a specific foodservice operation.
4. Forecasting is a prediction of future food needs and can be calculated using a number of formulas or models.
5. Production schedules and meetings are methods used to communicate production demand to the production staff.
6. Centralized ingredient assembly, portion control and yield analysis are methods of quality and cost control.

FOOD PRODUCTION

Production planning and scheduling are vital to the production of high-quality food and are important management responsibilities. The true test of the planning, however, is the production of food that is appealing to the clientele, prepared in the appropriate quantity, microbiologically safe, and within budgetary constraints. A knowledge of basic food preparation techniques and equipment usage will assist the foodservice manger in planning and achieving these goals.

The extent of actual preparation and cooking done on the premises depends on the type of foodservice system (conventional, commissary, ready-prepared, or assembly/serve), as explained in Chapter 2. Even in the conventional system, in which all or most of the food is prepared, the production methods and equipment needs vary with the type and size of the foodservice. For example, in some retirement homes in which all residents are served at the same, set time, it is necessary to prepare larger amounts of food at one time and with different equipment than in a restaurant where much of the food is prepared "to order" during an extended service period.

Quantity is the element that introduces complexity to food preparation in the foodservice system. Therefore, the foodservice manager should not only be knowledgeable about basic cooking methods but should understand the time-temperature relationships and quality control challenges inherent in quantity food production. The principles of food preparation in large quantity are much the same as that for small quantities except that some procedures differ because of the larger amounts of food involved. Mechanized equipment is essential for heavy processes and for time-consuming procedures, especially in the larger operations. Pie rollers, automatic bread and roll shaping equipment, steam-jacketed kettles with stirring paddles, timers on steam-cooking equipment, metering devices for measuring water, and high-speed vegetable cutters are examples of labor-saving equipment in use. The convection ovens and compartment steamers used in many foodservices reduce the time required for cooking. Nontransfer cooking, in which foods are cooked in the same pans used for serving, saves time and possible damage to the food quality through additional handling. Equipment is thoroughly reviewed in Chapter 11.

As discussed previously, a list of steps required in the transformation of raw food into acceptable finished products includes storage, thaw time, pre-preparation, preparation, assembly, and holding prior to serving. The extent of ingredient assembly, pre-preparations, and preparation of the food before being delivered to the production area depends on the size and physical layout of the foodservice operation. With some foods, such as fresh fruit and certain salads and sandwiches, the cooking procedure is omitted. However, most foods do require cooking and, at this point, quality control becomes critically important. Constant vigilance is required to make sure that food is cooked properly, that it is well seasoned, and that it is not held too long before service.

Objectives of Cooking in Food Production

The vast majority of food production involves at least some cooking. The basic objectives of cooking are as follows:

- Enhance the aesthetic appeal of the raw food product by maximizing the sensory qualities of color, texture, and flavor.

- Destroy harmful organisms to ensure that the food is microbiologically safe for human consumption.
- Improve digestibility and maximize nutrient retention.

It is the responsibility of the food manager, through planning and control, to ensure that these objectives are met each time a menu item is produced.

It is essential then that the food manager be knowledgeable about the physical and chemical properties of food and about the basic principles and techniques of food preparation. Appendix A is a concise summary of heat transfer and basic cooking methods. The appendix also includes tables on yields and common abbreviations used in food production. Other sources of food preparation information are listed at the end of Appendix A.

RECIPE DEVELOPMENT

Standardized Recipes

An important tool for production control is the *standardized recipe,* sometimes referred to as a *recipe formulation.* It enables the manager to predict quality, quantity, and the portion cost of the finished product, and it simplifies purchasing. Use of standardized recipes is helpful when training new or substitute production employees and makes management less dependent on the whims of and changes in personnel. Accuracy in the use of standardized recipes and in weighing and measuring ingredients takes the guesswork out of quantity food production. Clientele expect and should be able to depend on a food item being the same each time it is selected or served.

Standardized recipes are required if ingredients are assembled in a central area and are essential in a computer-assisted production system because the recipe is the trigger for other functions, such as forecasting. In a computerized system, each recipe and each ingredient has an identification number that is correlated with purchasing, inventory, and cost records. Production scheduling is simplified if recipes include all information needed to produce the menu item.

The standardized recipes must be *used* by the employees, too, if the system is to be effective. For example, if the daily cost report is calculated on the basis of standardized recipes provided by the production personnel, but employees change the amounts and ingredients, a discrepancy will appear that does not reflect the true financial picture.

Format. A suitable recipe form or format that provides all information needed for production of the menu item should be selected. An orderly arrangement of this information should be developed and the same general pattern followed for all recipes in the file. Each foodservice should decide on the format best suited to its operation and use this format consistently.

In most recipes, the ingredients are listed in the order in which they are used. A block arrangement, in which ingredients that are to be combined are grouped together, is helpful, and separating these groups with space or lines makes following recipes easier and faster. Listing the procedures directly across from the ingredients involved simplifies preparation and enables clear directions to be written in a minimum number of words. Figure 6.1 illustrates this suggested format. Some recipe writers like to number the steps in the procedure.

Certain information is essential, regardless of the form in which the recipe is written. The following are suggestions for material that should be included.

Recipe Title. The title should be printed in large type and either centered on the page or placed on the left or right side at the top of the page as shown in Figure 6.1. The recipe identification code for ease in filing is also placed here.

Yield and Portion Size. The total recipe yield may be given in measure, weight, number of pans, or number of portions. The portion size may be in weight, measure, or count. Yield and portion size calculations are presented later in this chapter.

Baking Time and Temperature. This information is listed at the top of the page so the preheating of the oven and the scheduling of baking can be determined without reading the entire recipe. Some recipe writers repeat the baking time and temperature in the instructions so the cooks can see them while working with the ingredients.

Ingredients and Quantities. The amounts of ingredients may be listed first, followed by the names of ingredients, as in Figure 6.1, or the ingredients may be given first with the quantities arranged in one or more columns to accommodate different yields (Figure 6.2). For the sake of accuracy, however, there should not be more than three ingredient amount columns on one card. Too many columns increase the chance of errors occurring when reading amounts and crowd the space needed to give complete directions for preparation.

Names of ingredients should be consistent. Descriptive terms are used to define clearly the kind and form of each ingredient. In some recipes, the term *before* the name of the ingredient designates the form as purchased or that the ingredient has been cooked or heated before using in the product. Examples are *canned* tomatoes, *frozen chopped* broccoli, *hot* milk, *boiling* water, and *cooked* turkey. The descriptive term is placed *after* the ingredient to indicate processing after the ingredient is weighed or measured—onions, *chopped*; eggs, *beaten*; or raw potato, *grated.* It is important in some recipes to designate AP (as purchased) or EP (edible portion). For example, 15 pounds (AP) of fresh broccoli would be 10 pounds (EP) or less. Whatever system is used, it should be consistent and understood by those using the recipes. Abbreviations should be consistent and easily understood, such as "qt" for quart or "lb" for pound. Tables A.4 and A.5 in Appendix A provide information on product yield and common abbreviations, respectively.

Procedures. Directions for preparation of the product should be divided into logical steps and are most effective when placed directly across from the ingredients involved. Procedures should be clear and concise so that employees can easily read and understand them. It is helpful if basic procedures are uniform in all recipes for

PASTA WITH VEGETABLE SAUCE

Yield: 50 portions or 2½ gal sauce   *Portion:* 6 oz sauce + 4 oz pasta

Ingredient	***Amount***	***Procedure***
Onions, chopped Olive oil	2 lb 1 cup	Sauté onion in oil until tender, using a steam-jacketed or other large kettle.
Oregano, dried, crumbled Basil, dried, crumbled Pepper, black Garlic powder Salt Bay leaves	¼ cup ½ oz (1/2 cup) 1 Tbsp 1 Tbsp 1 oz (1½ Tbsp) 2	Add spices to onion. Mix well.
Tomato juice Tomato paste	5 46-oz cans 1 lb 12 oz	Add tomato juice and paste to spices and onion. Heat to boiling. Reduce heat and simmer uncovered for 15–20 minutes. Remove bay leaves.
Zucchini, sliced Mushrooms, sliced	2 lb 8 oz 1 lb 8 oz	Add zucchini and mushrooms just before serving. Cook only until zucchini is tender.
Pasta Water, boiling Salt Vegetable oil (optional)	5 lb 5 gal 5 oz 3 Tbsp	Cook pasta according to directions. Serve 6 oz sauce over 4 oz pasta.

Approximate nutritive values per portion

Calories (kcal)	Protein (grams)	Carbohydrate (grams)	Fat (grams)	Cholesterol (mg)	Sodium (mg)	Iron (mg)	Calcium (mg)
265	8.2 (12%)	46.8 (69%)	5.7 (19%)	0	1023	3.4	47

Notes
- 2 cups finely chopped fresh basil may be substituted for dried basil.
- 4 oz. (2 cups) dehydrated onions, rehydrated in 3 cups water, may be substituted for fresh onions.

Variations
- **Italian Sausage Pasta.** Delete olive oil, salt, and zucchini. Brown 5 lb bulk Italian sausage in steam-jacketed kettle. Drain. Add onions to sausage and continue to cook until onions are tender. Add spices, tomato juice, and tomato paste. Simmer for 15–20 minutes. Add meat sauce to 6 lb 8 oz cooked pasta (approximately 3 lb AP) and mix gently. Be careful not to overcook pasta. Scale 12 lb per 12 × 20 × 2 inch pan. Sprinkle 1 lb shredded mozzarella cheese over each pan and place in low oven until cheese is melted. Suggested pasta combination: 1 lb (AP) rotini, 1 lb (AP) bow ties, 1 lb (AP) rigatoni.
- **Pizza Sauce.** Reduce olive oil to 4 oz. Delete zucchini and mushrooms. Increase tomato paste to 5 lb 8 oz and decrease tomato juice to 5¼ qt. Add 1 tsp fennel seed, 2 Tbsp sugar, 1 tsp paprika, and ¼ tsp cayenne. Spread 1 qt sauce on top of 18 × 26-inch pizza dough before adding toppings.
- **Sandwich Tomato Sauce.** Delete salt, zucchini, and mushrooms. Add 1 Tbsp sugar. Reduce olive oil to 1 Tbsp, onions to ½ cup, oregano to 1 tsp, basil to 1 Tbsp, pepper to ½ tsp, garlic powder to ½ tsp, bay leaf to 1, tomato juice to 1¼ qt, and tomato paste to 1½ cups. Yield: 50 1-oz servings.

Figure 6.1 Recipe format showing block arrangement. Nutritive value, notes, and variations are given at the end of the recipe.
Source: Reprinted with permission from *Food for Fifty,* 9th edition, by Grace Shugart and Mary Molt. Copyright © 1993 by Merrill/Prentice Hall.

Applesauce Cake

Desserts No. Ck–3
Portion: 2 × 2¾ in.
Cut 6 × 8

Oven temperature: 350 °F
Time: 30–35 minutes

Ingredients	*2 pans*	*3 pans*	*Procedure*
Shortening Sugar	1 lb 7 oz 2 lb 14 oz	2 lb 3 oz 4 lb 5 oz	Cream 5 min. on medium speed, with paddle.
Eggs	2 cups	3 cups	Add and beat 5 min. on medium speed.
Applesauce	2 qt + ½ c	3¼ qt	Add gradually on low speed. Beat 1 min. on medium speed after last addition. Scrape down.
Cake flour Salt Soda Cinnamon Nutmeg Cloves Raisins	2 lb 14 oz 4 tsp 1 oz 1 Tbsp 1½ tsp 1½ tsp 12 oz	4 lb 5 oz 2 Tbsp 1½ oz 4½ tsp 2¼ tsp 2½ tsp 1 lb 2 oz	Sift dry ingredients together and mix with raisins. Add to creamed mixture gradually on low speed. Beat 2 min., medium speed, after last addition. Scrape down once.
Total wt	13 lb 6 oz	20 lb 2 oz	Weigh into greased baking pans, 12 × 22 × 2 in., 6 lb 8 oz/pan.

Figure 6.2 Recipe format with columns for two quantities. Total weight of batter is helpful in adjusting recipes. Preparation steps may be numbered, if space permits.

similar products. For example, white sauce is basic to many other sauces and is an ingredient in many menu items. The procedures on the recipe should be worded the same in each recipe. Likewise, there are several basic procedures in baked products, such as those of creaming fat and sugar or of combining dry and liquid ingredients, that should be the same on all recipes using them.

Timing should be given for procedures in which mixers, steamers, or other mechanical equipment is used. For example, "cream shortening and sugar on medium speed for 10 minutes" or "cook on low heat until rice is tender and all water is absorbed, about 15 to 20 minutes."

Panning instructions should include the weight of product per pan to help in dividing the product equally into the required number of pans. For example: "Scale batter into two prepared 12 × 18 × 2-inch baking pans, 4 lbs 10 oz per pan." When layering ingredients in baking pans for a casserole-type entrée, it is helpful if the weight or measure of each layer is given. For example: "Place dressing, sauce, and chicken in two 12 × 20 × 2-inch counter pans, layered in each pan as follows: 4 lbs 8 oz dressing, 1½ qt sauce, 3 lbs chicken, 1¼ qt sauce."

Additional information that is not essential to the recipe but may be helpful such as substitution of ingredients, alternate methods of preparation, or comments about the appearance of the product, such as "These cookies puff up at first, then flatten out with crinkled tops," can be added as footnotes. Variations on a basic recipe usually are included at the end of the recipe, and may include tips on how to plate or garnish the product.

Decisions regarding the size and form of the recipe card or sheet, the format to be followed, and the manner of filing the recipes are contingent on the needs of the operator. Cards 4 × 6 inches and 5 × 8 inches are popular sizes, and heavy typing paper, 8½ × 11 inches, is used in some operations. In deciding on a size and format, keep in mind that the recipes will be used by cooks and other employees who will be busy weighing and mixing ingredients and may not be able to read a small crowded card easily. Recipes should be typed or printed and should be readable at a distance of 18 to 20 inches. Recipes that are used in the production or ingredient assembly areas should be placed in clear plastic covers to keep the copy clean.

In foodservices using a computer-assisted system, recipes are printed as needed and for the quantities required for the day's production. Because the printout is generated each time the recipe is used, it is considered a working copy and does not need a protective covering. An example of a computer-generated recipe is shown in Figure 6.3.

CHILI CON CARNE RECIPE CODE – 14-01-0-023-7 STATUS – STANDARDIZED 1

/ / -****** -*** ***

EQUIPMENT-STEAM KETTLE
COUNTER PANS-LIDS

RECIPE SOURCE–

04/16/86 10.20 AM 751124

NUMBER OF PORTIONS	68
PORTION SIZE / COST	.560 LBS. / $0.3044
MEAL PATTERN ALLOWANCE	1 PORTION
SUGGESTED SERVING UTENSIL	6 OZ LADLE
PAN SIZE	10X12X6
NUMBER OF PANS	
WEIGHT PER PAN	LBS.
HANDLING LOSS	6.00 PERCENT
MINIMUM BATCH	
MAXIMUM BATCH	
FORECAST UNIT	
*TOTAL RECIPE WEIGHT / COST	40.5 LBS. / $20.6992
TOTAL RECIPE VOLUME	

CODE	PERCENT	INGREDIENT	WEIGHTS AND MEASURES	AP/EP	STEP	PROCEDURE
0012113719	22.49	GROUND BEEF CHILI BULK	9.1 LBS 13.6 LBS	EP AP	A	1. COOK MEAT IN KETTLE UNTIL IT LOSES ITS COLOR. STIR OFTEN TO PREVENT LUMPS FROM FORMING. DRAIN OFF FAT.
0052001211	1.68	ONIONS CHOPPED FRESH	0.68 LBS		B	2. COMBINE WITH MEAT.
0052000826	0.02	GARLIC MINCED FRESH	1.0 TSP			
0082054100	0.34	SALT	0.14 LBS			
0082060401	0.57	CHILI POWDER	0.23 LBS			
0082093008	0.34	SUGAR GRANULATED	0.14 LBS			
0082060801	0.08	CUMIN GROUND	0.03 LBS			
0082063109	0.02	PEPPER BLACK GROUND	1.8 TSP			
0072011602	17.12	TOMATOES DICED CND	6.9 LBS		C	3. ADD TO MEAT MIXTURE.
0072011751	15.40	TOMATO PUREE CND	6.2 LBS			4. SIMMER 1 HOUR OR MORE. STIR
0072010363	33.90	BEANS RED CND	13.7 LBS			FREQUENTLY.
0000000001	8.05	WATER	3.3 LBS			5. TAKE UP IN SOUP PANS. COVER.

Figure 6.3 Example of computer-generated recipe.
Courtesy of Kansas State University Residence Halls.

The format for recipes in this type of system depends on the software purchased, so the format should be considered when comparing different software packages.

A recipe is considered **standardized** only when it has been tried and adapted for use by a given foodservice operation. Quantity recipes are available from many sources, such as cookbooks, trade journals, materials distributed by commercial food companies from their own experimental kitchens, and from other foodservice managers. Regardless of the source, each recipe should be tested and evaluated, then standardized and adjusted for use in a particular situation.

The first step in standardizing a new recipe is to analyze the proportion of ingredients and clarity of instructions, and to determine whether the recipe can be produced with the equipment and personnel available.

It is also important to assess the portion size defined for the original recipe to determine whether it is appropriate for the customer and financial objectives of the operation. For example, a recipe obtained from a restaurant would likely define portions in excess of what would be appropriate for a school or nursing home.

The recipe should then be tested. When doing so, make sure ingredients are weighed and measured accurately and that procedures are followed exactly. The yield, number, and size of portions as well as problems with preparation should be recorded.

Recipe Yield. Recipe yield is simply a measure of the total amount produced by a recipe. Recipe yield can be expressed in weight, measure, or count. For example, the yield of a soup recipe would be measured and expressed in quarts; a cake recipe would be measured and expressed in size and number of pans; and the yield of a cookie recipe is measured and expressed by count. Figure 6.4 is an example of a recipe evaluation form that may be used to document actual recipe yield. Along with yield analysis, the finished products should be evaluated for acceptability based on predetermined quality standards.

Quality Standards. Quality standards are measurable statements of the aesthetic characteristics of food items, and they serve as the basis for sensory analysis of the prepared product. Quality indicators include appearance, color, flavor, texture, consistency, and temperature. Figure 6.5 is an example of a score card for evaluating cakes and muffins and includes quality standards for shape, volume, color, texture, and flavor.

If the tested product is deemed suitable, the recipe is then adjusted to the quantities needed to meet production demand.

Recipe Adjustment

Three methods commonly used to adjust recipes are the factor method, the percentage method, and direct-reading measurement tables.

Factor Method. In the factor method, the quantities of ingredients in the original recipe are multiplied by a conversion factor, as explained in the following steps:

Recipe Evaluation Card

Please Return This Card to the Test Kitchen as Promptly as Possible.

Product ______________________ Residence Hall ______________

Date ______________

Quantity prepared ______________________________

Did you obtain yield as stated in recipe? ______________________

If not, what quantity was obtained? ______________________

Do you consider size of portion adequate? ______________________

If not, what change would you suggest? ______________________

____

Was product generally well accepted? ______________________

____

Further comments on recipe—for example—ease of using recipe; problems encountered; suggestions for changes in procedure, kind and/or amount of ingredients, etc.:

(Consult cooks for suggestions) ______________________

____

____

____

____

____

____

Reporting Supervisor

Figure 6.4 Recipe evaluation card.
Courtesy of Central Food Stores, University of Illinois.

Step 1 Convert all volume measurements to weights, when possible. For example, 3 cups of water weighs 1 pound 8 ounces. For ease in figuring, weights should be expressed in pounds and decimal components of pounds; 1 pound 8 ounces is 1.5 pounds. Add the weights of all ingredients.

Step 2 Divide the desired yield by the known yield of the original recipe to obtain the conversion factor. For example, if you have a recipe for pie crust for 12 two-crust pies that you wish to change to 66 two-crust pies, as shown in Table 6.1, divide the desired yield (66) by the known yield (12) to obtain the factor of 5.5.

Step 3 Multiply the amount of each ingredient in the original recipe by the factor, which in the example would be 5.5.

Score Card for Cake

Date ______________

Factor	*Qualities*	*Standard*	*Sample No.* 1	2	3	*Comments*
I. External appearance	Shape, symmetrical, slightly rounded top, free from cracks or peaks	10				
	Volume, light in weight in proportion to size	10				
	Crust, smooth uniform golden brown	10				
II. Internal appearance	Texture tender, slightly moist, velvety feel to tongue and finger	10				
	Grain, fine, round, evenly distributed cells with thin cell walls, free from tunnels	10				
	Color, crumb even and rich looking	10				
III. Flavor	Delicate, well-blended flavor of ingredients. Free from unpleasant odors or taste	10				

Directions for use of score card for plain cake:

Standard	10	No detectable fault, highest possible score
Excellent	8–9	Of unusual excellence but not perfect
Good	6–7	Average good quality
Fair	4–5	Below average, slightly objectionable
Poor	2–3	Objectionable, but edible
Bad	0–1	Highly objectionable, inedible

____
Signature of evaluator

Figure 6.5 Suggested score card for evaluating cakes or muffins.

Step 4 A check on the accuracy of the figures can be made by multiplying the total weight of ingredients in the original recipe by the factor and comparing with the total weight of ingredients in the new recipe. If the two are not the same, check the calculations before preparing the product.

Step 5 Change weights back to pounds and ounces and convert to quarts, cups, or other volume measures for ingredients that are more easily measured than weighed.

Step 6 Round off unnecessary or awkward fractions that would be difficult to measure or weigh as far as accuracy permits.

Percentage Method. In the percentage method, the percentage of the total weight of the product is calculated for each ingredient, and once this percentage has been established, it remains constant for all future adjustments. Recipe increases and decreases are made by multiplying the total weight desired by the percentage of each ingredient.

The percentage method is based on weights expressed in pounds and decimal parts of a pound. The total quantity to be prepared is based on the weight of each portion multiplied by the number of servings needed. The constant number used in calculating a recipe is the weight of each individual serving. A step-by-step procedure, as used at Kansas State University and reported by McManis and Molt (NACUFS J. 35, 1978), follows for adjusting a recipe by the percentage method:

Step 1 Convert all ingredients in the original recipe from measure or pounds and ounces to pounds and tenths of a pound. Make desired equivalent ingredient substitutions, such as frozen whole eggs for fresh eggs and powdered milk for liquid milk.

Step 2 Total the weight of ingredients in the recipe. Use edible portion (EP) weights when a difference exists between EP and as purchased (AP)

Table 6.1. Pastry—adjusting yield *from* 12 two-crust pies *to* 66 two-crust pies.

Ingredients	**Original Recipe 12 2-Crust Pies**	*Step 1* **Original Recipe in Weight**	*Step 2* **Original Recipe in Ounces**	*Step 4* **Each Amount in Step 2 Multiplied by Factor**	**New Recipe in Weight**	*Step 5* **New Recipe Rounded Weight and Measure**
Pastry flour	5 lb	5 lb	80	440.0	27 lb 8 oz	27 lb 8 oz
Salt	1¼ oz	1¼ oz	1¼	6.9	6.9 oz	7 oz
Shortening	3 lb	3 lb	48	264.0	16 lb 8 oz	16 lb 8 oz
Water	3 c	1 lb 8 oz	24	132.0	8 lb 4 oz	1 gal ½ c
Total weight		9 lb 9¼ oz	153¼	842.9	52 lb 11 oz	52 lb 11 oz

Step 3: Conversion Factor

$$\frac{66 \text{ (new)}}{12 \text{ (original)}} = \frac{11}{2} = 5\tfrac{1}{2} = 5.5$$

Total Weight in Step 2 × Factor
153¼ oz × 5.5 =
153.25 × 5.5 = 842.875 = 842.9 oz

Essential information needed:
Water: 1 c = 8 oz
4 c = 1 qt = 2 lb
4 qt = 1 gal = 8 lb

1 lb = 16 oz

Source: Procedure for adjusting recipe yield, Department of Institution Administration, Michigan State University.

weights. For example, the weight of onions or celery should be the weight after the foods have been cleaned, peeled, and are ready for use. The recipe may show both AP and EP weights, but the edible portion is used in determining the total weight.

Step 3 Calculate the percentage of each ingredient in relation to the total weight. Repeat for each ingredient. Use this formula:

$$\frac{\text{Individual ingredient weight}}{\text{Total weight of ingredients}} = \text{percentage of each ingredient}$$

Sum of percentages should total 100%.

Step 4 Check the ratio of ingredients, which should be in proper balance before going further. Standards have been established for ingredient proportions of many items.

Step 5 Determine the total weight of the product needed by multiplying the portion weight expressed in decimal parts of a pound by the number of servings to be prepared. To convert a portion weight to a decimal part of a pound, divide the number of ounces by 16 or refer to a decimal equivalent table (Table 6.2). For example, a 2-ounce portion would be 0.125 pound. This figure multiplied by the number of portions desired gives the total weight of product needed. The weight is then adjusted, as necessary, to pan size and equipment capacity. For example, the total weight must be divisible by the optimum weight for each pan. The capacity of mixing bowls, steam-jacketed kettles, and other equipment must be considered in determining the total weight. Use the established portion, modular pan charts, or known capacity equipment guides to determine batch sizes to be written.

Step 6 Add estimated handling loss to the weight needed. An example of handling loss is the batter left in bowls or on equipment. This loss will vary according to the product being made and preparation techniques of the worker. Similar recipes, however, produce predictable losses, which with some experimentation can be accurately assigned. The formula for adding handling loss to a recipe follows:

$$(100\% - \text{Assigned Handling Loss }\%)\ X = \text{Desired Yield}$$

$$X = \frac{\textit{Desired Yield}}{100\% - \text{Handling Loss }\%}$$

Example: Butter cake has a 1% handling loss. Desired yield is 80 lbs (or 600 servings).

$(100\% - 1\ \%)\ X = 80$ lbs

$.99\ X = 80$

$X = 80/.99$

$X = 80.80$ total lbs of ingredients to produce 80 lbs of batter

Step 7 Multiply each percentage number by the total weight to give the exact amount of each ingredient needed. Once the percentages of a recipe have been established, any number of servings can be calculated and the ratio of ingredients to the total will be the same. One decimal place on a recipe

Table 6.2 Ounces and decimal equivalents of a pound.

Ounces	Decimal Part of a Pound	Ounces	Decimal Part of a Pound
1/4	0.016	8 1/4	0.516
1/2	0.031	8 1/2	0.531
3/4	0.047	8 3/4	0.547
1	0.063	9	0.563
1 1/4	0.078	9 1/4	0.578
1 1/2	0.094	9 1/2	0.594
1 3/4	0.109	9 3/4	0.609
2	0.125	10	0.625
2 1/4	0.141	10 1/4	0.641
2 1/2	0.156	10 1/2	0.656
2 3/4	0.172	10 3/4	0.672
3	0.188	11	0.688
3 1/4	0.203	11 1/4	0.703
3 1/2	0.219	11 1/2	0.719
3 3/4	0.234	11 3/4	0.734
4	0.250	12	0.750
4 1/4	0.266	12 1/4	0.766
4 1/2	0.281	12 1/2	0.781
4 3/4	0.297	12 3/4	0.797
5	0.313	13	0.813
5 1/4	0.328	13 1/4	0.828
5 1/2	0.344	13 1/2	0.844
5 3/4	0.359	13 3/4	0.859
6	0.375	14	0.875
6 1/4	0.391	14 1/4	0.891
6 1/2	0.406	14 1/2	0.906
6 3/4	0.422	14 3/4	0.922
7	0.438	15	0.938
7 1/4	0.453	15 1/4	0.953
7 1/2	0.469	15 1/2	0.969
7 3/4	0.484	15 3/4	0.984
8	0.500	16	1.000

NOTE: This table is useful when increasing or decreasing recipes. The multiplication or division of pounds and ounces is simplified if the ounces are converted to decimal parts of a pound. For example, when multiplying 1 lb 9 oz by 3, first change the 9 oz to 0.563, by using the table. Thus, the 1 lb 9 oz becomes 1.563 lbs, which multiplied by 3 is 4.683 lbs or 4 lbs 11 oz.

is shown (e.g., 8.3 lbs) unless it is less than one pound, then two places are shown (e.g., 0.15 lb).

Tables 6.3 to 6.5 illustrate the expansion of a recipe for muffins from 60 to 340 servings.

Direct Reading Measurement Tables. The use of tables showing ingredient amounts for different numbers of portions saves time and simplifies recipe adjust-

Table 6.3 Original recipe for muffins *(yield: 60 muffins)*.

Ingredients	Amount
Flour, all-purpose	2 lbs 8 oz
Baking powder	2 oz
Salt	1 Tbsp
Sugar	6 oz
Eggs, beaten	4
Milk	1½ qt
Shortening	8 oz

ment. Conversion charts have been developed that give amounts in weight and measure in increments of 25 for 25 to 500 portions. Measurement tables are included in *Quantity Food Preparation: Standardizing Recipes and Controlling Ingredients* by Buchanan, and *Food for Fifty* by Shugart and Molt.

Adapting Small Quantity Recipes. Many quantity recipes can be successfully expanded from home-size recipes, but their development involves a number of carefully planned steps. Before attempting to enlarge a small recipe, be sure that it is appropriate to the foodservice and that the same quality can be achieved when prepared in large quantity and possibly held for a time before serving. Procedures should be checked because many home recipes lack detailed directions for their preparation. Before preparing the product, the extent of mixing, the time and temperature used in cooking or baking, and special precautions that should be observed and any other details that may have been omitted should be determined.

Enlarging the recipe in steps is more likely to be successful than increasing from a small quantity to a large quantity without the intermediate steps.

Suggestions follow for a step-by-step approach for expanding home-size recipes.

Step 1 Prepare the product in the amount of the original recipe, following exactly the quantities and procedures, noting any procedures that are unclear or any problems with the preparation.

Table 6.4 Percentage calculated on original recipe *(yield: 60 muffins)*.

Percentage	Ingredients	Measure	Pounds
35.79	Flour	2 lbs 8 oz	2.500
1.79	Baking powder	2 oz	0.125
0.67	Salt	1 T	0.047
5.37	Sugar	6 oz	0.375
6.27	Eggs	4	0.438
42.95	Milk	1½ qt	3.000
7.16	Shortening	8 oz	0.500
100.00	Total		6.985

6.985 lbs divided by 60 = .116 lb per muffin

Table 6.5 Expanded recipe for muffins *(yield: 340 muffins)*.

Percentage	Ingredients	Pounds
35.79	Flour	14.26
1.79	Baking powder	0.713
0.67	Salt	0.267
5.37	Sugar	2.14
6.27	Eggs	2.50
42.95	Milk	17.11
7.16	Shortening	2.85
100.00	Total	39.84

.116 lb per muffin × 340 = 39.44 lbs with 1% handling loss, 39.84 lbs batter needed, 39.84 lbs × % each ingredient = weight of ingredient

If using pounds and ounces, change decimal parts of pounds to ounces by using Table 6.2. If using measures for some ingredients, adjust to measurable amounts.

Step 2 Evaluate the product, using a written form such as that shown in Figure 6.4, and decide if it has the potential for the foodservice. If adjustments are necessary, revise the recipe and make the product again. Work with the original amount until the product is satisfactory.

Step 3 Double the recipe or expand to appropriate amount for the pan size that will be used and prepare the product, making notations on the recipe of any changes you make. For example, additional cooking time may be needed for the larger amount. Evaluate the product and record the yield, portion size, and acceptability.

Step 4 Double the recipe again, or if the product is to be baked, calculate the quantities needed to prepare one baking pan that will be used by the foodservice. If ingredients are to be weighed, home-size measures should be converted to pounds and ounces or to pounds and tenths of a pound before proceeding further. Prepare and evaluate the product as before.

Step 5 If the product is satisfactory, continue to enlarge by increments of 25 portions or by pans. When the recipe has been expanded to 100 or some other specific amount that would be used in the foodservice, adjustments should be made for handling or cooking losses. Handling loss refers to losses that occur in making and panning batters. About 3% to 5% more batter, sauces, and puddings are required to compensate for the handling loss. Cooking losses result from evaporation of water from the food during cooking. Soups, stews, and casseroles can lose from 10% to 30% of their water cooking. The actual yield of the recipe should be checked carefully. Mixing, preparation, and cooking times should be noted because these may increase when the product is prepared in large quantities. Preparation methods should be checked to see if they are consistent with methods used for similar products. An evaluation of the product should be made and its acceptance by the clientele determined before it becomes a part of the permanent recipe file.

Recipe Files

A master file of all recipes used by the foodservice should be maintained, with at least two complete sets available, one to be kept on permanent file in the office of the foodservice manager and the other to be available for production employees. Additional sets may be needed by other key personnel.

A system of classification and filing enables recipes to be located easily. Recipes can be coded by number or filed alphabetically under appropriate headings, such as appetizers, beverages, breads, desserts, eggs and cheese, fish, meat, pasta, poultry, salads, sandwiches, sauces, soups, and vegetables.

FORECASTING

The goal of forecasting is to estimate future needs using past data. Applied to foodservice, **forecasting** is a prediction of food needs for a day or other specific time period. Forecasting differs from tallying, which is a simple count of menu items actually requested or selected by the customers. Production planning begins with the menu and the production forecast. Other foodservice functions, such as purchasing, are triggered by the forecast. Sound forecasting is vital to financial management; it facilitates efficient scheduling of labor, use of equipment, and space.

Reasons for Forecasting

A great deal of lead time is needed to complete all phases of menu item preparation: purchasing, storage, thawing, pre-preparation, production, distribution, and final service. Forecasting serves as a means of communication with purchasing and food production staff to ensure that all of these stages are completed in a timely manner and that the final product meets standards of quality.

The purchasing representative needs to know how much food to order and when it needs to be available for use in the foodservice production area. The hot and cold food production staff(s) need to know how many servings of each menu item are needed, and in what form and for which service unit (e.g., cafeteria, vending, patients, catering) they are needed.

Accurate forecasting minimizes the chance of overproduction or underproduction—both of which have serious consequences. Without proper guidelines, production employees have a tendency to overproduce food for fear of running short of actual need. This can be a costly comfort measure. Leftover food is often held for later service or redirected to an alternate service unit such as the cafeteria or vending or catering services. Each choice is risky in that the food may not meet minimum quality standards at point-of-service, ultimately ending up in the garbage disposal.

Underproduction can be costly as well, and may result in customer dissatisfaction. To compensate for shortages, managers often substitute expensive heat-and-serve items, such as ready-prepared chicken cordon bleu. More serious than increased raw food costs, however, is the risk of upsetting a customer by providing him or her with a substitute menu item that was not ordered. Foodservice employees may get frustrated if food shortages occur too frequently, resulting in rushed, last-minute food preparation or delayed service.

In small health care organizations, such as long-term care facilities or hospitals, amounts to be produced can be determined by simple tally, especially if the patient census is stable and a nonselective menu is used. In large organizations with multiple service units, more sophisticated forecasting may be beneficial if there is wide variation in menu item demand. A simple tally system would be far too time consuming for these larger organizations. Regardless of the size and complexity of the foodservice organization, a good forecasting system is based on sound historical data that reflect the pattern of actual menu item demand in the foodservice operation.

Historical Data

Historical data are used to determine needs and to establish trends in all forecasting methods. To be of value, these data must be consistently and accurately recorded. Categories of data to collect vary depending on the type of foodservice organization, scope of services provided, and whether customers are allowed to select menu items. The following are a few examples of data categories for various organizations:

Restaurants/Cafeterias
- Customers served per meal
- Menu items selected per meal
- Beverage sales (types and amounts).

Schools
- Student enrollment
- Students purchasing full USDA meal
- A la carte items sold per lunch period
- Teachers and staff purchasing meals.

Hospitals
- Daily patient census
- Patients on therapeutic diets
- Daily patient admissions and discharges.

Vending Services
- Product placed in machine at each fill
- Total cash removed
- Food remaining in machine at each refill.

Figure 6.6 is an example of a form designed to document meal participation in a school lunch program. Over a period of time, a pattern of menu item demand or total meals served will emerge from the recorded data. This pattern, along with knowledge of pattern variance, will assist the production planner in making a valid estimate of future menu item needs. Factors influencing pattern variance include holidays, weather conditions, and special events.

DAILY PARTICIPATION RECORD
LUNCH PROGRAM
SCHOOL ____________ MONTH ____________

Days of operation	STUDENTS				ADULTS		2nd lunches	Absences (excluding kindergarten)
	Paid (1)	Reduced (3)	Free(2)	TOTAL	Program*	Nonprogram**		
1								
2								
3								
4								
5								
6								
7								
8								
9								
10								
11								
12								
13								
14								
15								
16								
17								
18								
19								
20								
21								
22								
23								
24								
25								
26								
27								
28								
29								
30								
31								
MONTHLY	TOTAL							

Average enrollment this month ________ (minus kindergarten)

Average enrollment – Average absences
= Average daily attendance

________ – ________ = ________

Total absences ÷ Days in session
= Average absences

________ ÷ ________ = ________

* Program adults: all foodservice workers. Adult meals are not reimbursable.

** Nonprogram adults: teachers, administrators, office workers, janitors, and any occasional visitors. Adult meals are not reimbursable.

Figure 6.6 Sample form for lunch participation in a school lunch program.
Courtesy of Beth Mincemoyer Egan, Manager–Food Services, Sun Prairie Public Schools, Sun Prairie, Wisconsin.

Criteria for Selecting a Forecasting Process

Careful planning and evaluation are essential in selection of the best forecasting method for a given foodservice operation. Numerous computer forecasting models have been developed during the past several years that have been a great aid to the foodservice manager. However, regardless of whether a manual or computer-assisted method is chosen, several factors should be considered before deciding on a forecasting system. These factors include cost, accuracy, relevancy, lead time, pattern of food selection, ease of use, level of detail, and responsiveness to changes. Table 6.6 is a summary of the considerations related to each of these factors.

Forecast Models

Types of forecasting models for use with manual or electronic systems include moving average, exponential smoothing, regression, and autoregressive moving average (Box-Jenkins). These models are mathematical descriptions of meals served or of menu item selection behavior. The information for the mathematical models is based on historical data and is expressed as an average of past service or selection behavior.

Table 6.6 Criteria for selecting a forecasting system

Factor	Considerations
Costs	Development, implementation, and system operational costs (e.g., data collection, analysis) are reasonable; that is, within budgetary guidelines Training and education for staff
Accuracy/relevancy	Past data and food selection patterns are relevant and accurately reflect current demand
Lead time	System allows adequate time for purchasing, delivery, and production Accounts for perishability of food items
Pattern of behavior	System can be adjusted for changes in menu item demand as a result of seasonality and consumer preference
Ease of use	Use of system is easily understood What knowledge and skills are required to operate system?
Level of detail	System can generate desired forecasts What is to be forecasted?
Responsiveness	System generates accurate information on a timely basis

Type of Model	Mathematics	Examples of Forecasting Using Actual Data								
Moving average	$F = \frac{\sum_{i=1}^{N} Ai}{N}$ F = Next forecast Σ = Summation i = Time unit of forecast (i.e., Mondays, Tuesdays, etc.) N = Number of observations A = Actual demand	Actual demand data	Week #	S	M	T	W	R	F	S
			1	104	112	128	142	133	133	117
			2	107	125	134	147	137	159	145
			3	137	155	155	150	150	153	134
			4	122	134	142	135	123	125	116
			5	108	128	134	140	147	146	121
		Week #5 forecast (N = 4)	5	118	132	140	144	136	143	128
		Forecast error	5	10	4	6	4	–11	–4	7

Figure 6.7 Moving average forecasting model.
Source: Adapted from Messersmith, A. M. and Miller, J. L.: *Forecasting in Foodservice.* Copyright © 1992. New York: John Wiley & Sons, Inc. Reprinted by permission of John Wiley & Sons, Inc.

The moving average and exponential smoothing models are commonly used in foodservices for production forecasting. Figure 6.7 illustrates the calculations for the moving average model and gives an example of how this method is used with past data from a small hospital. The number of customers served from a foodservice is generally different on each day of the week. For this reason, forecasts are calculated for intervals of seven days. For example, in a hospital, data collected on Mondays are used to forecast needs for future Mondays.

The *moving average* model is referred to as a time series method of forecasting and is easy to use. Using records from the past, a group of data is averaged and used as the first forecast. The next forecast is calculated by dropping the first number and adding the next. This process continues for all data available.

The *exponential smoothing* model is another time series model, similar to the moving average technique except that it accounts for seasonality of data and adjusts for forecast error. This results in a higher level of forecast accuracy. The simple exponential smoothing model predicts the next demand by weighting the data; more recent data are weighted more heavily than older data. The factor used to weight the data is referred to as *alpha.* Alpha is determined statistically and, in foodservice forecasting, is generally valued at 0.3. The purpose of alpha is to adjust for any errors in previous forecasts

Regression and *autoregressive moving average* models are sophisticated, statistical methods in which past data are analyzed to determine the best mathematical approach to forecasting. These methods generally require the assistance of a statistician and are used with computer-assisted forecasting systems. These two methods are beyond the scope of this textbook.

QUANTITIES TO PRODUCE

The forecast is the basis for determining quantities of menu items to be prepared and foods to be purchased or requisitioned from the storerooms. An estimate at the time the major food orders are placed is later adjusted, one to two days prior to the day of production, for more accurate decisions on amounts of food to be prepared. The amount of food needed is based on the number of persons to be served, portion size, and the amount of waste and shrinkage loss in the preparation of foods.

Recipes adjusted to the predicted number of portions needed provide much of this information. Most quantity recipes for noncomputer systems are calculated in modules of 50 or 100 or, in foods such as cakes or casserole-type entrées, to pan sizes and equipment capacity. For example, if a recipe produces two sheet cakes, which can be cut into 30 or 32 servings each, three cakes (or one and one-half times the recipe) would be required for 75 portions. When very large quantities are produced, the amount to prepare in one batch is limited to the capacity of the production equipment.

In foodservices that use a computer, recipes are printed daily and, when appropriate, provide amounts for the exact number of individual portions, or are adjusted to the number of pans or other modules required to serve the predicted numbers. To be effective, computer-assisted programs include recipes for all menu items offered, including fresh vegetables and fruits, salads, relishes, and meats, such as roast beef or baked pork chops. Quantities to purchase or requisition are readily available from these computer-generated recipes. In foodservices not having a computer-assisted system, standardizing and calculating recipes for more than one amount lessens the need for refiguring the quantities for each forecast.

A general procedure for determining amounts of meats, poultry, fruits, and vegetables follows:

Step 1 Determine the portion size in ounces.

Step 2 Multiply portion size by the estimated number to be served and convert to pounds. This is the edible portion (EP) required. EP may also be given in the standardized recipe.

$$\frac{\text{ounces} \times \text{number of portions}}{16\text{ oz}} = \text{number of pounds edible portion (EP) required}$$

Step 3 To determine the amount to order, divide the EP weight by the yield percentage (or the weight in decimal parts of a pound of ready-to-eat or ready-to-cook product from ONE pound of the commodity as purchased). The *USDA Food Buying Guide for School Food Service,* PA-1257, is a good source for yield information. A yield guide is included in Appendix A.

$$\frac{\text{EP weight}}{\text{yield}} = \text{amount to order}$$

Step 4 For foods to be purchased, convert the amount needed to the most appropriate purchase unit (e.g., case, crate, or roast). If the food is to be used for other menu items, combine the amounts and then convert to purchase units.

As an example, if three-ounce portions of fresh asparagus are needed for 50 people, you would calculate the amount to purchase in the following way:

1. $\dfrac{3 \text{ oz} \times 50 \text{ portions}}{} = \dfrac{150 \text{ oz}}{16 \text{ oz}} = 9.375$ lbs of EP needed
2. $\dfrac{9.375 \text{ lbs needed}}{0.53 \text{ lb yield from 1 lb as purchased}} = 17.68$ lbs to purchase
3. Convert to purchase unit, 18 lbs to 20 lbs

In a computer system, these figures would be calculated automatically from the forecast, portion size, and yield data.

PRODUCTION SCHEDULING

Once the needed quantity of food is known and recipes are obtained from the file or from the computer, supplies are requisitioned or ordered, as previously discussed. The next step in production planning is scheduling of food preparation. Careful planning ensures the efficient use of employee time and equipment, and a minimum of production problems. Foods that are ready for service at the scheduled time without undue holding are superior in quality to those prepared too early.

Production planning and scheduling require a knowledge of the steps through which a product must go and the time required for each, as well as the steps that can be completed early without affecting the quality of the food. Most menu items go through part or all of the following steps:

- *Storage:* dry, refrigerated, freezer
- *Thaw time*
- *Assembly:* weighing or measuring ingredients
- *Pre-preparation:* vegetable cleaning, peeling, cutting, chopping, preparing pans
- *Preparation:* mixing, combining ingredients, panning
- *Cooking:* baking, frying, broiling, steaming, simmering
- *Finishing:* setting up salads, portioning desserts, slicing meats, packaging food for freezing
- *Storage prior to serving:* heated, refrigerated, frozen.

Up to the point of final cooking, many steps can and should be scheduled early, possibly the day before. Each recipe should be broken down into production stages to determine which steps can be done in advance. Preparation of casserole-type entrées should be scheduled so that the cooking of all recipe components is completed in time for final assembling and baking; for example, browning the beef cubes, dicing and cooking the vegetables, and preparing the topping for beef pot pie; or cooking ground beef, tomato sauce, and pasta for a casserole.

For some foods, it may be better to assemble the product in batches as needed for final cooking and serving. Combining the entire amount may reduce the flexibility of the individual ingredients if there is an overproduction. For example, the ground beef could be used for tacos, the pasta for a salad, and the tomato sauce in chili. These would then be "new" menu items rather than "leftovers." In commercial cooking, where food may be prepared in relatively small quantities because of the uncertain number of portions needed or if the menu includes a large selection of items, the finished product may be cooked to order or in amounts compatible with the anticipated patronage.

Cheese balls, croquettes, and similar foods must be shaped and chilled for several hours before deep-fat frying. Other procedures, such as breading of chicken, cutlets, or fish prior to cooking; shaping of meat balls; and filling of deviled eggs are examples of time-consuming procedures that can be scheduled early. Some menu items or ingredients can be prepared two to three days ahead, but care must be taken not to schedule early preparation if the quality or safety of the food is endangered. If frozen foods are to be defrosted, time must be allowed for thawing in the refrigerator.

When scheduling the cooking of foods in quantity, allowances must be made for the time required for heat to penetrate. For example, it will take longer for heat to reach the center of a pan with 50 servings of scalloped potatoes than a casserole serving six. Sheet cake baked in an 18 × 26-inch pan needs more cooking time than the same amount of batter baked in 9-inch layer cake pans. Four beef roasts in an oven require more cooking time than just one. Size of container, then, and oven load affect the total cooking time.

Final cooking should be scheduled so that only food needed for immediate service is cooked at one time. Batch cooking of vegetables in small amounts, baking of pans of entrées at intervals, and continuous deep-fat frying "to the line" or "to order" are examples. Some cooks are more comfortable if all food is cooked at once, but batch cooking should be scheduled and employees encouraged to recognize that the quality of the prepared food is better when it has not been held for long periods.

Due to variations in the complexity of menu items from day to day, production should be planned several days ahead to distribute the work load evenly. An increasing number of foodservices are using advance production to equalize workloads. Food preparation is scheduled during slack periods, and the food is frozen for later use. Such a program should not be attempted, however, without adequate freezing and refrigeration equipment and without some knowledge of foods that freeze well and the effect of storage length on different foods. Proper cooling, packaging, and freezing are especially important to the quality and safety of the finished products. The foods, of course, should be prepared under close supervision so that adequate sanitation and processing precautions are observed. If the product will be stored for any length of time, a system of inventory control of the frozen prepared foods should be implemented.

Production Schedules

A production schedule, sometimes called a production sheet or work sheet, lists in detail the items to be produced for the current day's menu plus any advance preparation needed. This schedule should include, as a minimum, the menu items, amounts to prepare, cooking times, special instructions for preparation and serving, and the name of the employee assigned to prepare each item. Some production schedules include space for recording quantities of food prepared and amounts left after the serving is completed. This information is important to the menu planner or production supervisor who must plan for incorporating these foods into the menu for the next meal or the next day, or to freeze for later use. Also added on some production schedules is information concerning weather, holidays, or special events that might assist in explaining the quantities prepared and served.

A single schedule for a small foodservice may include preparation and cleaning assignments for all employees, with production times set for specific times of the day to assist employees in planning their work. In a large, more complex operation, schedules are made for each department and usually are quite detailed. Figure 6.8 is an example of a production schedule that also serves as a production record. If cycle menus are used, time is saved by setting up orders and production schedules for each cycle as a "package" to be used with the cycle. Adjustments, of course, may be needed if the menu or the number of persons to be served changes.

Production Meetings

A meeting with appropriate staff and employees to discuss the menu and production plans heightens the effectiveness of the written production schedule. Such meetings generally do not need to be long, but they should be held regularly and at a time when activity in the production area is at a minimum. At meeting time, the menu can be explained and special instructions given for the items as needed. Employees also have an opportunity to discuss the schedule and any production problems they may anticipate.

No amount of paperwork can replace the human element in food production. Food must be prepared by people, and no matter how carefully plans are made and how many instructions are written, a manager must follow through to be sure that the menu as served measures up to the menu as planned.

PRODUCTION CONTROL

Ingredient Assembly

Central assembly of ingredients for food production has been found to be cost effective in many operations. In this system, the ingredients needed for recipes for the day's production and for advance preparation are weighed, measured, and assembled in a central ingredient room or area. If pre-preparation equipment and low-temperature storage are available, certain other procedures, such as peeling, dicing, and chopping of vegetables; breading and panning of meats; opening of canned goods; and thawing of frozen foods, can be completed in the ingredient room. The extent of responsibilities depends on the space, equipment, and personnel available.

PRODUCTION SCHEDULE

Meal Count ______

Weather ______

DATE AREA

Special Events ______

PERSON	MENU ITEM	QUANTITY TO PREPARE	ACTUAL YIELD	DIRECTIONS	TIME SCHEDULE	LEFT OVER AMOUNT	RUN OUT TIME	SUBSTITUTION	CLEANING ASSIGNMENT

PRE-PREPARATION:

PERSON	MENU ITEM	QUANTITY	PERSON	MENU ITEM	QUANTITY

Form 4.1

F.S. Form 60

Figure 6.8 Production schedule for a large residence hall food center. Information useful for production forecasting is also recorded on this form.
Courtesy of Kansas State University Residence Halls.

After ingredients have been weighed or measured and the pre-preparation completed, each ingredient is packaged in a plastic bag or other container and labeled. The ingredients for each recipe are assembled and delivered, with a copy of the recipe, to the appropriate production unit. In some operations, the assembled ingredients are distributed when needed according to a predetermined schedule.

There are many advantages to an ingredient room, the term used most often for this production facility. Increased production control, improved security, more consistent quality control, and more efficient use of equipment, especially if pre-preparation is included in this area, are possible with central ingredient assembly. Because cooks are not involved in the time-consuming job of weighing and measuring ingredients, their time and skills can be used more effectively in production.

There are some disadvantages, the main one being the lack of flexibility. For example, the ingredients must be weighed the day before, or earlier in some cases, which does not provide for last-minute changes in menus or quantities needed. Cooks may feel restricted by not being able to add their own touches to the food they are preparing. This attitude usually changes when they are satisfied that the recipes as written will result in a product of the desired quality and once the cooks have confidence in the person who weighs the ingredients.

Personnel and Equipment. Accuracy in measuring ingredients contributes to the acceptability of the finished product, so it is important that the ingredient room personnel be well qualified and that they be provided with adequate equipment.

Personnel assigned to the ingredient room must be able to read and write and perform simple arithmetic with at least minimal ease and accuracy. These traits, as well as honesty and dependability, are desirable for this person or persons to have. Safety precautions and sanitation standards should be stressed in their training.

Weighing is the quickest, easiest, and most accurate means of measure in most cases, so good scales are essential. A scale that accurately weighs up to 25 pounds is usually adequate and, if more than that is required, the ingredients are divided in two or three lots for easier handling. Some foodservices have separate scales for very small amounts such as spices. If scales that accurately weigh small amounts are not available, volume measurement may be preferable for those foods. Scales that are calibrated to pounds and tenths of a pound instead of pounds and ounces are useful with a system that uses a computer for calculating recipes.

The equipment needed depends on the functions performed in this area. Suggested minimum equipment for weighing and measuring includes the following:

- Worktable, six to eight feet long, with one or two drawers,
- Counter scales, with gradations of one ounce minimum to 25 pounds,
- Mobile storage bins for sugar, flour, and other large-volume staples,
- Shelving for bulk staples and spices,
- Mobile racks for delivery of foods to production areas,
- Refrigeration (and freezer if frozen foods are distributed),
- Sink and water supply,
- Can opener,
- Trash containers,

- Counter pans with lids if canned foods are opened,
- Trays for assembling ingredients,
- Rubber spatulas,
- Measuring utensils (gallon, quart, pint, cup measures; measuring spoons),
- Scoops for dipping flour and sugar,
- Packaging materials (paper and plastic bags, paper cups), and
- Masking tape and marking pens to label ingredients.

If vegetable preparation is also done in the ingredient room, the following additional equipment is needed:

- Double or triple sink,
- Waste disposal,
- Peeling, slicing, and/or dicing equipment,
- Cutting board,
- Assorted knives and sharpening equipment, and
- Plastic tubs or bags for cleaned products.

Portion Control

An early step in recipe standardization is that of making a decision on the size of portions to be offered. Standardized portions are important not only in the control of costs, but also in creating and maintaining consumer satisfaction and goodwill. No one likes to receive a smaller serving than other customers for the same price.

Food is portioned by weight, measure, or count and begins with the purchase of foods according to definite specifications so that known yields can be obtained from each food. Portioned meats, fish, and poultry; fresh fruits ordered by size (count per shipping box); canned peaches, pears, pineapple slices, and other foods in which the number of pieces is specified are examples. Also helpful in portion control is the purchase of individual butter and margarine pats and individually packaged crackers, cereals, and condiments.

A knowledge of common can sizes is also helpful in portion control. For example, No. 10 cans are common in large-volume operations. They come packed six to a case and each can contains approximately 12 cups of product. Table 6.7 provides a summary of common can sizes.

During food production, portions are measured by scoop or dipper or are weighed on portion scales. For example, the recipe for meatballs may call for dipping the mixture with a size 16 dipper (or scoop), which results in a ¼-cup or 2-ounce portion. The numbering system for scoop sizes is based on the number of scoops per quart. Table 6.8 shows approximate dipper and ladle sizes. Dippers range from size 6 (10 tablespoons/6 ounces) to size 100, which holds a scant 2 teaspoons. Muffins are also measured by dippers, but bread and roll types of dough are weighed. A roll cutter provides equal portions without the necessity of weighing each roll. Each recipe should indicate the yield expected in number of portions, total weight or measure, and the size in weight or measure of each serving.

The appropriate utensil and its size for serving the product should be indicated on the recipe. Ladles, which are used for serving sauces, soups, and similar foods, are

Table 6.7 Common can sizes.

Can Size (industry term)[1]	Average net weight of fluid measure per can[2]		Average volume per can		Cans per case	Principal products
	Customary	Metric	Cups	Liters	Number	
No. 10	6 lb (96 oz) to 7 lb 5 oz (117 oz)	2.72 kg to 3.31 kg	12 to 13⅔	2.84 to 3.24	6	**Institutional size:** Fruits, vegetables, some other foods.
No. 3 Cyl	51 oz (3 lb 3 oz) or 46 fl oz (1 qt 14 fl oz)	1.44 kg or 1.36 L	5¾	1.36	12	Condensed soups, some vegetables, meat and poultry products, fruit and vegetable juices.
No. 2½	26 oz (1 lb 10 oz) to 30 oz (1 lb 14 oz)	737 g to 850 g	3½	0.83	24	**Family size:** Fruits, some vegetables.
No. 2 Cyl	24 fl oz	709 mL	3	0.71	24	Juices, soups.
No. 2	20 oz (1 lb 4 oz) or 18 fl oz (1 pt 2 fl oz)	567 g or 532 mL	2½	0.59	24	Juices, ready-to-serve soups, some fruits.
No. 303	16 oz (1 lb) to 17 oz (1 lb 1 oz)	453 g to 481 g	2	0.47	24 or 36	**Small Cans:** Fruits and vegetables, some meat and poultry products, ready-to-serve soups.
No. 300	14 oz to 16 oz (1 lb)	396 g to 453 g	1¾	0.41	24	Some fruits and meat products.
No. 2 (vacuum)	12 oz	340 g	1½	0.36	24	Principally vacuum pack corn.
No. 1 (picnic)	10½ oz to 12 oz	297 g to 340 g	1¼	0.30	48	Condensed soups, some fruits, vegetables, meat, fish.
8 oz	8 oz	226 g	1	0.24	48 or 72	Ready-to-serve soups, fruits, vegetables.

[1] Can sizes are industry terms and do not necessarily appear on the label.

[2] The net weight on can or jar labels differs according to the density of the contents. For example: A No. 10 can of sauerkraut weighs 6 lb 3 oz (2.81 kg); a No. 10 can of cranberry sauce weighs 7 lb 5 oz (3.32 kg). Meats, fish, and shellfish are known and sold by weight of contents.

Source: Food Buying Guide for School Food Service, PA-1331, U.S. Department of Agriculture, 1984.

Table 6.8 Dipper (or scoop) equivalents.

Dipper No.*	Approximate Measure	Approximate Weight	Suggested Use
6	10 T (2/3 c)	6 oz	Entrée salads
8	8 T (1/2 c)	4–5 oz	Entrées
10	6 T (3/8 c)	3–4 oz	Desserts
12	5 T (1/3 c)	2½–3 oz	Muffins, salads, desserts
16	4 T (1/4 c)	2–2¼ oz	Muffins, desserts
20	3⅕ T	1¾–2 oz	Sandwich fillings, muffins, cup cakes
24	2⅔ T	1½–1¾ oz	Cream puffs
30	2⅕ T	1–1½ oz	Large drop cookies
40	1½ T	¾ oz	Drop cookies
60	1 T	½ oz	Small drop cookies, garnishes
Ladles	⅛ c	1 oz	Sauces, salad dressings
	¼ c	2 oz	Gravies, sauces
	½ c	4 oz	Stews, creamed foods
	⅔ c	6 oz	Stews, creamed foods
	1 c	8 oz	Soup

These measurements are based on level dippers and ladles.
*Portions per quart.

sized according to capacity (1 ounce/ ⅛ cup to 8 ounces/1 cup). Although spoons are used for serving some foods, they are not as accurate as ladles unless the employees have been instructed in their use and know how full to make the spoon. For cakes and other desserts baked in a pan, instructions for cutting should be included. Many foodservices have pie markers or other marking devices to ensure equal portions of baked products.

Employees should know the number of servings expected from a certain batch size and should be familiar with the size of the portion. In addition to the information included on recipes, a list of portion sizes for all foods should be made available to employees either in an employees' manual or posted in a convenient location (see Figure 6.9).

PRODUCT EVALUATION

As mentioned earlier in this chapter, product evaluation is part of the initial testing phase of a new recipe and is important for quality control. Product evaluation or sensory analysis is actually an ongoing process to ensure that the yield expectations and quality standards established during the recipe standardization process are met each time a menu item is produced.

Many foodservice organizations conduct sensory analysis just prior to meal service. This analysis is best done by a team or panel of persons knowledgeable about

Figure 6.9 Sample of breakfast portion guide.

Menu Item	Portion	Utensil
Meats		
Bacon	2 strips	Tongs
Canadian bacon	2 oz. slice	Tongs
Sausage, links (16 per lb.)	2 each	Tongs
Mixed fruit	½ c. (4 oz)	Spoodle, slotted
Juices, pre-portioned	4 oz.	N.A.
Breads		
Toast	2 slices	Tongs
Sweet rolls	1 each	Tongs
Coffee cakes, 18 × 26″ pan	6 × 10 cut	Spatula
Biscuits	1, 2 oz.	Tongs
Muffins	1, 2½ oz.	Tongs
Pancakes	3, 1 oz. each	Spatula
Hot cereal	¾ c. (6 oz)	Ladle
Dry cereal, pre-portioned boxes	1 each	N.A.
Eggs		
Scrambled	¼ cup	#16 scoop
Omelette	3 oz. each	Spatula

product standards and trained to judge quality characteristics in the interest of the consumers. Figure 6.10 is an example of a sensory analysis form used in a small long-term facility.

SUMMARY

Management's responsibility to serve high-quality food starts with the setting of standards and ensuring that employees are aware of them. The use of standardized recipes, good quality ingredients, and proper supervision of food production are vital to quality control.

Basic to production planning and scheduling is the forecast, which is a prediction of the number of persons to be served for a meal or a day and, for a selective menu, an estimate of the number who will choose each menu item. Quantities of food to prepare are based on the predicted number of servings needed and the portion size to be offered. This information, plus special instructions for preparation and work assignments, are included in a production schedule.

Once standardized recipes are implemented, the manager must design, monitor, and control procedures to ensure that preestablished quality standards are met. Centralized ingredient assembly, portion control, and sensory analysis are examples of quality control methods.

Heartland Country Village—Foodservice Department
Sensory Analysis Sheet

Date: __________ Day: __________ Cycle: __________ Cook: __________

FOOD ITEMS	TEMP. °F*	QUALITY DESCRIPTION⁺
Breakfast		
Hot Cereal		
Eggs		
Other		
Milk		
Lunch/Dinner		
Entree		
Entree alternate		
Starch		
Vegetable		
Vegetable alternate		
Salad		
Dessert		
Modified diet items		
1.		
2.		
Milk		
Supper		
Soup		
Sandwich		
Casserole		
Vegetables		
Alternates		
1.		
2.		
Modified Diet Items		
1.		
2.		
Milk		

* Hot food *held* at 150° – 180°; Cold Food 35° – 50°. See recipes for proper end stage T°.

⁺ Consider flavor, texture, appearance, then indicate a rating of good, fair, poor.

Comments: (Indicate action taken to correct food items of fair or poor quality)

Figure 6.10 Daily sensory analysis form.

REVIEW QUESTIONS

1. What impact does the type of foodservice system (i.e., conventional, commissary, ready-prepared, and assembly/serve) have on the production function of a foodservice operation?
2. What are the objectives of cooking? How might these objectives influence recipe development in various settings such as day care centers, nursing homes, hospitals, and restaurants?
3. Why are standardized recipes important in foodservice operations?
4. Write two measurable quality standards for a baked product such as muffins or cookies.
5. Calculate the quantity of lettuce to purchase when 1 pound AP equals .76 EP and 100¾-ounce servings are needed.
6. What information should be included on a production sheet?
7. What are the advantages of an ingredient assembly center in a foodservice operation?
8. Describe the use of historical data in the process of forecasting.
9. Describe the relationship between the purchasing and production functions in a foodservice operation.
10. What factors should be considered when scheduling production?
11. Why are portion control standards important? How does portion control relate to the standardization of recipes?

SELECTED REFERENCES

Buchanan, P. W.: *Quantity Food Preparation: Standardizing Recipes and Controlling Ingredients.* Chicago, Ill.: American Dietetic Association, 1983.

Byers, B. A., Shanklin, C. W., and Hoover, L. C.: *Food Service Manual for Health Care Institutions,* 1994 edition. Chicago, Ill.: American Hospital Publishing, Inc., 1994.

Greathouse, K. R., and Gregoire, M. B.: Variables related to selection of conventional, cook-chill, and cook-freeze systems. J. Am. Diet. Assoc. 1988; 88:476–478.

Knight, J. B., and Kotschevar, L. H.: *Quantity Food Production, Planning, and Management,* 2nd ed. New York: Van Nostrand Reinhold, 1989.

Kotschevar, L. H.: *Standards, Principles, and Techniques in Quantity Food Production,* 4th ed. New York: Van Nostrand Reinhold, 1988.

Messersmith, A. M., and Miller, J. L.: *Forecasting in Foodservice.* New York: John Wiley & Sons, 1992.

Nettles, M. F., and Gregoire, M. B.: Operational characteristics of hospital foodservice departments with conventional, cook-chill, and cook-freeze systems. J. Am. Diet. Assoc. 1993; 93:1161–1163.

Nutrient losses in quantity food preparation. Hosp. Food Nutr. Focus. December 1994; 11(4):4–7.

Pennies per portion and hundreds of dollars per day. Hosp. Food Nutr. Focus. October 1994; 11(2):4–6.

Shugart, G., and Molt, M.: *Food for Fifty,* 9th ed. Englewood Cliffs, N.J.: Merrill/Prentice Hall, 1993.

7

Assembly, Distribution, and Service

OUTLINE

Foodservice managers have the responsibility of making certain that after food is prepared, it is safely held, transported, delivered, and served to consumers. Therefore, the goals of a delivery and service system should include the following:

- Maintain quality food characteristics.
- Ensure microbial safety of food.
- Serve food that is attractive and satisfying to the consumer.

In addition, the system should be designed and selected for optimal use of available resources: labor, time, money, and space.

This chapter provides information for the foodservice manager so that he or she can make sound decisions about delivery and service systems. The information is appropriate whether the manager is evaluating a currently existing system or preparing to select a new system for a particular situation. Chapter content includes factors that affect selection of a system, equipment needs for delivery-service functions, and a review of various styles of service.

KEY CONCEPTS

1. Numerous options and alternatives exist to assemble, distribute, and serve meals.
2. System selection for assembly, distribution, and service is contingent on many factors including the type of foodservice system and the production process used by the organization.
3. Tray assembly is used in many high-volume foodservice operations.
4. The time and temperature relationship is a critical factor to consider in food delivery systems.

METHODS OF ASSEMBLY, DELIVERY, AND SERVICE

Modern technological research and development related to foodservice have brought many advances in methods of delivery and service of food and in the equipment used for those processes. These developments resulted in large part from the production systems discussed in Chapter 2 and from the complexity of modern-day foodservice operations. With the increased time and distance between production and service, the potential for loss of food quality has also increased. Newer delivery and service methods have been designed to protect against such loss.

Most menu items are at peak quality at the time production and cooking are completed. It is not possible to serve food at that precise time in many foodservice systems because of the need to assemble, transport, and deliver meals for service. Equipment that maintains foods at proper temperatures for best quality and ensures safety of the foods in transit is a necessity. Methods of delivery and service that involve the shortest possible time and distance are best able to help achieve the desired goal.

Methods—Delivery and Service as Subsystems

The term *distribution* or *delivery* refers to the transportation of ready-prepared foods from production to place of service; *service* involves assembling prepared menu items and distributing them to the consumer. The equipment required for both delivery and service is an essential part of these subsystems. Whereas delivery and service *are* subsystems in the overall foodservice system, they are small systems within themselves and are referred to here as "systems."

Basically, there are two major *on-premise* delivery systems: centralized and decentralized. Although some people think of these two terms only in relation to tray service to hospital patients, they are used in relation to other types of foodservices as well.

Centralized Delivery-Service System. In the centralized method, prepared foods are portioned for individual service and meals assembled at a central area in or adjacent to the main kitchen. The completed orders are then transported and distributed to the customer. This is typical of over-the-counter service in fast-food restaurants, of table or counter service in restaurants that utilize waiters or waitresses to transport and deliver meals to the patrons of room service in hotels or motels, and of banquet service where plates are filled in a central location and transported by various means to the dining areas for service. This method is used also in many hospital and health care facilities. Foods are portioned and plated, and trays for individual patients are assembled in the central serving room. Completed trays are then transported by various means to the patients throughout the facility. Soiled trays and dishes are returned to the central area for washing.

Centralized delivery-service systems are prevalent today because of the close supervision and control of food quality, portion size, assurance of correct menu items on each tray or order, and correct food temperatures at point of service that this system affords. Also, it requires less equipment and labor time than does the decentralized method. If the number of people to be served is large, however, the total time span required for service may be excessively long.

Decentralized Delivery-Service System. In the decentralized system, bulk quantities of prepared foods are sent hot and/or cold to serving pantries or ward kitchens located throughout the facility, where the reheating (if needed), portioning, and meal assembly take place. Thus, instead of one central serving area, there are several smaller ones close to the consumers. Often, these pantries have facilities for limited short-order cooking of eggs and toast and for coffee making. Refrigerators, ovens for reheating, temperature-holding cabinets, and a counter or conveyor belt for tray assembly are also equipment provided in these serving pantries. Dishwashers can be provided for warewashing in the ward kitchens; or soiled dishes and trays can be returned to the central area for washing, which eliminates the need for duplication of dishwashing equipment in each pantry. If dishes are washed in the central area, the clean dishes must be returned to the pantries for use for the next meal. It is time and energy consuming to transport dishes twice each meal, soiled and clean, from and to the pantries. Over a period of time, this may be more expensive than providing dishwashing facilities for each serving unit.

Decentralized service is considered most desirable for use in facilities that are low and spread out in design or in any facility where there are great distances from the main kitchen to the consumers. It is expected that foods will be of better quality and retain desired temperatures more effectively if served near the consumer rather than plated in a central location and transported to distant locations within the facility. Also, with many serving units rather than one, meals for all consumers can be distributed within a reasonable time span.

Types of foodservice that utilize the decentralized system include large hospitals and medical centers; industrial plants with several serving units or with mobile carts that are used to travel throughout the plant to serve workers at their workstations; hotels providing room service from serving pantries on various floors, and banquets from a serving kitchen within the facility.

Costs and values of centralized versus decentralized methods should be studied and carefully weighed before deciding on which one to adopt. Both can be successfully utilized in specific situations and with given conditions.

ASSEMBLY

Assembly is the fitting together of prepared menu items to complete an entire meal unit. Assembly can occur at a number of points along the sequence of process steps depending on the type of foodservice organization and the production system used. Restaurants, for example, assemble hot meals at the centralized production point and serve the meal immediately and directly to the waiting customer. Institutions, on the other hand, use tray assembly systems for speed efficiency. This method of assembly is common in organizations such as health care facilities, schools, and airlines where large numbers of meals must be served at specific times.

Tray Assembly

Two major systems are used to assemble meal trays. In one, food is assembled at a central location, usually the production kitchen, using a trayline, and then various distribution methods are used to deliver the trays to units. Figures 7.1, 7.2, and 7.3 illustrate various trayline configurations. The second system transports food in bulk to units where it is assembled or plated as individual meals. This is referred to as decentralized assembly and service.

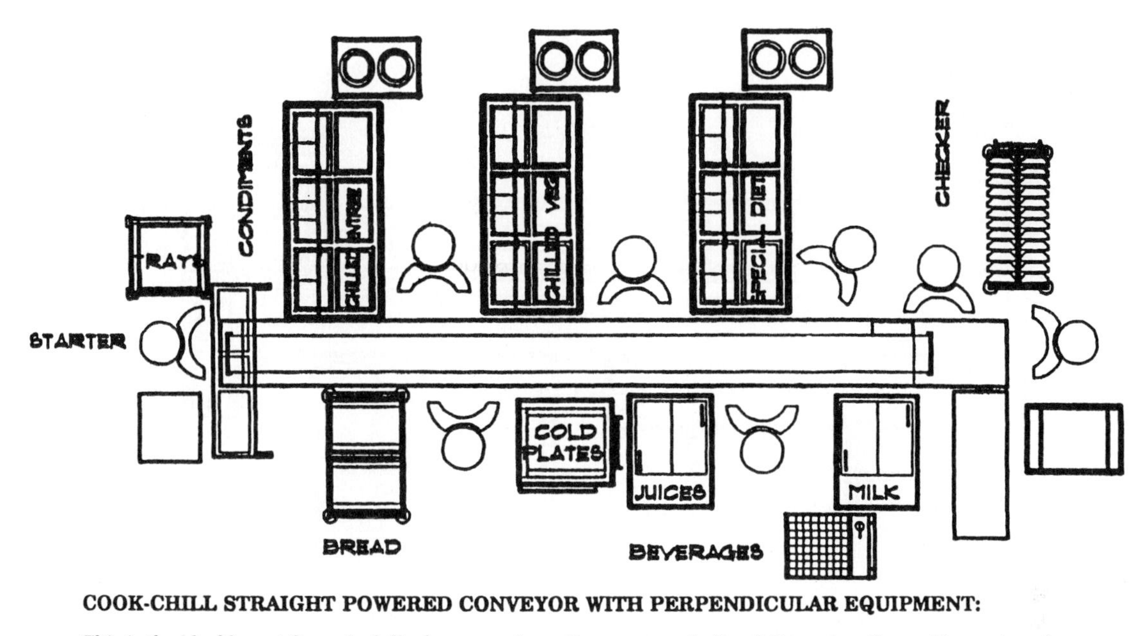
CONDIMENTS
CHILLED ENTREE
CHILLED VEG
SPECIAL DIET
CHECKER
STARTER
COLD PLATES
JUICES
MILK
BREAD
BEVERAGES
COOK-CHILL STRAIGHT POWERED CONVEYOR WITH PERPENDICULAR EQUIPMENT:
This is the ideal layout for cook chill when space is available. Speed is not a problem when traying up for a cook chill system, as the tray is going back into a refrigerated space, before delivery to retherm. The system shown was for a restaurant style menu and a Crimsco convection retherm in a floor pantry.

Figure 7.1 Cook/chill trayline.
Courtesy of Crimsco, Inc., Kansas City, Missouri.

FACTORS AFFECTING CHOICE OF DISTRIBUTION SYSTEMS

Every organization has its own requirements for delivery and service based on the type of foodservice system, the kind of foodservice, the size and physical layout of the facility, style of service used, skill level of available personnel, economic factors related to labor and equipment costs, quality standards for food and microbial safety desired, timing necessary for meal service, space requirements versus space available for foodservice activities, and the energy use involved.

No one factor alone can be considered when deciding on a delivery-service system, because most of the factors interact with, and have an influence on, the others. They must be regarded as a whole when a choice is made.

Type of Foodservice System

The system used determines to some extent its own needs for delivery and service. Of the four types of foodservice systems discussed in Chapter 2, the commissary system, for example, is the only one requiring delivery trucks to take prepared foods to satellite serving units. In both the conventional and ready-prepared systems, foods are prepared on the premises, and the assembly/serve system has meals that are purchased and delivered by the vendor to the service facility. Table 7.1 illustrates the assembly, delivery, and service components of each type of foodservice.

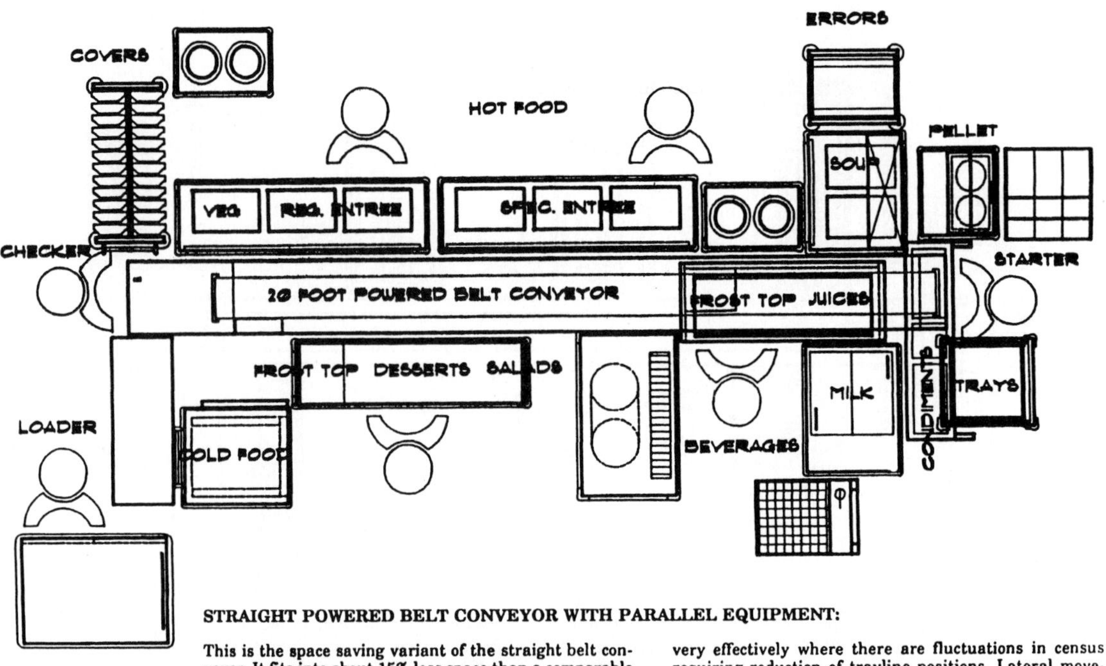
ERRORS
COVERS
HOT FOOD
PELLET
SOUP
VEG.
REG. ENTREE
SPEC. ENTREE
CHECKER
STARTER
20 FOOT POWERED BELT CONVEYOR
FROST TOP JUICES
FROST TOP DESSERTS SALADS
MILK
CONDIMENTS
TRAYS
LOADER
COLD FOOD
BEVERAGES
STRAIGHT POWERED BELT CONVEYOR WITH PARALLEL EQUIPMENT:
This is the space saving variant of the straight belt conveyor. It fits into about 15% less space than a comparable circular conveyor. It will cost out at about one-half the price of a comparable circular conveyor. It may be used very effectively where there are fluctuations in census requiring reduction of trayline positions. Lateral movement of staff means that trayline need not stop as frequently.

Figure 7.2 Parallel cook/serve trayline.
Courtesy of Crimsco, Inc., Kansas City, Missouri.

As noted previously, menu items processed in the commissary are either held in bulk or portioned before storage. Three alternatives for storage following food production are frozen, chilled, or hot-held. Each method requires different equipment.

Bulk foods may be placed in 12-inch × 20-inch counter-size pans for freezing so that they may be reheated and served from the same pan. Or, if they are to be transferred to serving units in the chilled or hot state instead of frozen, they are placed in heavy containers with lids that clamp on securely. Otherwise, spillage may result during transport to the foodservice facility.

Carriers (Figure 7.4, for example) to hold the portioned food in their containers are filled at the commissary. At scheduled times each day, carriers are loaded onto a truck for transfer to the service unit. In many cases, the driver is responsible for unloading the truck and taking the food carriers to the storage or service area as required. Empty carriers from the previous delivery are collected and returned to the commissary on the delivery truck.

The fleet of trucks required by the commissary depends on geographic distances to be traveled and number of deliveries to be made by each truck driver. Timing can be crucial, especially in those situations where the food is delivered hot just at meal times. School foodservices or college residence hall service may utilize the hot food delivery system to a greater extent than other foodservices because they may not have finishing kitchens. Distances for hot foods to be transported should be short. Timing is also essential for delivery of meals to airports for loading onto planes before takeoff. Heavy penalties are imposed for deliveries not meeting airline schedules.

Delivery of frozen foods requires well-insulated carriers to maintain food in the frozen state during the time it is being transported. If the service facility has adequate space for holding frozen food, there is little problem with delivery time, since meals can be sent a day or two ahead. If there is no such storage space, delivery timing must be correlated with meal periods and time for rethermalizing and assembling the menu items.

At this point, foods are on the premises, and the procedures for delivery and service within the facility may be the same for all four systems.

Kind of Foodservice Organization

The type of organization determines to a large extent the delivery and service system requirements. Those where large numbers of people must be served quickly, such as schools, colleges, and industrial plants, usually provide cafeterias for meal service. Fast-food restaurants serve foods as quickly as possible, too, but with over-the-counter service or drive-thru windows or drive-in service by car hops.

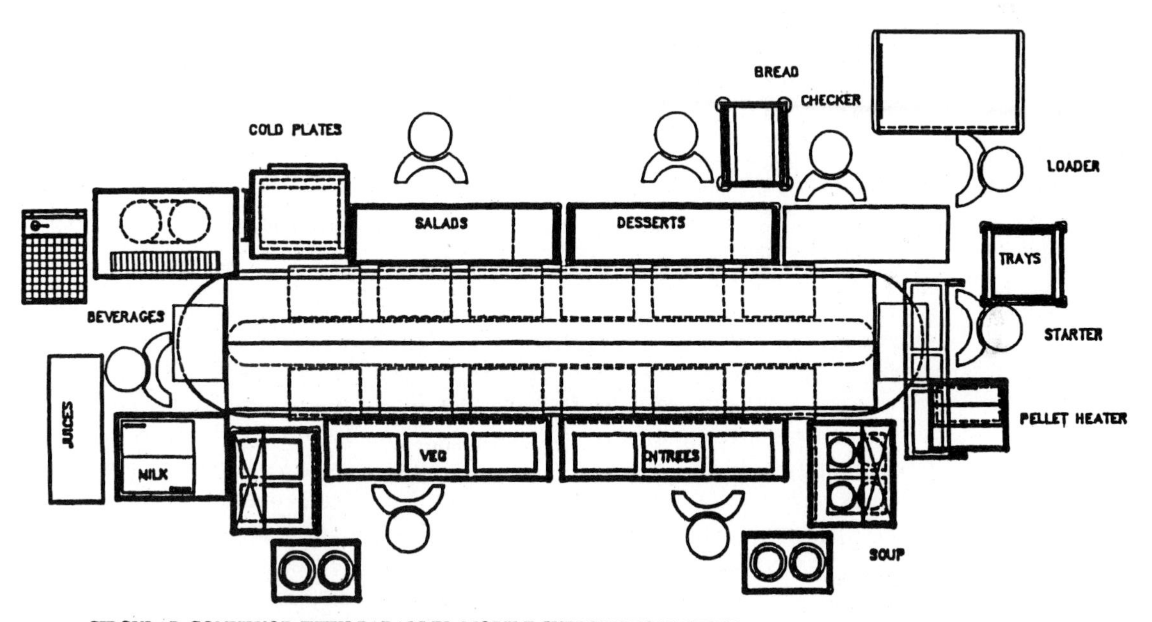

CIRCULAR CONVEYOR WITH PARALLEL MOBILE SUPPORT EQUIPMENT:

This is a significantly more expensive solution to the trayline layout problem. By using mobile support equipment a flexible approach is possible. The flexibility of the mobile support equipment means that as your census and menu changes, you can re-lay your trayline to meet new conditions. The parallel layout of the hot and cold equipment means staffing of positions also vary with the census, up or down. This layout is the second most space efficient.

A less desirable type circular conveyor is the giant one with built-in work stations both hot and cold. It is approximately 25% more expensive than the circular conveyor with mobile support equipment. It is totally inflexible.

Figure 7.3 Cook/serve circular trayline.
Courtesy of Crimsco, Inc., Kansas City, Missouri.

Table 7.1 Sequence of process steps for entrée menu items using alternate food production and storage methods in foodservice systems.

Process Steps	Cook/Freeze[AB]	Cook/Chill[AB]	Cook/Hot-Hold[AC]	Heat/Serve[D]
Ingredient and menu item storage	Ingredients enter foodservice	Ingredients enter foodservice	Ingredients enter foodservice	Commercially prepared shelf-stable menu items enter foodservice
Preparation	Preparation	Preparation	Preparation	
Heating	Initial heating	Initial heating	Final heating	
Postheating storage	Freezing and frozen storage Thawing	Chilling and chilled storage	Hot-holding storage	Room temperature storage
Portioning and assembly	Portioning and assembly	Portioning and assembly	Portioning and assembly	Portioning and assembly
Storage during distribution	Cold-holding and distribution	Cold-holding and distribution	Hot-holding and distribution	Cold-holding and distribution
Final heating	Final heating	Final heating		Final heating
Service	Service	Service	Service	Service

[A] Used in commissary foodservice systems.
[B] Used in ready-prepared foodservice systems.
[C] Used in conventional foodservice systems.
[D] Used in assembly/service foodservice systems.

Source: Adapted with permission from M. E. Matthews: *Microbiological Safety of Foods in Feeding Systems.* Published by the National Academy Press, Washington, DC, 1982.

Hospitals and nursing homes cater to the foodservice needs not only of their patients but also of the employees, professional staff, and visitors. This calls for tray service for patients who are bedfast, perhaps dining room service for ambulatory patients in some care centers, cafeteria service for staff, employees, and visitors, and often vending machines provided as a supplemental service for between-meal hours.

Airlines have special delivery-service needs also. Food must be transported by delivery truck from commissary to airport for loading onto the planes for service after takeoff (see Figure 7.5). Meals may be transported preplated hot or cold for reheating in the galley of the plane. Passengers are served meals or snacks on trays at their seats by flight attendants.

Table service restaurants may use differing styles of service (see later section titled "Style of Service"), but all employ servers to carry meals from kitchen to the guests. In restaurants in which customers serve themselves, *cafeteria* or *buffet service,* employees replenish the food and may serve beverages to the guests' tables.

The large hotel or motel may have several types of service within its facility, including a counter or coffee shop for fast meals and table service dining rooms. Some may be more "exclusive" and expensive than others, so more formal types of service may be used. Because many hotels cater to conventions and group meetings, banquet service is also offered. Room service is available in most hotels, which calls for a different means of delivery and service, such as servers using trays or tables on wheels to take meals to guests in their rooms.

Meal service aboard cruise ships parallels that of hotels or motels, except that there is no banquet service. In the main dining room, guests are served at one or two sittings, depending on facilities and number of guests aboard. In addition to regular meals, specialty foods such as pizza and short-order menu items may be served in snack bars, around the pool, or in a lounge. Midmorning coffee or bouillon, afternoon tea, and elaborate midnight buffets necessitate various delivery-service methods. Foodservice is provided almost around the clock.

Size and Physical Layout of Facility

The size and building arrangement of the facility are other factors to consider when determining a delivery system. Some restaurants, for example, may be in a high-priced downtown location and, thus, are generally narrow and several stories in height in order to utilize valuable land to its best advantage. In this case, the bakery

(a)

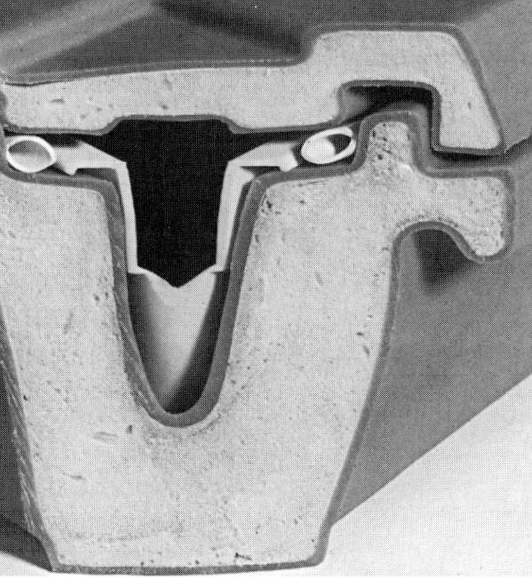

(b)

Figure 7.4 (a) An example of a carrier for transporting prepared hot food. This model has two snap-in heat packs for efficient heat distribution that extends holding time at a desired temperature. National Sanitation Foundation International (NSFI) approved. (b) Cutaway shot of same insulated carrier. Note the one-piece polyethylene, double-wall construction; thick polyurethane insulation; and the design features for heat retention. Courtesy of Cambro Manufacturing Company, Huntington Beach, California.

Figure 7.5 Marriott In-Flite Service supplies more than 150 carriers and is typical of foodservice delivery to airlines.
Courtesy of H. H. Gill, president, Gladieux Corporation, Toledo, Ohio (now merged with Marriott Corporation).

may be on one level, preparation and cooking units on another, and dishwashing on still another, with all being on different levels from the dining room. This calls for a well-coordinated system of mechanized conveyors, subveyors, and elevators to deliver food quickly to the place of service.

Hospitals and health care facilities can be constructed as high-rise buildings or low, rambling facilities with miles of corridors. Different systems are required for each to ensure tray delivery to the patient within a reasonable time. The distance and the routing from production to service areas are points for consideration.

Style of Service

Whether or not the style is *self-service* such as cafeteria, buffet, vended, or pick-up by the consumer; *tray service,* either centralized or decentralized; *waiter or waitress service* for table, counter, or drive-in facilities; or *portable meal service* with meals delivered to home or office throughout an industrial plant, each has different equipment and delivery needs. (See later section on "Equipment Needs.")

Skill Level of Available Personnel

Labor needs and required skills vary for different types of delivery systems and for the equipment used in each type of system. When planning to alter the current delivery system or select a new one, the foodservice manager must assess the current skills and availability of the foodservice employees. Judgment must be made on the skills needed to operate a new system and on the learning ability of the employees. A training program should then be designed to ensure that employees are well trained in the use, care, and safety features of all equipment and delivery procedures.

Economic Factors

Labor and equipment required in the various delivery-service systems must be costed and evaluated in relation to budgeted allocation. Unless adequate funding is available, the foodservice would not, for example, be able to install automated electronic delivery equipment. Economic factors play a part in deciding where and how frozen or chilled foods should be reheated, assembled, and served. Decentralized service requires duplication of assembling and serving and, sometimes, dishwashing equipment, as well as personnel for the many pantries or ward kitchens throughout the facility, and so may be more expensive to install and operate than the centralized service. Cost comparisons of the numerous types of carts and trucks for transporting food should precede the selection of a specific delivery and serving system.

Quality Standard for Food and Microbial Safety

Management establishes standards for food quality and safety, then selects equipment for heating, holding, and transporting food to achieve those standards. How hot should the food be when served to the consumer? How can that temperature be maintained through delivery and service? How hot must foods be at the time of serving and portioning to aid in achieving the desired standards?

Considerable research has been conducted to find answers for these questions; a summary is reported by a group of North Central Regional Researchers. Studies relate to the four foodservice systems, microbial safety, nutrient retention, and sensory qualities. Microbial quality of menu items studied is dependent on the type of food, quality of raw ingredients, batch size, type of equipment used for cooking, and position of menu items in foodservice equipment, among others. The management of time and temperature relationships throughout all stages of product flow in every foodservice system is considered of major importance. (See Chapter 3 for more on microbial safety.) Time and temperature relationships are also important in nutrient retention and on sensory qualities of food products. Managers should be knowledgeable about these factors to improve their foodservice.

Timing Required for Meal Service

The time of day desired or established for meals is another factor influencing the choice of a delivery-service system. For example, if 1200 people are to be served at a 7 P.M. seated banquet, all food must be ready at once and served within a few minutes to all of the guests. Many serving stations and adequate personnel for each station are prerequisites for achieving the time objective. Preheated electric carts can be loaded with the preplated meals a short time before service and then taken to various locations in the dining room for service from the carts to guests. An alternate method is to place the plates as they are served on trays and carried by servers to the dining room. This will require several trips from serving area to dining room, thus, it takes more time than when carts are used.

If only a few people need to be served at one time, as in a restaurant where customer orders come to the kitchen over a period of a few hours, food is cooked to order, or in small batches, and held for short periods of time.

In school foodservice, many children are ready for lunch at the same time. To avoid long waits in a cafeteria line, however, a staggered meal period can be scheduled, which allows various grades to be dismissed for lunch at 5- or 10-minute intervals. Another option, if space allows, would be multiple serving lines.

Large hospitals have the problem of serving their many patients within a reasonable meal period time span. Should all be served at approximately the same time as may be possible with decentralized service? Or is a one- to two-hour time span acceptable as may be required with centralized service? Various systems meet specific needs. The dietitian must work with nursing staff to ensure that patient care is not disrupted and that quality food and service are provided.

Space Requirements or Space Available

Allocation of space for departments and their activities is determined at the time of building construction. The delivery-service system preferred should be stated early in the facility planning process so adequate space will be available for those foodservice activities. Any later remodeling to change to a different system can be disruptive and expensive, if it's possible at all.

Decentralized systems require less space in the main kitchen area but more throughout the facility for the serving pantries than do centralized systems. In hospitals with centralized service, tray assembly equipment, as well as trucks or carts, take up considerable space. Based on the number and size of the transport carts or trucks, the space needed for their storage when not in use can be calculated. Added space must be allocated for moving the carts through the facility with ease.

Energy Usage

A concern for energy use and its conservation plays a role in deciding on a delivery-service system. Systems that utilize a large number of pieces of electrically powered or heated equipment are more costly to operate than those that use the "passive" temperature retention equipment, such as insulated trays or pellet-heated plates.

A report of a research study by Franzese (1984) in 66 New York City hospitals showed that 30% of the hospitals changed their meal delivery systems during the period between 1979 and 1983. There was a trend toward a decrease in decentralized and microwave systems and an increase in insulated tray meal delivery systems that use hot food assembly at a central tray line and transport portioned food in trays designed to maintain food temperatures during transport to the patient. The "active" hot food systems, using hot/cold carts and rethermalizing food previously prepared, decreased in use. Reasons given for this "switch" to centralized from decentralized service and to the hot food assembly with "passive" temperature retention equipment were to decrease electrical energy consumption, to decrease labor costs, to increase the manager's control of food and supplies, and to improve the quality of patient food.

Concern for energy conservation diminished during the 1980s when management instead focused on a labor shortage in many parts of the United States. Labor savings seemed to be the most marketable equipment feature during these lean labor years. Today, energy awareness is on the increase again and energy savings are an important consideration in delivery system and equipment selection.

EQUIPMENT NEEDS

Delivery and service of food in institutions necessitates the use of specialized equipment for each step of the procedure: reheat if necessary, assemble, transport, distribute, and serve. Every foodservice system has its own requirements. Manufacturers work closely with foodservice directors to design pieces of equipment that best fill those specific needs. Equipment for delivery and service may be classified in several ways:

- *In general:* fixed or built in, mobile, and portable.
- *For a specific use:* reheating, assembling, temperature maintenance, transporting, and serving.
- *For each of the four foodservice systems:* conventional, commissary, ready-prepared, and assembly/serve.

A brief description of general and specific classification follows for an understanding of the various delivery-service systems in their entirety. For more detailed information about this equipment, see Chapter 11.

General Classification of Delivery-Service Equipment

Fixed or Built-In Equipment. Equipment that is fixed or built in should be planned as an integral part of the structure at the time a facility is being built.

One such system is the automated cart transport or monorail. This has its own specially built corridor for rapid transit, out of the way of other traffic in the building. It is intended for use by all departments since it is so expensive to install. It can transport items in a few seconds from one part of the building to another and is desirable because of its speed. An alternate plan for tray delivery may be needed if a power failure should occur, which could incapacitate the automated tray delivery system.

Other fixed equipment includes elevators; manual or power-driven conveyors for horizontal movement as for tray assembly; and subveyors and lifts (dumbwaiters) to move trays, food, or soiled and clean dishes to another level within the facility.

Mobile Equipment. Mobile equipment is equipment that is moved on wheels or casters. This includes *delivery trucks* for off-premise use to transport food from a commissary or central kitchen to the meal sites, and for "Meals-on-Wheels" delivery to home or offices.

Another type is *movable carts and trucks,* either hand-pushed or mechanized, for on-premise transport of either bulk food for decentralized service or preplated meals for centralized service. Such carts are available in many models, open or closed, insulated or not, temperature controlled for heated or refrigerated units, or combinations of both (see Figure 7.6 for an example). Some movable carts are designed to accommodate the served plates of hot food for banquet service, others are designed for entire meals assembled on trays for service in hospitals, and still others for bulk quantities of food. Assembly equipment and galley units are also mobile, instead of built in, which permits flexibility of arrangement. An example of a galley is shown in Fig. 7.7.

Portable Equipment. Included in this category are items that may be *carried*, as opposed to mobile equipment that is moved on wheels or casters. For delivery and service, equipment such as pans of all sizes and shapes, many with clamp-on lids to prevent spillage in transit, and hand carriers (also called totes) are commonly used.

Figure 7.6 Crimsco hot/cold tray cart designed with two convection ovens and two refrigerated units to keep foods above 140°F and below 40°F for service.
Courtesy of Crimsco, Inc., Kansas City, Missouri.

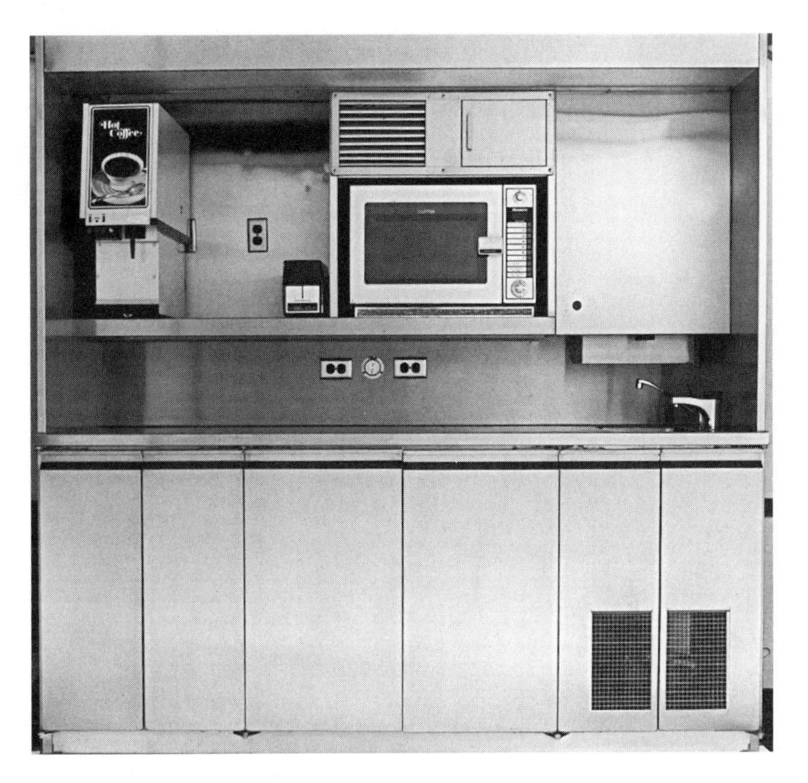

Figure 7.7 CYBEX Model 8430 Galley Station, a 5-foot × 7-foot unit for patient meal services, has a sink, an undercounter refrigerator/freezer, an ice maker/dispenser, storage space, and workcounter.
Courtesy of Crimsco, Inc., Kansas City, Missouri.

Totes may be insulated to retain temperature of foods for short-time transport or delivery (see Figure 7.8).

Also, a variety of plates and trays can keep preplated foods at proper temperatures for service. When these are used, unheated carts can be employed for transporting meals to consumers. Common types of plates and trays include pellet disc and insulated trays.

Pellet Disc. A metal disc (pellet) is preheated and at mealtime is placed in a metal base. Individual portions of hot food are plated and placed over the base and then covered. Either china or disposable dishes can be used. This hot metal pellet generates heat and keeps the meal at serving temperature for approximately 40 to 45 minutes. (Figure 7.9 is a modification of the pellet disc where the entire unit is heated.)

Insulated Trays with Insulated Covers. Insulated trays are designed with a variety of configurations for the differing types of dishes used for the menu of the day. Thermal, china, or disposable dishes can be used. After the food is served, the dishes are placed on the tray and covered with the insulated cover. No special carts are needed to transport these trays because they are nesting and stackable, and, of course, no temperature-controlled units are necessary. Some insulated tray systems are designed to create "synergism"; that is, when stacked properly, the cold and hot sections of each tray work together in a column to maintain proper temperatures.

Figure 7.8 Portable totes are insulated for temperature maintenance. This model has an optional insulated shelf to separate cold food (below) from the hot food.
Courtesy of Cambro Manufacturing Co., Huntington Beach, California.

The combined temperatures of the individual sections exceed the sum of the individual temperatures. Meals in these insulated trays retain heat quite well for short periods of time, such as during transport and distribution. Note that many foods retain heat much better than others under any circumstances. For example, a hearty entrée such as beef stew retains heat much better than a single serving of green beans. Portion size also influences how well and for how long a food item will maintain desirable temperature. Figures 7.10 and 7.11 illustrate two types of tray systems designed to maximize food quality and heat retention.

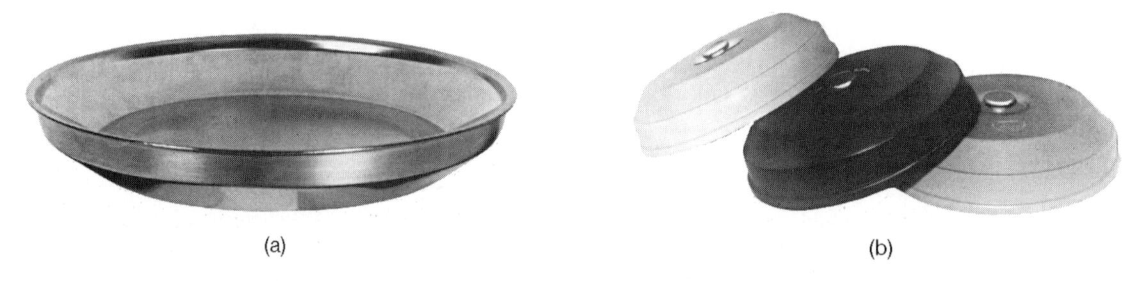

Figure 7.9 (a) Unitized base of pellet system is constructed of two 20-gauge stainless steel shells that are hermetically sealed. Bases are heated in specially designed heater/dispenser before filled dinner plate is set in and covered. (b) Cover for unitized pellet.
Courtesy of Seco Products, St. Louis, Missouri.

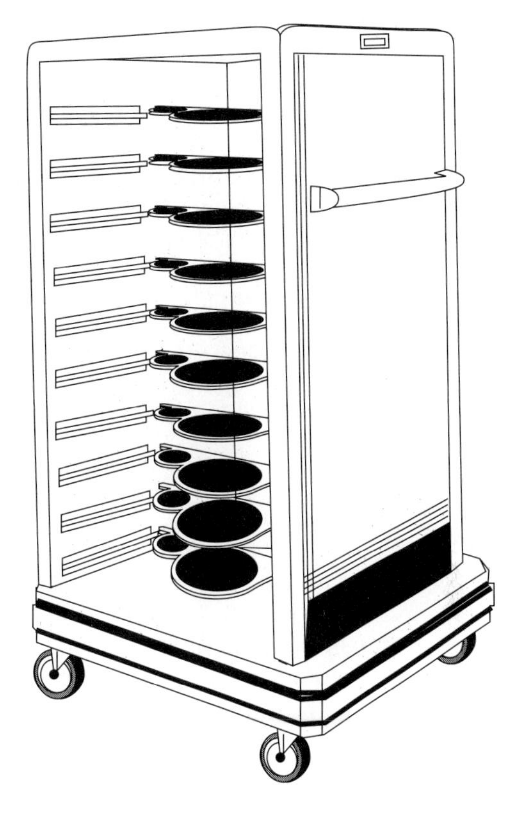

Figure 7.10 Aladdin rethermalization cart. Meals are assembled cold; assembled trays are placed in carts, on specially designed conduction heating elements. Carts are held in walk-in refrigerators until 35 minutes before meal service, when meals are heated within the cart.
Courtesy of Aladdin Synergetics, Inc., Nashville, Tennessee.

Equipment for Specific Uses

Reheating Frozen or Chilled Foods. Foods prepared, cooked, and then frozen or chilled for later service must be reheated at serving time. This may be done in the central serving area or in serving pantries throughout the facility. Equipment used for reheating in both cases is the same and includes convection ovens, conduction (conventional) ovens, microwave ovens, infrared ovens, and the integral cabinets described in the previous section. Also used are immersion equipment (for food in pouches) such as steam-jacketed kettles or tilting frypans. Microwave ovens are the fastest for single portions, but unless a fleet of these ovens or a tunnel-type microwave is available, reheating a large number of meals can take a long time. Convection ovens with the forced-air heat can reheat many meals at one time, the number depending on size model selected. Frozen foods usually are tempered in the refrigerator before reheating to reduce time for bringing foods to serving temperature. With any rethermalization system, the objective is to heat the food product to service temperature and to retain nutrient content, microbial safety, and sensory quality.

Meal Assembly. The assembling of meals for service is an important step in the delivery-service system. Methods vary for various types of establishments and the activities involved must be suited to the specific needs of each.

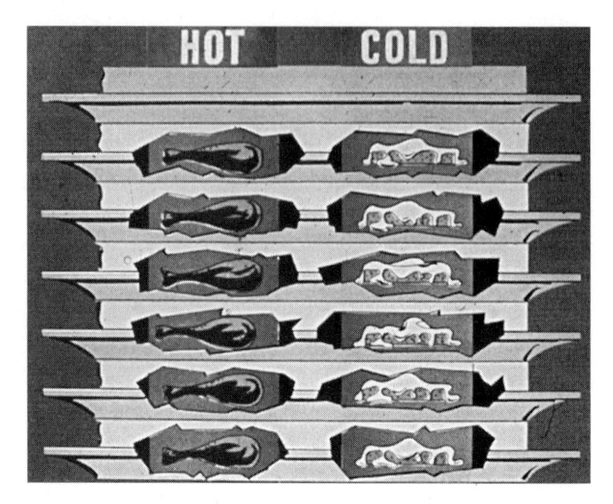

Figure 7.11 Cutaway section of one-piece insulated tray servers. They are stacked for delivery and designed for natural thermal retention—hot foods on trays are stacked above hot food, cold foods above cold. An insulated cover protects the top and bottom of stack, and the system provides proper serving temperatures. Lightweight stacks are easy to carry for delivery on the premises or off. Courtesy of Aladdin Synergetics, Inc., Nashville, Tennessee.

Meal assembly requires that the various menu items that make up a meal be collected and put in one place. This may require equipment as simple as a convenient table or counter for bagging or plating the foods cooked behind the counter in a fast-food restaurant. In a table service restaurant, servers may pick up the cold foods at one or more stations and hot foods from the chef's station and assemble them all on a tray for service.

The most complex type of assembly is that of tray service for many patients or other consumers. Trays, dishes, silverware, and food are prepositioned along a conveyor belt. Employees are stationed to place a specific item or items on the tray as it passes along. A patient menu or diet card precedes the tray and indicates which menu items should be placed on the tray. Conveyors of various types are commonly used for this purpose. All must be sized to the width of the trays used. The simplest is a manual or self-propelling conveyor with rollers that move trays when they are pushed from one station to the next. Others are motor driven. Power-operated conveyors can be set at varying speeds for moving trays along the belt automatically (see Figure 7.12). Conveyors may be mobile or built in.

Temperature Maintenance and Holding. Foods prepared and ready for service often must be held for short periods until needed, while being transported to another area for service, or during the serving period itself. Equipment for this short-time holding includes refrigerated and heated storage units of many types. Note that heated storage cabinets will *not* heat the food, but will, when preheated, maintain for short periods the temperature of the food as it was when it was placed in them.

Heated or refrigerated cabinets may be built in, pass through from kitchen to serving area, or from mobile carts and trucks of all types, some designed with both refrigerated and heated sections. Movable refrigerated units are often used for banquet service. Salads and desserts can be preportioned and placed in the production area and held until moved to the banquet hall at service time. Likewise, hot foods for large groups can be portioned and placed in preheated carts close to serving time but held until all plates are ready to be served at the same time.

Infrared lamps are also used to keep certain foods hot on a serving counter during the serving period.

Transportation and Delivery. Equipment for transportation and delivery is described earlier in this chapter under "Mobile Equipment" and "Portable Equipment." Open or closed noninsulated carts, including the monorail, are used to transport meals served on pellet- or capsule-heated dishes, or placed on insulated trays with covers. Temperature maintenance carts with heated and refrigerated sections or insulated nonheated carts are used to transport meals preplated on regular dishes and placed on noninsulated trays. Other carts are designed with heated wells and compartments for bulk amounts of soup, vegetables, meats, and so forth, as well as for cold ingredients and other food items for meal assembly in another location.

Roll-in refrigerator units serve as transport equipment also with preplated salads and desserts set up in the production area and moved later to the dining areas. Similarly, other mobile serving equipment, such as banquet carts and buffet or catering tables, can serve the dual functions of transporting and serving. Some catering carts for snack items, soups, sandwiches, and beverages are used to take foodservice to workers in plants or office buildings. Insulated totes are inexpensive yet effective means for home delivery of meals.

Many methods and pieces of equipment are available for transporting food from the kitchen to the consumer. The manager must identify the specific needs of the organization when choosing among them. Consideration must be given to the total number to be served; the distance to be traveled between production and service areas; layout of the building with routes including doors, ramps, and elevators involved; and the form of food to be transported: hot, cold, bulk, or preplated.

Serving. Cafeteria counters of varying configurations and with sections for hot and cold foods, buffet tables with temperature-controlled sections and sneeze

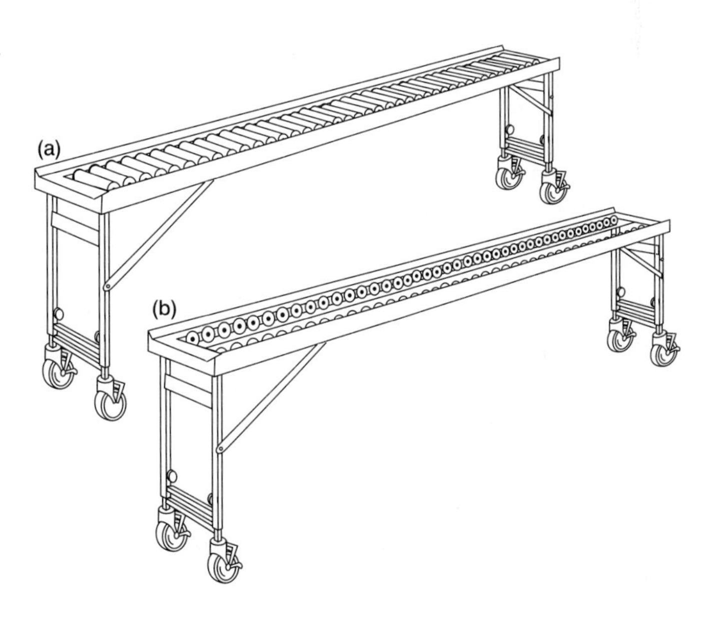

Figure 7.12 (a) Roller-type conveyor and (b) skate wheel conveyor. Both are mobile, stainless steel construction, adjustable in height, and NSFI approved. Courtesy of Precision Industries, Inc., Miami, Florida.

guards, and vending machines all provide a means for self-service. Various methods for tray service have been described. The diagram of alternate plans for tray service shown in Figure 7.13 provides a summary of them.

For dining room table service, trays or carts are used to carry the assembled menu items ordered to the guests. Serving stations, small cabinets often located in or near the dining room, are equipped with table setup items such as silverware, napkins, and perhaps water, ice, glasses, coffee, and cups. This speeds service and reduces the distance traveled to serve guests. Other specialized serving equipment is noted under "Styles of Service."

STYLES OF SERVICE

There are many styles of services, differing widely among institutions, but all with the common objective of satisfying the consumer with food of good quality, at the correct temperature for palatability and microbial safety, and that is attractively served.

The style of service selected, appropriate for a particular type of foodservice operation, should contribute toward reaching those objectives. Also, the style must be economically compatible with the goals and standards of the organization. The basic types or styles of service discussed follow:

1. *Self-service:* cafeteria—traditional, free flow, or scramble; machine vended; buffet, smorgasbord, salad bar; and drive-thru pick-up.
2. *Tray service:* centralized or decentralized.
3. *Waiter-waitress service:* counter, table—American, French, Russian, family, banquet; and drive-in.
4. *Portable meals:* off-premise or on-premise delivery.

Self-Service

The simplest provisions for foodservices involve guests or customers carrying their own food selection from place of display or assembly to a dining spot. The best known example of self-service is the cafeteria, although buffet service with its variations, smorgasbord and salad bars, and vending are also popular.

Cafeteria. Cafeterias are of two types. The *traditional* cafeteria is one in which employees are stationed behind the counter to serve the guests and encourage them with selections as they move along a counter displaying the food choices. There are many configurations for counter arrangement, from the straight line to parallel or double line, zigzag, and U-shaped. In each case, however, the patrons follow each other in line to make their selections.

The traditional self-service is used in colleges and other residences, cafeterias open to the public, school lunchrooms, in-plant foodservices, and commercial operations. The emphasis is on standardized portions and speedy yet courteous service.

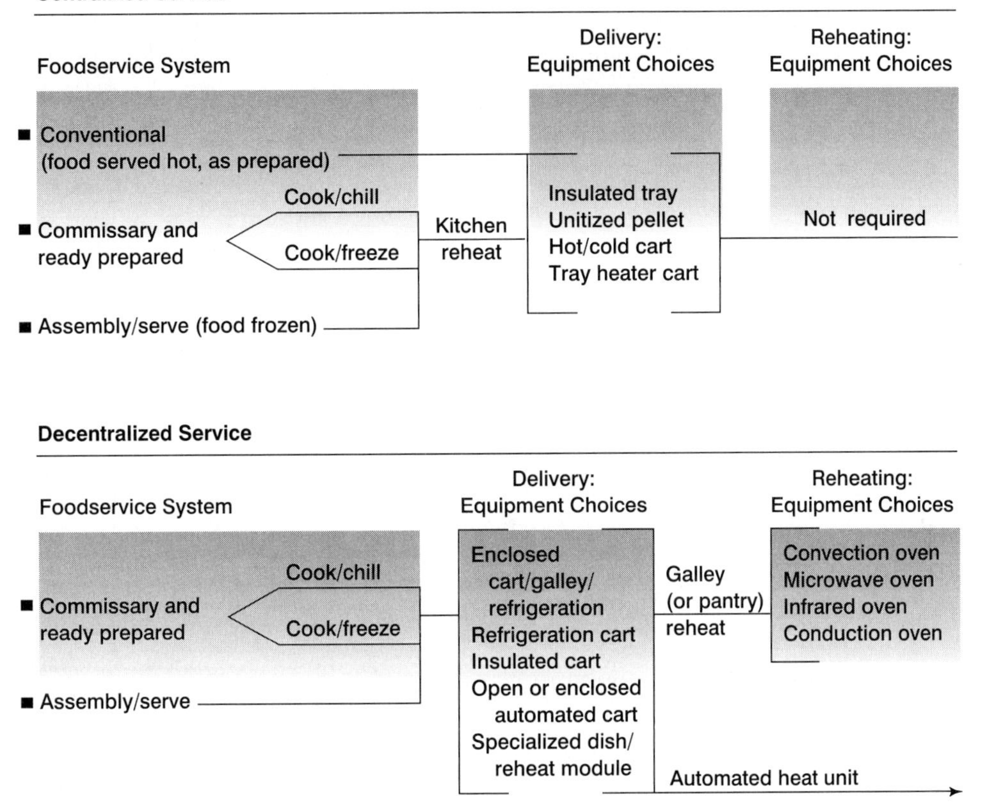

Figure 7.13 Alternative systems for food service to patients.
Adapted from Herz et al. *Analysis of Alternative Patient Tray Delivery Concepts.* Natick, Mass.: U.S. Army Natick Research & Development Command, 1977.

The rate of flow of people through the cafeteria line varies according to the number of choices offered and patron familiarity with the setup.

The second type of cafeteria is known variously as the *hollow square, free flow,* or *scramble system*. In this, separate sections of counter are provided for various menu groups, such as hot foods, sandwiches and salads, and desserts. The sections are usually placed along three sides of the serving room, and customers flow from the center to any section desired. This may seem confusing for the first-timer, but it does provide speed and flexibility by eliminating the need to wait in line for customers ahead to be served. Also, it relieves the pressure on those who do not wish to hurry in making decisions. To be successful, it is necessary to have repeat business and a mechanism for controlling the number of people who enter at one time.

Machine Vended. The history of vending dates back as far as 215 B.C. in Greece, but *food* vending began in this country centuries later with penny candy and gum machines. Other items such as cold drinks and coffee soon were dispensed from vending machines. Today, a complexity of menu items, including complete meals, is available through vending machines. Some contain heating elements to cook or reheat foods before dispensing them. Others are refrigerated, or low temperature controlled for holding frozen foods, such as ice cream.

Machine-vending foodservice skyrocketed in use and popularity in the 1950s and 1960s as it met a demand for speedy service and meant that foodservice was available 24 hours a day, seven days a week. Its popularity continues, and today vending is accepted as an important component of the foodservice industry, especially as a means for supplementing other styles of service. Schools, residence halls, hospitals, industrial plants, office buildings, and transportation terminals in particular have used this mode of service for coffee breaks, after-meal-hour snacks, and, in some, as the sole means of providing meals.

Food for the vending machines can be prepared by the institutions using them, or by an outside vending company that delivers fresh foods at frequent intervals and keeps the machines supplied and in good working order. Fast turnover of the food and good supply service are requisites for the safety and success of vended foods. Also, the foods offered must be fresh and displayed attractively. Glass-front display cases are far superior to metal-front machines for this purpose. Cleanliness and adherence to city health and sanitation codes are essential. Cooperative efforts by those concerned with packaging, production, merchandising, transportation, storage, and sanitation have brought about improvements in the quality and variety of the food offered and will continue to do so in the future. Another innovation is the "talking" machine, which "communicates" some brief statement to the customer.

Buffet. Buffet service, such as the smorgasbord and the popular salad bar, provides a means for dramatically displaying foods on a large serving table. Guests move around the table to help themselves to foods of their choice. Selections usually are numerous, and eye appeal is an important factor in the foods offered. Foods that hold up during the meal hour and the proper equipment to keep this food hot or cold as desired are essential to the success of this type of service. For aesthetic appeal and to comply with health regulations, displayed food must be protected against patron contamination. Portable sneeze guards placed around the foods give some protection as customers serve themselves. Figure 7.14 is an example of one type of mobile food bar with sneeze guard.

Drive-Thru Pick-Up. This type of service, popular with fast-food establishments to speed customer service, is a variation of the drive-in service. Customers drive through the restaurant grounds in a specially designated lane, make their food selection from a large menu board posted outside, and call their order in through a speaker box (usually next to the menu board). By the time they reach the dispensing window, their order has been assembled and packaged for pick-up.

Tray Service

Meals or snacks that are assembled and carried on a tray to individual consumers by an employee is a type of service provided for those unable to utilize other dining

Figure 7.14 Portable food bar with sneeze guard and separate heating controls, adaptable for many uses.
Courtesy of Restaurant Equippers, Columbus, Ohio.

facilities. Airlines, hospitals, nursing homes, and other health care facilities use this method. For persons who are ill or infirm, attractively appointed trays served by pleasant-mannered employees do much to tempt their appetites and help restore health. Airline travelers also appreciate attractive meals served to them in flight and often judge the airline company by the quality of its food and service.

The two types of tray delivery service in hospitals, centralized and decentralized, are described earlier in this chapter. After trays have been transported to serving pantries, they are carried to patients by an employee either of the nursing or dietary department. Good cooperation between the two departments is a prerequisite to coordinate timing for prompt delivery. Delays in getting trays to patients can cause loss of temperature and quality of the food and, thus, a major objective of foodservice is not attained.

Meals for airline foodservice are supplied by the airline's own commissary or on contract from caterers. Foods generally are preplated at the commissary and loaded, either hot or cold, into insulated holding cabinets that are delivered to planes at flight time. Meals loaded hot must be served soon after takeoff; those chilled are reheated on the plane in either small convection or microwave ovens. The length of the flight may determine the method to be used because adequate time must be allowed for reheating all meals. Flight attendants assemble the meals in the galley, placing the hot entrée (if one is served) on a tray with cold foods and hand carry the trays or wheel them down the aisle on carts to service passengers where they sit.

Beverages are distributed in a similar manner. Many airlines have expanded the use of cold plates, sandwiches, and snacks on shorter flights, but all continue to use good merchandising techniques to make food attractive and satisfying and so help meet the competition for business among airlines.

Waiter-Waitress Service

Counter. Lunch counter and fountain service are perhaps the next thing to self-service in informality. Guests sit at a counter table that makes for ease and speed of service and permits one or two attendants to handle a sizable volume of trade. Place settings are laid and cleared by the waiter or waitress from the back of the counter, and the proximity of the location of food preparation to the serving unit facilitates easy handling of food. The U-shaped counter design utilizes space to the maximum, and personnel can serve many customers with few steps to travel.

Table Service. Most restaurants and hotel or motel dining rooms use more formal patterns of service in addition to the counter service, although both employ service personnel. Many degrees of formality (or informality) can be observed as one dines in commercial foodservice establishments around the world. Generally, the four major styles of service classified under table service are American, French, Russian, and banquet.

American service is the one generally used in the United States, although all styles are employed to some degree. A maître d', host, or hostess greets and seats the guests and provides them with a menu card for the meal. Waitresses or waiters place fresh table covers, take the orders, bring in food from the kitchen serving area, serve the guests, and may also remove soiled dishes from the tables. Busers may be employed to set up tables, fill water glasses, serve bread and butter, and remove soiled dishes from the dining room. Checkers see that the food taken to the dining room corresponds with the order and also verify prices on the bill before it is presented to the guest. Characteristic of this type of service is that food is portioned and served onto dinner plates in the kitchen.

French service (synonymous with "fine dining") is often used in exclusive, elegant restaurants. In this style, portions of food are brought to the dining room on serving platters and placed on a small heater (*rechaud*) that is on a small portable table (*gueridon*). This table is wheeled up beside the guests' table and here the chief waiter (*chef de rang*) completes preparation, for example, boning, carving, flaming, or making a sauce. The chief waiter then serves the plates, which are carried by an assistant waiter (*commis de rang*) to each guest in turn. This style is expensive, because two professionally trained waiters are needed to serve properly and service is paced slowly. The atmosphere is gracious, leisurely, and much enjoyed by patrons because of the individual attention they receive.

Russian service is the most popular style used in all of the better restaurants and motel dining rooms of the world. Due to its simplicity, it has replaced, to a high degree, the French style that seems cumbersome to many. In Russian service, the food is completely prepared and portioned in the kitchen. An adequate number of portions for the number of guests at the table are arranged on serving platters by the chef. A waiter or waitress brings the platters, usually silver, with food to the dining room along with heated dinner plates and places them on a tray stand near the guests' table. A dinner plate is placed in front of each guest. The waiter then carries the platter of food to each guest in turn and serves each a portion, using a spoon and a fork as tongs in the right hand and serving from the left side. This is repeated until all items on the menu have been served. Although this service is speedy, only requires one waiter, and needs little space in the dining room, it has the possible disadvantage that the last person served may see a disarrayed unappetizing serving platter. Also, if every guest orders a different entrée, many serving platters would be required.

Banquet service, unlike other types discussed, is a preset service and menu for a given number of people for a specific time of day. Some items, such as salads, salad dressings, butter, or appetizers, may be on the table before guests are seated. Either the American style or the Russian style of service is used.

For American style, dinner plates filled in the kitchen are transported to the guests in one of several different ways; preheated carts are filled with numerous plates and taken to the dining room before guests arrive. Service personnel remove plates from these carts to serve their guests. Another way is for each server to obtain two dinner plates from the serving station, and with one in each hand, go, as a group, to the dining room and serve one table completely. Several trips back and forth are required to finish this service. Still another method is to use busers to carry trays of dinner plates to the dining room, place them on tray stands, and return for another load. Service personnel, working as a team in the dining room, serve the plates as busers bring them in. The head table is served first; then the table farthest from the serving area is served next, so that each succeeding trip is shorter. All guests at one table are served before proceeding to the next table.

The Russian style is used at banquets as described for restaurant service. Sixteen to 20 guests per server is a good estimate for banquet service.

Family style is often used in restaurants or residences of various types. Quantities of the various menu items, appropriate for the number of guests at the table, are served in bowls or platters and placed on the dining table. Guests serve themselves and pass the serving dishes to the others. This is an informal method that is popular for Sunday "fried chicken dinner specials" and in Chinese restaurants for foods that are to be shared, family style. Family-style service is used in some long-term care facilities in an effort to create a home-like atmosphere.

Drive-In Service. Drive-in service requires waiters or waitresses (carhops) to serve patrons who drive up to the restaurant and remain in their parked cars. This type of service was most popular in the 1950s and 1960s. Today patrons seem to prefer going into a restaurant instead of eating in their cars. Most drive-ins provide dining space inside. Although many still conduct business as they did in the past and enjoy success doing so, others have changed to the drive-thru customer pick-up method described earlier.

Table Settings and Serving Procedures

Many foodservice operations including restaurants, catering services, long-term care facilities, and hospitals are emphasizing formal or fine dining in at least some of their service options. For example, a nursing home may offer formal dinners for special events such as birthdays, holidays, or anniversaries. A catering service may offer formal sit-down dining, family-style dining, or a buffet. Proper table settings are defined for each type of service.

According to *Emily Post's Etiquette* (1984) few absolute rules apply to table setting but the end result must achieve geometrical balance; the centerpiece should be in the center, places should be set at equal distances, and utensils should be balanced.

The amount of space allotted to each guest or customer is referred to as the cover. Amounts vary but 24 to 36 inches of table edge space should be allowed for each guest. Dinner plates and the handles of flatware should be placed 1 inch from the edge of the table. This is referred to as the set line. Figure 7.15 illustrates a correct place setting for a luncheon.

Waitress-waiter training manuals are excellent resources in which to find proper serving techniques. The following are a few of the traditional service rules used in formal dining:

1. Using his or her left hand, the waitperson serves the food from the left-hand side of the guest.
2. The dinner plate is placed directly in front of the guest.
3. The waitperson uses his or her right hand to serve beverages from the right-hand side of the guest.
4. Dishes and beverage containers are removed from the left.

Portable Meals

Off-Premise Delivery. One example of off-premise service is delivering meals to the homes of elderly, chronically ill, or infirm individuals not requiring hospitalization. This plan, sometimes called Meals-on-Wheels, attempts to meet the need for nutritious meals for those persons who are temporarily disabled, or for the elderly who may live alone and are unable to cook for themselves. In communities where such a plan is in operation, meals are contracted and paid for by the individual in need of the service or

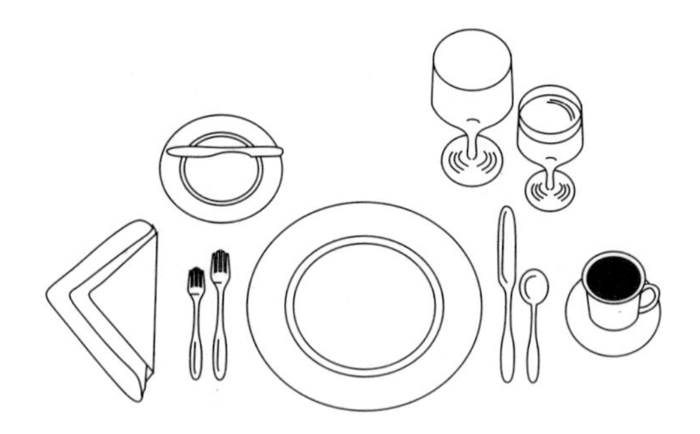

Figure 7.15 Luncheon place setting.

by some federal or community agency or volunteer organization for persons unable to pay. Desirably, the menus are planned by a dietitian working cooperatively with the organization providing the meals. Food may be prepared be restaurants, hospitals, colleges, or other foodservices and delivered by volunteer workers. Preplated meals are covered and loaded into some type of insulated carrier to ensure food safety while in transit and to maintain desired temperatures until delivered to the home.

A similar service is provided by caterers for workers in office buildings or to customers in their homes by pizza restaurants or others, but on a profit-making basis.

On-Premise Delivery. Another example of portable meals often used in some industrial plants is the distribution of foods to workers at their workplace by mobile carts that move throughout the plant. Carts are equipped with heated and refrigerated sections for simple menu items such as soup, hot beverages, sandwiches, snack items, fruits, and pastries. Workers pay the cart attendant as selections are made. This provides a time-saving service for employees who might have long distances to go to a central cafeteria in a large plant during a short meal period.

An alternative type of portable service is utilized by some companies not having foodservice facilities: A mobile canteen is provided by a catering firm and driven each day to the yard of the plant. Workers go outside to buy their meals from the canteen truck.

Although variations of these basic styles of service can be found in today's innovative foodservice systems, the types discussed here should provide understanding of the most commonly used service systems.

SUMMARY

The delivery and service of food after it has been prepared are important aspects of the total foodservice system. Consumer satisfaction depends in large part on the pleasing presentation of carefully prepared, assembled, and transported food in every type of foodservice operation.

Foodservice managers should be cognizant of the major goals of delivery and service systems. These goals are to maintain quality food characteristics including desirable temperatures, ensure microbial safety, and present food attractively. In addition, the system selected should save worker steps and energy, reduce labor time and costs, and lessen worker fatigue.

Factors affecting the selection of a particular delivery system, either centralized or decentralized, as well as the appropriate equipment needed include the type of foodservice system (conventional, commissary, ready-prepared, or assembly/serve); the type of organization, such as school, hospital, commercial, or other; the size of physical facilities and amount of space available; the style of service to be used; the skill level of personnel; the labor and equipment costs involved; the quality standards required and desired; the timing for meal service; and the energy usage involved.

The style of service used whether self-serve, tray, or waiter-waitress service must be appropriate for the type of operation and for attaining its goals. Training the workers to use correct serving procedures and to present the food to the consumers in a pleasing and courteous manner is also an essential element in achieving a successful foodservice operation.

REVIEW QUESTIONS

1. Define the terms *delivery* and *service* as they relate to foodservice in institutions.
2. Discuss the advantages and disadvantages of centralized and decentralized delivery-service systems.
3. Describe the impact of the skill level of available personnel on the choice of delivery-service systems.
4. What is the difference between energy usage of "passive" and "active" temperature retention equipment?
5. Describe the various equipment options that can be employed for heat retention in a centralized hospital delivery system.

SELECTED REFERENCES

Bryner, R. A.: *Hospitality Management: An Introduction to the Industry,* 7th ed. Dubuque, Iowa: Kendall-Hunt Publishing Company, 1995.

Dowling, R. A., Cotner, C. G.: Monitor of tray error rates for quality control. J. Am. Diet. Assoc. 1988; 88:450–453.

The Effective Trayline. Kansas City, Mo.: Crimsco Inc., 1991.

Franzese, R.: Food services survey shows delivery shift. Hospitals. August 16, 1984; 58:61.

Friedland, A.: Rethermalization carts. Food Mgmt. 1995; 30:44.

Khan, M.: *Foodservice Operations.* Westport, Conn.: AVI Publishing Company, 1987.

Post, E. L.: *Emily Post's Etiquette: A Guide to Modern Manners,* 14 ed. (pp. 689–706). New York: Harper & Row Publishers, 1984.

Spears, M. C.: *Foodservice Organizations: A Managerial and Systems Approach,* 3rd ed. Englewood Cliffs, N.J.: Prentice Hall, 1995.

Transporting at safe temperatures. Foodservice Equipment & Supplies. June 1994, pp. 55–56.

The Trayline Doctor. Kansas City, Mo.: Crimsco Inc., 1995.

3

Physical Facilities

8

Cleaning, Sanitation, and Facility Safety

Providing a clean, safe foodservice facility is basic to achieving a successful operation and ensuring the health and well-being of both employees and customers. Also, a clean, safe environment contributes to the aesthetic satisfaction that guests derive from dining and gives a feeling of personal security to all.

Sanitation and safety are closely related environmental factors, to be considered when planning a facility and followed in its daily operations. Maintaining high standards of cleanliness and making sure that the workplace is free from hazards are management responsibilities.

To help maintain minimum sanitation and safety standards, certain regulations have been established and enforced by city, state, and federal law. Agencies such as the U.S. Public Health Service (USPHS) and state and city health departments, and organizations such as the National Sanitation Foundation International (NSFI), Occupational Safety and Health Association (OSHA), and the National Safety Council recommend standards. Results of research conducted at universities, hospitals, and food, chemical, and equipment manufacturing companies provide data for recommending or setting standards. These groups, individually or cooperatively, work with the foodservice industry to prepare, distribute, and interpret pertinent information in publications and exhibits. Some present seminars, classes, and programs for various foodservice groups and the public. All are aimed at informing concerned persons of the rules, regulations, and standards to be met and how to achieve them.

Standards of cleanliness and sanitation will be only as high as those established and enforced by the foodservice director. That person must create the philosophy of good sanitation and impart a sense of urgency about the matter to the employees. This is best accomplished through an ongoing training program for the foodservice workers. This assumes that foodservice directors themselves have had some training and are knowledgeable about sanitary procedures and practices advocated in legislative measures. In many cities, a formal training program in food protection and safety is mandatory before foodservice operators can obtain a required food sanitation certificate. In other cities, voluntary or mandatory certification programs for managers, as well as workers, have been initiated by state restaurant associations or by the state health department.

The purpose of this chapter is to review the principles of cleaning and sanitation as they relate to warewashing and maintenance of facilities. Worker safety is also included in this chapter because it closely relates to the care and maintenance of an operation. The chapter concludes with guidelines on how to design, implement, and monitor a safety program.

KEY CONCEPTS

1. The design and layout of a foodservice operation have a significant impact on cleaning and sanitation management.
2. The foodservice manager must understand the terminology and principles of cleaning and sanitation in order to design, implement, and manage an effective cleaning and sanitation program.
3. Dishwashing is a cleaning and sanitizing process that can be done manually or by machine.
4. Facility cleaning must be planned, organized, and scheduled to ensure that equipment and facilities are cleaned appropriately and in a timely fashion.
5. Pest control is an integral part of a cleaning program.
6. OSHA is a federal agency whose primary objective is to ensure safe and healthful working conditions for working men and women.

SANITARY DESIGN OF FACILITIES

Numerous factors influence the management of a sanitation program in any given foodservice operation—not the least of which is the design of the facility itself. Many of the factors related to sanitary design are covered in Chapter 10, Facilities Planning and Design. A few of the factors are reviewed here because they are an integral part of sanitation management.

The Physical Plant

Built-In Features. Many features that facilitate easy cleaning and maintenance of the foodservice should be built in at the time of construction. For example, the coving, a curved, sealed edge between the floor and wall, eliminates sharp corners that would be difficult to clean.

Water Supply. The adequacy and quality of the water supply available to the institution are of prime concern in the operation of a sanitary foodservice. It is relatively easy to check on the water and its safeness from the reports of the department of health. The relative hardness of the water will have a bearing on the effectiveness of detergents and other chemicals.

Trash and Waste Removal. Provision for the daily removal of trash and any other wastes not eliminated by mechanical means is essential for good sanitation. Until such disposal is made, garbage should be stored temporarily in garbage cans with tight-fitting lids, in a well-ventilated area. The containers should be scrubbed daily, rinsed, and sanitized.

Sanitary Facilities: Rest Rooms and Locker Rooms. Adequate sanitary facilities for employees are basic requirements that must be provided in the physical plant of any food or related service. Most states have laws specifying the kind and number of facilities to be provided.

Hand-washing facilities must be provided in an area adjacent to the toilet room and be supplied with running hot and cold water, soap, and paper towels or air-drying blowers. The hot water supply for hand washing should not exceed 120°F (lower than required for dishwashing), otherwise persons could be burned. Toilet and wash room should have self-closing doors and completely screened outside windows or built-in ventilation fan(s).

Another provision for employee cleanliness and comfort is locker rooms, one for each sex, with ample space for changing clothes. Minimal equipment includes individual lockers, with locks for storage of uniforms, and benches to sit on while changing.

In the kitchen work area, drinking fountains and additional hand-washing facilities, separate and apart from food preparation sinks, are required. They should be placed so that they are readily available to employees in various parts of the kitchen.

Only if a lavatory is close to the workstation will most employees wash their hands frequently. Otherwise, they tend to use food preparation or pot washing sinks, or go without washing soiled hands, both of which are unsanitary practices. Sinks that are used for washing equipment or utensils should not be used for hand washing.

Equipment Design and Placement

Each piece of equipment that will come in contact with food should be so designed and constructed that the food contact surfaces are nonabsorbent (i.e., stainless steel), continuous, smooth, and free of open seams, cracks, chipped places, exposed junctions, and sharp corners. All junctions should be rounded or coved. The food contact surface should not only be readily cleanable, but also readily accessible for cleaning.

The NSFI has been instrumental in establishing sanitation standards for materials used in foodservice equipment and for equipment design, construction, installation, and maintenance. These standards, based on scientific research, are agreed on by joint committees made up of qualified representatives from research, industry, education, and the public health profession. Publications are available from the foundation in which minimum standards are clearly and specifically defined and are adaptable to various geographic areas. Attention will be called to some of the standards in this text, but each foodservice director should request specific standards as needed from the NSFI.

Manufacturers who meet the standards are privileged to use the seal of approval of the NSFI (see Figure 8.1) on their products, and the purchaser is assured of acceptable design, materials, construction, and performance if the equipment is properly installed and operated. Recommended standards give much attention to the proper conditions for installation and use of equipment as well as construction in the factory.

The placement of the pieces of large fixed equipment should be determined thoughtfully to allow the worker space for necessary activity and to avoid accidents and spillage. Adequate space beneath and behind heavy equipment is necessary for mopping and cleaning if such items are not built into the floor or wall with proper fittings and joinings. Figure 8.2 is an illustration of equipment of simple design with a good work arrangement that ensures that high sanitation standards can be maintained.

CLEANING AND SANITATION

The terms *cleaning* and *sanitizing* are frequently assumed to be one and the same when in fact there are important differences. Cleaning is the physical removal of visi-

Figure 8.1 National Sanitation Foundation International's seal of approval.

Figure 8.2 Simple design and good construction of equipment facilitate ease of maintenance and contribute to the beauty of a modern foodservice. Note that here the equipment is built into the floor or is mounted on wheels.

ble soil and food from a surface. Sanitizing is a process that reduces the number of micro-organisms to safe levels on food contact surfaces such as china, tableware, equipment, and work surfaces. Sanitized surfaces are not necessarily *sterile,* which means to be free of micro-organisms.

Cleaning and sanitizing are resource-intensive functions in any foodservice operation. These functions require time, labor, chemicals, equipment, and energy. Careful design and monitoring of the cleaning and sanitizing functions result in optimal protection of employees and customers. Mismanagement of these two functions can result in:

- Injury to employees and customers,
- Waste of chemicals and money, and
- Damage to equipment and facilities.

It is essential that managers understand the principles of cleaning and sanitizing and the many factors that influence these processes.

Principles of Cleaning

Cleaning is a two-step task that occurs when a cleaning compound (or agent) such as a detergent is put in contact with a soiled surface. Pressure is applied using a brush, cloth, scrub pad, or water spray for a long enough period of time to penetrate the soil so it can be easily removed during the second step of rinsing. Many factors influence the effectiveness of the cleaning process. Table 8.1 is a summary of these factors. Each of these factors must be considered when making a cost-effective selection of detergents and other cleaning compounds such as solvents, acids, and abrasives.

Detergents. The selection of a compound to aid in cleaning the many types of soil and food residues is a complex one because so many compounds are available from which to choose. An understanding of the basic principles involved in cleaning will assist the foodservice manager in making this decision.

Detergents are defined as cleansing agents, solvents, or any substance that will remove foreign or soiling material from surfaces. Specifically listed are water, soap, soap powders, cleansers, acids, volatile solvents, and abrasives. Water alone has some detergency value, but most often it is the carrier of the cleansing agent to the soiled surface. Its efficiency for removing soil is increased when combined with certain chemical cleaning agents.

The three basic phases of detergency are penetration, suspension, and rinsing. The following actions and agents are required for each phase:

1. *Penetration:* The cleaning agent must penetrate between the particles of soil and between the layers of soil and the surface to which it adheres. This action,

Table 8.1 Factors that influence the cleaning process.

Factor	Influence on Cleaning Process
1. Type of water	Minerals in hard water can reduce the effectiveness of some detergents. Hard water can cause lime deposits or leave a scale, especially on equipment where hot water is used, such as in dishmachines and steamtables.
2. Water temperature	Generally, the higher the temperature of the water used for cleaning, the faster and more efficient the action of the detergent; however, $\leq$ 120°F is recommended (and in some cases mandated) as higher temperatures can result in burns
3. Surface	Different surfaces, especially metals, vary in the ease with which they can be cleaned
4. Type of cleaning compound	Soap can leave a greasy film. Abrasives such as scouring powders can scratch soft surfaces. Many cleaning agents are formulated for specific cleaning problems. Lime removal products are an example.
5. Type of soil to be removed	Soils tend to fall into one of three categories: protein (eggs), grease or oils (butter), or water soluble (sugar). Stains tend to be acid or alkaline (tea, fruit juice). Ease of cleaning depends on which category the soil is from and the condition of the soil (e.g., fresh, baked-on, dried, or ground-in.

known as *wetting*, reduces surface tension and makes penetration possible. Agents are water, soaps, and synthetic detergents, which are rather fragile suds formers.

2. *Suspension:* An agent is required to hold the loosened soil in the washing solution so it can be flushed away and not redeposited. Agents, which vary according to the type of soil, include the following: *For sugars and salts,* water is the agent because sugars and salts are water soluble and are easily converted into solutions. *For fat particles,* an emulsifying action is required to saponify the fat and carry it away. Soap, highly alkaline salts, and nonionic synthetics may be used. *For protein particles,* colloidal solutions must be formed by *peptizing* (known also as sequestering or deflocculating). This action prevents curd formation in hard water; otherwise, solvents or abrasives may be needed.

3. *Rinsing:* The agent used must remove and flush away soils and cleaners so they are not redeposited on the surfaces being washed. Clean, clear hot water is usually effective alone. With some types of water, a *drying* agent may be needed to speed drying by helping the rinse water drain off surfaces quickly. This eliminates alkaline and hard water spotting, films, and streaks on the tableware or other items being cleaned.

The development of polyphosphate detergents has provided a wide variety of highly satisfactory cleaning compounds. Film deposits from precipitation, poor rinsability, and harshness to hands are no longer problems, and the selection of cleaners can be made to meet the needs for particular uses in the institution foodservices; for instance, a suds-producing hand wash with high wetting action would be satisfactory as such but entirely unsatisfactory for machine dishwashing where a nonsuds-producing detergent is needed.

Selection of the right detergent for the job of cleaning in any situation is determined in large measure by the hardness of the water. The sequestering of the lime and magnesia of hard water by the polyphosphates produces a clear, not muddy, solution with insoluble precipitates, as is the case when some of the phosphates and silicates are used.

Foodservice sanitation is concerned mainly with china, crockery, glass, and metal surfaces. Common soils to be removed are saliva, lipstick, grease, and carbohydrate and protein food particles that may adhere to dishes, glassware, silverware, cooking utensils, worktable tops, floors, or other surfaces. Some types of food soils such as sugars, starches, and certain salts are water soluble. The addition of a wetting agent to hot water will readily remove most of these simple soils. The soils that are insoluble in water, such as animal and vegetable fats and proteins, organic fiber or carbon residues, and mineral oil, are more difficult to remove. Abrasives or solvents may be necessary in some cases to effect complete cleanliness.

The use of a "balanced" detergent or one with a carefully adjusted formula of ingredients suitable for the hardness of the water and the characteristics of the soil is advised in order to produce the best results. The properties of the detergent must cause complete removal of the soil without deposition of any substance or deleterious effect on surfaces washed.

Detergents for *dishwashing machines* must be a complex combination of chemicals that will completely remove the soil in a single pass through a high-speed machine. It must soften the water, solubilize and emulsify greases, break down proteins, suspend soil, protect the metal of the machine, increase wetting action, and counteract minerals in the wash water. Other characteristics desired in some situations are defoaming action where excess sudsing is a problem, and chlorination action where a chlorine-type detergent is used to remove stains and discolorations.

It is the phosphate in dishwashing compounds that reacts with the minerals in the water. Many different types of phosphates are used, from the crude ones, such as trisodium phosphate, to the more refined, such as pyrophosphates; but the most efficient is the highly refined polyphosphate ingredient that completely cancels the mineralization of the water and "conditions" water in the wash tank for the cleansing task.

Managers responsible for purchasing dishwashing and other cleaning compounds should be aware of the chemicals they contain and the suitability for use in a particular situation. The effectiveness of the product instead of the price should be the primary factor in the selection of any cleaning compound. In evaluating the cost of a rinse agent, for example, consider the concentration required (parts per million) for effective drying in relation to the price to get a true usage cost.

The quality of service offered by the company selling detergents is an important factor in selecting a particular brand. Sales representatives know their products' capabilities and can give good assistance in selection of items for a foodservice's specific needs. For example, the salesperson should know as much or more about the operation of various dishwashing machines than do the managers. Since the company representative is accustomed to making quantitative and qualitative tests under differing circumstances and conditions, that person can wisely recommend the amounts of a compound to use. Since it is to the company's advantage to have its products do the best job possible, the representative will not want to recommend using too much because it may be as ineffective as too little.

Dishwashing machines should be equipped with automatic detergent dispensers. Also, a water softener more than pays for itself and is recommended for use in hard water areas.

Solvent Cleaners. Solvent cleaners, commonly referred to as degreasers, are necessary to clean equipment and surface areas that get soiled with grease. Ovens and grills are examples of areas that need frequent degreasing. These products are alkaline based and are formulated to dissolve grease.

Acid Cleaners. Tough cleaning problems such as lime buildup on dishwashing machines and rust on shelving are treated with acid cleaners. There are a number of these products from which to choose and they vary depending on the specific purpose for the product.

Abrasives. Abrasive cleaners are generally used for particularly tough soils that do not respond to solvents or acids. These products must be used carefully to avoid damage to the surface that is being cleaned.

Principles of Sanitation

Immediately after cleaning, all food contact surfaces must be sanitized. Heat and chemical sanitizing are the two methods of sanitizing surfaces effectively.

Heat Sanitizing. The objective of heat sanitizing is to expose the clean surface to high heat for a long enough time to kill harmful organisms. Heat sanitizing can be done manually or by a high-temperature machine. The minimum temperature range necessary to kill most harmful micro-organisms is usually 162°F to 165°F. Table 8.2 summarizes minimum washing and sanitizing temperatures for manual and machine methods.

Chemical Sanitizing. A second method of effective sanitizing is through the use of chemicals. The primary reason for choosing this method over heat sanitizing is the savings that are realized in energy usage. Chemical sanitizing is achieved in two ways. The first is by immersing the clean object in a sanitizing solution of appropriate concentration and for a specific length of time, usually one minute. The second method is by rinsing, swabbing, or spraying the object with the sanitizing solution. The rinsing and spraying methods can be done manually or by machine. Careful management of sanitizers is important for several reasons including these:

- The sanitizer becomes depleted over time and must be tested frequently to ensure that the strength of the solution is maintained for effective sanitizing. Test kits are available from the manufacturer.
- The sanitation solution can get bound up by food particles and detergent residues if surfaces are inadequately rinsed, leaving the sanitizer ineffective.

It is essential that the manager work closely with the manufacturer's representative to ensure that chemical sanitizers are used appropriately. The three types of commonly used chemical sanitizers in foodservice operations are chlorine, iodine, and

Table 8.2 Minimum washing and sanitizing temperatures for heat sanitation.

	Wash	Sanitize
Manual	110°F	170°F
Machine, (spray types)		
1. single tank, stationary rack, single temperature machine	165°F	165°F
2. single tank, conveyor, dual temperature machine	160°F	180°F
3. single tank, stationary rack, dual temperature machine	150°F	180°F
4. multitank, conveyor, multi-temperature machine	150°F	180°F

(1) Some local regulations may mandate stricter standards.
(2) Minimum time for exposure to heat is 1 minute.
(3) 195°F is the maximum upper limit for heat sanitation for manual or machine methods, as higher temperatures cause rapid evaporation and therefore inadequate time for effective sanitation.

Source: From the *1995 Food Code*, U.S. Department of Commerce.

quaternary ammonium compounds (quarts). The properties of these sanitizers are summarized in Table 8.3.

DISHWASHING

Dishwashing (sometimes referred to as warewashing) requires a two-part operation, that is, the cleaning procedure to free dishes and utensils of visible soil by scraping or a water flow method, and the sanitizing or bactericidal treatment to eliminate the health hazard. Dishwashing for public eating places is subject to rigid regulations.

The two groups of equipment and utensils that are commonly considered for discussion under dishwashing are kitchen utensils, such as pots, pans, strainers, skillets, and kettles soiled in the process of food preparation, and eating and drinking utensils, such as dishes, glassware, spoons, forks, and knives.

Kitchen Utensils

Mechanical pot and pan washing equipment is relatively expensive; therefore, in many foodservices this activity remains a hand operation. A three-compartment sink is recommended for any hand dishwashing setup (see Figure 8.3). Public health authorities recommend that sinks used for manual washing and sanitizing operations be of adequate length, width, and depth to permit the complete immersion of the equipment and utensils.

The soil may be loosened from the utensils by scraping and then soaking them in one compartment of the sink, well filled with hot water, previous to the time of washing. A forced-flow pump system unit, such as that shown in Figure 8.4, facilitates the cleaning of pots and pans. It can be installed onto the end of any sink and is relatively inexpensive to install and operate, and highly effective in the loosening and removing of cooked-on food.

After the surface soil has been removed from the utensils, the sink is drained and refilled with hot water to which a washing compound is added. The utensils are washed in the hot detergent solution in the first compartment; rinsed in the second compartment, the outlet and inlet of which are adjusted so as to keep the water level constant if hot water is kept running throughout the process; and sanitized in the third compartment.

There are several methods for sanitizing both dishes and utensils. One recommended method is by immersing them for at least one minute in a lukewarm (at least 75°F) chlorine bath containing a minimum of 50 parts per million (ppm) available chlorine. Dishes and utensils must be thoroughly clean for a chlorine rinse to be an effective germicidal treatment. Another method of sanitizing hand-washed dishes or utensils is immersion in clean soft water of at least 170° for 1 minute. Figure 8.5 illustrates a sink arrangement for the wash, rinse, and sanitize phases of hand dishwashing. (Note that some states require a "prewash" phase in addition to the standard three phases.) Utensils also may be successfully sanitized by subjecting them to live steam in an enclosed cabinet after washing and rinsing. The hot, clean utensils should be air dried before being stacked upside down on racks or hung for storage.

Table 8.3 Properties of commonly used chemical sanitizers.

	Chlorine	Iodine	Quaternary Ammonium
Minimum Concentration			
For immersion	50–100 parts per million (ppm)	12.5–25.0 ppm	200 ppm
For power spray or cleaning in place	100–200 ppm	25–50 ppm	
Temperature of Solution	75°F/23.9°C+	75°–120°F/23.9°–48.9°C Iodine will leave solution at 120°F/48.9°C	75°F/23.9°C+
Time for Sanitizing			
For immersion	1 minute	1 minute	1 minute; however, some products require longer contact time; read label
For power spray or cleaning in place	Follow manufacturer's instructions	Follow manufacturer's instructions	
pH (detergent residue raises pH of solution, so rinse thoroughly first)	Must be below pH 10	Must be below pH 5.5	Most effective around pH 7 but varies with compound
Corrosiveness	Corrosive to some substances	Noncorrosive	Noncorrosive
Response to Organic Contaminants in Water	Quickly inactivated	Made less effective	Not easily affected
Response to Hard Water	Not affected	Not affected	Some compounds inactivated but varies with formulation; read label. Hardness over 550 ppm is undesirable for some quats
Indication of Strength of Solution	Test kit required	Amber color indicates effective solution, but test kits must also be used	Test kit required. Follow label instructions closely

Source: From *Applied Foodservice,* 4th ed., Copyright © 1992 by the Educational Foundation of the National Restaurant Association.

Figure 8.3 Three-compartment stainless steel pot and pan sink with sliding platform. Note the rounded corners for ease of cleaning and good sanitation.
Courtesy of S. Blickman, Inc.

Dishes, Glassware, and Silverware

Washing of these items can be accomplished by hand or by use of mechanical dishwashers. Steps for these are outlined in Figures 8.6 and 8.7. In either case, *prewashing* or *preflushing,* which applies to any type of water scraping of dishes before washing, is recommended to prevent food soil in the wash water. The usual types of water scraping equipment include (1) a combination forced water stream and food waste collection unit built into the scraping table, by use of which dishes are rinsed under the stream of water before racking, (2) a hose and nozzle arrangement over a sink for spraying the dishes after they are in racks, and (3) a prewash cabinet

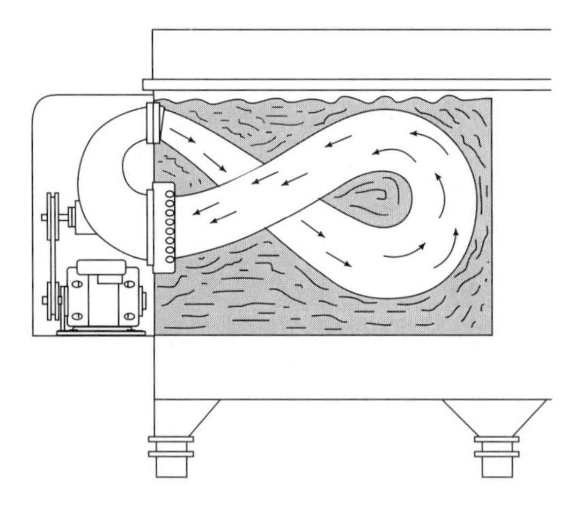

Figure 8.4 A pump-forced flow of water loosens food particles from cooking utensils in a pot and pan sink. The pump (at left) can be easily installed on a sink.
Courtesy of Kewanee Washer Corporation.

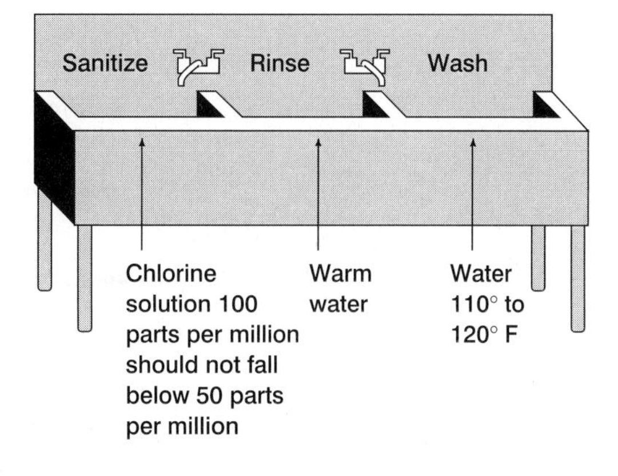

Figure 8.5 Sink arrangement and USPHS-recommended temperature and sanitizing method for hand dish- or panwashing.

through which the racks of soiled dishes pass and are jet sprayed to remove food particles prior to their entering the wash section of the dishwashing machine. The prewash cabinet can be built in as part of the larger model machines or, in small installations, may be a separate unit attached to the wash machine in such a way that the water used is the overflow from the wash tank. The prewash water should be at a temperature of 110°F to 140°F to provide for the liquefying of fat and the noncoagulation of protein food particles adhering to dish surfaces. The installation and use of a prewash system lessens the amount of organic waste and the number of microorganisms entering the wash tank, removes fat that might otherwise result in suds formation, reduces the number of wash water changes, cuts the costs for detergents, and results in cleaner dishes.

Many dishwashing machines are on the market. Various types are discussed in Chapter 11, but Figure 8.8 illustrates the general principle of *how* dishes are washed in a single-tank, spray-type machine. Large machines have divided tanks (Figure 8.9) so that the wash and rinse waters are kept separate and the dilution of the wash water is less rapid.

After the prerinse, the dishes are loaded into racks or on conveyor belts in such a way that food-contact surfaces will be exposed to direct application of the wash water with detergent and to the clean rinse waters. Cups, bowls, and glasses must be inverted and overcrowding or nesting of pieces avoided if dishwashing is to be effective. Wash water shall not be less than 120°F, and if hot water is the sanitizing agent, the rinse water shall be 180°F. Figure 8.10 is an example of a dish machine temperature documentation form. The pressure of the rinse water must be maintained at a minimum of 15 pounds per square inch (psi) but not more than 25 psi to make the sanitizing effective.

China, glassware, and silver can be washed in the same machine, but it is preferred wherever possible to subject glasses to friction by brushes so that all parts of the glass are thoroughly cleaned, which means the use of a special machine designed for that purpose (Figure 8.11). This is especially important in soda fountains, bars, and similar establishments where glasses are the primary utensils used. To prevent water spotting, it is advisable to use a suitable detergent for the washing of silver and also a drying agent with high wetting property in the final rinse water to facilitate air dry-

Manual Dishwashing

Important Steps (in Doing the Job)	*Key Points*[a]
1. Get ready (materials and equipment)	Sinks Hot water Washing powders (chlorine, or other bactericidal treatment, if used) Scraper Garbage can Drying racks
2. Scrape dishes and prewash	Use scraper Garbage in can or disposer
3. Wash dishes	Each piece separately Hot water, 110–120°F Use detergent (washing powder)
4 Rinse	Place in basket Set in hot rinse water
5. Sanitize	Place basket in vat In hot water (170°F for one min or 212°F for 30 sec)[b] 2 min in chemical solution (chlorine of approved strength)
6. Dry	Lift out basket Place on drain board Air dry
7. Store	Cups and glasses bottoms up Stack dishes on mobile carts In clean protected place
8. Clean vats	Stiff brush Washing powder
9. Use separate baskets for dishes, cups, and glasses	
10. Silver may be air-dried or by use of clean towel	
11. Fingers should not touch surfaces which come in contact with food or drink	
12. All multiservice eating and drinking utensils must be thoroughly cleaned after each usage	
13. Single-service containers must be used only once	

[a] Key points are those things which will make or break the job, injure the worker, or make the job easier to do.
[b] This standard may vary with local or area regulations.

Figure 8.6 Job breakdown details procedure for manual dishwashing.

Machine Dishwashing

Important Steps (in Doing the Job)	*Key Points*[a]
1. Get ready (materials and equipment)	Sort dishes, cups, glassware, and silver Water temperature: 140°F wash, 170°F rinse[b] Check washing powder and dispenser, wash and rinse sprays
2. Scrape and prerinse	Use brush to scrape off garbage
3. Rack dishes, etc.	Place in separate racks Do not pile dishes, cups, etc. Cups and glassware bottoms up One kind at a time, separately
4. Place racks in machine	Every dish, etc., under spray
5. Wash	Start machine Turn on wash spray 140° to 160°F Do not hurry machine if it is manually operated Keep washtank water clean
6. Rinse	10 sec 170°F[b]
7. Dry	Remove racks Allow dishes, etc., to dry in racks on drain table or if flight type machine, do not remove dishes before they reach end of conveyer
8. Store	Fingers should not touch surfaces which come in contact with food In a clean, dry place above floor Away from dust and flies Dishes stacked Cups and glasses bottoms up—leave in racks in which they are washed
9. Clean machine	Take out scrap trays and clean Clean wash sprays Use clean water Add new washing powder as required

[a] Key points are those things which will make or break the job, injure the worker, or make the job easier to do.
[b] This means temperature of 170° at the dish, and requires 180°F water in the line. This standard may vary with local or area regulations.

Figure 8.7 Job breakdown details procedure for machine dishwashing.

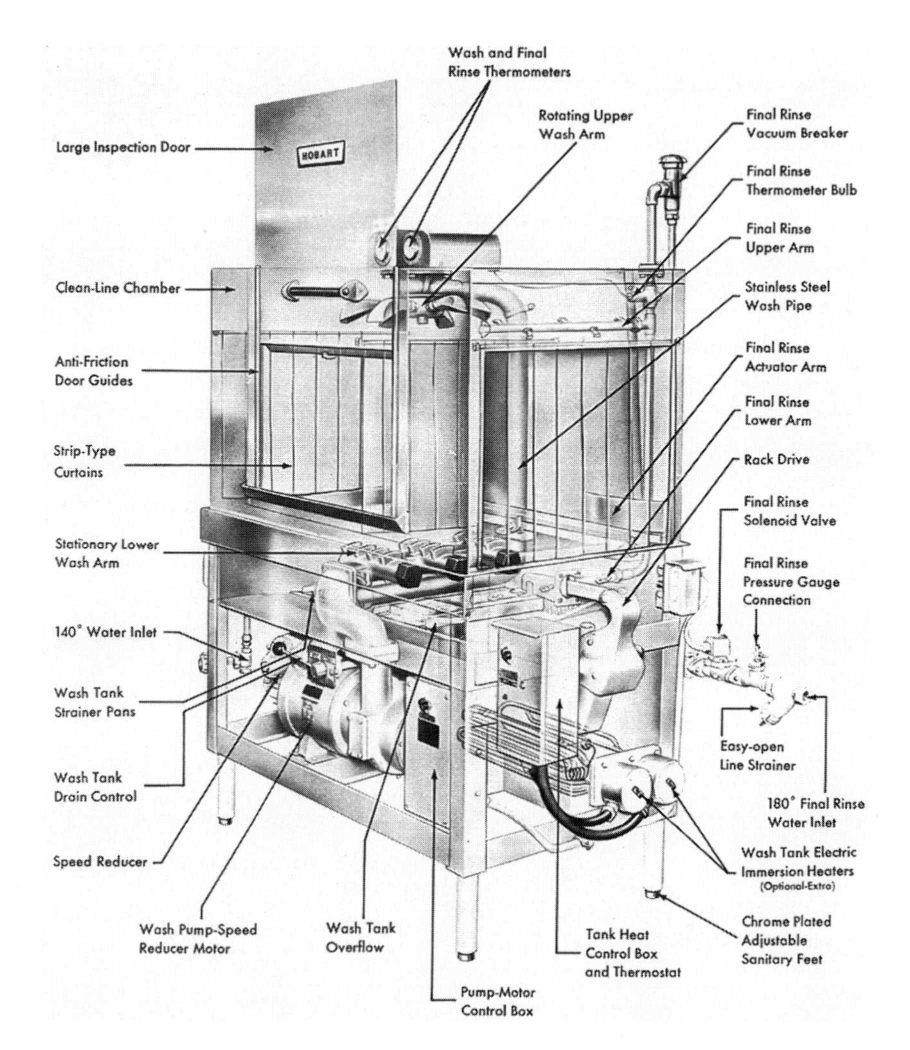

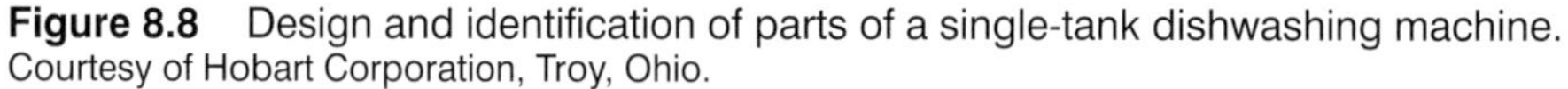

Figure 8.8 Design and identification of parts of a single-tank dishwashing machine. Courtesy of Hobart Corporation, Troy, Ohio.

ing. The introduction of a drying agent with low foam characteristics into the sanitizing rinse promotes rapid drying of all types of tableware.

Provision for the storage of clean glasses and cups in the racks or containers in which they have been washed reduces the possibility of hand contamination.

Some machines are designed for a chemical solution rinse rather than the high-energy use of the 180°F temperature water. In this case, the rinse water used with the chemical sanitizer shall not be less than 75°F or less than that specified by the manufacturer. Chemicals used for sanitizing should be dispensed automatically to make sure that the proper amount and concentration are used.

Dishes can be dried by hot air blast within the machine or allowed to air dry. If a hot water rinse is used, dishes should be hot enough to air dry within 45 seconds before they are removed for storage.

All steps require energy except air drying. For this reason, the low-temperature models are preferred by many operators. To minimize energy use, only fully loaded racks or conveyors should be put through the machine.

The NSFI (see the "For More Information" section at the end of this chapter) offers acceptable standards for wash and rinse cycles of three types of dishwashing machines: (1) single-tank, stationary rack, hood, and door types, (2) single-tank, conveyor type, and (3) multiple-tank conveyor type with dishes in inclined position on conveyor or in rack. These standards can be obtained and used as a check on specifications of various makes of dishwashers and by managers for ensuring that the specified conditions such as water temperature and pressure are met. Employees, too, should be trained to follow proper procedures in the use and care of the dishwashing machine, or hand washing of dishes if no machine is available.

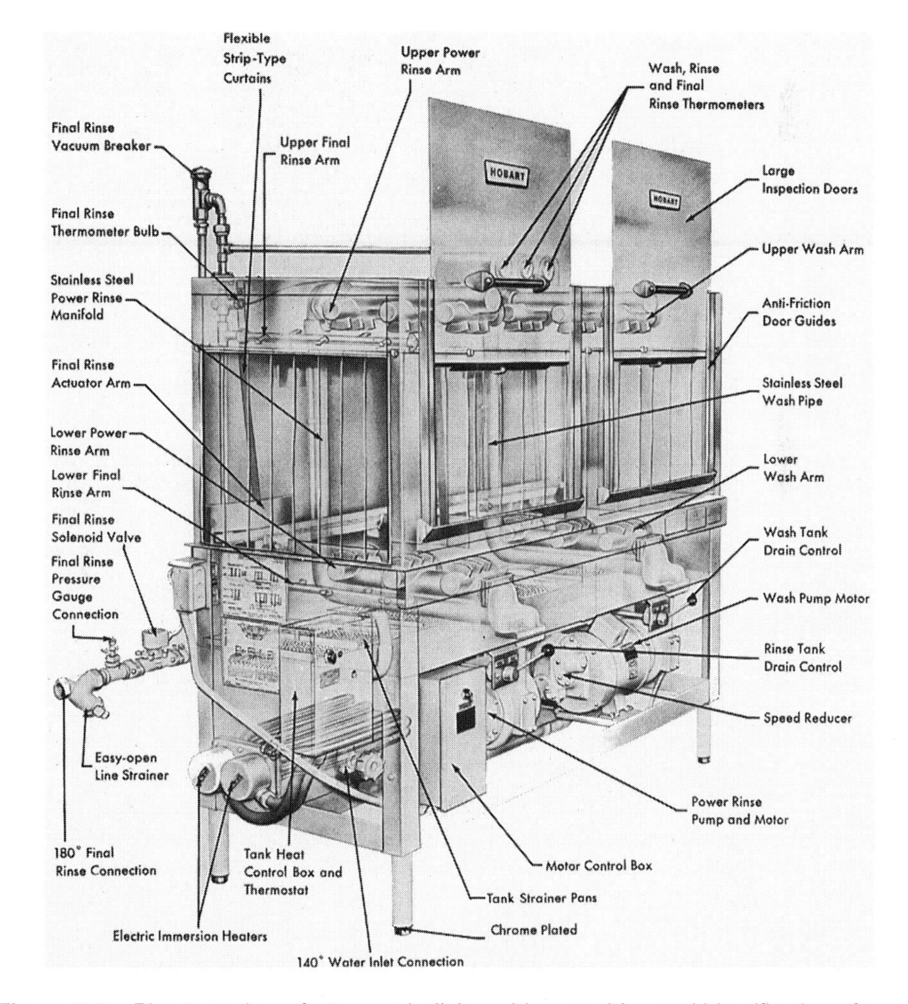

Figure 8.9 Phantom view of a two-tank dishwashing machine and identification of parts. Courtesy of Hobart Corporation, Troy, Ohio.

Heartland Country Village				Dishwasher Temperatures		
Month	Breakfast		Lunch		Dinner	
	Wash	Rinse	Wash	Rinse	Wash	Rinse
1						
2						
3						
4						
5						
6						
7						
8						
9						
10						
11						

Figure 8.10 Temperature documentation form.

Any machine can fail in its function if it is not kept clean and properly maintained, and dishwashing equipment is no exception. Corrosion or lime deposits in nozzles can alter the jet or spray materially. Also, detergent sanitizers can be inactivated by contact with soiled surfaces and lose their power of penetration. The removal of microbial contamination is necessary; otherwise, the washed surfaces of dishes will have deposited on them bacterial populations and soil proportionate to that in the washing solution.

Good maintenance includes frequent examination and lubrication where needed by a qualified maintenance person to ensure the continuing satisfactory operation of motors, nozzles, pumps, thermostats, thermometers, and all moving parts of a dishwashing machine.

The installation of elaborate equipment, however, offers no real security for good sanitation, since the efficiency of the machines depends almost entirely on the operator, the availability of an adequate supply of hot water at the proper temperature and pressure, the selection and concentration of the detergent used for the hardness of the water, and the length of time the dishes are subjected to treatment. In the small, hand-operated, single-tank machines, the process and length of washing time are under the control of the operator and are followed by the rinsing process, also under manual control. Other machines have automatic controls that regulate the length of times for washing and rinsing. Thermometers that record the temperatures of both wash and rinse waters and thermostatic controls, except for the final rinse, are included as standard parts of dishwashing machines. Booster heaters with temperature controls are available and necessary to provide the sanitizing rinse temperature, because the 180°F water in the pipe lines of a building would be a hazard to personnel safety. The installation of electronic detergent dispensers makes it possible to maintain optimum detergent concentration in the wash water and sanitizing chemical rinse in low-temperature machines. Each of these mechanical aids is most helpful in reducing human error and ensures clean, properly sanitized dishes and pots and pans.

FACILITIES CLEANING AND MAINTENANCE

The total housekeeping and maintenance program of a foodservice department must be planned to reflect concern for sanitation as "a way of life" if this philosophy phrased by the NSFI is to pass from words to reality. Food sanitation results can be obtained through establishing high standards, rigid scheduling of assignments that are clearly understood by the workers, ongoing training, proper use of cleaning supplies, provision of proper materials and equipment to accomplish tasks, and frequent meaningful inspections and performance reviews.

Organization and Scheduling

The organization of a plan for housekeeping and maintenance begins with a list of duties to be performed daily, weekly, and monthly. Most organizations believe that

Figure 8.11 Glasswasher installed for convenient use in a back-counter unit.

"sanitation is a part of every person's job" and that daily cleaning of the equipment and utensils used by each person is that person's responsibility.

General cleaning of floors, windows, walls, lighting fixtures, and certain equipment should be assigned to personnel as needed. Do not implement a policy of "clean if you have time." These tasks can be scheduled in rotation so a few of them are performed each day; at the end of the week or month, all will have been completed and the workers then repeat the schedule. Figure 8.12 gives an example of such a schedule. Each of the duties on the schedule list must be explained in detail on a written work sheet or "job breakdown" for the employee to follow. This description is the procedure that management requires to be used in performing each task. The job

Typical Job Assignments for Heavy-duty Cleaner

Monday	Filter grease in snack bar
	Clean left side of cafeteria hot-food pass-through
	Clean all kitchen windows
	Clean all kitchen table legs
	Vacuum air-conditioner filters; wipe exterior of air conditioner
	Wash all walls around garbage cans
	Complete high dusting around cooking areas
	Clean outside of steam kettles
	Wash kitchen carts
	Clean cart-washing area
Tuesday	Snack bar: Wash inside of hood exhaust
	Clean all corners, walls, and behind refrigerator
	Empty and clean grease can
	Wash garbage cans
	Main range area: Clean sides of ovens, deep-fat fryers, grills, drip pans, and hood over ovens
Wednesday	Clean two refrigerators in cooks' area
	Clean right side of cafeteria hot-food pass-through
	Clean kettles, backs of steamers, and behind steamers
	Clean walls around assembly line and pot room
Thursday	Clean all ovens in cooks' area, bottoms of ovens, and between ovens and stoves
	Clean long tables in cooks' area, including legs and underneath
	Clean and mop storage area
Friday	Clean stainless steel behind kettles and steamers
	Clean main range and tops of ovens
	Clean legs of assembly line tables
	Clean vents in all refrigeration equipment
	Clean cart-washing area

Figure 8.12 Certain weekly or monthly tasks are scheduled, a few each day, for balanced workload and completion of all such tasks within the time frame of the rotation period.

How to Clean a Food Slicer

Equipment and supplies needed:

Three cloths:
- One to wash
- One to dry
- One to apply rust preventative

One-gal container for detergent
One container with sanitizer
One table knife

Cleaning products needed:

Hand detergent
- In amount needed to make one gal of solution
- Usual proportion: 1 oz to 1 gal of water

Sanitizer:
- Usual proportion: 2 oz chlorine to 1 gal water

Rust preventative:
- In amount needed to moisten cloth for application of thin film to specified metal surfaces

Approximate time: 20 min
Frequency of cleaning: Daily, after each use.
Approximate cost: Labor ______________
Supplies ______________

What To Do	*How To Do It*
1. Remove parts	1. *a.* Remove electric cord from socket. *b.* Set blade control indicator at zero. *c.* Loosen knurled screw to release; remove meat holder and chute. *d.* Grasp scrap tray by handle; pull away from blade; remove. *e.* Loosen bolt at top of knife guard in front of sharpening device; remove bolt at bottom of guard; remove guard. *f.* Remove two knurled screw nuts under receiving tray; remove tray.

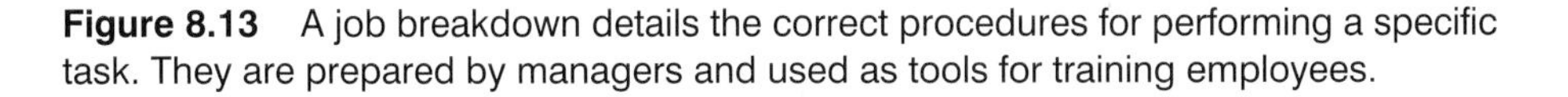

Figure 8.13 A job breakdown details the correct procedures for performing a specific task. They are prepared by managers and used as tools for training employees.

breakdown includes the name of the task, tools, equipment, and materials to be used, and the step-by-step list of *what to do* and *how to do it*. Figure 8.13 is an example of such a job breakdown.

In addition to establishing procedures, a time standard for accomplishing each task is important. Based on studies of the actual time required for performing the same tasks by several different workers, an average time standard can be set. This is used to determine labor-hour requirements for each department within the foodservice and also provides management with data to establish a realistic daily workload.

Equipment. Heavy-duty power equipment is available to foodservice managers to aid in keeping the facility clean and properly maintained. Mechanical food waste disposals are indispensable in most foodservices and eliminate the need for garbage cans, which may become unsightly and difficult to keep clean. Disposals are located where food waste originates in quantities such as in vegetable and salad preparation units, the main cooking area, and the dishwashing room. In the last, a disposal can be incorporated as a part of the scraping and prewash units of the dishwashing machine.

Compactors and can and bottle crushers (Figure 8.14) reduce appreciably the volume of trash, including items such as disposable dishes and tableware, food cartons, bags, and crates. (See Chapter 11 for details about this equipment.)

Scrubbing/waxing machines are used for cleaning both kitchen and dining room floors and are available in many sizes and models. Adequate drains in the kitchen floor are essential for proper sanitizing, and should be cleaned regularly.

Care of equipment used in food preparation, storage, and service is an essential part of the maintenance program to ensure good sanitation. All containers and utensils must be cleaned thoroughly after *each* use. This is especially true of meat grinders and slicers, cutting boards, and knives in order to prevent any cross-contamination and ensure safety against *Clostridium perfringens, Campylobacter jejuni,* and salmonellae.

The thorough cleaning and sanitizing of stationary equipment are more difficult but quite as necessary as is the cleaning of dishes and small portable equipment. No piece of large equipment should be purchased unless the operating parts can be disassembled easily for cleaning purposes. Dishwashing machines, mixers, peelers, slic-

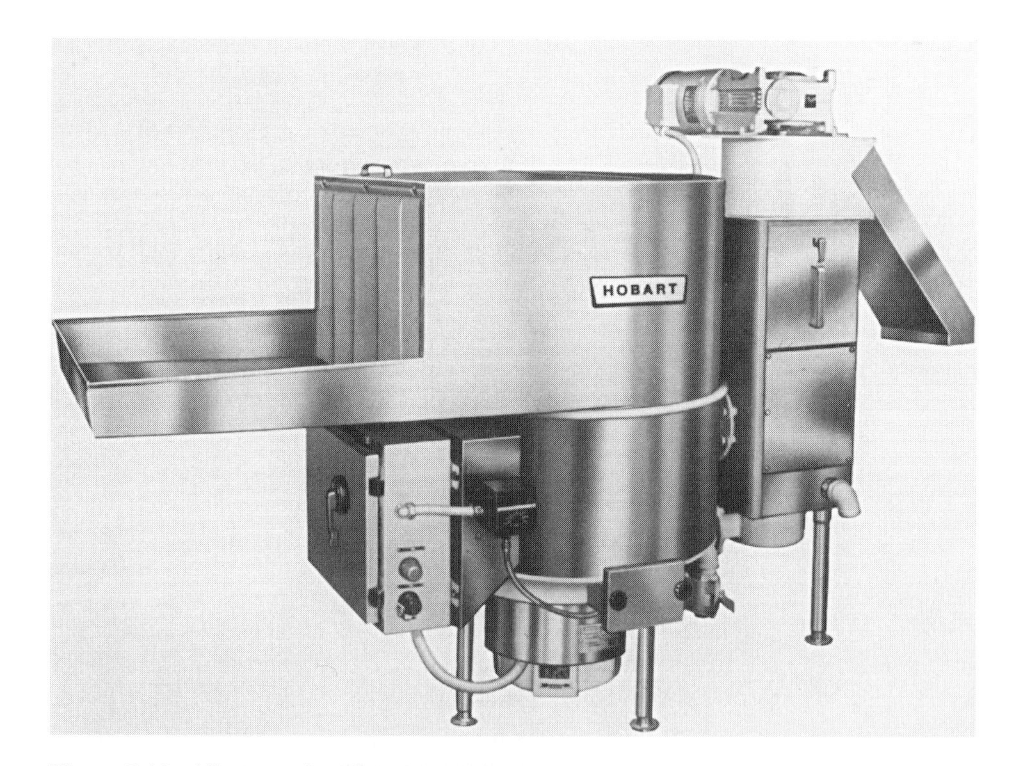

Figure 8.14 Waste equipment system.
Courtesy of Hobart Corporation, Troy, Ohio.

ing machines, and stationary can openers are also examples of equipment that should be cleaned after each use. The standard practices for hand dishwashing should be followed in the routine cleaning of such equipment. An example of the detailed daily care necessary for one piece of equipment is shown in Figure 8.13.

Preventive Maintenance

Preventive maintenance is a documented program of routine checks or inspections of facilities and equipment to ensure the sanitary, safe, and efficient operation of a foodservice department. It includes regular cleaning and maintenance such as oiling motors on mixers, and any repairs that may become evident during the inspection process. This program is usually done in cooperation with the maintenance or plant engineering department.

Regular cleaning, for example, of counter tops, floors, etc., needs to be done daily and is usually assigned as part of an employee's regular daily duties. Other cleaning tasks that need to be done less frequently must be scheduled and assigned as needed; for instance, daily, weekly, monthly. Examples include washing walls, and cleaning hoods and filters. Some large operations have cleaning crews that are responsible for these tasks. In smaller operations, however, the manager must decide on a way to distribute these tasks fairly among the employees. All of these tasks must be written as a master cleaning schedule that includes minimally what each task is, when it should be done, and who should do it. Master schedules must be supplemented with specific cleaning tasks and employees must be trained on the proper cleaning procedures.

The maintenance of facilities and equipment is usually the responsibility of the maintenance or plant engineering department. Each piece of equipment is inspected by a representative from the department on a routine basis. The foodservice manager develops a list or file of all equipment including name, identification number, purchase date, and installation and repair information for each piece of equipment. Then, together with the maintenance department, a schedule of inspection and routine repair is developed. Detailed records of repairs and costs are kept and used to determine when a piece of equipment needs to be replaced. Figure 8.15 is an example of an equipment record card.

Pest Control

The importance of rodent and insect control cannot be emphasized enough. Rats, mice, flies, roaches, grain insects, fruit flies, and gnats all facilitate the transmission of communicable disease; therefore, it is essential for any foodservice to try to effect complete elimination of resident pest infestations and then to correct conditions within the establishment so that such pests cannot gain entrance in the future.

Two conditions, food and a place to "harbor" or hide and live, are required for these pests to survive. Adherence to strict rules for proper food storage and maintenance of high standards for cleaning the nooks and corners, such as drawers in cooks' tables, around sink pipes and drains, as well as the general overall sanitation and cleaning program, provide good preventive maintenance against pests.

EQUIPMENT RECORD CARD

Equipment: ____

Manufacturer: ____

Model No.: ____________________ Serial No.: ____________________

Capacity: ____________________ Attachments: ____________________

Operation: Electric [ ] Gas [ ] Steam [ ] Hand [ ]

Purchased from: ____________ [ ] New [ ] Used Cost: $ ____________

Purchase date: ________ Guarantee: ________ Warranty: ________

Routine maintenance: ________________________(daily, weekly, monthly)

Date	*Description of Repairs*	*Cost*

Figure 8.15 Equipment record for preventive maintenance.

Many roaches and insects gain entrance to a building on incoming foodstuffs and packages, which makes their control difficult. Their reproduction is rapid, and they thrive in the warm, damp hiding places afforded in many foodservices. Screens to help keep out flies, covered trash and garbage cans, closed cracks and crevices in walls and around equipment and areas around pipes, and clean storerooms are preventive measures to try to block the entrance and reduce the hiding places of such pests. The use of certain residual insecticides is effective treatment when there is no danger of polluting food, whereas the use of less toxic insecticides is recommended for contact spraying.

Ratproofing the building to make it impossible for rodents to gain entrance is the best preventive measure for ensuring it will be free of rodents. This means the closing of openings as small as one-half inch in diameter, placing rat guards on all wires both inside and outside of pipes leading into the building, and careful joining of the cement walls and foundations of the building. Trapping and the use of rodenticides are part of a rodent-control program and are used either inside or outside the building. However, the most effective rodenticides are also the most dangerous to humans and pets; therefore, they must be used with care and caution.

Constant attention and alertness to signs of pests and an effective program for their destruction by a trained person within the organization or an outside agency are usually required. Specialized entomological services can be scheduled, as often as once a month. The effectiveness of such an effort depends on its scope, regularity, and intelligent administration of a cleaning program and proper care of foodstuffs to eliminate the environmental factors conducive to the harboring of pests.

Checks and Inspections

Maintaining high standards of sanitation is essential in *all* foodservice establishments regardless of type or size. Consumers expect and demand a clean facility. In fact, it is one of the first criteria they use for judging an eating place. The best way to ensure that proper sanitation procedures and high standards of sanitation are followed and achieved is to develop a departmental sanitation program. A cooperative effort is necessary to carry out an effective program. By setting high departmental standards and conducting routine self-inspections, management can be assured that sanitation regulations are met. It is critical that management take corrective action on sanitation deficiencies in order for the program to be fully effective.

All foodservice operations are regulated by local, state, or federal agencies. The purpose of these agencies is to administer and enforce regulations and standards for food protection. The major federal agency involved in setting and enforcing standards is the Food and Drug Administration (FDA). The FDA has the responsibility for developing model codes to be adopted by state and local health departments. The Model Food Service Sanitation Ordinance developed in 1976 is used as the basis for setting standards and codes. Official inspections are conducted on a periodic, monthly, or annual basis depending on the type of foodservice and size of foodservice operation. Figure 8.16 provides an example of a health department's inspection form.

SAFETY

Physical safety of workers and customers alike is a major concern of foodservice administrators. A work environment free of hazards that cause accidents and a dining facility in which customers are safe and secure should be aims of all managers.

The Occupational Safety and Health Act, which became effective April 28, 1971, makes it illegal *not* to have a safe establishment. It is administered by the U.S. Department of Labor (see the "For More Information" section at the end of this chapter). The act mandates action on the part of management to ensure safe and healthful working conditions for all of the nation's wage earners. It states, among other things, that each employer has a duty to furnish the employees with a place of employment that is safe and free from any hazards that can cause serious physical harm or death. The organization set up to enforce this act has the authority to inspect any place of business and to penalize those who do not comply with the provisions of the law. Managers must strictly comply to correct specific potential hazards and furnish written records of any accidents that have occurred.

Two OSHA standards of particular concern to foodservice operators are the Hazard Communication Standard (HCS) and the bloodborne pathogens standard. The HCS, also recognized as the "right to know," requires that employers develop and implement a program to communicate chemical hazards to all employees. An inven-

State of Oregon
OREGON STATE HEALTH DIVISION
Department of Human Resources
Environmental Health Services
FOOD INSPECTION REPORT

FAC. TYPE | CTY. NO. | ESTAB. NO. | INSPECT TIME | TRAVEL TIME

ESTABLISHMENT NAME: ____________

STREET ADDRESS: ____________

CITY/COUNTY: ____________

NAME OF LICENSEE: ____________

INSPECTION DATE
MO. DAY YR.

Semi-Annual ______ Complaints ______

Critical Items ______ Reinspection ______

SCORE: ____________

Public Notice of Restaurant Sanitation

☐ EXCEEDED THE SANITATION STANDARDS

☐ MET THE SANITATION STANDARDS

☐ FAILED TO MEET THE SANITATION STANDARDS

AN EVALUATION OF SANITATION ON YOUR PREMISES HAS THIS DAY BEEN MADE AND YOU ARE HEREBY NOTIFIED OF THE VIOLATIONS FOUND. SUCH VIOLATIONS MAY RESULT IN DENIAL, SUSPENSION, OR REVOCATION OF YOUR LICENSE; OR CLOSURE OF THE RESTAURANT MAY RESULT FROM UNCORRECTED CRITICAL VIOLATIONS OR FAILURE TO MAINTAIN THE MINIMUM ACCEPTABLE SCORE. YOU MAY OBTAIN A CONTESTED CASE HEARING FOR ANY DENIAL, SUSPENSION, REVOCATION OR CLOSURE BY CONTACTING THE LICENSING AGENCY. SUCH HEARINGS ARE IN ACCORDANCE WITH ORS CHAPTER 183.

ITEM NO./PT. OAR/ORS	SPECIFIC PROBLEM	REQUIRED CORRECTION & TIME LIMIT

OPERATOR ____________ SANITARIAN ____________

Figure 8.16 An example of one state's sanitation inspection report form for foodservice establishments.

No.	Item	Points
	FOOD SOURCE	
01	Approved source; unadulterated; no spoilage (OAR 32-026)	5
02	Approved milk dispensing; original container for shellfish (OAR 32-026)	1
	FOOD PROTECTION	
03	Potentially hazardous food meets temperature requirements during storage, preparation, display, service, transportation (OAR 32-024,-026,-028)	5
04	Dependable, spirit stem thermometers provided and located (OAR 32-024)	1
05	Potentially hazardous food properly thawed, cooked, preheated before hot-holding, rapidly reheated and cooled (OAR 32-024,-026)	4
06	Unwrapped and potentially hazardous food not reserved; self-service foods effectively protected (OAR 32-026,-028)	4
07	Food protection during storage, preparation, display, service, transportation (OAR 32-028)	2
08	Handling of food (ice) minimized (OAR 32-026,-024)	2
09	In use, food (ice) dispensing utensils properly stored (OAR 32-022)	1
	PERSONNEL	
10	Personnel with infections and sores restricted (ORS 624.080)	5
11	Hands washed and clean, good hygienic practices, no smoking in food preparation and service areas (OAR 32-030)	5
12	Clean clothes, hair restraints, no unauthorized personnel (OAR 32-030,-028)	1
	FOOD EQUIPMENT AND UTENSILS – SANITIZING	
13	Food (ice) contact surfaces of equipment; designed, constructed, maintained, installed, located (OAR 32-018)	2
14	Non-food contact surfaces of equipment; designed, constructed, maintained, installed, located (OAR-018)	1
15	Dishwashing facilities; designed, constructed, maintained, installed, located, operated (OAR 32-020)	2
16	Accurate thermometers, chemical test kits provided (OAR 32-020)	1
17	Pre-scrape, soak, wash, rinse water; clean, proper temperature (OAR 32-020)	2
18	Sanitization rinse; clean, temperature, concentration, exposure time; equipment, utensils sanitized (OAR 32-020)	4
19	Wiping cloths; clean, use restricted, stored in sanitizing solution (OAR 32-020)	1
20	Food contact surfaces of equipment and utensils clean, free of abrasives, detergents (OAR 32-020)	4
21	Non-food contact surfaces of equipment and utensils clean (OAR 32-020)	2
22	Protected storage, handling of clean equipment/utensils (OAR 32-022)	1
23	Single-service articles, storage, display, no re-use (OAR 32-022)	1

No.	Item	Points
	WATER, SEWAGE AND PLUMBING	
24	Water source approved, safe, at least 20 P.S.I. (OAR 32-014)	5
25	Approved sewage and waste disposal (OAR 32-023)	4
26	Sewage disposal violations which do not directly or indirectly effect food service operations inside the restaurant (OAR 32-023)	2
27	Plumbing installed, maintained (OAR 32-023)	1
28	Cross-connection, back siphonage, backflow (OAR 32-023)	5
	TOILET AND HANDWASHING FACILITIES	
29	Number, convenient, accessible, designed, installed (OAR 32-012,-16)	2
30	Clean, enclosed, self-closing doors, fixtures good repair, hot and cold water, hand cleanser, sanitary towels/hand drying devices, provided, proper waste receptacles, handwashing signs posted (OAR 32-023,-012,-016)	2
	GARBAGE AND REFUSE DISPOSAL	
31	Containers or receptacles, covered; adequate number insect/rodent proof, frequency, clean (OAR 32-023)	1
32	Outside storage area enclosures properly constructed, clean (OAR 32-023)	1
	INSECT AND RODENT, ANIMAL CONTROL	
33	Approved and effective control of insects/rodents, outer openings protected, no birds, turtles, other animals (OAR 32-006,-028)	4
	FLOORS, WALLS AND CEILINGS	
34	Floors; installed, constructed with sealed junctures, smooth, easily cleanable, good repair (OAR 32-002)	1
35	Walls and ceilings; installed, constructed; washable up to splash level in food preparation, utensil washing and toilet areas (OAR 32-004)	1
36	Floors and junctures; clean, dustless cleaning methods (OAR 32-002)	2
37	Walls, attached equipment, ceilings; clean (OAR 32-004)	2
	LIGHTING AND VENTILATION	
38	Uniform lighting provided in food preparation and storage areas (OAR 32-008)	1
39	Adequate, designed ventilation systems in food preparation, dishwashing and restroom areas; filters cleaned (OAR 32-010)	1
	HOUSEKEEPING	
40	Toxic items properly stored, labeled, used (OAR 32-028,-032)	5
41	Premises maintained free of litter, unnecessary articles, cleaning/maintenance equipment properly stored (OAR 32-032)	1
42	Complete separation from living/sleeping quarters, personal items separate (OAR 32-032)	1
43	Clean, soiled linen properly stored (OAR 32-032)	1

Other Areas (do not affect grade)

____ Employes certified in First-aid for choking victims (ORS 624.130)

____ Smoking area comply with ORS 433.845 (effective July 1, 1983)

____ Food Sanitation Training ____________

100% - 92% and no uncorrected critical violations:
EXCEEDED THE SANITATION STANDARDS

91% - 60% and no uncorrected critical violations:
MET THE SANITATION STANDARDS

59% or below:
FAILED TO MEET THE SANITATION STANDARDS

All 4 and 5 point critical violations must be rectified immediately with an approved alternative procedure and must be corrected within the designated time period. Previously cited 1 and 2 point items which have not been corrected are considered repeat violations and will accumulate 1 or 2 penalty points respectively each time they are observed on semi-annual inspections. Failure to correct critical violations as described above or two consecutive complete inspection sanitation scores below 60% may result in the closure of the restaurant or the revocation, suspension or denial of the license.

Figure 8.16 *Continued.*
Courtesy of Oregon State Health Division.

tory of all chemicals used by the operation must be maintained, and they must be properly labeled. The manufacturer must supply, for each chemical, a Material Safety Data Sheet (MSDS) that identifies the chemical and includes a hazard warning. Figure 8.17 is a sample of an MSDS form. The bloodborne pathogen standard requires that all employees be made aware of potentially infectious materials that they may be exposed to while on duty. Examples of pathogens include the hepatitis B virus and

Material Safety Data Sheet May be used to comply with OSHA's Hazard Communication Standard, 29 CFR 1910.1200. Standard must be consulted for specific requirements.	**U.S. Department of Labor** Occupational Safety and Health Administration (Non-Mandatory Form) Form Approved OMB No. 1218-0072
IDENTITY *(As Used on Label and List)*	*Note: Blank spaces are not permitted. If any item is not applicable, or no information is available, the space must be marked to indicate that.*

Section I

Manufacturer's Name	Emergency Telephone Number
Address *(Number, Street, City, State, and ZIP Code)*	Telephone Number for Information
	Date Prepared
	Signature of Preparer *(optional)*

Section II — Hazardous Ingredients/Identity Information

Hazardous Components (Specific Chemical Identity; Common Name(s))	OSHA PEL	ACGIH TLV	Other Limits Recommended	% *(optional)*

Section III — Physical/Chemical Characteristics

Boiling Point		Specific Gravity (H_2O = 1)	
Vapor Pressure (mm Hg.)		Melting Point	
Vapor Density (AIR = 1)		Evaporation Rate (Butyl Acetate = 1)	
Solubility in Water			
Appearance and Odor			

Section IV — Fire and Explosion Hazard Data

Flash Point (Method Used)	Flammable Limits	LEL	UEL
Extinguishing Media			
Special Fire Fighting Procedures			
Unusual Fire and Explosion Hazards			

(Reproduce locally) OSHA 174, Sept. 1985

Figure 8.17 Sample of a Material Safety Data Sheet (MSDS).

the human immunodeficiency virus (HIV). For the foodservice manager, this means educating employees on the risks of and proper procedures for entering patient rooms or cleaning food trays in the dishroom that may be contaminated with hazardous matter.

The National Safety Council, although not a regulatory agency but a nonprofit service organization, is devoted to safety education. Through its research, reports, and printed materials available to the public, the council provides valuable assistance to managers of numerous types of businesses, including foodservice.

Worker Safety

The provision of a safe workplace through a well-designed facility (see Chapter 10) with equipment facilities that meet federal, state, and local standards is a first step toward ensuring worker safety. However, safety is more than a building with built-in safe features. Safety can never be *assumed*, because accidents can and do occur. Managers and employees must work together on a safety awareness program to attain a good safety record.

"Accidents don't happen; they are caused"—and they can be prevented. The National Safety Council has defined an accident as any suddenly occurring, unintentional event that causes injury or property damage. An accident has become a symbol of inefficiency, either human or mechanical, and usually represents a monetary loss to the organization. The company not only loses the productivity of the injured individual but also incurs indirect costs such as medical and insurance expenses, cost of training new workers, waste produced by inexperienced substitute workers, administrative costs for investigating and taking care of accidents, and cost of repair or replacement of broken or damaged equipment. Not only from the humanitarian standpoint, but also from the economic, should dietitians and other foodservice managers be aware of the advantages of good safety measures. All should seek ways to improve working conditions and employee performance that will reduce accidents with their resulting waste, and maintain low accident frequency and severity rates. *Severity rate* is computed by the number of working days lost because of accidents, and *frequency rate* by the number of lost-time accidents during any selected period, each multiplied by 1,000,000 and the result divided by the total number of hours worked during the same period. National Safety Council statistics rank the food industry about midway among all industry classifications in terms of severity rates. However, in terms of frequency rate, it is nearly twice as high as the average for all industries reporting.

Foodservice managers must organize for safety and develop a wholesome regard for safe procedures among the entire staff.

Safety Program

Specific topics for a safety campaign may be centered around the "three Es" of safety: engineering, education, and enforcement.

The *engineering* aspect refers to the built-in safety features of the building and equipment, and the manner in which the equipment is installed to make it safe to use. Encased motors, safety valves on pressure steamers, easily manipulated spigots on urns, and guards on slicing and chopping machines are examples of safety features. A maintenance program to keep equipment in good working order is the responsibility of management, as are all other phases of providing a safe environment.

A study of traffic patterns in kitchen and dining areas and the placement of equipment and supplies in locations to avoid as much cross traffic as possible, and the arrangement of equipment within a work unit to provide for logical sequence of movement without backtracking are a part of the engineering phase of the safety program.

Education for safety is a never-ending process. It begins with the establishment of firm policies regarding safety, which then should be discussed with each new employee during the orientation period. "Safety from the first day" is an appropriate slogan for any organization.

Since safety is an integrated part of every activity, it should be taught as a component of all skills and procedures. Written procedures for tasks to be performed by each employee must include the safe way of doing each task, and the written outline then used to train the employee in the correct steps to follow. These written, step-by-step procedures provide a follow-up, on-the-job reference for the employee, and can be used by managers as a check against employee performance.

Safety education, however, is more than training each employee in the procedures for a particular job. An ongoing group program based on *facts* about safe and unsafe practices keeps employees aware of safety. The National Safety Council, the Bureau of Vital Statistics, various community safety councils, and trade and professional organizations can provide statistics and materials for planning such a program. Data obtained from records kept on accidents *within* the organization are invaluable and more meaningful than general statistics.

A form for reporting accidents should be provided by management (see Figure 8.18 for an example). These written records should include the type of accident, kind of injury that has occurred and to whom, when it occurred, the day and hour, and where it took place. In foodservices, most accidents occur at rush hours when it is especially difficult to take care of the injured, find replacement help, and continue efficient customer service. This fact alone should provide incentive enough for the manager to do all that is possible to promote safety.

An analysis of the causes of accidents provides further data for preventing them. Causes may be classified into "unsafe acts" and "unsafe conditions." Usually it is found that unsafe acts outnumber unsafe conditions three to one. From this, there is an immediate indication of the need for proper training to reduce accidents.

In the foodservice industry, falls cause the largest number of food-handling accidents, usually due to greasy or wet floors, with cuts second, and burns and strains from lifting next in order. Falls and strains result in the greatest loss of time from the job and monetary loss to the institution.

It is management's responsibility to ferret out the reasons, remove the hazards, and then train the employees to prevent recurrence of the same accident. Good housekeeping procedures, such as storing tools and materials in proper places and keeping aisles and pathways clear, optimum lighting of work areas, prompt repair of broken tools and equipment, replacement of worn electrical cords, and proper care and removal of broken china and glassware, are only a few of the things that can be done

Board of Education
Office of Assistant Supervisor of Home Economics
in Charge of Cafeterias

Department Report of Personal Injury Involving
the Employees of the City Schools

School ____________________ Date ____________________

Name ____________________ Title ____________________

Address ____________________ Date of accident ____________________

Sex ____________________ Time of accident ____________________

State fully in your own words how accident occurred: ____________________

____

____

Exact part of person injured and extent of injuries: ____________________

____

____

Probable period of disability ____________________

Was medical attention necessary? ____________________

Name and address of Physician ____________________

____

Give location where accident occurred ____________________

Will employee lose any time? ____________________

If not able to work, give probable date of recovery ____________________

Are there any indications of permanent injury? ____________________

State monthly salary of employee ____________________

How long in our employ? ____________________

IMPORTANT: Fill out report in *duplicate* on day of accident and mail immediately to above office.

____
Employee

Figure 8.18 Example of a typical accident report form; to be filled in and filed immediately.

to correct unsafe conditions. Employees should be encouraged to report to the manager any unsafe conditions they may notice. A simple form could be developed and made available to the employees for such reporting. Having the information in writing is helpful to the manager who must then follow up to correct the situation.

The possibility of fires is an ever-present threat in foodservice establishments, making it essential that all employees follow good procedures in use of equipment and cooking techniques. Further, they should know the location of fire extinguishing equipment, and how to use it. Directions for and practice in the use of fire extinguishers, fire blankets, and other first aid equipment, necessities in every institutional kitchen, are included in training meetings, for supervisory personnel particularly. Information about the various types of fire extinguishers and which should be used for grease, paper and wood, and other types of fires is important. Figure 8.19 gives this information. Group training in precautionary procedures to be followed in everyday work and instructions on what to do in case of an accident should be part of the overall safety program. Be certain that all employees know where and to whom to report an accident and that the phone numbers to use for emergencies are posted on or near the telephone.

Many aids are available to foodservice managers to use in setting up a training program. The National Restaurant Association's *Safety Operations Manual* is an excellent resource. The National Safety Council has posters, pamphlets, and other materials available for use in training sessions. These are invaluable sources of information and illustration for foodservice managers. Clear eye-catching posters that create favorable impressions and serve as reminders of good, safe practices are effective supplements to other types of training. The safety rules given in Figure 8.20 may be used as topics for training sessions. However, each foodservice organization should establish its own similar list of safety rules to be adhered to in its own department.

The third "E" in the overall safety campaign is *enforcement*. This represents the follow-up or constant vigilance required to prevent carelessness and to make certain

KNOW YOUR FIRE EXTINGUISHERS

saif
State Accident Insurance Fund

	WATER TYPE				FOAM	CARBON DIOXIDE	DRY CHEMICAL			
							SODIUM OR POTASSIUM BICARBONATE		MULTI-PURPOSE ABC	
	STORED PRESSURE	CARTRIDGE OPERATED	WATER PUMP TANK	SODA ACID	FOAM	CO2	CARTRIDGE OPERATED	STORED PRESSURE	STORED PRESSURE	CARTRIDGE OPERATED
CLASS A FIRES — WOOD, PAPER, TRASH HAVING GLOWING EMBERS (A — ORDINARY COMBUSTIBLES)	YES	YES	YES	YES	YES	NO (BUT WILL CONTROL SMALL SURFACE FIRES)	NO (BUT WILL CONTROL SMALL SURFACE FIRES)	NO (BUT WILL CONTROL SMALL SURFACE FIRES)	YES	YES
CLASS B FIRES — FLAMMABLE LIQUIDS, GASOLINE, OIL, PAINTS, GREASE, ETC. (B — FLAMMABLE LIQUIDS)	NO	NO	NO	NO	YES	YES	YES	YES	YES	YES
CLASS C FIRES — ELECTRICAL EQUIPMENT (C — ELECTRICAL EQUIPMENT)	NO	NO	NO	NO	NO	YES	YES	YES	YES	YES
CLASS D FIRES — COMBUSTIBLE METALS (D — COMBUSTIBLE METALS)	SPECIAL EXTINGUISHING AGENTS APPROVED BY RECOGNIZED TESTING LABORATORIES									
METHOD OF OPERATION	PULL PIN-SQUEEZE HANDLE	TURN UPSIDE DOWN AND BUMP	PUMP HANDLE	TURN UPSIDE DOWN	TURN UPSIDE DOWN	PULL PIN-SQUEEZE LEVER	RUPTURE CARTRIDGE-SQUEEZE LEVER	PULL PIN-SQUEEZE HANDLE	PULL PIN-SQUEEZE HANDLE	RUPTURE CARTRIDGE-SQUEEZE LEVER
RANGE	30'-40'	30'-40'	30'-40'	30'-40'	30'-40'	3'-8'	5'-20'	5'-20'	5'-20'	5'-20'
MAINTENANCE	CHECK AIR PRESSURE GAUGE MONTHLY	WEIGH GAS CARTRIDGE ADD WATER IF REQUIRED ANNUALLY	DISCHARGE AND FILL WITH WATER ANNUALLY	DISCHARGE ANNUALLY -RECHARGE	DISCHARGE ANNUALLY -RECHARGE	WEIGH SEMI-ANNUALLY	WEIGH GAS CARTRIDGE-CHECK CONDITION OF DRY CHEMICAL ANNUALLY	CHECK PRESSURE GAUGE AND CONDITION OF DRY CHEMICAL ANNUALLY	CHECK PRESSURE GAUGE AND CONDITION OF DRY CHEMICAL ANNUALLY	WEIGH GAS CARTRIDGE-CHECK CONDITION OF DRY CHEMICAL ANNUALLY

Figure 8.19 Types of fire extinguishers.

General Restaurant Safety Rules
(please post)

- Report *every* injury *at once,* regardless of severity, to your Supervisor for first aid. *Avoid delay.*
- Report all *unsafe conditions,* broken or splintered chairs or tables, defective equipment, leaking radiators, torn carpeting, uneven floors, loose rails, unsafe tools or knives, broken china and glass, etc.
- Understand the *safe way* to perform any task assigned to you. If in doubt, see your Supervisor. Never take unnecessary chances.
- If you have to move over-heavy objects, ask for help. *Do not overlift.* When lifting any heavy object, keep your back straight, bend your knees and *use your leg muscles.* Your back has weak muscles and can easily be strained.
- Aisles, passageways, stairways must be kept clean and free from obstructions. Do not permit brooms, pails, mops, cans, boxes, etc., to remain where someone can fall over them. Wipe up any grease or wet spots from stairs or floors or ramps *at once.* These are serious falling hazards.
- Walk, do not run, in halls, down ramps or stairs, or around work areas. Be careful when passing through swinging doors.
- Keep your locker clean and the locker top free from all loose or discarded materials, such as: newspapers, old boxes, bottles, broken equipment, etc.
- Wear safe, sensible clothes for your work. Wear safe, comfortable shoes, with good soles. Never wear thin-soled or broken-down shoes. *Do not wear high-heeled shoes for work.* Ragged or over-long sleeves or ragged clothing may result in an injury.
- If you have to reach for a high object, use a ladder, not a chair or table or a makeshift. There is no substitute for a good ladder. *Never overreach.* Be careful when you have to reach high to fill coffee urns, milk tanks, etc.
- Horseplay or practical jokes on the job are forbidden.
- Do not argue or fight with fellow employees. The results are usually unpleasant and dangerous.
- Keep floors clean and dry. Pick up any loose object from the floor immediately to prevent someone from falling.
- Do not overload your trays. Trays should be loaded so as to give good balance. An improperly loaded tray can become dangerous.
- Dispose of all broken glass and china immediately. Never serve a guest with a cracked or chipped glass or piece of china. Check all silverware.
- Take sufficient time to serve your guests properly. Too much haste is liable to cause accidents to your guests and to yourself. *Haste makes waste.*
- Remove from service any chair, table or other equipment that is loose, broken or splintered so as to prevent injury.
- *Cashiers.* Close cash registers with back of hand. Do not permit fingers to hang over edge of drawer.
- Money is germ-laden. Keep your fingers out of your hair, eyes, and mouth after handling. Wash hands carefully before eating. Report the slightest cut or sore *at once* for treatment.
- Help *new employees* work safely on the job. Show them the right way to do the job—the safe way.

Figure 8.20 Safety rules—a basis for safety training programs.

that the rules and prescribed procedures are observed. Enforcement can be accomplished in many ways. In some organizations, safety committees are set up among the employees, who observe and report unsafe conditions and practices. Membership on this committee may be rotated so that everyone will be personally involved in a campaign against accidents. In other places, contests among departments serve as incentives for keeping accident rates low. Honor rolls for accident-free days or months help call attention to safety records. If possible, one person in each organization should have the overall responsibility for developing and supervising the safety program, after being specifically trained for the task.

Probably the most effective overall enforcement plan, however, is a periodic inspection of the department by someone on the supervisory staff. The use of a checklist as a reminder of all points to be observed is helpful. Any foodservice manager could develop a form for use in a specific operation. The comprehensive checklist illustrated in Figure 8.21 includes both food safety and sanitation and may serve as a model for developing a checklist for a specific department.

Customer Protection

Customers of foodservices deserve the same careful concern given employees in regard to safety. They expect and should have assurance that the food served will be safe for consumption and that the facility for dining is also safe. This includes everything from a safe parking area that is well lighted and free of any stumbling "blocks" to furniture that is in good condition and will not cause snags or splinters. The flooring must be kept in good repair to prevent tripping and falls, and any spillage should be wiped up at once so that no one will slip and fall.

Dining rooms should be adequately lighted and ample aisle space provided between tables so that diners can see to make their way through the room without tripping.

Servers must be well trained in correct serving procedures so they will not spill any hot food on the customers or anything on the floor that could cause accidents. Any spillage must be cleaned up at once.

Managers are liable for accidents that may occur on the premises. Lawsuits could result that are costly as well as detrimental to the reputation of the establishment. Therefore, extreme care should be taken so that none will occur.

SUMMARY

It is the responsibility of the foodservice manager to design, implement, and monitor a program of cleaning and sanitation for his or her operation. Program design begins with an understanding of principles and factors that influence the cleaning and sanitation tasks. These principles and factors must be considered when managing the major cleaning and sanitation functions, which include dishwashing and facilities maintenance.

The steps to safety in any foodservice include awareness, involvement, and control. The first step is *awareness* on the part of managers for the need to provide a safe environment for employees and patrons, and to assume the responsibility for and positive

Check Sheet
Safety in the Kitchen

Rating Scale: 5—1; 5 points is highest and 1 point the lowest.

Burns
1. Are handles of pans on the stove turned so the pans cannot be knocked off? ______
2. Are flames turned off when removing pans from stove? ______
3. Are dry pot holders used for lifting hot pans? ______
4. Are fellow workers warned when pans are hot? When pans of hot food are to be moved? ______
5. Is steam equipment in proper working order to avoid burns from leaks? ______
6. Is hot water regulated at proper temperature so it will not scald? ______
7. Are lids lifted cautiously and steamer doors opened slowly to avoid steam burns? ______

Cuts
1. Are broken dishes and glasses promptly cleaned up and disposed of in special container provided? ______
2. Are knives stored in the slotted case provided for them? ______
3. Are knives left on the drain board to be washed, and not dropped into the sink? ______
4. Is the safety hood put over the slicer after each use and cleaning? ______
5. Is the can opener in good repair so it cuts sharply and leaves no ragged edges? ______
6. Are safety devices provided on slicers and choppers? ______

Electricity
1. Are electric cords in good repair? ______
2. Are sufficient outlets provided for the equipment in use? ______
3. Are hands always dry before touching electrical equipment? ______
4. Are there extra fuses in the fuse box? ______

Falls
1. Are spilled foods cleaned up immediately? ______
2. Are corridors and stairways free from debris? ______
3. Are articles placed on shelves securely so they will not jar off? ______
4. Are step ladders sturdy and in good repair? ______
5. Are brooms and mops put away properly after use and not left out against a wall or table to trip someone? ______
6. Are hallways well lighted and steps well marked so no one will trip? ______

Fires and Explosions
1. Are gas pipes free from leaks? Have they been checked by the gas company? ______
2. Are matches kept in a covered metal container? ______
3. Are fire blankets and extinguishers provided? ______
4. Has the fire extinguisher been checked in the last month? ______
5. Is the first-aid box fully supplied? ______
6. Is hot fat watched carefully, and is cold fat stored away from flame? ______

Please Report Immediately Any Fires or Accidents to the Food Manager
or Dial ________________ to Report a Fire

Figure 8.21 Example of a kitchen safety check sheet to be used by managers to help identify unsafe practices.

attitude toward accident prevention. *Involvement* includes initiating a safety education program or campaign that keeps employees safety conscious. A training program that indoctrinates employees with the philosophy of working safely and instructs them in how to do so is a major part of being involved. Seeking employee suggestions about safe procedures and forming safety committees in which employees participate are other forms of involvement. *Control* is the process of insisting on safety, checking on safety codes and meeting them, analyzing accident records as a basis for improvement, and, above all, good consistent supervision of employee work. This assumes that the institution has established safety policies, written procedures for job performance, and adopted a procedure for reporting and handling accidents that are known to all in the organization.

Benefits of a safety program include a reduction in accidents, improvement in employee morale, patron satisfaction and feeling of security, and fewer workers' compensation claims, resulting in reduced costs and better financial performance for the foodservice. The objective is to keep injuries to a minimum and the working force at maximum efficiency.

REVIEW QUESTIONS

1. Which organizations establish and enforce sanitation and safety standards?
2. What is OSHA? What is its influence on foodservice operation?
3. Compare and contrast the terms *clean* and *sanitized.*
4. What are the advantages of developing and implementing a cleaning program?
5. What are the most frequent causes of accidents in foodservice operations? How can they be prevented?
6. What is meant by the term *preventive maintenance*? Why is it important in a foodservice operation?
7. What factors should be considered when deciding between a high-temperature or chemical dishwashing machine system?
8. What is a Material Safety Data Sheet?

SELECTED REFERENCES

Burkholder, V. R.: Halting kitchen hazards. School Food Service J. October 1989, pp. 58–63.

Lydecker, T.: Don't get burned; fireproof your restaurant. Restaurants USA. August 1992, pp. 20–23.

National Restaurant Association: *Right to Know: A Foodservice Operator's Guide to the OSHA Hazard Communication Standard Program.* Chicago, Ill.: National Restaurant Association, 1988.

National Restaurant Association, The Education Foundation: *Applied Foodservice Sanitation,* 4th ed. New York: John Wiley & Sons, 1992.

Reed, G. H.: Sanitation in food service establishments. Dairy, Food Environ. Sanitation. August 1992, pp. 566–567.

Spertzel, J. K.: Safety is no accident. School Food Service J. April 1992, pp. 50–52.

Somerville, S. R.: Safety is no accident. Restaurants USA. August 1992, pp. 14–19.

FOR MORE INFORMATION

National Restaurant Association
Education Foundation
250 S. Wacker Drive, Suite 1400
Chicago, IL 60606

National Safety Council
1121 Spring Lake Drive
Itasca, IL 60143-3201

National Sanitation Foundation International
P.O. Box 130140
Ann Arbor, MI 48113-0140

Occupational Safety and Health Administration
Publications Office
200 Constitution Ave NW, Room N3101
Washington, DC 20210

9

Environmental Management

OUTLINE

Efforts made by foodservice operators to conserve energy and water and to minimize waste can have a significant impact on more global efforts to protect the environment and to preserve natural resources.

The first Earth Day celebrated on April 22, 1970, was marked by nationwide demonstrations and political speeches drawing attention to our heavy dependence on nonrenewable sources of energy and to the fact that the supplies of these sources were dwindling rapidly. *The Limits of Growth,* a book published by the Club of Rome in 1972, intensified the importance of the message by predicting the year in which resources would be completely depleted. The authors based their predictions on conservative estimates of known reserve growth rate, population, food production, available capital, and land use on a global scale for a closed system, "spaceship" Earth. The predicted depletion dates for resources included: petroleum, 2020; natural gas, 2019; aluminum, 2025; zinc, 2020, copper, 2018; and iron, 2143. Although there is some disagreement about these dates, no one debates the fact that these resources will eventually run out and that strategies must be developed to cope with increasing scarcity as total depletion approaches.

Many advances have been made in our efforts to shift from oil, natural gas, coal, and nuclear to renewable sources of energy such as geothermal, hydroelectric, biomass, wind and solar. With energy being one of the biggest overhead costs in a foodservice operation, these new sources of energy hold much hope. Utility rates have, in most cases, paralleled the general inflation rate and foodservice operators have faced increasing energy usage during the past several years. One of the primary causes of this escalation of energy use is the demand from customers and employees for comfort. No matter what type of foodservice operation, air-conditioned dining rooms and kitchens have become the norm. In many areas, an air-conditioned working environment has become a union bargaining position.

In addition to the depletion of natural resources, the world faces other environmental issues such as soil, water, and air pollution, global warming, acid rain, deforestation, and the generation of ever-increasing amounts of waste.

As in all American businesses, the foodservice industry is constantly facing competitive challenges. In the sixties, low cost was the competing challenge. Flexibility and quick response to customer demands were the challenges of the seventies. Quality was the emphasis in the 1980s. And now as the 1900s come to a close, the entire world is entering an era of environmentalism, "zero discharge," and "total pollution management." The impetus for this focus on the environment has been brought about by increasing public pressure, skyrocketing cleanup costs, rising criminal and civil liabilities, and stringent laws and regulations. Environmental excellence is becoming a number one priority of corporate management.

A staggering statistic is the conservative estimate that of the 20,000 pounds per person of active materials (food, fuel, forest products, and ores) extracted in the United States every year, only 6% becomes durable goods. The other 94% is wasted within a few months of extraction. Each person in this country creates 4.3 pounds of garbage per day, which adds up to 200 million tons of garbage being created each year ("Billionaires in the future will be created out of trash," Financial News, April 28, 1995).

Reductions in pollution and waste can lead to a healthier environment and a healthier economy. The Institute for Local Self-Reliance in Washington, DC, estimates that nine jobs are created for every 15,000 tons of solid waste that are recycled into a new product, whereas only two jobs are created for every 15,000 tons incinerated, and only one job for every 15,000 tons sent to landfills. Each environmental issue, however, is complex because each has economic, sociological, political, and ecological ramifications that must be considered. Effective solutions will require the collaboration of those in agriculture, manufacturing, the packaging industry, foodservice operators, policy makers, and consumers.

Foodservices are actively engaged in solid waste management programs that include source reduction and recycling of virtually every waste product generated in the operation. Some examples of these are discussed in this chapter.

KEY CONCEPTS

1. An effective energy management program requires the constant participation of every employee in the organization.
2. Every operation in the foodservice should be analyzed to determine its potential for energy reduction.
3. Energy conserving equipment should be purchased and installed wherever and whenever possible.

4. Air-conditioning costs may be reduced by remotely locating sources of heat, purchasing insulated equipment, using heat pump water heaters, using radiant heat barriers, and cutting out units during off-peak periods.
5. Solid waste management is an ethically, legally, and economically mandated priority of foodservice management today.
6. The first step in an integrated solid waste management program is source reduction.
7. Recycling reduces waste handling costs, dependence on scarce natural resources, manufacturing energy costs, amount of material sent to landfills, and the potential pollution of nature.
8. Foodservices may recycle virtually all of the waste products that they generate including paper, plastic, aluminum, steel, glass, polystyrene, and food.

CONSERVATION OF NATURAL RESOURCES

Ways in which to cut energy costs have received much attention. Foodservice operators have found that targeting unnecessary energy usage and incorporating techniques to reduce energy consumption result in a more efficient operation overall. Water conservation has received less attention. However, as more and more communities suffer from periodic water shortages, this precious resource will become the focus of more effort.

Energy Conservation

An energy management program requires the constant participation of every employee in the operation. Inservice training and incentive programs should be set up to ensure the cooperation of all involved.

Using large pieces of equipment at less than full capacity is one of the most common energy wasters. This includes but is not limited to dishwashers, ovens, griddles, fryers, ranges, and steam-jacketed kettles, which can be operated with partial loads and/or left on between loads.

Routine maintenance and cleaning of equipment are essential components of an energy reduction program. Weak or broken door springs on ovens and refrigerators may reduce efficiency by 35%. Carbonized grease and cooking residue on griddle plates can reduce cooking efficiency by 40%.

Utility companies often offer free equipment service adjustment, energy audits, and assistance in establishing effective energy management programs. An example follows of information that may be obtained from a utility company.

HOW MUCH ENERGY DOES IT SAVE . . . to use OPEN BURNER RANGES instead of Hot Tops?

You are way ahead in conserving fuel and reducing operating costs when you use OPEN BURNER RANGES. Tests show dramatic savings in fuel consumption when OPEN BURNER RANGES are compared to Hot Top ranges.

- Similar quantities of water boil in up to *one-third less time* on the OPEN BURNER RANGE.

- Boiling similar quantities of water requires up to *55% less fuel* on the OPEN BURNER RANGE.
- Hot Tops must be preheated. Preheating takes 30 to 60 minutes. The gas flame on the OPEN BURNER RANGE comes on instantly when you need it. *No preheating necessary!*

Tests show additional energy can be conserved no matter what kind of range top you have.

- Covering pans reduced energy consumption
  —on OPEN BURNER RANGES by up to 20%.
  —on Hot Tops by up to 35%.
- Heating larger quantities of food can be done more efficiently than heating smaller quantities of food.
  —on OPEN BURNER RANGES, Btu consumption per pound was reduced up to 19% when the quantity of water heated was doubled.
  —on Hot Top ranges, Btu consumption per pound was reduced up to 20% when the quantity of water heated was doubled.

Other energy conservation suggestions for use of equipment in food service establishments follow:

- Ventilation systems, air conditioning, heating:
  Make-up air for hoods; use thermostatically controlled unheated and unrefrigerated air.
  Use heat recovery systems in hoods; heat exchangers for hot water and/or comfort heating.
  Use evaporative coolers (swamp coolers) to comfort-condition kitchen air (do not use refrigerated air).
  Size air-conditioning units and comfort heaters accurately for climate area; limit size of heaters for kitchen area to take advantage of heat from cooking equipment.
  Use economizer cycle systems (use of outside air when cool enough to eliminate need for refrigerated air).
  Place air conditioning and furnace filters in an easily accessible location to ensure frequent scheduled cleaning or replacement.
  Insulate heating and air-conditioning ducts adequately and completely, using two inches of insulation.
  Zone and wire air-conditioning and furnace units to permit zone control of unoccupied areas.
  Install covered and locked thermostats, 68°F for heating, 78°F for cooling.
  Use time clocks to decrease utility consumption by mechanical equipment in off-peak periods.
  Keep filters and extractors clean.
- Water heating:
  Locate water heater in close proximity to major use.
  Insulate all hot water lines.
  Size water heating equipment accurately; do not undersize or oversize; use quality equipment.
  Install spring-loaded faucet valves or spring-loaded food controls to limit hot water waste.

Use quality valves to minimize dripping faucets and repair all leaks promptly.
Consider solar-assisted and/or waste heat exchanger water heating systems to preheat water.
Use single-system, high-temperature water heating equipment and automatic mixing valves.
Use water softening equipment to soften water in areas where water is hard. Reduce water temperature where possible.

- Dishwashing:
  Size dishwasher to handle average maximum requirements.
  Install easily accessible switch to permit shut down of equipment in slack periods. Consider chemical dishwasher for small establishments.
- Cooking equipment:
  Be selective in specification of equipment offering greatest efficiency and flexibility of use.
  Careful planning can save on operating and initial equipment costs. Do not overestimate equipment requirements. Specify thermostatically controlled equipment whenever possible.
  Preheat just before use and turn off when not in use.
  Keep equipment clean for most efficient operation.
  Use the correct size equipment at all times.
- Ranges: Specify open top burners—they require no preheat and offer maximum fuel efficiency as compared to center-fired and front-fired hot tops, or even heat tops. Open burners reduce the air cooling load because there is minimal heat radiation when cooking operation is completed.
- Convection ovens: Versatile and perform most baking/roasting operations in shortest period of time.
- Steamers: Self-contained (boiler); high production at minimal operating cost.
- Grooved griddle: Replaces the underfired broiler; minimizes air pollution problems; operates much more efficiently than underfired broilers and places less of a load on air cooling systems; generally has greater cooking capacity.
- Broilers: Underfired—minimize specifying—reasons stated in grooved griddles. Overfired—preferred over underfired. More efficient, faster, no pollution problems. Compartment over broiler provides use for waste heat as plate warmer, finishing foods, browning, cheese melting (may be used for these purposes in lieu of salamander).
- Salamanders/cheese melters: Specify those that use infrared ray radiation; they reach full operating temperature within seconds, are efficient, can be turned off when not in use.
- Braising pans/tilting skillets: A versatile volume production piece of equipment that can serve many cooking operations—fry, boil, braise, roast, steam, food warmer. Consider caster equipped pans and installation of additional gas outlets near serving lines, banquet facilities, and such to obtain maximum utilization of this equipment and reduce gas consumption.
- Fryers: Floor fryers provide maximum production capabilities, have self-contained power oil filter units for ease and speed of filtering, which prolongs oil life. Consider inclusion of multiproduct programmed computers; they are avail-

able built into the fryer and will provide consistently high-quality fried products with novice operators; lowers labor, food, and oil costs. Specify more than one size of fryers; full production capacity and standby or nonpeak period smaller capacity fryers; save on initial equipment cost; fuel costs and oil costs.

Standard operating procedure in the past has been that the first person in the kitchen in the morning turns everything on from the salamanders to the broilers and the last person to leave at night turns everything off. Utility costs can be dramatically reduced by simply shutting off equipment when it is not being used. The disadvantage of this procedure is that some pieces of equipment require a considerable amount of time to reach the desired cooking temperature. New technology is available to handle this problem.

Systems may be installed in the gas supply line that allow for the firing of gas cooking equipment on demand. One such system installed in a grill restricts gas flow to 20%—just enough to keep the burners warm. When the grill is needed, the cook hits a button to restore full gas flow, which takes 20 to 30 seconds. The system can be programmed to cook items from 2 to 10 minutes. At the end of the timed cooking period, the grill automatically shuts down. This system not only saves on gas usage costs but reduces the amount of heat released into the kitchen, which in turn saves on air-conditioning costs. Employee safety is also improved by reducing the number of high heat surface areas on which one can be inadvertently burned. In addition, food waste from overcooking is eliminated by the timing device.

Many electric utility companies now provide an analysis of air-conditioning demand and special cooling control programs. These programs monitor temperatures and compressors, cut out units during off-peak periods, and keep temperatures in the comfort zone.

Moving heat producing equipment out of the air-conditioned or ventilated area is another energy-saving practice. Refrigerator condensing coils and compressors generate a surprising amount of heat, and they can be moved outside the building or into a basement. The heat generated from an ice machine can be eliminated from a service area by remotely locating the ice machine and pumping ice to each of the stations where it is needed.

Heat pump water heaters use the heat generated by all of the cooking equipment in the kitchen to warm water and at the same time return cool, dehumidified air to the kitchen, lowering the air temperature and humidity. The cool air may be directed to the general kitchen area, to a specific area, or outside when cooling is not needed. This type of water heater is four to six times as efficient as conventional gas water heaters and three times more efficient than conventional electric water heaters, does not require a flue or fireproof enclosure because no combustion is involved.

Manufacturers of cooking equipment are putting more insulation into their equipment to keep the kitchen cooler and finding that an additional benefit is that the insulated equipment is more efficient. Additional examples of energy efficient equipment are high efficiency gas burners and infrared heating in fryers, both of which produce far more Btus from the same amount of gas; and electrical induction heating in fryers and grills, which keeps the kitchen cooler and is more efficient than traditional heating methods.

Radiant heat barriers (RBs) are materials installed in the attics or on the undersides of roofs to reduce summer heat gain and winter heat loss and thus to reduce building heating and cooling energy usage. Usually made of thin sheets of a highly reflective material such as aluminum, RBs radiate back the heat leaving the building during heating season and reduce air-conditioning cooling loads in warm or hot climates.

Turning off lights in areas not being used and using daylight for ambient illumination can reduce the lighting load during peak demand hours. To use "free" daylight, the task is to admit the sun's rays in a way that makes this "free" energy truly usable as light.

Cash rebate programs are popular inducements for operators to install energy-saving equipment. Usually these rebates are a percentage of the equipment's installed cost. Most utilities have special rates for off-peak hours. Making use of those hours can lower utility costs.

All of the previously mentioned energy-saving suggestions will also require a commitment on the part of the management team to continuously develop, communicate, and monitor energy-saving strategies. The National Restaurant Association and the Federal Energy Administration recommend that foodservice operations organize energy management programs in the following manner:

1. Assign responsibility for energy conservation to a committee comprised of members representing all areas of the company's operations and chaired by a manager committed to the program.
2. Conduct an energy audit to determine baseline data on current operating costs, energy consumption, and operating practices.
3. Develop an energy conservation plan based on the energy audit with specific goals and strategies including the improvement of employee practices in all areas of energy use and the acquisition of energy-saving equipment.
4. Measure the results by comparing baseline to postimplementation results.
5. Maintain or modify the plan as needed based on feedback and results achieved.

Water Conservation

Water conservation programs in foodservice operations should be developed in the same manner as those for energy conservation. Simple practices such as turning off faucets completely, running dishwashers at full capacity, using low-flow toilets in rest rooms, recycling "gray water" (wash water and other waste water that goes down sink drains) for watering exterior landscaping, and serving water to customers only when requested can reduce water usage and result in cost savings for the operation.

SOLID WASTE MANAGEMENT

An urgent need exists to reduce the amount of **municipal solid waste** (MSW), which is by definition the solid waste produced at residences, commercial and industrial establishments, and institutions and excludes construction/demolition debris and automobile scraps. The issue has economic, political, ecological, and sociological ramifications.

The cost for hauling away solid waste continues to rise. For example, the Los Angeles Unified School District Foodservice generates approximately 60,000 tons of trash annually—70% is from the meal program, 29% is paper, and 1% is glass and metal. The cost of disposing of this trash in landfills is $34 per ton or $1.3 million a year.

Legislation has been passed in several states such as that in California where Assembly Bill 939 mandated that companies reduce solid waste output 25% by 1995 and 50% by the year 2000. Mandatory recycling regulations, including landfill bans, are in place in more and more communities across the country.

In foodservice operations, an **integrated solid waste management system** should be employed. By definition, an integrated solid waste management system is the "complementary use of a variety of waste management practices to safely and effectively handle the municipal solid waste stream with the least adverse impact on human health and the environment" (U.S. EPA, 1989). The goals of such a system are to reduce air and groundwater pollution, to reduce the volume of waste, and to extract energy and materials safely prior to final disposal. The hierarchy of integrated solid waste management is:

- Source reduction including reuse of materials,
- Recycling of materials including composting,
- Waste combustion with energy recovery, and
- Use of landfills.

Source reduction is "the design and manufacture of products and packaging with minimum toxic content, minimum volume of material, and/or a longer useful life" (U.S. EPA, 1989). Source reduction has been identified as a priority by many. This includes the elimination of single-use containers and double packaging, the phasing out of metal containers, the banning of packaging that is not recyclable, sanitizing glass and plastic containers for storage purposes, and donating leftover food to programs for the homeless.

The closing of landfills and geometric increases in waste disposal costs have led to the need to reduce trash volume. A variety of equipment options are available to accomplish this goal. Most pay for themselves within a year of purchase with the savings they make possible. Among some commonly used waste management tools are:

- Cardboard crushers,
- Garbage disposals in all sinks to keep trimmings out of the trash,
- Pulper extractor systems, which shred and sanitize garbage (see Figure B.26 in Appendix B),
- Trash "crushers" including can compactors and glass "smashers" (see Figures B.24 and B.25), and
- Polystyrene "melters."

Next to source reduction, recycling is critical for these reasons:

- It conserves scarce natural resources for future generations.
- It reduces the quantity of waste materials sent to landfills since landfill space in many locales will be exhausted shortly if present trends continue.
- It reduces energy costs in manufacturing since using recycled materials often requires less energy and releases less air pollutants than does the use of raw materials.

- It reduces waste that is dumped in oceans, streams, forests, and deserts.
- It prevents the contamination of groundwater sources caused by putting hazardous materials down drains.

Many companies have appointed recycling coordinators and formed recycling committees and teams. Their responsibilities include implementing the recycling program, training staff and customers, encouraging involvement, communicating regarding issues and concerns, and overseeing the program on a daily basis.

Nonprofit organizations, such as the Steel Recycling Institute, offer their services to the foodservice industry. Representatives work with public recycling officials, haulers, scrap dealers, and foodservice organizations to help raise awareness of and implement steel recycling programs.

Every foodservice setting produces some steel waste. The most common type is the #10 can, however smaller cans and lids from plastic and glass containers are also found. All steel products can, and should be, recycled because the steel industry needs old steel to produce new steel. Approximately 66% of steel is now recycled in the United States. Recycling foodservice steel waste provides the steel industry with a much needed resource, reduces material sent to a landfill, helps save energy, and conserves precious domestic natural resources.

The steps in recycling steel cans are as follows:

Step 1 Rinse the cans to remove most of the food particles. This is required for basic sanitation reasons because the cans usually sit for sometime before being picked up. To reduce the amount of water consumed in this process, cans may be rinsed in leftover water that has been used to wash pots and pans or they may be put through the dish machine in available empty spaces.

Step 2 Flatten the cans to reduce volume for efficient storage and economical transportation. Manually this may be done by removing the bottom in the same manner as the lid and stepping on the body of the can. Mechanically, machines are available that will flatten all sizes of cans with the bottom intact. Steel lids and bottoms have sharp edges and should be stored in a can until it is full. The top of this full can may then be taped or crimped shut and transported to storage.

Step 3 Recycle the cans through local options. Dockside recycling may be accomplished by having the company's waste hauler provide a container for steel recyclables. A ferrous scrap processor or independent recyclers will also provide this service. Or used steel cans be delivered to a scrap yard or recycling facility.

Many large foodservices use a piece of equipment that rinses the unopened can, opens it, empties the contents into the desired container, rinses the empty can, flattens it, and delivers it to a Dumpster.

Foodservice operations that use great quantities of glass bottles, such as bars, employ equipment designed to "disintegrate" glass containers. By crushing the bottles to a "gravelly" consistency that is not sharp and ecologically correct, the volume of glass waste is reduced to 1/12 of its original size. The crushed glass may be recycled and the operator has a cleaner, safer working environment. Additional advantages accrue from reductions in empty bottle storage space, waste removal costs, and number of required trips to the Dumpster.

Polystyrene, commonly called foam or Styrofoam, was the target of environmental activists in past years because it was manufactured from chlorofluorocarbons (CFCs), which reduce the earth's protective ozone layer. Currently, polystyrene is manufactured by an injection process using hydrocarbons, a by-product of the oil refining industry, which have no effect on the ozone layer. The manufacture of polystyrene from hydrocarbons does contribute to air pollution as did the "burning off" of the by-product when it was not used.

The National Polystyrene Company's (NPRC's) facilities accept used and baled polystyrene. The bales are broken apart and loaded onto a conveyor leading to a grinder. The little pieces of polystyrene are washed and ground. After drying, the polystyrene is sent to an extruder where it is melted, extruded into strands, and cut into pellets. NPRC sells the postconsumer polystyrene pellets to manufacturers who produce items such as school and office supplies, construction materials, protective foam packaging, video and audio products, egg cartons, and sandwich clamshells.

Paper, plastic, and other dry fibrous materials are being turned into building material for low-cost housing. The multipurpose product is cost effective and construction time is minimal compared to other building alternatives.

A reusable, recyclable entrée dish made out of resin has recently been developed for United Airlines. The dish can be used up to 20 times before being reground into resin flakes to become part of new dishes.

The Hilton Hotels have found that putting food waste and by-products into the compost heap not only helps the environment but saves money as well. As part of a comprehensive focus on environmentalism, the hotel chain has implemented employee-driven environmental programs including purchasing only recycled products where possible and recycling paper, glass, cardboard, aluminum, and food waste. One hotel in the chain converts 15 tons of wet garbage a day into 1 ton of fertilizer and a rich soil conditioner and then sells this compost to golf courses and horticulturists in the area. The compost heap generates temperatures of 160°F, which is adequate to destroy any pathogenic bacteria. The challenge is to control the odor. Using a combination of fans and biologically active filters a 95% reduction in noxious odors has been attained. The final product looks like peat moss and smells like a rich, soil-like, humus peat moss. A nutrient analysis reveals a rich nitrogen, potassium, and phosphorus product.

Much work has been done to research the feeding of food by-products and food wastes to cattle and sheep. The advantages of this idea are that the wastes are diverted from landfills, nutrient density of animal diets can be increased, ration costs can be reduced, and profits for farmers may be increased. The challenges of such a program are that the by-products or wastes must be carefully matched to the animals' requirements, transportation and processing must prevent spoilage without adding to costs, and moisture content must be reduced. In one operation, food wastes are processed through a pulper and packed in 30- to 40-gallon tubs. The food waste is mixed with ground waste paper, cracked corn, and a nonprotein nitrogen source. The mixture is then stored in a storage silo. The recycled newsprint lowers the moisture content, and the corn and nitrogen provide energy for the fermentation process and raise the crude protein content to an acceptable level.

A simpler "food waste to animal fodder" recycling program is used by some communities and foodservice operations. By allowing pig farmers to pick up food waste dockside or curbside, everyone saves.

The purchase of products made from recycled material whenever feasible should be practiced in order for recycling to be an effective method of waste management. Some governmental agencies require that a certain percentage of all paper products purchased be made from recycled materials.

The final alternatives in the integrated waste management system are incineration, which reduces the volume of solid waste and can produce energy, and landfilling, which is the least desirable option.

SUMMARY

For both the foodservice operator and the environment, real savings may be obtained from the better control of energy use. Cutting overhead costs in order to boost profits is a far better alternative to raising menu prices in this economic climate. The foodservice management team needs to develop strategies for their particular operation that will accomplish energy savings without compromising the quality of products and service provided to customers.

The deterioration of the global environment, which supports life, is accelerating. Human use of resources and energy is largely responsible. Using less need not mean a resulting decline in quality of life and may result in the creation of jobs.

As Margaret Mead so wisely put it, "Never doubt that a small group of thoughtful, committed people can change the world. Indeed, it is the only thing that ever has." It is now an ethical imperative that management make the environment a corporate commitment that each employee understands. Implementing an integrated solid waste management system is a foodservice practice that both preserves natural resources and protects the environment.

REVIEW QUESTIONS

1. Identify some of the most common energy "wasters" in a foodservice operation.
2. Describe how preventive maintenance and routine cleaning can reduce energy usage.
3. Compare and contrast energy use on an open burner range versus a hot top range.
4. In what ways do utility companies help foodservices to reduce energy use?
5. List and describe several ways in which air-conditioning costs can be controlled.
6. Why is solid waste management such a concern for today's foodservice manager?
7. Give several specific examples of source reduction to control foodservice waste.

8. List and describe several solid waste management "tools" employed in the foodservice industry.
9. Describe the steps in the recycling of polystyrene.
10. What is necessary to recycle #10 food cans? Why is it particularly important that foodservices recycle this product?

SELECTED REFERENCES

Amdahl energy reduction and recycling programs saved $6 million in 1994, while improving air quality and helping to conserve scarce landfill space. Business Wire. March 30, 1995.

American Dietetic Association: Position of The American Dietetic Association: Environmental issues. J. Am. Diet. Assoc. 1993; 93(5):589–591.

Associated Press: Pigs dig in to help recycle trash. The New York Times. February 20, 1994; 1:49.

Bay, J.: Tips on choosing a recycler. Food Process. 1995; 56(2):82.

Beasley, M. A.: Waste management: Winning support. Food Mgmt. 1992; 27:46.

Birnesser, D. J., Moore, L. H., and Kaye, T. P.: Development of a regional integrated solid waste system. Public Works. 1993; 124(4):52.

Bottle Disintegrator. Marietta, Ga.: Glass Recycling Inc., 1995.

Bryant, F.: Energy team focuses on lighting, comfort challenges of 24-hour-a-day operation. Energy Users News. 1994; 19(1):62.

Cummings, L. E., and Cummings, W. T.: Foodservice and solid waste policies: A view in three dimensions. Hosp. Res. J. 1991; 14:163–171.

Devi, S. L.: Hilton Village's conservation program lauded. Honolulu Star Bulletin. July 21, 1993.

Facilities Operation Manual. Washington, D.C.: National Restaurant Association, 1986.

Federal Energy Administration, Office of Energy Conservation and Environment. *Guide to Energy Conservation for Food Service.* Washington, D.C.: Government Printing Office, January 1977.

Ferris, D. A., Shanklin, C. W., and Flores, R.A.: Solid waste management in foodservice. Food Technol. 1994; 48:110–116.

Hoelting, F.B.: Redefining the food chain: Recycling food waste into animal feed. Am. School and Univ. 1993; 66(2):36N.

Hotels: Industry finding ways to save money. Greenwire. September 9, 1994.

How much can you save with a radiant barrier? Consumer's Res. Magazine. 1994; 77(7):21.

Kaufman, W.: Next step in recycling chain is food. National Public Radio *Morning Edition.* March 17, 1994.

Knisely, J.: Daylighting design fits into an energy-conscious market. EC&M Elec. Construction and Maintenance. 1994; 93(11):7.

Lecard, M.: Cheap chills: Ultra-efficient refrigerators. Sierra. 1994; 79(1):26

Los Angeles Unified School District Recycling Handbook. Los Angeles: LAUSD, 1995.

Lustigman, A.: How green is it? Scientific Certification Systems Inc.'s new environmental safety standard. Sporting Goods Bus. 1993; 26(8):107.

Makower, J.: Good, green jobs. Sacramento: California Department of Conservation, 1995.

Managing Solid Waste: Answers for Foodservice Operators. Washington, D.C.: National Restaurant Association, 1991.

Mann, D. C.: Conserving limited resources: Property management. J. Property Mgmt. 1994; 59(4):51.

Mans, J.: Let the utilities pay for equipment: Cost-saving technique for electric utilities. Dairy Foods Magazine. 1993; 94(5):49.

Meadows, D.L.: *The Limits to Growth: A Report for The Club of Rome's Project on the Predicament of Mankind.* New York: Universe Books, 1972.

Nelson, K.L., and Randazzo, M.: Restaurant installs grill controls at no cost, earns positive cash flow. Energy Users News. 1994; 19(1):6.

Neth, M.: Domizil takes ecological approach from ground up. Hotel and Hotel Mgmt. 1993; 208(10):18.

New dish reduces waste and saves money. Food Eng. November 1993, p. 60.

Patterson, P.: Energy-saving tricks reduce overhead expenses. Nation's Rest. News. 1993; 27(17):56.

PR Newswire: Billionaires in the future will be created out of trash. Financial News. April 28, 1995.

Recycling Steel Cans from Food Service Facilities. Pittsburgh, Pa.: Steel Recycling Institute, 1994.

Reis, D. A., and Betton, J. E.: The environment and its effect on today's management. Industrial Mgmt. 1992; 34(1):26.

Renewable energy in the southland. Los Angeles Times. April 21, 1995, p. B2.

Smith, A.: Green building better for the environment. Austin Business News. 1993; 13(37):1.

Smith, C.: Environmentalism is checking in. Los Angeles Times. September 8, 1994, p. 5D.

Somerville, S.: Short cuts: Twenty quick ways to slash kitchen costs. Rest. Hosp. April 1995, p. 46.

Toolen, T.: New Jersey pigs eat Philly's garbage. Rocky Mountain News. May 1, 1994, p. 28A.

U.S. Congress, Office of Technology Assessment. *Facing America's Trash Crisis: What Next for Municipal Solid Waste?* Washington, D.C.: Government Printing Office, October 1989.

U.S. Environmental Protection Agency, Office of Solid Waste and Emergency Response. *Characterization of Municipal Solid Waste in the United States: 1990 Update. Executive Summary.* Washington, D.C.: Government Printing Office, June 13, 1990.

U.S. Environmental Protection Agency, Office of Solid Waste. *The Solid Waste Dilemma Task Force.* Washington, D.C.: Government Printing Office, February 1989.

Vasanthakumar, N.: Benchmarking for environmental excellence. Industrial Mgmt. 1995; 37(1):1995.

Walshe, N.: Savings schemes: Energy bill reduction. Food Manufacture. 1994; 69(9):23.

Willig, J. T.: *Environmental TQM,* 2nd ed. New York: McGraw-Hill, 1994.

10

Facilities Planning and Design

OUTLINE

Definitions and Goals
Preliminary Preparation for Facility Planning
  Trends Affecting Foodservice Design
  Information on Developments in Design and Equipment
  Regulatory Considerations
  Special Considerations for Specific Types of Foodservices
Steps in the Planning Procedure
  The Prospectus
  The Planning Team
  Feasibility Study
  Menu Analysis
  Architectural Features
  Budget/Cost Relationship
Design Development
  Space Allowances and Relationships
  Schematic Drawing
Work Areas
  Mechanics of Drawing
  Designing by Computer
  Architect's Blueprints
  Specifications and Contract Documents
  Bids, Contracts, Construction, and Inspection
Summary

Facility planning and design are among the responsibilities of foodservice managers. Their involvement can range from planning a new foodservice facility to remodeling or making minor changes within an existing facility.

The concepts presented in this chapter apply to all planning projects regardless of size or scope. However, managers must identify their own goals and needs, work to maximize the project's attributes, and plan for and around any constraints that exist. Providing an appropriate, efficient facility for the production and service of high-quality, attractive food is the desired outcome of all foodservices.

KEY CONCEPTS

1. Preliminary planning for a foodservice design project should include study of the current trends in foodservice design, innovations in equipment and design, regulatory codes and operating licenses required, and specific needs for various types of foodservices.
2. The first step in a facility design project is to prepare a *prospectus*, which is a written description of all aspects of the project under consideration.
3. The *planning team* may include any or all of the following members: the owner or administrator, foodservice manager, architect, foodservice design consultant, equipment representative, business manager, builder/contractor, maintenance/mechanical engineer.
4. The menu is the key to equipment needs, which in turn determines space requirements for the equipment.
5. Decisions made on architectural features are important for determining project cost, ease of cleaning, good sanitation, safety, adequate type and amount of lighting and temperature control for high productivity, and noise reduction for a more pleasant work environment.
6. All initial expenditures must take into consideration the project budget and also such factors as operating costs, life expectancy, conformance to sanitary standards, and provision of comfort for employees and customers.
7. The first step in design development is to determine optimum space allowances and draw a flow diagram showing the location of the work units.
8. In the schematic drawing, equipment is drawn to scale in each work unit with required traffic aisles and work spaces included.
9. The American with Disabilities Act mandates some general guidelines for implementing reasonable accommodations in the workplace and dining areas for persons with disabilities.
10. The seven major work areas in foodservice departments are receiving, storage and issuing, pre-preparation, preparation, serving, warewashing, and support services.

DEFINITIONS AND GOALS

To understand the planning process thoroughly, foodservice managers and others involved in a design project need to know certain definitions and examples of terminology, as used in this chapter, and also the goals to be achieved. Definitions include the following:

- *Physical:* Pertains to material existence measured by weight, motion, and resistance. Thus, anything taking up space in a facility must be accounted for and fit the available space.
- *Design:* Refers to the broad function of developing the facility, including site selection, menu, equipment requirements, and other planning functions that will guide the project into reality.
- *Layout:* Refers to the process of arranging the physical facilities, including equipment, such that operational efficiency is achieved. This involves a design drawn on paper to show walls, windows, doors, and other structural components. After this outline drawing is complete, required work areas are designated on the plan, then the equipment and other facilities are arranged and drawn onto the plan.

Foodservice managers must be involved with the development of all aspects of the design plan to ensure that the facility is properly coordinated and functional. Although other professionals will design the electrical, water, and plumbing systems, as well as the lighting, heating, ventilation, and structural components of the building, the foodservice manager must provide input on the specific needs of the foodservice facility. The finished project plan results in either success or failure for the organization involved!

PRELIMINARY PREPARATION FOR FACILITY PLANNING

Before attempting to develop a final design, foodservice managers need to prepare themselves for the tasks ahead, which include the following:

- Study trends that affect foodservice design.
- Learn what is new in design and equipment.
- Obtain and read copies of regulatory codes and required operating licenses that have a bearing on foodservice design and operation.
- Become knowledgeable about special requirements for specific types of foodservices

Thus, they will have the information needed for making a worthwhile contribution to the overall planning team and the background to make sound decisions as the project progresses.

Trends Affecting Foodservice Design

Changes in Patterns of Dining Out. More people than ever before are eating meals away from home. Depending on the economy, however, the types of foodservices patronized will be affected. The foodservice industry is responding to this trend by making changes in the style of foodservices, types of food served, and prices charged. All these factors, in turn, influence a facility's design.

Change in Desired Menu Items. Continuing changes in customer preferences for types of foods and meals eaten away from home also affect the design requirements of a foodservice facility. A concern for physical fitness and well-being, for example, changed the menus of most foodservices. Menus now offer lighter, more healthy food selections and limited desserts. This changes the equipment needed and the space requirements and, thus, affects the remodeling or new construction to accommodate preparation of foods customers prefer.

Economic Factors. Costs of wages, food, and utilities can influence selection of a type of foodservice and its design. For example, as employees' wages increase, automation of equipment (e.g., robots) and the purchase of convenience foods become more common. In addition, as costs for food and the energy to prepare it continue to rise, the foodservice design must provide for efficient operation. Managers must strive to maximize the operation's efficiency through the best possible design and **layout** of work centers; for example, by minimizing the distances a worker must cover to transport and distribute food.

Flexibility of Use. One trend in facility design is to make the existing space adaptable either for multiple uses or for meeting future demands. This may be accomplished in part by selecting wheel-mounted and **modular** equipment or by using portable units. The key is to choose equipment that is uniform in size, movable, and adaptable for numerous work activities.

Energy Conservation. With the current ever-rising energy costs, the trend is to design and equip foodservice facilities to save energy to the greatest extent possible. Equipment manufacturers are producing equipment that gives a high yield of energy for the work accomplished. The energy used by specific pieces of equipment is stated in the specifications, and comparisons should be made by the foodservice manager before selecting a particular make or model.

Other energy conservation trends are toward better insulation, heat recapture for other uses, and recirculation of heat. Solar heat designs are used in some areas, especially for restaurants, and may be a future trend of other types of foodservices. (See Chapters 9 and 11 for details on energy-conserving equipment.)

Built-In Safety, Sanitation, and Noise Reduction. In planning the total facility, the safety of the employees, safety of food, and overall sanitary conditions are considerations in new designs. These may be achieved by the type of floor covering, ventilation, building materials, lighting, and equipment selected, and the method of their installation. Ease of cleaning reduces labor costs, and materials and designs chosen for their safety features help reduce accidents. All make for an attractive, safe working environment for the employees. Many of these features reduce noise and worker fatigue, and hence result in greater productivity.

Information on Developments in Design and Equipment

Visits to new or remodeled facilities of the same type you are planning and talks with managers of those facilities may garner new ideas and serve as a means to obtain firsthand information. Those with recent building experience usually are pleased to share workable ideas, mistakes that were made, and suggestions for improvement.

Obtaining catalogs and specification sheets from various equipment companies for comparative purposes and determining equipment space needs is essential. A file of such reference materials will be invaluable during work on the project. Equipment company representatives can be excellent sources of information for learning what is new and workable in various situations. Design consultants can also be contacted with any specific questions that arise.

Trade journals should be reviewed for articles on planning and design. Information gained can contribute new ideas and helpful suggestions to the planner. If the project is to remodel a facility, present staff and employees may have excellent suggestions resulting from their own work experience. Giving them an opportunity to express opinions is a valuable resource and should be mutually beneficial.

Regulatory Considerations

Foodservice managers need to know which federal, state, and local laws, codes, and regulations will affect their building or remodeling project. These regulations have to do with zoning restrictions; building standards, including those to accommodate persons with disabilities; electrical wiring and outlets; gas outlets and installations; health, fire, and safety codes; sanitation standards that govern water pollution and waste disposal systems; and installation of heavy-duty equipment.

Regulations have been established by agencies and organizations such as state, county, and local health and engineering departments, the American Gas Association, Underwriters' Laboratories, and the National Sanitation Foundation International by federal legislation such as the Occupational Safety and Health Act and the Americans with Disabilities Act. Copies of these codes may be obtained by writing to or visiting the appropriate agency. Large libraries also usually have copies of the codes and regulations.

Other professional persons will assist in the identification and application of regulatory codes and standards. These people make up a part of the planning team discussed later in this chapter.

Building permits are required but, in most instances, will not be issued for foodservice projects until health department officials have reviewed and approved the plans. Breit (1991) notes that it is expedient, therefore, to contact a local health department official and work closely with that person as plans are developed, so that approval is ensured.

Special Considerations for Specific Types of Foodservices

A brief review follows of some special considerations to keep in mind when planning a specific type of foodservice.

Commercial Facilities. Restaurants catering to downtown shoppers and businesspeople prefer a location near a busy intersection. Their customers, who often have a limited lunch period, may be those within a 10-minute walk of the restaurant. Because rents for prime downtown sites are likely to be high, effective use of every square inch of space is a top planning priority. Many such restaurants are built vertically with several levels for their various functions. Coordinating these activities with a good transport system between floors is a unique planning challenge. Suburban restaurants typically draw patrons from a larger area, making adequate parking the first essential. In addition, the location should be easily accessible and highly visible to approaching motorists. Shopping centers, which not only attract large numbers of customers but also provide ample parking, are considered desirable locations for commercial foodservices.

Hotels and motels usually have coffee shops located in visible locations, with entrances from both the street and the lobby. However, the main dining, party, and banquet rooms are frequently less visible, with access only through the lobby. For these facilities, basic food items are often prepared in a central kitchen. Finishing or banquet kitchens then should be located adjacent to the various serving areas.

Schools and Universities. School foodservices are preferably located on the first floor, convenient to the central hallway. The area should be well ventilated to allow cooking odors to dissipate rather than to permeate classrooms. Dining areas in some schools may have to double as a study hall or gymnasium, which would present a different planning situation.

Many large city school systems utilize a central production kitchen, or commissary, for food production for all schools in the system. Often, these are cook/chill or cook/freeze systems, which require specialized cooking equipment and good transportation systems and schedules. With this system, individual schools need only limited equipment for finishing off baking, reheating certain items, and serving the food.

Colleges and universities provide many and varied types of foodservices to accommodate the needs of the entire campus community. Residence halls may have their own kitchen and dining rooms or, if there are several halls on campus, may have a central production unit for certain items, such as baked goods, or for pre-preparation of produce or meats. The trend is to have a choice of menu items, usually served cafeteria style.

Peak workloads at the three meal periods may necessitate duplicate pieces of large equipment and adequate work space for personnel. Student unions usually offer a variety of types of foodservices, including large banquet halls for seated service, short-order units, cafeterias, and possibly small dining rooms for special meals. Some colleges and universities have invited various fast-food companies to operate one of their units on campus to satisfy student requests for that type of food. Each type of foodservice requires different space and equipment, making planning for these facilities a challenge.

In-Plant Facilities. The *in-plant* or *industrial foodservice area* should be central in location, allowing for ready access from as many places in the plant as possible. Every provision should be made to expedite service so that all workers can be accommodated quickly during a fairly short lunch break. Mobile units and vending operations can be used in remote areas of large plants or in those too small to justify the space and expenditure for kitchen equipment, management, and labor. Adequate passageways for such carts are essential.

Homes for Children and Adult Communities. Homes for children and retired persons make use of the residence hall and cottage-type of plan. Because a quiet, home-like atmosphere is desirable for this type of situation, the facilities are often located away from the busy streets of the city and surrounded by private grounds.

Hospitals and Health Care Centers. Facility planning for hospitals and other health care centers must provide for the needs of staff, employees, visitors, and guests as well as the patients. The decision as to type of service to use, centralized or decentralized, has to be made at the outset of planning, because space and equipment requirements differ greatly for each. A central kitchen ordinarily provides food for these groups with one dining room/cafeteria that serves everyone except bed patients. Sometimes small, private dining rooms can be planned for official catering functions. For after hours, vending machines can be installed to supplement regular meal service. Adequate passageways for transporting patients' meals on carts and trucks, as well as space for cart storage, are other special considerations. Elevators or lifts designated solely for foodservice use will expedite meal service to patients. Office space for clinical dietitians in large hospitals is another planning consideration. If off-premise catering, Meals-on-Wheels, or other services are to be provided, adequate space for them must also be included in the facility plan.

Correctional Facilities. Kitchen and dining rooms for correctional facilities present a planning challenge different from most other types of foodservices. Since inmates often serve as foodservice workers, the basic design consideration is to provide for personal safety, security, and protection against sabotage. The foodservice manager should have a full view of all operations. Therefore, the office should be centrally located in the kitchen, raised above the floor level, and safety glass windows should be in place on all four sides. For security, all cabinets should be open with no drawers, and secure locks should be provided for all storage areas. Hall (1989), a registered dietitian, recommends a storage warehouse built outside but adjacent to the kitchen so deliveries for daily use can be made easily, thus eliminating the need for large storage areas in the kitchen and reducing chances of theft. She also suggests that the serving area be designed to prevent face-to-face interaction between servers and inmate "customers" that could create confrontations. A partitioning wall from the ceiling down to within approximately 24 inches of the front of the serving counter achieves this objective while still allowing easy viewing for selection of foods. The dining area is best divided into small units seating 100 to 125 persons for control of potential riots.

Other planning considerations may include the delivery of food to some inmates in their cells. The choice of centralized service with food portioned and served onto trays in the kitchen for distribution or decentralized service with bulk food delivery to serving areas throughout the facility for service is one to be made by the planning team.

Generally, kitchens and dining areas in any type of foodservice facility should provide maximum convenience and accessibility for customers. For efficiency's sake, locating dining rooms adjacent to kitchens is preferable. Foodservices are best located on the first floor to obtain the best lighting, ventilation, and outdoor views. Basement-level locations can have a poor psychological effect on both patrons and employees if the area is dark and unattractive. The disadvantages of foodservices that are located above the first floor are inaccessibility to patrons and problems related to bringing in supplies and removing trash and waste. The physical environment has much to do with the success of any foodservice design.

STEPS IN THE PLANNING PROCEDURE

After preliminary study to prepare for the facility design project, completing the following developmental steps will lead to a completed layout design:

- Prepare a prospectus (a program or planning guide).
- Organize a planning team.
- Conduct a feasibility study.
- Make a menu analysis.
- Consider the desired architectural features: building materials, floors, walls, lighting, heating, cooling, ventilation, refrigeration, plumbing.
- Consider (and adjust if necessary) the costs versus money available relationships.

Upon completion of these preliminaries, the design development process can proceed.

The Prospectus

The **prospectus** is a written description that details all aspects of the situation under consideration and helps other professionals on the planning team understand the exact needs of the foodservice department. It should contain the elements that will affect and guide the proposed design and also present a clear picture of the physical and operational aspects of the proposed facility or remodeling project. Usually, it is based on questions such as these:

- What type of foodservice is planned?
- What is the foodservice to accomplish?
- Which major type of food production system will be used?
- How many people and what age groups are to be served? How many must be served at one time?
- What will be the hours of service? style of service?
- What is the menu pattern?
- In what form will food be purchased? how often?
- What storage facilities will be needed? amount of refrigerated and amount of freezer space?
- What equipment and what capacity for each piece will be required to prepare and serve the menu items?
- What are desirable space relationships?
- How will safety precautions be incorporated in the plan? sanitary measures?
- What facilities must be planned for persons with disabilities?
- What energy sources are most economical? available?
- What activities will be computerized?

The prospectus usually contains three major sections:

1. The *rationale* includes title, reason or need for project, and its goal, objectives, policies, and procedures.
2. *Physical and operational characteristics* include architectural designs and features, all details about the menu, food preparation and service, employee and customer profiles, and anticipated volume of business.
3. *Regulatory information* includes built-in sanitation, safety, and noise control features, and energy and type of utility usage desired.

Rationale. The title, goal, objectives, policies and procedures, and a statement of need for the project are, perhaps, the most difficult components to define. The following definitions and examples should help make the exercise easier:

- *Title:* Description of the plan. Narrow the title to reflect the actual scope of the design that is proposed. Example: Design for a warewashing area of the Coastal Restaurant Foodservice.
- *Goal:* State the single outcome of the project. Example: To develop a central warewashing area that will process all dishes, utensils, and pans of the foodservice.
- *Objective:* Specific statements that indicate what is necessary to achieve the goal. Example: The warewashing area will (1) utilize no more than 36 square feet of floor space, (2) be operated with no more than four persons, and (3) operate with minimum energy usage.
- *Policy:* A definite course or method of action selected from among alternatives and in light of given conditions to guide and determine present and future decisions. Example: All dishes, utensils, and pans will be washed and stored within 45 minutes of use.
- *Procedure:* A particular way of accomplishing something. Example: Conveyor belts will be used to carry dirty dishes to the warewashing area. Or, scraping,

racking, and washing of dirty utensils and pans and storing clean ones should be accomplished with 80% automation.

The statement of need for the project may be simple or complex, depending on the project; for example: "The foodservice dry and refrigerated storage areas need to be expanded 60% to accommodate an increase in meal census from 500 per day to 1200 per day as a result of a recent building addition."

Physical and Operational Characteristics. Physical characteristics relate to architectural or design features, such as building style appropriate to the type of food to be served. Mexican foods, for example, often call for Mexican or Spanish architecture. A necessary design feature to identify in a remodeling project could be an existing support pillar, an elevator shaft that cannot be moved, or a desired solar heating system. These must be identified at this stage because they could affect other considerations, such as the style of the roof and types of windows.

Operational data refer to activities that take place in the foodservice department. The types of food on the menu are the key concern in the planning stage. Further, the form in which food will be purchased—fresh, canned, and/or frozen—and the approximate quantities of each must be estimated with some accuracy. This information helps planners determine the amount and kind of storage space required. Food preparation methods to be used, including on-premise pre-preparation, tell the planners what equipment is needed and the amount of space that is required to accommodate the equipment.

The major operational characteristic, or type of foodservice system, is basic to all design planning. Space requirements for an assembly/serve system are quite different from those of a conventional system. Other decisions to make in advance are whether to use centralized or decentralized service and the method of delivery and service to be employed.

Other operational characteristics are the hours of service, anticipated volume of business (both total and per meal period), and the number of diners to be seated at any one time. These data help determine the required size for the dining area.

A customer profile, including age, size of group, and mobility, helps determine the probable dining space required per person. An employee profile includes the number of employees, the number of shifts and employees on duty per shift, the employees' sex (to plan locker and rest room space), and each work position as it relates to standing, sitting, walking, pushing carts, and so forth. Special considerations to meet the needs of persons with disabilities are included as well. This information is essential so that adequate space for work and movement of people and equipment will be allocated.

Regulatory Information. This section identifies the standards of safety, sanitation and cleanliness, noise control, and waste disposal that the design must meet. Also included are standards established by the Americans with Disabilities Act to provide for employees and patrons with disabilities. Guidelines for selection of the type of utilities to be used and energy constraints are also stated.

Because every project is unique, the various sections of a prospectus will not always include all of the same data. Just those that pertain to the situation are necessary. For example, a dry and refrigerated storage design does not need a customer profile. Instead, employee and equipment characteristics would be the focus, in addition to information regarding the menu, food items, and safety and sanitation regulations.

The key in writing the plan is to include all pertinent technical data and to always comment on how the data presented will affect the proposed design. The seating space in a cocktail lounge, for example, will depend on the size of the tables and chairs selected. The person who writes the prospectus and later helps develop the design should be a professional foodservice manager with the knowledge and authority to make decisions about the anticipated menu, space, and equipment needs. That person must also be able to provide other operational data required by planning team members.

The Planning Team

When the project plan is complete, it is time to organize a team to develop the design plan. The expertise required of those on the team will vary depending on the extent of the project, its objectives, and its size. Typical planning team members include the following types of persons:

1. Owner or administrator (the person with authority to spend money for the project and give final approval to carry it out),
2. Foodservice manager,
3. Architect,
4. Foodservice design consultant,
5. Equipment representative,
6. Business manager,
7. Builder/contractor, and
8. Maintenance engineer/mechanical engineer.

Not all team members are involved in all planning stages. Certain members are included at intervals throughout design development. Generally, the owner or administrator and foodservice manager will co-plan the initial design and bring in other team members for planning meetings at appropriate times during the project's development.

The planning team chooses a floor plan, selects materials, and writes specifications cooperatively. However, team members need to check the plans many times before submitting final proposals to builders and to equipment vendors for bids. It is essential to include every detail and be so specific that no part of the architectural features, equipment layout, and specifications is left to chance or misinterpretation.

Feasibility Study

A feasibility study—the collection of data about the market and other factors relating to the operation of the proposed facility—justifies the proposed project, helping to ensure that the project is worth pursuing. This study follows the prospectus outline, with data being collected for each major category. Since each project is unique, categories vary according to need. For instance, the feasibility study for a new restaurant would include research on the proposed site, potential customer profiles and community growth, building trends, competition in the area, and possible revenue-generating sources, such as catering. For a small remodeling project in an existing building, the feasibility study would focus more on operational details than on community and competitive information.

Because the financial commitment for most projects is so large, cost information for the building or for remodeling, and for equipping it, is an essential part of the feasibility study. Many people can assist in this effort, but one person should be the coordinator. This person may be the business or financial manager who also interprets the cost data for the planning team.

Data sources for the feasibility study may include

1. Payroll, production, cash register, and inventory records;
2. City, county, state, and national regulations, obtained from the respective agencies involved or from library copies of those documents; and
3. Statistics regarding trends, average costs, and customer information obtained from trade journals and independent studies.

The feasibility study is a critical component of a project plan. If done well, funds are more likely to be made available, allowing the project to proceed. Resources for a restaurant feasibility study are shown in Table 10.1.

Menu Analysis

An important step in preliminary planning is identifying the type of menu to be served (see Chapter 4 for menu types) and the various food preparation methods required for that menu type. This is the key to equipment needs, which in turn determines the space requirements for the equipment.

Menu, foodservice system, and style-of-service decisions are the major foodservice planning components. The menu affects equipment design and layout, as well as personnel skills and staffing levels required. For example, if the menu and menu pattern contain no fried foods, frying equipment need not be included in the design and no cooks will be needed to perform this task.

The prospectus should include a sample of several days' menus and a menu pattern. The pattern specifies meal categories or courses, while the menu identifies the respective preparation methods required. From this sample menu, the foodservice manager analyzes the variables involved in producing menu items, such as type of storage needed, portion size, total number of portions, batch size, processing required, utensils needed, necessary work surfaces, and type of equipment required (see Table 10.2). The estimated time when a batch is needed and when preparation is complete is also helpful in deciding whether equipment could be shared or whether duplication is necessary. The manager also evaluates the menu for production, service, acceptability, and feasibility. At this point, menu changes can be made to balance equipment use, workload, and acceptability.

Architectural Features

During a project's planning phase, the planning team considers certain architectural features such as building style and materials; types of floors, walls, ceilings, and noise

Table 10.1 Restaurant feasibility study resources.

Sources	Available Information
Chamber of commerce, economic development authority, or city planning office	Population trends (historical and projected) Age, occupation, income level, ethnic origin, and marital status by census tract Retail sales, food and beverage sales Maps Consumer shopping habits and patterns Employment and unemployment statistics Major employers and industries Planned commercial, residential, and industrial developments Demographic and socioeconomic trends Area master plan
Housing and community development	Residential occupancy and housing Urban renewal projects Property values
Building commission	Building permits data Planned construction developments
Zoning commission	Planned uses for zoned areas Building height, signage, parking, and construction restrictions
Transportation department	Road traffic patterns and counts Proposed traffic developments Types of public transportation and routes
Department of revenue	Income and other related sales taxes Real estate assessments
Convention and visitors bureau	Number of visitors by month and spending habits/statistics Conventions—size, type, frequency, and duration Recurring festivals and fairs
Utility companies	Estimates of gas and/or electric expenses for the proposed restaurant
Local newspapers and magazines	Restaurant critiques Dining guides Planned commercial developments

Source: Courtesy of National Restaurant Association.

reduction components; lighting; heating and cooling; ventilation; built-in refrigeration; and plumbing. Not only is making a decision on these features essential for determining project cost, but also for ensuring ease of cleaning, good sanitation, safety, adequate type and amount of lighting and temperature control for high productivity, and noise reduction methods for a more pleasant work environment. Because certain refrigeration units are usually built-in, the number and location of such units must be determined before construction or remodeling begins.

Some of the components previously mentioned are included here as a basic information review.

Table 10.2 Menu analysis.

Menu Item	Storage Type	Portion Size	Table Portions	Batch Size	Process Required	Utensils	Work Surface	Large Equipment	Holding Equipment
Spinach salad	Refrigerated	1 ounce	350	100	Wash Trim Drain	Knife Drain pan	Sink Counter	Sink Counter	Refrigerator
Beef patty	Freezer	4 ounces	300	50	Grill	Spatula	Counter Utility cart	Grill	Warming oven

A sample of a menu analysis to be made for all menu categories. When similar menu items are used, such as several salad dressings that require the same production and service treatment, only one analysis need be made. This information is the basis for determining equipment and space needs for designing a foodservice facility.

Building Style and Materials. Type of foodservice operation, its geographic location, and its menu markedly influence the type of architecture and materials used. Material selection for any building depends on the type of architecture planned, the permanence desired, the location, and the effect of local weather conditions on the materials. The building engineer and the architect will know the characteristics of the various building materials and help select the most suitable in relation to cost.

The restaurant shown in Figure 10.1 is a fine example of modern design. It is built of local stone and situated to take advantage of a scenic location and provide diners with an outstanding view.

Floors. Floors have to meet certain utility, durability, and resiliency requirements. They should be impervious to moisture, grease, and food stains, as well as nonslippery and resistant to scratches and acids, alkalies, and organic solvents. Floors should be durable enough to withstand the wear from heavy traffic characteristic of large food units and, in kitchens, to support the weight of heavy-duty equipment.

What is the best flooring material for a foodservice unit? Opinions vary. Hard surfaces tire employees and may cause accidents by causing persons to slip or fall when the floor is wet. However, this type of flooring is highly resistant to wear and soil, comparatively easy to maintain, and permanent. Rough or abrasive slip-resistant tile is safer but more difficult to clean and, therefore, less hygienic. Norwich (1991) reported that some flooring manufacturers are now producing specially constructed resilient sheet vinyl that is smooth on the surface, easy to clean, and seems more slip resistant than hard tiles (e.g., quarry). Sheet vinyl flooring is easier to stand on than tile and has no grouting to clean. Although this type of flooring has potential for future use in foodservice installations, at present, it seems too soft to support the weight of heavy-duty equipment. Quarry tile seems to be the preferred flooring in most institutional kitchens today. Any floor surface that is to comply with health department requirements must be **coved** 6 inches up at all walls and equipment bases. In addition, floors must be installed so that they are sloped to drain at various parts of the kitchen for ease in cleaning.

Walls, Ceilings, and Noise Reduction. The type of wall and ceiling materials selected for kitchen and dining areas can contribute to the overall aesthetic value

(a)

(b)

Figure 10.1 (a) The curved facade of the Top of the Falls restaurant overlooks the whole Niagara Falls panorama. (b) The multilevel design of the restaurant interior offers every customer a spectacular view.
Courtesy of Herbert H. Gill, President, Gladieux Corporation, Toledo, Ohio.

and sanitary conditions, as well as help to reduce the noise level. As with floors, wall and ceiling materials that are durable and easily cleaned will meet health department regulations and reduce labor time required for cleaning. In addition to wall and floor joinings being coved, all other corners and angles used in installations should be rounded to make cleaning easier and prevent chipping. All pipes and wiring conduits should be concealed in walls.

The amount of natural and artificial light available helps determine the wall finish for a given room. Various colors and textures of materials reflect and absorb different amounts of light. Consider this aspect in relation to the amount of light desired.

Several materials are suitable for walls in foodservice kitchens. Ceramic tile is probably the most suitable material because of its durability and cleanability, although it is expensive to install. However, over its expected life span, the cost of ceramic tile is comparable with other materials.

Fiberglass reinforced panels (FRP), a plastic-like paneling, is quite durable, available in several colors, and less expensive to install than ceramic tile. A minimum-quality material for kitchen wall surfaces is wallboard painted with washable enamel. Since it lacks durability, it is not suitable for use in wet areas, such as around sinks and in warewashing rooms.

A desirable arrangement is to cover walls with ceramic or other glazed tile to a height of 5 to 8 feet where food and water splashes occur. The remainder of the wall may be smooth-finished washable enamel or semigloss paint.

Stainless steel is another material highly suited for kitchen walls; however, due to its very high cost, its use is often limited to cooking areas. It is also quite reflective and may cause glares from the lighting.

Ceiling heights vary widely, with kitchens typically averaging 14 to 18 feet. Kitchen and dining room ceilings should be acoustically treated and lighter in color than the walls. The use of sound-absorbing materials, such as draperies and carpeting, tends to minimize local noises in dining rooms. There are many acoustical ceiling materials to choose from for kitchen use. They must resist deterioration from rapid temperature and humidity changes and from corrosive cooking fumes. In addition, those that have a low reflectance value, are resistant to fire, and are washable are most suited for new or remodeled foodservice kitchens.

Sound-absorbing materials are used not only as surface finishes in construction but also as insulators. Vents, radiator pipes, and water pipes may act as carriers of sound and the most effective means of noise prevention is their careful and thorough insulation with sound-absorbing material. Because of later inaccessibility and prohibitive costs, it is most important that this precaution be taken in the original construction.

Figure 10.2 shows an example of sound reduction treatment in a noisy area. Features such as automatic lubrication of the so-called noiseless power equipment that keeps it in quiet working condition, rubber-tired carts, rubber collars on openings in dish-scraping tables, and ball-bearing glide table drawers help to minimize noise in the kitchen.

Lighting. The amount and kind of lighting required for a foodservice represent a long-term investment and merit the assistance of technical experts in the field. However, the lighting's adequacy, efficiency, and suitability are far more important concerns than its installation cost.

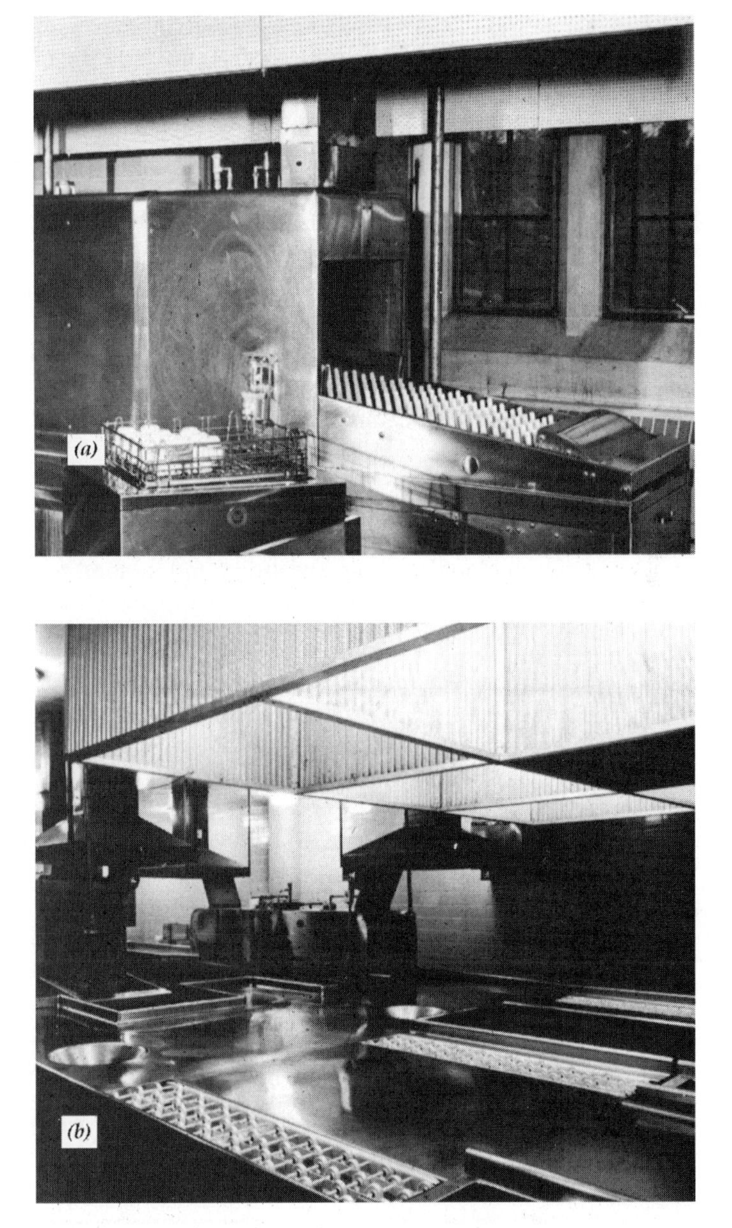

Figure 10.2 Special sound-absorbing ceiling treatments above dishwashing area. (a) Removable perforated acoustical panels hung from ceiling. (b) Box-like corrugated aluminum sheets filled with 1 inch of fiberglass insulation suspended from ceiling in metal frame.

The design should allow for as much natural light as possible. Natural light not only makes food look more appealing, but it can also reduce operating expenses. In addition, natural light exerts a positive psychological effect on workers and guests. Because it is not possible to rely totally on natural light, it is desirable for foodservice managers to have some knowledge of lighting and its requirements when working with lighting experts.

The amount or intensity of light, the kind and color of light, and type of fixtures and their placement combine to create good lighting. The reflective values of walls, ceiling, and other surfaces also affect lighting. Light intensity is measured in footcandles obtained from light meter readings, and the number of required footcandles per square foot depends on the work to be done. The general guidelines shown in Table 10.3 are helpful for planning.

Planners should choose the light fixtures and their placement during the project's design phase so outlet and switch locations can be identified. Fixtures should harmonize with the architectural plan and be placed to provide the recommended illumination level and balance for dining areas as well as for food storage, preparation, and serving areas. Studies have shown that proper workplace lighting can increase employee productivity by 3% to 4%, a significant amount in overall efficiency.

Lighting systems may be either indirect, direct, or a combination of the two. In indirect systems, about 90% to 100% of the light is directed upward, whereas in direct systems, a corresponding amount is directed downward. Luminous ceiling lighting gives an evenly dispersed light that creates the effect of natural sunlight and is desirable in kitchen areas. Brightness, though, should be low enough to prevent glare or cause reflections on shiny surfaces that may affect workers' eyes. Light fixtures should be positioned to prevent employees from standing in their own shadows while working. Good lighting reduces eye strain and general worker fatigue, and is conducive to accuracy in work, as well as to good sanitation and safety in the workplace.

Heating, Ventilation, and Air Conditioning. The heating, ventilation, and air conditioning (HVAC) system provides comfortable temperatures for employees and guests. An architect, working with an HVAC specialist, is best qualified to specify a system of sufficient capacity for the facility in question. Foodservice presents a somewhat different problem from other building uses because cooking processes generate heat, moisture, and food odors.

Air conditioning means more than air cooling. It includes heating, humidity control, and circulating, cleaning, and cooling of the air. Systems are available with controls for all features in one unit. The system may be set up to filter, warm, humidify, and circulate the air in winter and, by adding cooling coils and refrigeration, maintain a comfortable summer temperature. Dehumidification may be necessary in some climates.

The placement of air ducts is important to prevent direct blasts of cold air onto those in the room. Figure 10.3, which shows a retirement home dining room that seats 400 people, illustrates several points. The ceiling is designed with wood slats

Table 10.3 Guidelines for achieving effective lighting.

Light Intensity (in footcandles)	Activity
10 to 20	Walkways (halls and corridors)
15 to 20	Dining
15 to 35	Rough work
35 to 70	General work or food display area
70 to 150	Reading recipes, weighing and measuring ingredients, inspecting, checking, and record keeping

backed by sound absorption material. This permits enough sound to create a friendly atmosphere, but reduces the sound level enough to allow people with hearing aids to be comfortable. The lighting is excellent, with natural light on three sides and adequate, well-spaced ceiling lights. Ducts located beside the right-hand row of lights supply conditioned air. The suspended umbrellas add an aesthetic touch.

Satisfactory kitchen ventilation typically consists of an exhaust fan system, built into a hood placed over cooking equipment, to eliminate cooking odors, fumes, moisture, and grease-laden vapors. In the absence of direct air conditioning, cool outdoor air may be drawn into the kitchen by fans to reduce the temperature and increase circulation of air, making body surfaces feel cooler.

Although air conditioning may be considered expensive to install and operate, employee productivity is estimated to increase 5% to 15% in such a controlled environment. As a result, planners should carefully consider what type of temperature control system is most appropriate for their climate and facility.

Built-In Refrigeration. The smooth, efficient operation of foodservice departments will be enhanced by planning the adequate kind and amount of refrigeration. The foodservice manager should have some knowledge of the principles of refrigeration, types of systems used, and how to determine space needs for the facility being planned. Walk-in, built-in, and reach-in refrigerators and freezers are essential. At this stage of planning, however, only the permanent built-in types need to be considered. Reach-in and portable types are discussed in Chapter 11.

Figure 10.3 Distinctive treatment of a retirement residence dining room.

Mechanical refrigeration is the removal of heat from food and other products stored in an enclosed area. The system includes the use of a *refrigerant* (chemical) that circulates through a series of coils known as the *evaporator*. It begins as a liquid in the coils then is vaporized and pressure is built up by the heat it absorbs from the food. This starts a *compressor*, which pumps the heat-laden gas out of the evaporator and compresses it to a high pressure. The compressed gas flows to a *condenser*, which is air or water cooled, the heat is released, the gas is reliquefied, and the cycle is ready to repeat when the temperature in the refrigerator or freezer becomes higher than desired.

It is desirable for a refrigerant to have a low boiling point, an inoffensive odor, high latent heat, and a reasonable cost, as well as to be nontoxic, nonexplosive, nonflammable, noncorrosive, stable, and not harmful to foods. Most important, with today's concern over the depletion of the ozone layer in our atmosphere, refrigerants used must be "ozone friendly." Many of the old, standard refrigerants, such as Freon (F12 and F22), are the harmful chlorofluorocarbons (CFCs) that are responsible for the depleting ozone layer. They are therefore being phased out of use. Less harmful are the hydrochlorofluorocarbons (HCFCs), which are now being used by compressor manufacturers for refrigerants in their compressors. According to Richardson, HCFC 22 is 90% safer than CFC 12. He estimates that by the year 2000, all CFCs will be entirely eliminated from use in any foodservice. Thus, the foodservice industry is doing its share to help alleviate the worldwide concern for stress on the ozone layer.

Refrigeration systems may be central, multiple unit, and single unit. In a *central* system, one machine supplies refrigeration in an adequate amount for all cooling units throughout the building. This system is rarely used because of the problem of trying to maintain desirable refrigeration in all the different units and because, in case of a breakdown, *all* refrigeration is gone. In a *multiple* or parallel system of refrigeration, there is a compressor for a series of coolers, the compressor being of proper capacity to carry the load required to maintain the desired temperature in the series of coolers. A *single* unit is the self-contained refrigerating system used in the reach-in types.

The location and space allocations for built-in units require careful planning. Generally, they are placed close to the delivery area to minimize distance to transport items received into refrigerated storage. They also need to be close to the preparation units that most frequently use the products stored in them. Three separate walk-in refrigerators are recommended as a minimum, one for fresh produce, one for dairy products and eggs, and one for meat and poultry. Each food group requires a different temperature for optimum storage. Walk-in freezers may also be planned.

Many factors influence the amount of refrigerated and freezer space needed:

1. *The size of the establishment:* Since permanent walk-in units that are smaller than about 8 × 10 feet are uneconomical to install, small facilities may use reach-in rather than walk-in units.
2. *The kind of foodservice system used:* Systems with cook/chill methods require a large amount of refrigerated space, whereas cook/freeze and assembly/serve systems require primarily freezer space.

3. *The frequency of deliveries:* Establishments that are close to markets and receive daily deliveries would require less storage space than those foodservices located in remote areas where deliveries are infrequent.
4. *The form in which food is purchased:* If primarily frozen foods are purchased, naturally more freezer space will be required than if all fresh or canned goods are used.

The total space required may be estimated by measuring the size of units of purchase (e.g., cases, bags, crates) and multiplying each by the number of units to be stored at one time. This will give the total cubic feet of space required, which will be divided into the separate food items to be stored together. Most walk-in units are 7 or 8 feet high. Aisle space in the refrigerator must be wide enough for trucks or carts to enter. Width of shelving is based on the items to be stored; 2 to 3 feet is the usual width. Space for the insulation of the walk-ins also has to be included, a minimum of 3 inches on all sides for built-in refrigerators and 5 to 8 inches for built-in freezers.

Floors of walk-ins should be made of strong, durable, easily cleaned tile that is flush with the adjoining floor to permit easy entry and exit of food on trucks or carts. Wall surfaces should be washable and moisture resistant. Each unit should be equipped with an internal door-opening device and a bell as a safety measure. An exterior wall-mounted recorder to show the box's inside temperature saves energy by eliminating the need to open the door to check temperatures. (See the "Schematic Drawing" section in this chapter for more information on location of refrigerated storage space.)

Plumbing. Although architects and engineers plan the plumbing for a facility, foodservice managers must be aware of and able to describe the foodservice's need for kitchen and dishroom floor drains and proper drains around steam equipment; the desired location for water and steam inlets and for hand-washing sinks in work areas and rest rooms; the water and steam pressure needs for equipment to be installed; and adequate drains to sewer lines for waste disposal equipment.

Electricity. Food managers are responsible for providing information on the needed location of electrical outlets and the voltage requirements of all equipment to be used in the facility. Equipment manufacturers' specifications list power requirements for their equipment. These must be compatible with the building's power supply or else the equipment will not operate at peak efficiency or will overload the wiring.

The mechanical engineer on the planning team details the electrical specifications, based on the foodservice's requirements, including the wattage and horsepower of the facility's equipment. Hospitals and health care facilities may require special electrical receptacles for food carts used to deliver patients' meals. Because these carts will be moved to various locations, compatible receptacles must be installed at all points of use. Figure 10.4 shows an overhead electrical raceway installation of receptacles for carts in a hospital kitchen.

All pipes and wiring going into a kitchen should be enclosed and out of sight. A modular utility distribution and control system offers many advantages compared with fixed and permanent installations. The entrance of and controls for all utilities

Figure 10.4 Overhead electrical raceways accommodate 10 patient food carts. Connection plates are equipped with receptacles and point-of-use circuit breakers.
Courtesy of Louisiana State University Medical Center Hospital, Shreveport.

are centered in one end-support column or panel of the system. All pipes and wiring are enclosed, but controls for both operation and quick-disconnect are on the outside of the panel within easy reach. Water, steam, gas, and electrical outlets may be installed as desired in panels extending from the one-point control column along a wall; to a center room unit, directly behind equipment; or from above as shown in Figure 10.5. Utility distribution systems are usually custom designed. A wall-type unit may house electrical wiring, plumbing assemblies, gas piping, and contain controls for water-wash cleaning for the exhaust ventilator. A fire control system for protection of cooking equipment may be located in the exhaust ventilator, which is located above the utility distribution system.

Budget/Cost Relationship

Because unlimited budgets are rare, studying the costs involved in any facility design project is inevitable. Planners usually establish a predetermined budget, which the project's total cost cannot exceed. Yet, the quality and features that foodservice managers select for a facility may well affect its operating costs. A detailed financial analysis may reveal that a higher initial expenditure for top-quality design and fixtures will result in lower operating costs during the project's anticipated life—its life-cycle cost—than a less expensive design.

Building and construction costs are affected by many interrelated factors, including the prevailing prices for labor and materials, quality and quantity of items selected, and the building's overall design. It may be helpful to think of these three factors—cost, quality, and quantity—as a triangle. If the project's budget is a fixed amount, it may be necessary to restrict quantity, quality, or both. However, if a predetermined amount of space is the top priority for the facility, planners must anticipate financing a building or site large enough to accommodate the required size. Alternatively, if planners assign top priority to the quality of fixtures and equipment, they must be flexible regarding the project's cost and quantity factors.

The particular design selected for the building or foodservice department will impose certain operating costs, especially those for labor. A well-planned arrangement on one floor minimizes the distance food and people must travel and permits good supervision. Compact work units, with the proper equipment easily accessible to workers, tend to reduce steps, motion, and fatigue, helping to minimize labor and operating costs. In poorly planned facilities, it is not uncommon for employees to spend at least 10% of their time locating and assembling utensils and supplies. Some assessments indicate that in an efficiently planned department, only the dietitian or foodservice supervisor, storeroom clerk, dishroom supervisor, pot and pan washer, and janitor would need to leave their work areas.

It is also important to include the total costs for cleaning materials, utilities, building and equipment depreciation, and the amount of equipment needed. Such cost will vary directly with the amount of space allocated to the foodservice department.

Furnishings and other equipment should contribute to efficient operation and reflect the best design, materials, and workmanship to conform to established sanitary standards. The degree of comfort for both guests and employees depends on the provisions made for them during the project's planning phase. They are particularly sensitive to amenities such as air conditioning, type of lighting, sound deaden-

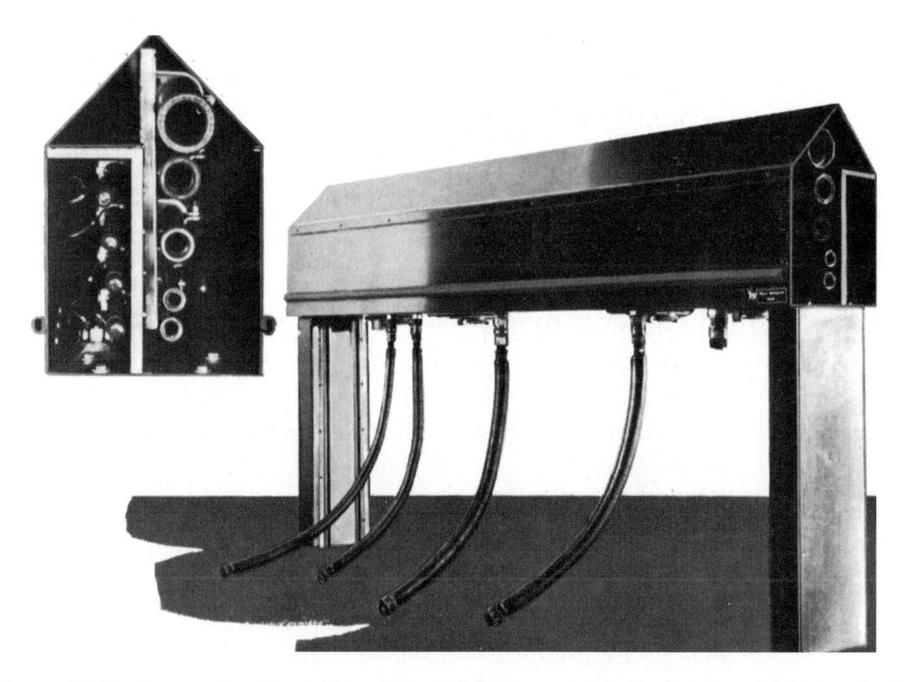

Figure 10.5 An overhead modular utility distribution and control system facilitates flexibility in arrangement, as well as ease of maintenance.
Courtesy of Avtec Industries, Inc.

ing, artistic incorporation of color and design, comfortable chair and work surface heights, and clean, well-ventilated rest rooms.

The facility's cost directly influences what can be done with a fixed budget. However, the material in the remainder of this chapter assumes that adequate funds are available for foodservice planning on a moderate scale.

DESIGN DEVELOPMENT

After completing preliminary preparations, the feasibility study, menu analysis, prospectus writing, and cost considerations, the foodservice manager needs to develop a design and layout plan. Providing adequate facilities for all anticipated activities, incorporating the ideas that planning members generate, and considering the facility's future growth are important aspects of design development.

A logical sequence for developing a design and for completing a foodservice facility follows:

1. Determine space allowances. Draw a flow diagram showing the space relationships of the work units and routes for supplies and workers.
2. Prepare a schematic design to scale, to show space allowances and relationships and placement of equipment, for consideration by the planning team before the architect begins preparing blueprints. Revise as needed.
3. Prepare and submit the architect's complete set of blueprints and contract documents, including specifications, to reliable interested contractors, builders, engineers, and equipment representatives for competitive bids.
4. Formulate contracts with accepted bidders.
5. Inspect construction, wiring, plumbing, finishing, and the equipment and its installation, as specified in blueprints and contracts. This is the responsibility of the architect and contractor.

Space Allowances and Relationships

Determining the amount of floor space and how to divide it for foodservice activities varies with every operation. Each needs adequate space to prepare and serve the planned number of meals. Yet, allowing too much space can result in inefficiency and lost time and effort.

The prospectus and menu analysis specify the number and kind of activities to be performed. For each activity, such as vegetable preparation, cooking specific menu items, and service methods to be used, the required equipment is listed. From the manufacturers' equipment catalogs, the size and, thus, the space requirements for each model to be purchased can be found. The space for equipment plus adequate aisle space represents a fair estimate of the total area required.

One commonly used procedure to determine kitchen space requirements begins with a calculation of the amount of space needed for the dining room. Fairly accurate estimates for dining areas can be calculated if the type of service and number of persons to be seated at one time are known. Likewise, seating capacity can be determined by using the generally accepted number of square feet per seat for different kinds of foodservices. Variations from the following suggestions will depend on the sizes of tables and chairs and whether a spacious arrangement is desired:

School lunchrooms	9 to 12 sq ft per seat
Hotel and club banquet rooms	10 to 11 sq ft per seat
Commercial cafeterias	16 to 18 sq ft per seat
Industrial and university cafeterias	12 to 15 sq ft per seat
Residence halls	12 to 15 sq ft per seat
Restaurants and hotels (table service)	15 to 18 sq ft per seat
Lunch counters	12 to 20 sq ft per seat

The kitchen size is often about one-third to one-fourth of the dining room area. This is a rough estimate at best because so many variables are involved. For example, dining room and kitchen space requirements are entirely different for a fast-food restaurant and a school cafeteria serving the same number of persons per meal period. The restaurant's turnover rate may be three customers per hour for each seat during a 3-hour meal period; thus, the restaurant would need to prepare food in small batches. In the school cafeteria, one-half of the group may be seated at one time with the total number served during a 50-minute period; therefore, larger quantities of food would be prepared and ready to serve the students. As a result, the restaurant kitchen would most likely be considerably smaller than the school kitchen with its larger capacity equipment.

Hospital foodservices confront a unique situation for space determination because only one-third to one-half of the total number of meals served are eaten in the dining room; patients are usually served in bed. Consequently, hospital kitchen space requirements are large, relative to dining areas, so the quantity and variety of food needed for patient, staff, employees, and guests can be prepared and assembled.

Flow Diagram of Space Relationships. Designing the floor plan begins with a diagram showing the flow of work, food, and supplies for one procedure to the next in logical sequence. To find the shortest, most direct route is the goal. The assembly-line concept provides for efficient operations by creating a continuous work flow for the tasks of receiving, storing, issuing, preparing, cooking, and serving the food, while minimizing traffic lines, backtracking, and cross traffic. After food has been served and consumed, the direction reverses to remove soiled dishes and trash. Figure 10.6 shows a typical foodservice flow diagram with desirable work area relationships. Only those work units required in a specific planning project need be shown. Because many foods are now purchased ready-to-cook, certain preparation units may be unnecessary in some kitchens. For example, since most foodservices no longer purchase carcass meat or wholesale cuts of meat, the meat pre-preparation unit has been eliminated entirely in those facilities.

The relationship of one work unit to another is also a consideration; that is, deciding on which work units need to be close to each other, which ones adjacent to other areas of the building, and which must be located near an outside door. Figure

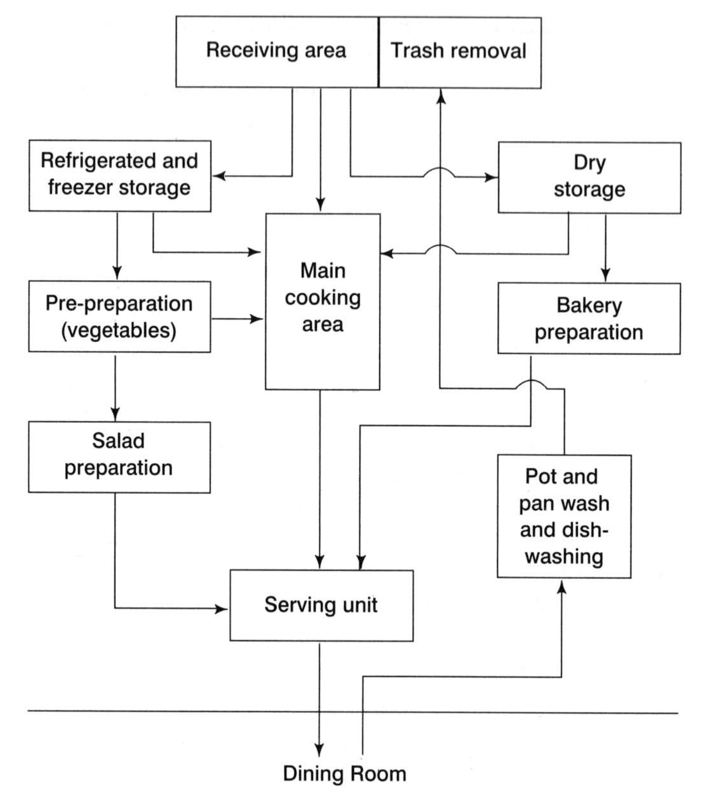

Figure 10.6 Flowchart diagram showing desirable work area relationships and progression of work from receiving goods to serving without backtracking and with little cross traffic.

10.7 shows the relationship of areas in a medium-size facility using the conventional foodservice system. As can be noted, the main cooking unit is the central area of most kitchens with supporting units feeding to or from it. Further discussion of desirable relationship of units is given under the "Work Areas" section that follows.

Schematic Drawing

Translating a flow diagram into a preliminary floor plan schematic is the next step in design development. The floor plan is a sketch or sketches of possible arrangements of the work units, with equipment drawn to scale within the allocated space. The required traffic aisles and work space also have to be included. Some general guidelines and a brief description of various work areas and their basic equipment needs are given on the following pages.

General Guidelines. Several considerations should be noted when planning a foodservice facility. The main traffic aisles should be a minimum of 5 feet wide, or wide enough to permit carts or hand trucks to pass without interfering with each other or with the workers in a unit. Aisles between equipment and worktables must have at least a 3-foot clearance; 3.5 to 4 feet are required if oven doors are to be opened or contents from tilting kettles must be removed in the aisle space. Usually one or two main aisles go through a kitchen with aisles into work areas that are parallel or perpendicular to the main aisle but are separate from them. (See Figure 10.8 for an example of an efficient cook's unit.)

The **Americans with Disabilities Act** (ADA) protects the rights of those with disabilities to enjoy and have access to employment, transportation, public accommodation, and communications. It has two main sections, one dealing with employment (see discussion of this part of the act in Chapter 13) and the other with public accommodation. The provisions detailed in the act are voluminous, and anyone wishing to ensure compliance must become familiar with them.

The ADA, which went into effect July 26, 1992, for companies with 25 or more employees (July 26, 1994, for companies with 15 or more), mandates some general guidelines for implementing "reasonable accommodation" to make the workplace and dining area accessible to persons with disabilities. The ADA applies to almost every public facility, and to new construction and alterations in existing facilities. Accommodations may include installing ramps, widening doors, and lowering shelves and counters. Aisles must be at least 36 inches wide (preferably 42 inches) to accommodate persons with wheelchairs. Figure 10.9 gives dimensions of dining and serving room space requirements for wheelchair accommodation in order to comply with the ADA regulations.

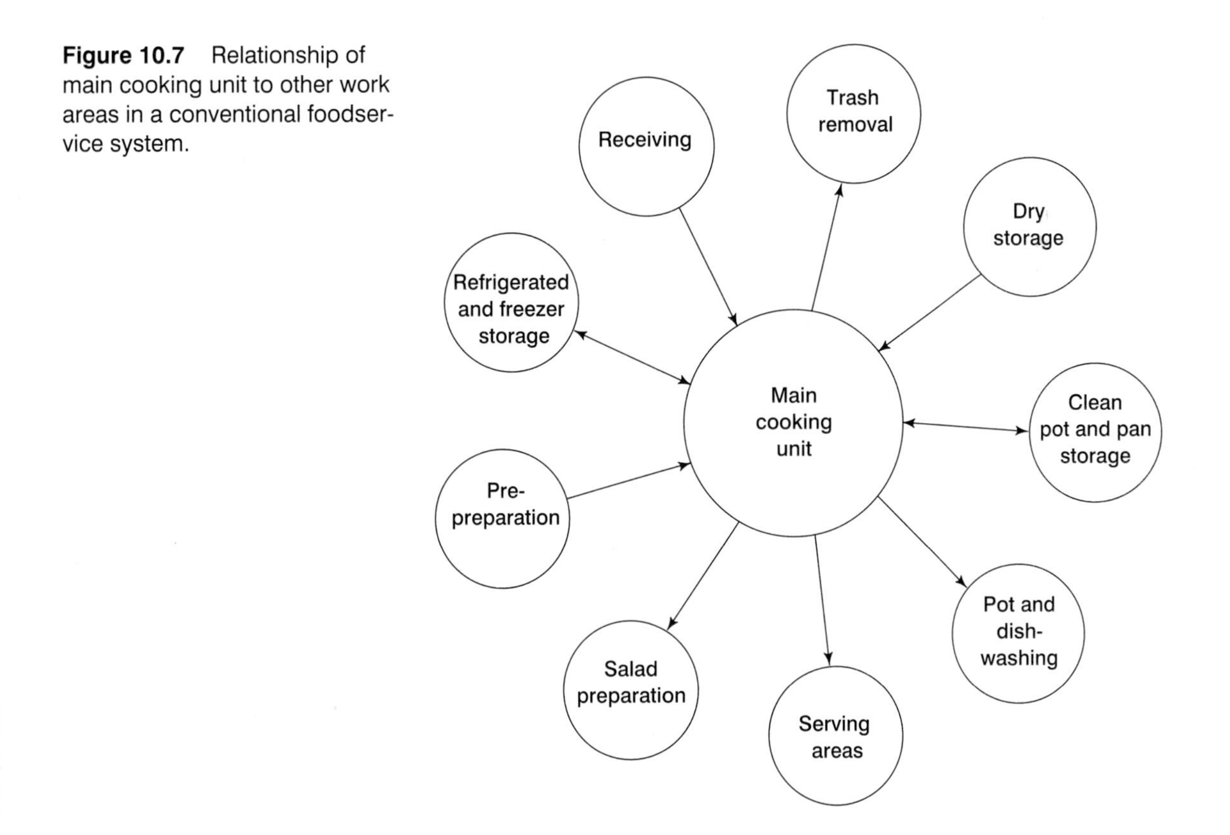

Figure 10.7 Relationship of main cooking unit to other work areas in a conventional foodservice system.

Figure 10.8 Compact, efficient kitchen arrangement with serving and dishwashing units in one room. A power-driven conveyor for soiled dish return is shown on the right.

Checklists for determining compliance with some of the ADA regulations are given in Figures 10.10 and 10.11. These lists serve as aids in assessing a facility for compliance and for future building programs. Taking the actions outlined in these checklists will not necessarily ensure compliance with the ADA; however, they can be used as tools to identify and eliminate potential problem areas. The diagrams of space requirements, shown in Figure 10.9, help interpret some of the requirements of this act. Refer to the act itself for complete regulations.

A minimum of 4 linear feet or worktable space is recommended for each preparation employee, but 6 feet is preferable. Work heights are generally 36 to 41 inches for standing and 28 to 30 inches for sitting positions (see Figure 10.12).

Tools and equipment require adequate storage space that is located at the place of use. Sinks, reach-in refrigerators, and space for short-term storage of supplies should be located in or near each of the work areas so employees at one location will have everything needed to perform their work. This includes space for racks to store clean pots and pans. Hand-washing facilities and drinking water should also be in a convenient location for all personnel.

Rectangular or square kitchens are considered the most convenient. The length of a rectangular-shaped kitchen should be no more than twice its width for best efficiency. Employees will save steps if the dining room entrance is on the longer side of a rectangular kitchen. Figure 10.13 shows another efficient arrangement for some restaurants, with a square dining space and the kitchen occupying a smaller space in one corner. The dining area is on two sides, and entrances to the kitchen can be

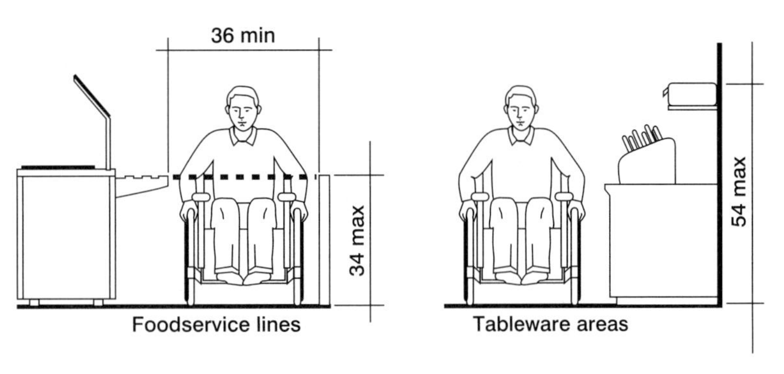

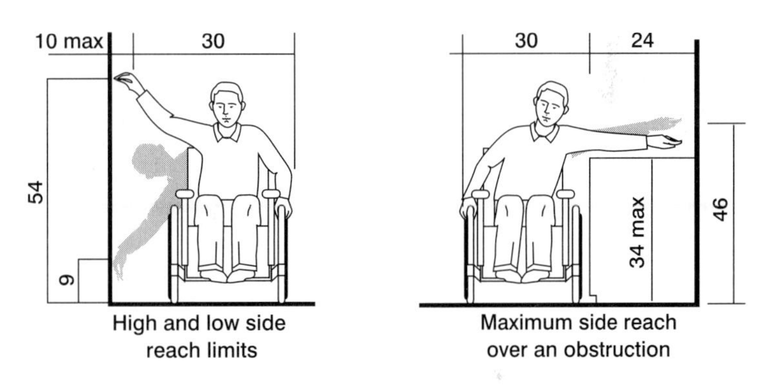

Forward reach

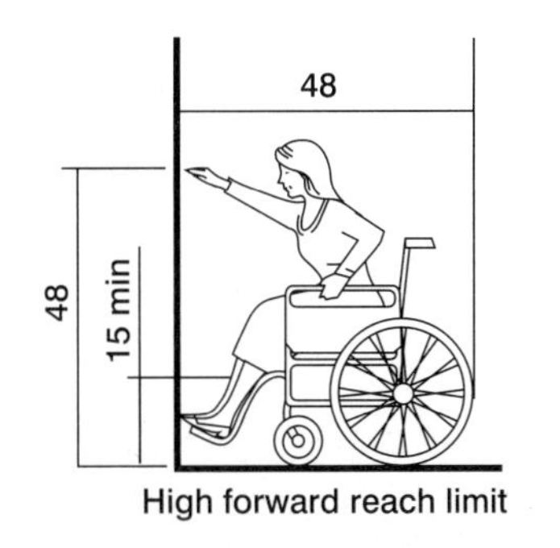

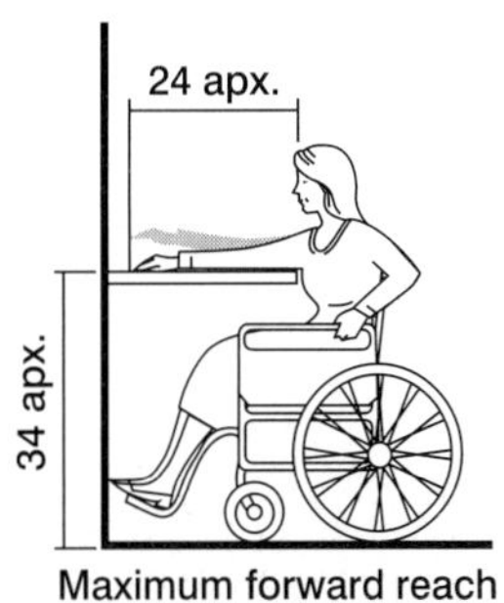

Figure 10.9 ADA requirements for space to accommodate wheelchair patrons in foodservice facilities.
Courtesy of Liberty Northwest Insurance Corporation, Portland, Oregon.

REMOVAL OF BARRIERS

(a) *General.* A public accommodation shall remove architectural barriers in existing facilities, including communication barriers that are structural in nature, where such removal is readily achievable (i.e., easily accomplishable and able to be carried out without much difficulty or expense).

(b) *Examples.* Examples of steps to remove barriers include, but are not limited to, the following:

1. installing ramps
2. making curb cuts in sidewalks and entrances
3. lowering shelves
4. rearranging tables, chairs, vending machines, and other furniture
5. lowering telephones
6. installing flashing alarm lights
7. widening doors
8. installing offset hinges to widen doorways
9. eliminating a turnstile or providing an alternative accessible path
10. installing accessible door hardware
11. installing grab bars in toilet stalls
12. rearranging toilet partitions to increase maneuvering space
13. insulating lavatory pipes
14. installing a raised toilet seat
15. installing a full-length bathroom mirror
16. lowering the paper towel dispenser in a bathroom
17. creating a designated accessible parking space
18. removing high pile, low density carpeting

Figure 10.10 Compliance with the ADA—barriers checklist.

located on each side. During slow periods of service, one section or side of the dining room can be closed off with a folding partition. When there is sufficient business, both sides can be used.

Routing servers counterclockwise through the kitchen or patrons through a cafeteria line is more efficient than a clockwise arrangement, at least for right-handed people. This way the right hand of the patron or employee is closest to the food to be selected.

Large kitchens usually have specialized work areas, each with its own equipment and short-term storage facilities. For efficiency in work and to reduce the noise level in the kitchen, these work areas may be divided with semipartitions, walls 5 to 5.5 feet high. Thus, there is separation of work but air circulation in the kitchen is not blocked by ceiling-high partitions.

In smaller kitchens, the work areas may merge and hence equipment can be shared by employees. For example, the cook and the salad worker may share an electric mixer that could be located at the end of the cook's table but close to the salad preparation area. This requires careful planning of work schedules so both workers will not need the equipment at the same time.

BARRIERS CHECKLIST

Building Access

1. Are 96″ wide parking spaces designated with a 60″ access aisle?
2. Are parking spaces near main building entrance?
3. Is there a "drop off" zone? at building entrance?
4. Is the gradient from parking to building entrance 1:12 or less?
5. Is the entrance doorway at least 32 inches?
6. Is door handle easy to grasp?
7. Is door handle easy to open (less than 8 lbs pressure)?

Building Corridors

1. Is path of travel free of obstruction and wide enough for a wheelchair?
2. Is floor surface hard and not slippery?
3. Do obstacles (e.g., phones, fountains) protrude no more than 4 inches?

Rest rooms

1. Are rest rooms near building entrance/personnel office?
2. Do doors have lever handles?
3. Are doors at least 32″ wide?
4. Are rest rooms large enough for wheelchair turnaround (51″ minimum)?
5. Are stall doors at least 32″ wide?
6. Are grab bars provided in toilet stalls?
7. Are sinks at least 30″ high with room for a wheelchair to roll under?
8. Are sink handles easily reached and used?
9. Are soap dispensers and towels no more than 48″ from floor?

Figure 10.10 *Continued*
Courtesy of Liberty Northwest Insurance Corporation, Portland, Oregon.

WORK AREAS

Seven major types of work may occur in foodservice departments: (1) receiving, (2) storing and issuing of dry and refrigerated foods, (3) pre-preparation, (4) preparation/cooking, (5) food assembly/serving, (6) warewashing (e.g., dishes, pots and pans), and (7) supporting services such as administration and janitorial work and employee/storage areas such as the locker and rest rooms, and storage for extra china, linens, paper goods, and supplies. The number of work areas to plan for a specific foodservice depends on the type of operating system to be used (see Chapter 2), the volume of business, types of menu items to be prepared, and the form in which food will be purchased.

The *receiving area* includes an outside platform or loading dock, preferably covered, and adjacent floor space, large enough to check in, examine, weigh, and count food, and to check invoices when they are delivered. The floor of the platform should be equal to the height of a standard delivery truck bed and on the same level as the building's entrance. The suggested minimum width is 8 feet, and the length is dictated by the number of trucks that are to be unloaded at any one time.

ADA Compliance Checklist

The following checklist will aid in assessing the current level of compliance and assist in future accessibility issues. Any changes in an establishment must be readily achievable. This means that the task must be easily accomplished and able to be carried out without much difficulty or expense.

Passenger Arrival
- ☐ Adequate space
- ☐ Ramp to entry
- ☐ Proper width of walk
- ☐ Door opening clearance
- ☐ No obstructions

Parking
- ☐ Special stalls
- ☐ Access to building by level path

Walks
- ☐ Minimum 48″ wide
- ☐ Firm, nonslip surface
- ☐ Curb cuts at streets, driveways, parking lots
- ☐ 5% maximum grade
- ☐ Free of obstructions
- ☐ Level platforms at doors

Ramps
- ☐ Maximum 8⅓% grade
- ☐ Free of grates
- ☐ Landings at 32″ high, extended 12″ beyond ramp
- ☐ Well illuminated
- ☐ Firm, nonslip surface
- ☐ Level approaches
- ☐ Guardrails on walls

Entrances
- ☐ One major entrance for wheelchair
- ☐ Level approach platform

Doors
- ☐ 32″ wide clear opening
- ☐ Thresholds flush with walk or floor
- ☐ Vestibules with 6′6″ separation
- ☐ Handles maximum 42″ high
- ☐ Closers with time delay
- ☐ Closers with 8 lbs maximum pressure
- ☐ Single effort with 8 lbs maximum pressure
- ☐ Kickplates 16″ high
- ☐ Vision panels at 36″ above floor maximum

Corridors, Public Spaces, Work Areas
- ☐ Corridors minimum 60″ wide
- ☐ Floors on common level
- ☐ Nonslip floor materials
- ☐ Recessed doors when opening to corridor
- ☐ Noncarpeted circulation paths

Stairs
- ☐ Minimum 42″ wide
- ☐ Nonprojecting nosings
- ☐ Level, differentiated approaches
- ☐ Handrails 18″ beyond top/bottom
- ☐ Maximum 7″ risers
- ☐ Nonslip treads
- ☐ Handrails 32″ high
- ☐ Well illuminated

Figure 10.11 ADA compliance checklist.
Courtesy of Liberty Northwest Insurance Corporation, Portland, Oregon.

Space should also be allowed for hand trucks, platform scales, and a desk or work space for the receiving clerk for checking off items delivered. Large institutions that process their own meat need to include an overhead track with hooks for carcass meat. This track would extend from the loading platform through to the meat department's refrigerators.

The exterior door must be wide enough (6 feet is common) to accommodate hand trucks, large cartons, and any large pieces of equipment that are to be installed in the kitchen. A glass-walled office facing the loading dock that is equipped with a double-faced platform scale is efficient for a clerk in the office to check weights of goods being delivered and received.

The *storage areas* should be close to the delivery entrance so goods will not have to be moved far to be stored. Space needed for canned foods, staples, and grocery items is known as "dry stores." This area should be easily accessible to the bakery and the cook's units in particular. Dry storerooms must be cool and well ventilated. Other requirements are moisture-proof floors, screened windows, metal-slatted shelves for case goods, and tightly covered storage bins for items such as cereals, rice, and condiments. Wooden, or polypropylene, mobile pallets should be provided for stacking sacks of flour, sugar, and similar products to keep them off the floor. These pallets should be mobile for ease in cleaning the floor. Space should be arranged to accommodate carts and hand trucks. A desk and files should be included for keeping inventory records, either by computer or manually. Scales are a necessity. Lockable double doors or a wide single door should open to the preparation areas.

Walk-in refrigerators and freezers must be provided for perishable foods. Reach-in refrigerators located in the work units used for daily supplies and leftovers are usually not considered as storage. Refrigerated storage areas should be as close as possible to the receiving platform and accessible to the work unit that will use it most frequently.

The amount of storage required depends on the frequency of deliveries, daily or less frequent, and the form of food purchased. Also, the extent of the menu and the variety of foods offered will influence the amount and kind of storage required. Restaurants may also require space for storing wines and liquor.

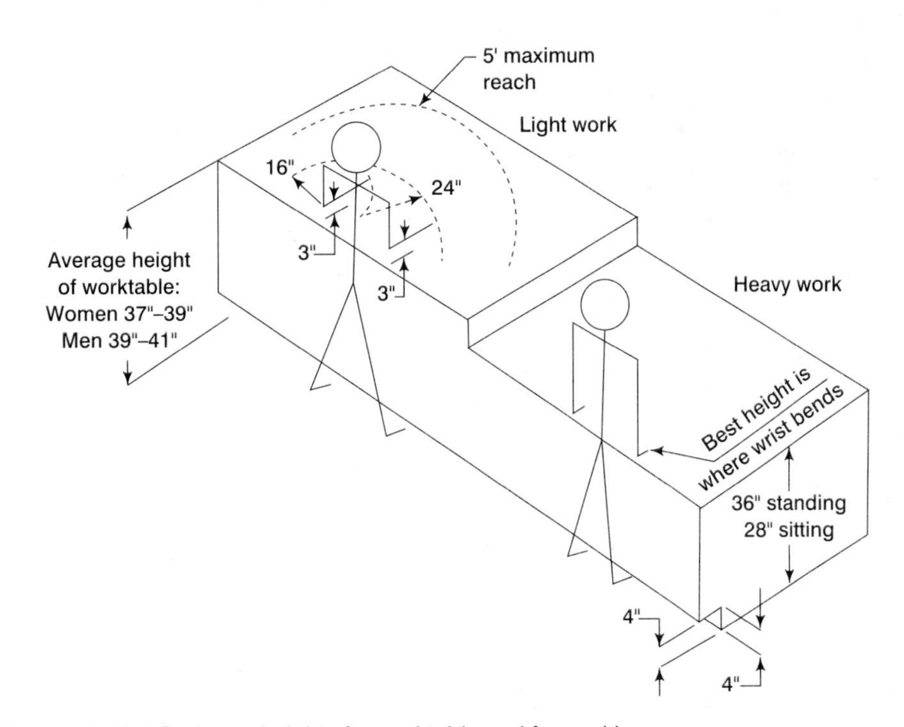

Figure 10.12 Optimum heights for worktable and for working area.
Courtesy of Arthur C. Avery.

Figure 10.13 Efficient kitchen/dining room arrangement.

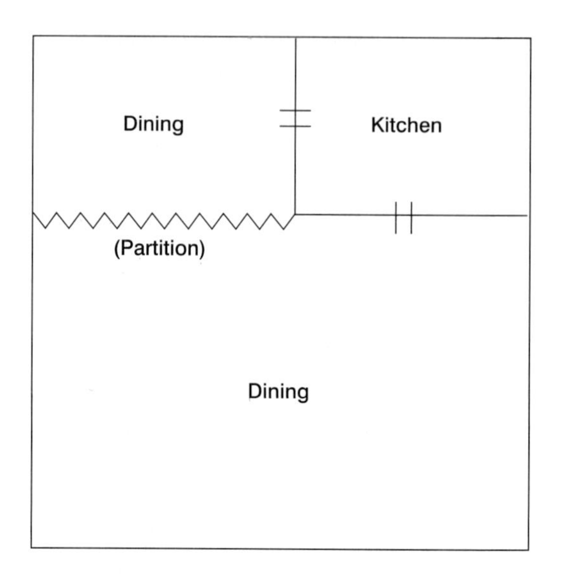

Cleaning supplies must be stored separate and apart from all foods, helping to ensure that none of those poisonous chemicals will be mistakenly issued as a food product. Also, additional space must be allocated for extra stocks of paper goods and reserves of china, glassware, linens, towels, uniforms, and aprons.

A *central ingredient room*, if used, will be located adjacent yet connected to the storage areas. Adequate table or counter space for weighing, measuring, and counting ingredients and ample aisle space for carts carrying assembled ingredients to the production units are basic requirements for this room.

The *preparation area* for meat, fish, and poultry includes butcher blocks, an electric saw and grinder, sinks, storage trays, and refrigerators. The overhead tracks for bringing in carcass meat from the delivery area that were mentioned earlier would lead to this unit.

For many foodservices, however, this unit is almost a thing of the past except in very large facilities. The trend toward buying prefabricated and pan-ready meats, poultry, and fish decreases the need for this once necessary work unit.

The *vegetable preparation area* should be located near the refrigerated storage and the cooking and salad areas (see Figure 10.6). The usual vegetable preparation area is equipped with a chopper, a cutter, a two-compartment sink, worktables, a cart, knives, and cutting boards. If a peeler is needed, it may be either a pedestal or a table model, placed to empty directly into a sink. Figure 10.14 shows three possible arrangements for this unit.

Two separate sinks should be provided to permit unhampered use. Food waste disposals are placed in the drainboard to the sink or on a worktable near the end of the sink, or space for a garbage can may be provided, often under an opening cut into the worktable or drainboard.

Because the vegetable preparation area is often responsible for pre-preparing some items for the salad unit, ample space for many workers may be needed.

Tables that are 30 to 36 inches wide and 6 to 8 feet long are adequate, permitting employees to work on either side for most types of preparation. Providing at least one table low enough for employees to sit at comfortably to perform certain tasks is advisable. Figure 10.15 shows a close-up view of part of a vegetable preparation unit in a large, central production kitchen. Workers are slicing potatoes for use by the cook's unit. Note the electrical circuit posts on the left side to accommodate heavy-duty equipment requirements.

The main *cooking area* is the hub of the kitchen, which is usually located in or near the center of the kitchen. It is most efficient when adjacent to the vegetable preparation area, the storage rooms, and behind or near the serving area (see Figures 10.6 and 10.7). The equipment needs are entirely dependent on the amount and type of foods to be cooked on the premises. The usual for a conventional type of production method would include ovens, broilers, fryers, steam equipment, mixers with attachments, and cook's tables with a sink, pot and pan storage racks, and overhead utensil racks. Ranges may also be used although, in many cases, they have been replaced with specialized pieces of equipment such as pressure steam cookers for batch cookery, tilting fry pans, meat roasting ovens, convection ovens, and grills. These may be more energy efficient and generate less heat than the ranges.

The grouping of equipment varies according to the size and shape of the kitchen. However, steam equipment is usually installed together in a row with the appropriate floor drains in front. Grills and broilers for short-order cooking should be closest to the serving unit but not next to deep-fat fryers. Fire danger is great when the

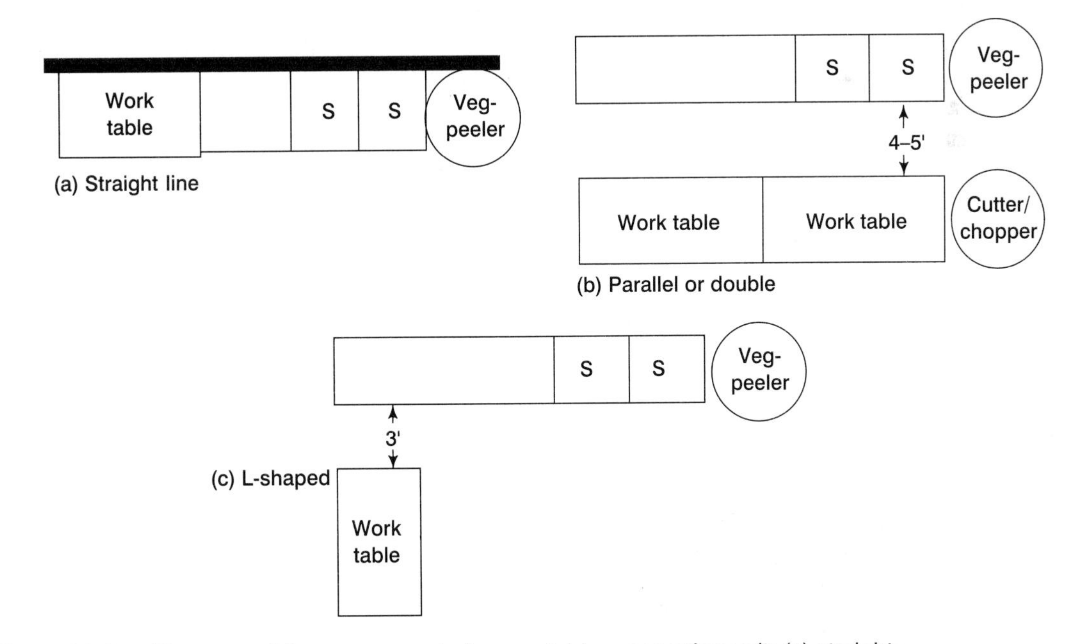

Figure 10.14 Three possible arrangements for vegetable preparation unit: (a) straight line, (b) parallel, and (c) L-shaped.

Figure 10.15 Vegetable pre-preparation for the cook's unit in a large, central production kitchen. Courtesy of Bob Honson, Director, Nutrition Services, Portland, Oregon Public Schools.

intense heat from grills and broilers is close to hot fat that may splatter. Figure 10.16 illustrates four possible arrangements for the cook's unit. Note the amount of space required for each arrangement, including a 12- to 18-inch cleaning space between back-to-back rows of cooking equipment.

Ceiling-mounted hoods with separately vented exhaust fans, which extend 1 foot down over all cooking equipment, help ventilate the kitchen by removing odors, smoke, moisture, and fumes. Hoods also facilitate the installation of direct lighting fixtures to illuminate cooking surfaces.

Water outlets at each point of use, such as a swing-arm faucet between each pair of steam-jacketed kettles, above or beside a tilting frypan, or over the range area, are a great convenience and a timesaver for the cooks. A cook's table, located directly in front of the cooking equipment, may contain a small hand sink at one end and an overhead rack for small utensils. A rack for storing clean pots and pans should be easily accessible to the cook's unit, the pot and pan sink, and the power washer, if one is used. Much of the equipment in the cook's area can be wall or wheel mounted for ease in cleaning.

The *salad area* is generally located at one side or at the end of the kitchen, as close as possible to the serving unit and to the product walk-in refrigerators. The unit requires a liberal amount of worktable space and refrigeration for set-up salads. In cafeterias, it is most efficient to have the salad preparation area located directly behind the salad counter on the cafeteria line. A pass-through refrigerator allows kitchen workers to place the trays of set-up salads in the refrigerator from their side, and the counter workers remove them as needed.

Mobile refrigerated units are available for banquet or special party use so that set-up salads may be refrigerated in them until meal time, and then moved directly to the dining area for service. For short-order salad making, a refrigerated table is a convenience for storing salad ingredients between times of use. In a hospital, nursing home, or restaurant that is built on more than one floor, easy access to service eleva-

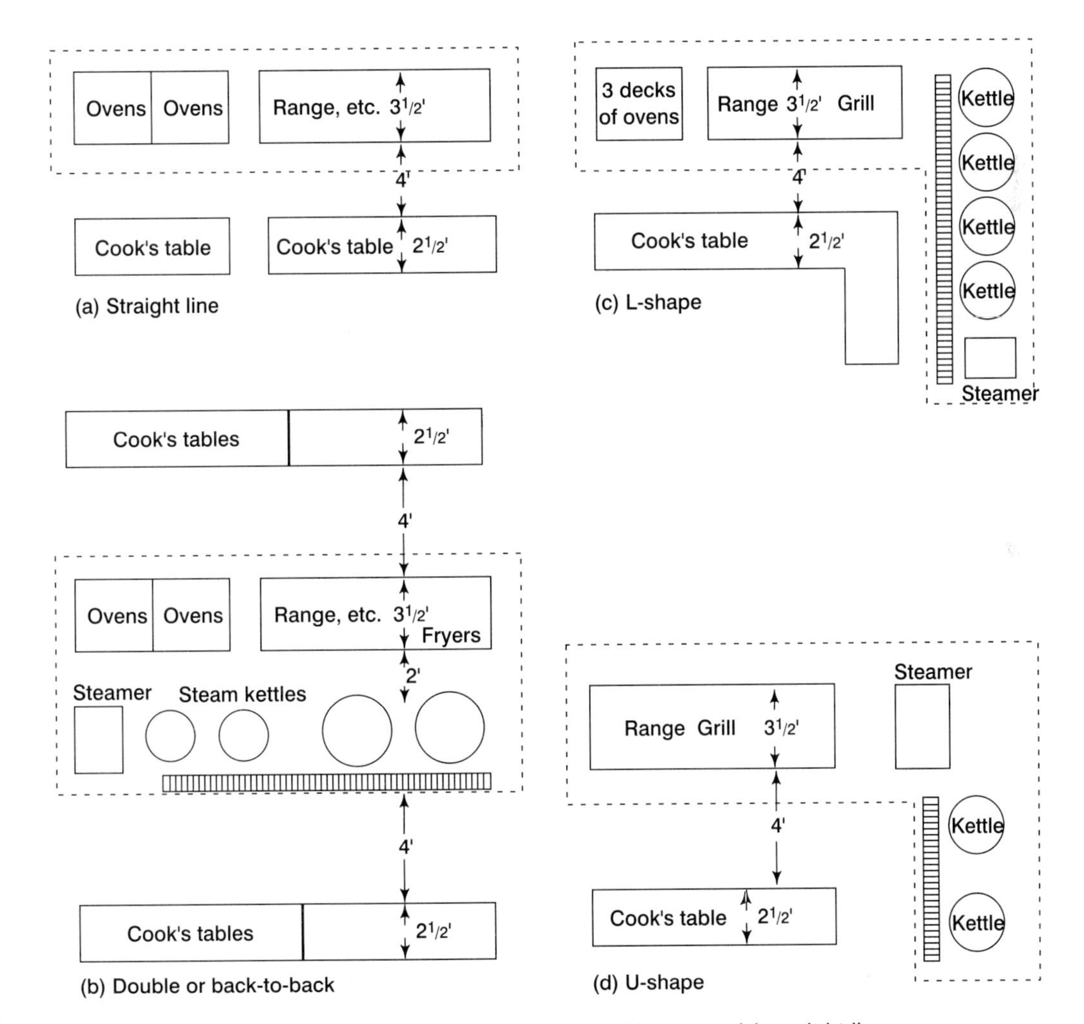

Figure 10.16 Four suggested arrangements for a main cooking area: (a) straight line, (b) back-to-back or double, (c) L-shape, and (d) U-shape. Note the amount of aisle space and total floor space required for each.

tors, subveyors, or dumb waters enables made-up salads to be delivered in good condition. Figure 10.17 shows some possible arrangements for the salad unit.

The *bakery and dessert preparation area* operates as a fairly independent unit. Having little direct association with the other preparation areas, it may be separated from them. Because the quality of the products from this unit is not as dependent on time and temperature as are meats, vegetables, and salads, the bakery need not be as close to the serving area as the other units.

Equipment for a typical bakeshop unit includes baker's table with roll-out bins, ovens, pan storage and cooling racks, mixers, steam-jacketed kettle, dough divider and roller, pie crust roller, and reach-in or small walk-in refrigerator. Large bakery units may include dough mixer, proof box, dough troughs, and reel ovens. Small operations may not have a separate bakery unit, but placing a baker's table near the cooking unit allows the equipment to be shared.

Routing of work and placement of equipment should be in a counterclockwise arrangement for greatest efficiency. The finished product should be on the side closest to the serving unit for shortest transport distance. Performance of tasks should

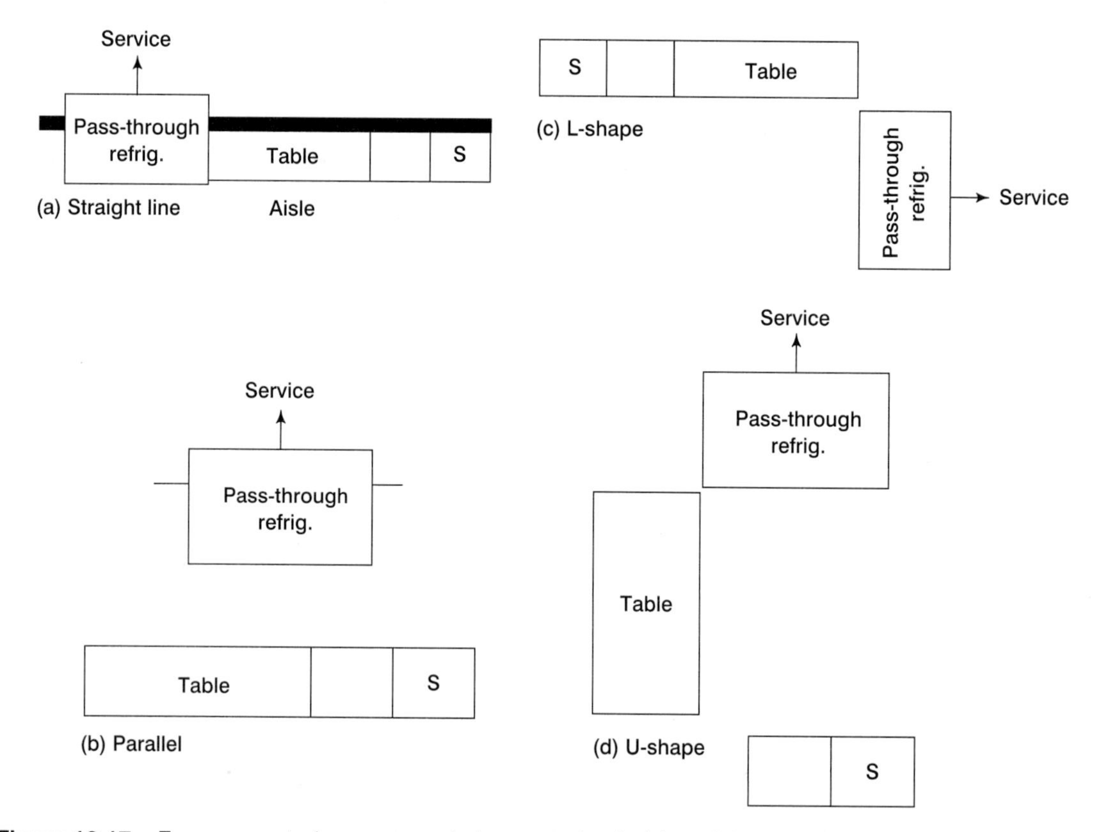

Figure 10.17 Four suggested arrangements for a salad unit: (a) straight line, (b) parallel, (c) L-shape, and (d) U-shape. Note that equipment is arranged so work progresses from right to left (preferred progression) in all except (c).

proceed in a direct line from one function to the next without any backtracking or crisscrossing of workers.

If frozen desserts such as ice cream and ices are to be made on the premises rather than purchased, a separate room with specialized equipment to handle these products will have to be provided. It must meet strict sanitation codes and requirements established for production of frozen dessert items.

The *assembly/serve area* may be at various preparation centers in the kitchen where servers pick up their orders for table service or for assembling trays for hospital tray service. The latter requires a tray line as described in Chapter 6. Separate serving rooms may adjoin the kitchen and, in some facilities, serving pantries may be located throughout the building. Cafeteria counters located between the kitchen and dining room can be of many different configurations. The length and number of counters needed depend on the number of persons to be served, the number of menu items offered, and the desired speed of service. Speed of service can be increased if the counter is designed for customer movement from right to left or counterclockwise, making it easier for right-handed persons to pick up and put food on their trays.

Serving counter designs depend on the amount of available space. Counters can be arranged in a straight line, in a parallel or double line with a serving station in between, in zigzag sections, or in a hollow square. Whatever configuration is selected, the design should permit speedy service and prevent long waits for patrons, as well as keep labor at a minimum. To speed the flow, silverware, napkins, condiment bars, and beverage dispensers are often placed in the dining room area, encouraging patrons to move away from the serving area more rapidly. A revolving counter section, placed near the wall to the kitchen, is ideal for displaying cold items, such as salads, sandwiches, and desserts.

The hollow square arrangement (sometimes called the "scramble" or "supermarket" system) may be constructed with a center island for trays, silverware, and napkins, and with serving counters on three sides. With this design, patrons enter the square, pick up a tray and utensils, and move to any section of the counter that they desire without standing in line to wait for others to make their choices. A typical arrangement is illustrated in Figure 10.18.

Cafeteria counters are usually custom made so that the desired length and design can be obtained. Sections for hot food, once heated with steam, are now mostly electrically heated and have thermostat-controlled units. Hot foods may be placed on the counter in the pans in which they were cooked, if the counter openings are the same size as the pans. This hot unit should be located as close as possible to the kitchen cooking area. Heated, pass-through holding cabinets can be installed into the wall between the kitchen and serving room close to the hot-food section to facilitate supplying foods to the counter.

The arrangement of food items on a cafeteria line may be in a logical sequence; that is, in the order that the food would be eaten. Schools usually prefer this arrangement so that students will choose the most nourishing items first and desserts last. For commercial cafeterias, however, a psychological arrangement might be more profitable; for example, the most eye-appealing items, such as salads and desserts, are placed first for greater selection, and hot foods placed near the end of the counter. Counter units can be mobile to provide flexibility in arrangement.

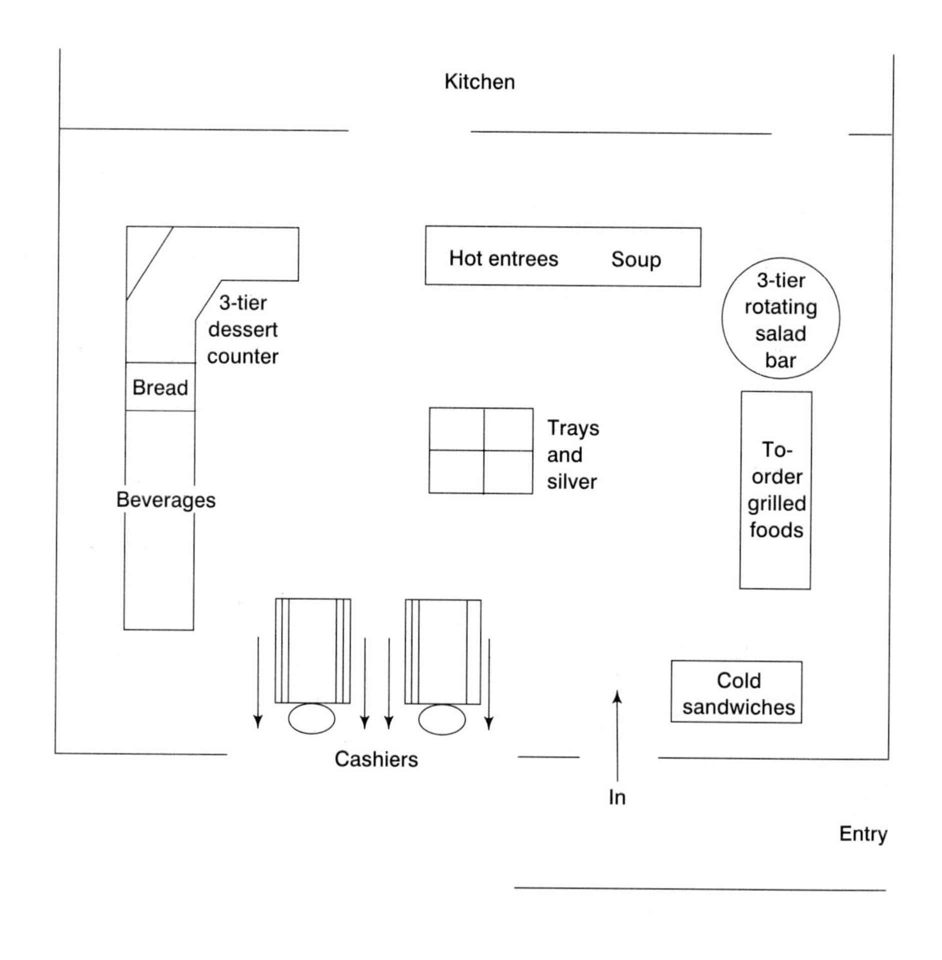

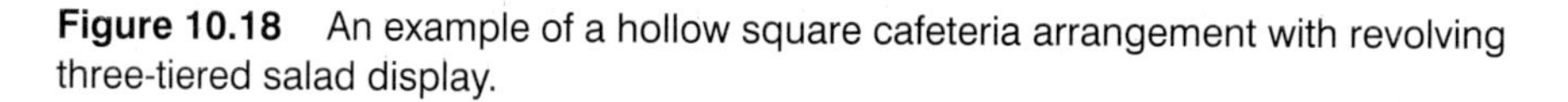
Figure 10.18 An example of a hollow square cafeteria arrangement with revolving three-tiered salad display.

The size of serving pantries, such as those in hospitals, depends on whether centralized or decentralized service is used. Refer to Chapter 7 for details of the equipment and space requirements for these two service systems.

Warewashing includes dishes, silverware, glassware, trays, and pots and pans. Each of these is discussed individually. The *pot and pan* washing area should be located near the cooking and bakery units because most of the soiled pots and pans come from those units. The area should not be in a main aisle or traffic lane. It is often at the end or back of the cook's unit or in an alcove allocated for this purpose.

Equipment needs include a three-compartment sink: one for soaking and washing, one for rinsing, and one for sanitizing (with drainboards). Racks for clean pots and pans are also needed and, in some facilities, a mechanical pot and pan washer. Hand washing of pots and pans may be aided by a manually guided power scrubber or a pump-forced flow of water to loosen food from pans as shown in Chapter 8, Figures 8.8 and 8.9. After hand washing, pots and pans may be sanitized in a steam cabinet or run through the dish machine.

Large foodservices, particularly hospitals with serving carts for meal delivery to patients, may need space for a room-sized cart and pan washer as shown in Figure 10.19.

Dishwashing areas should be compact, well lit, and well ventilated. It is desirable to locate this unit away from the dining room because of the noise. If this is not possible, surrounding the area with acoustical material will help muffle the sounds. Two examples of such installations are shown in Figures 10.2a and 10.2b. Mechanical conveyors save time and money by transporting soiled dishes from the dining area to the dishwashing room. The location of the dishwashing area should be such that the return of soiled dishes will not interfere with the routine of service or cross through work units.

The process of dishwashing is described in Chapter 8. The design of, and space for, the dishwashing area must allow for the smooth flow of dishes through the processes of sorting, scrapping, washing, rinsing, drying, and storage. The overall arrangement of the area and the size and type of dishwashing machine to be selected depend on

(a) (b)

Figure 10.19 (a) Insulated rack, cart, and pan washer with "drip off" area in front of the door. (b) Mechanical pot and pan washer for corner installation.
Courtesy of Alvey Washing Equipment, Division of Alvey Industries, Cincinnati, Ohio.

the number of pieces to be washed, the speed with which they must be returned for reuse, and the shape of the available space. Arrangements for a dishwashing area may be straight line, L-shaped, U-shaped, open square, platform, or closed circle. The straight-line type is often installed near a side wall in small operations. The U-shaped arrangement is compact and efficient for small spaces, whereas the open square might be preferable for a larger facility and could easily accommodate a glass washer. Machines are designed for either right- or left-hand operation, although the usual flow direction is from right to left. Figure 10.20 shows one layout plan for a small space. Figure 10.21 shows a closed-circle, or fast-rack conveyor, arrangement that mechanically moves racks continuously to the soiled dish end of the machine. Although this arrangement requires about the same amount of floor space as a straight-line type, it is more compact and can be operated by fewer employees.

Any dishwashing layout should be arranged far enough away from the walls to permit workers to have easy access. At least a 4-foot aisle is desirable on either side of the dishwasher.

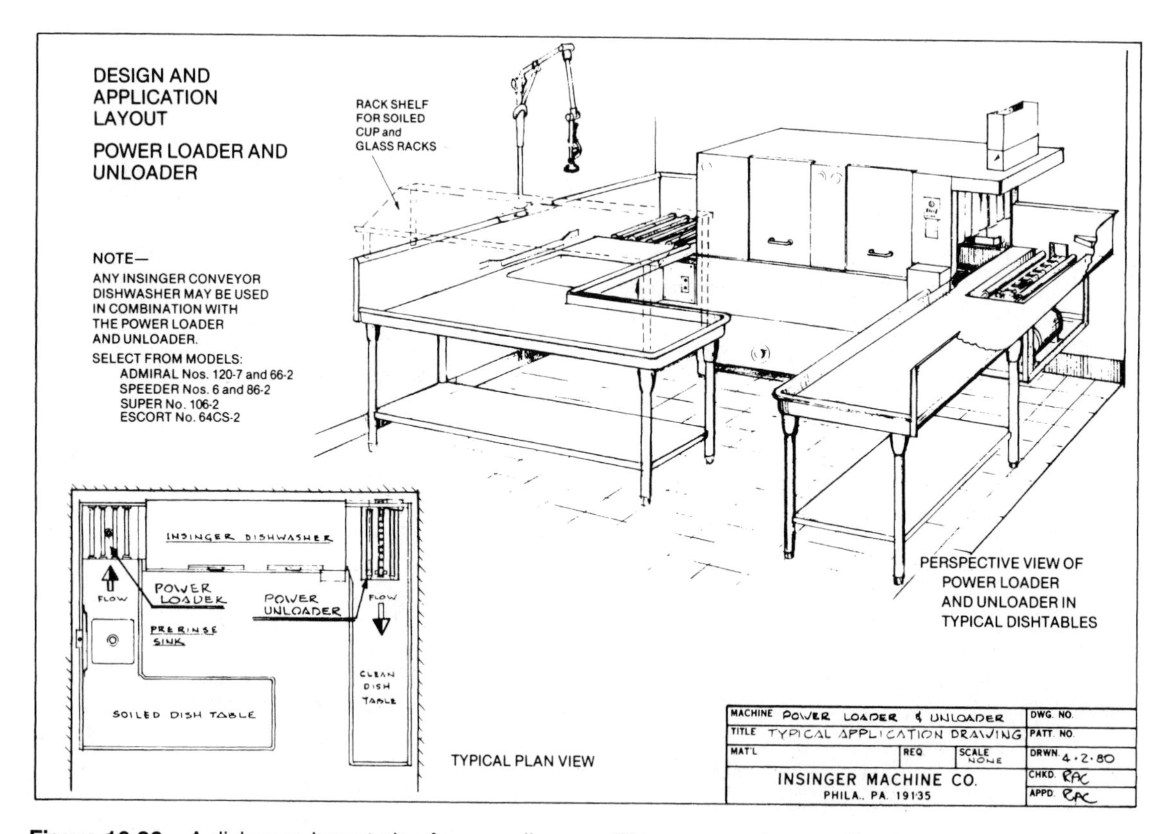

Figure 10.20 A dishroom layout plan for a small space. This company's power loader and unloader enables many facilities to use larger dish machines in a small area. The unloaders eject dish racks at a right angle to the machine, which is a space-saving feature. Courtesy of Insinger Machine Co., Philadelphia.

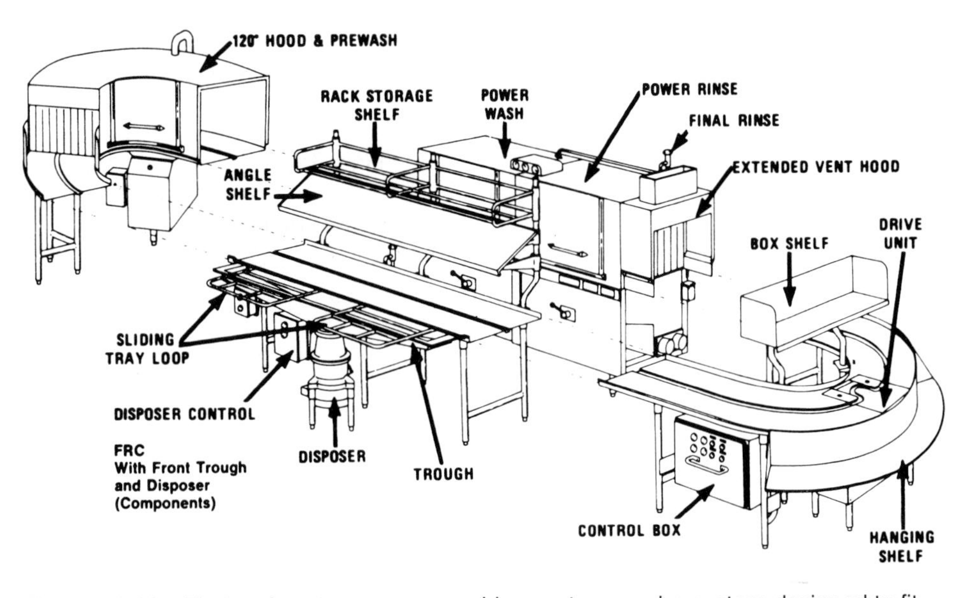

Figure 10.21 Fast-rack conveyor warewashing system can be custom designed to fit the space and needs of the situation.
Reprinted courtesy of Hobart Corporation, Troy, Ohio.

Equipment for the dishwashing unit may include a prewash arrangement, the dish machine, possibly a glasswashing machine, soiled and clean dish tables, waste disposals, storage carts, and carts or conveyors to transport dishes to and from this area. The usual division of space allocated to dish tables is 60% for soiled dishes and 40% for clean. (See Appendix B for further details on dishwashers.)

Prewashing or preflushing equipment includes a unit built into the dish machine, a hose and nozzle, or a forced water spray as illustrated in Figure 10.22. Note that the forced water spray method utilizes more water than the other methods; therefore, it is not a desirable choice if water conservation is a concern. The hose and nozzle can be near the machine, but the forced spray should be far enough away so that dishes can be easily racked following the preflush. Food waste disposals can be installed with either type. A method for returning emptied racks to the soiled dish table for reuse should also be provided. (See the power-driven conveyor in Figure 10.8.)

A booster heater to increase the temperature of the usual 120°F to 140°F water used throughout the building to the 180°F required for the sanitizing rinse water for the dish machine should be installed near the machine. Some dishroom layouts may include an oversized sink for washing serving trays that are too large to go through the dish machine. If many trays are to be washed (as in hospitals serving patients in bed), a special machine designed for washing trays would be desirable. Such a machine is shown in Appendix B.

Proper ventilation of the dishwashing area is essential. A hood-mounted exhaust fan should be installed over the unit, or rustproof, watertight exhaust ducts, which are vented directly to the outside, may be attached directly to the machines to remove steam and hot air.

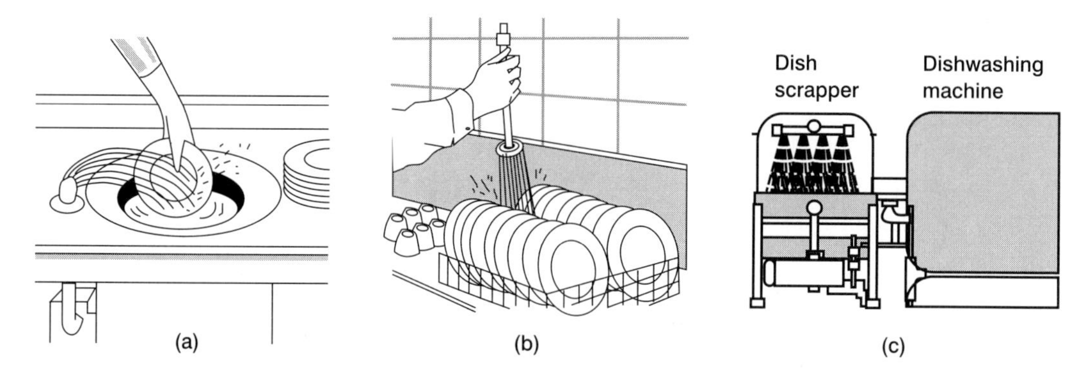

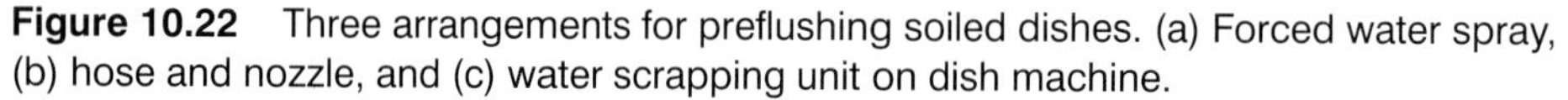
Figure 10.22 Three arrangements for preflushing soiled dishes. (a) Forced water spray, (b) hose and nozzle, and (c) water scrapping unit on dish machine.

Supporting/auxiliary services must not be forgotten when planning a facility. Space for *employees'* rest rooms, lockers, showers, and hand-washing facilities are to be included. The number of toilets and other amenities is determined by the number of workers of each sex on duty at any one time and by the Health Department's standards and codes. Requirements of the ADA must be met when planning these facilities as well.

Office space for the foodservice management staff is preferably located so the staff has a view of the kitchen and the work going on. This may be accomplished in part by using glass walls or large windows for the office. The number of persons who will need desks, files, chairs for visitors, and aisle space will determine the size of the offices. Those staff members not directly supervising food production may have offices in an area adjacent to the kitchen.

Janitors' closets for storage of mops, brooms, and cleaning materials, as well as a large low sink for washing mops, require consideration in planning a foodservice facility. An area equipped with a steam hose, often located near the back door, may be required for sanitizing food carts and trucks, especially in hospitals that have many such items to clean. This should be a separate area with curbing around it and it should be equipped with floor drains.

Trash and food waste storage and removal space is necessary if disposal facilities are not available in the building. Many buildings have their own incinerator for burning trash, central compactors to compress trash and cans, and preparation unit waste disposals. When such facilities are not available, both garbage and trash must be collected and held for frequent removal. A cooled room near the back entrance may be provided for the daily storage of garbage, but when feasible, unit or central disposers incorporated into the system are more desirable and efficient.

The *dining room* is generally a part of the total foodservice design plan. For greatest efficiency, it is located adjacent to the kitchen or serving area, sometimes opening off the cafeteria. Dining rooms that are quiet, well lit, and well ventilated are conducive to the enjoyment of food and hospitality. The size of the dining room was discussed in the "Space Allowances and Relationships" section.

Equipment for dining areas includes the tables, chairs, and small serving stations. Two- and four-seat tables that can be combined to accommodate larger groups are typical of most public dining rooms. Tables in school foodservices are larger to conserve space, but difficult for waiter or waitress service and less satisfactory for socializing. The size of the tables to be used, the type and size of chairs, and the number of people to be seated at one time are basic to determining space needs. Also, space between tables and aisle space must be added; minimum space between chair backs is 18 inches after guests are seated. Main traffic aisles of 4.5 to 5 feet are recommended. Public dining rooms should accommodate patrons who may be in wheelchairs or who use walkers, and so may need wider aisles. (See Figure 10.9 for details.)

Folding partitions that are decorative as well as functional may be used to close off part of the dining room for special groups or when all of the room is not in use. Customer rest rooms are located close to the dining area for convenience and security. See Chapter 11 for more on dining room furnishings.

The planning team, supplied with this information, should now have a conference to discuss ideas. They should reject or discard features and/or components of the plan until agreement is reached on what shall be included and the boundaries for the project. If it is a remodeling project, the team decides how much can be done and, perhaps, what has to be left undone. Decisions on quantity and quality within the confines of the budget will be made. Only with this agreement among all team members will each be fully committed to the project and continue to devote work time and provide the expertise needed to bring the project to a successful conclusion—to bring the menu and customers together through a planned system of time and motion.

Mechanics of Drawing

The actual drawing of a plan to scale requires certain tools and techniques. Paper with a ¼-inch grid is a convenient size with which to work (usual scale is ¼ inch to 1 foot) and yet also provides a good scale for visually depicting the layout. (If a ⅛-inch scale is used, buy ⅛-inch squared paper and so on.) A pen and India ink, or a heavy black ink pen, a good ruler, preferably an architect's ruler with various scales marked on it, and some tracing paper and masking tape are other needed supplies.

An outline of the size and shape of the space allocated is first drawn to scale with pencil on the squared paper. When the location of doors and windows has been decided on, these are marked off on the outline. Then, the outline of the space is inked in, using proper architectural symbols for walls, doors, and windows as illustrated in Figure 10.23.

The next step is to obtain a set of templates, to-scale model drawings, of each piece of equipment to be used. They must be to the same scale as the floor plan. Label each template with the name and dimensions of the piece of equipment it represents; see Figure 10.24. Sometimes a different color is used for each work unit. Templates should include overall measurements of features that require space, such as the swing of door openings, control boxes or fittings, and any installation needs as specified in equipment catalogs. The templates are then cut out, placed on the floor plan, and moved about until a good arrangement is found. Templates may then be secured to the plan with a bit of rubber cement (for easy removal if changes are made).

Figure 10.23 Architectural symbols used on blueprints to show placement and arrangement of various types of doors and windows.

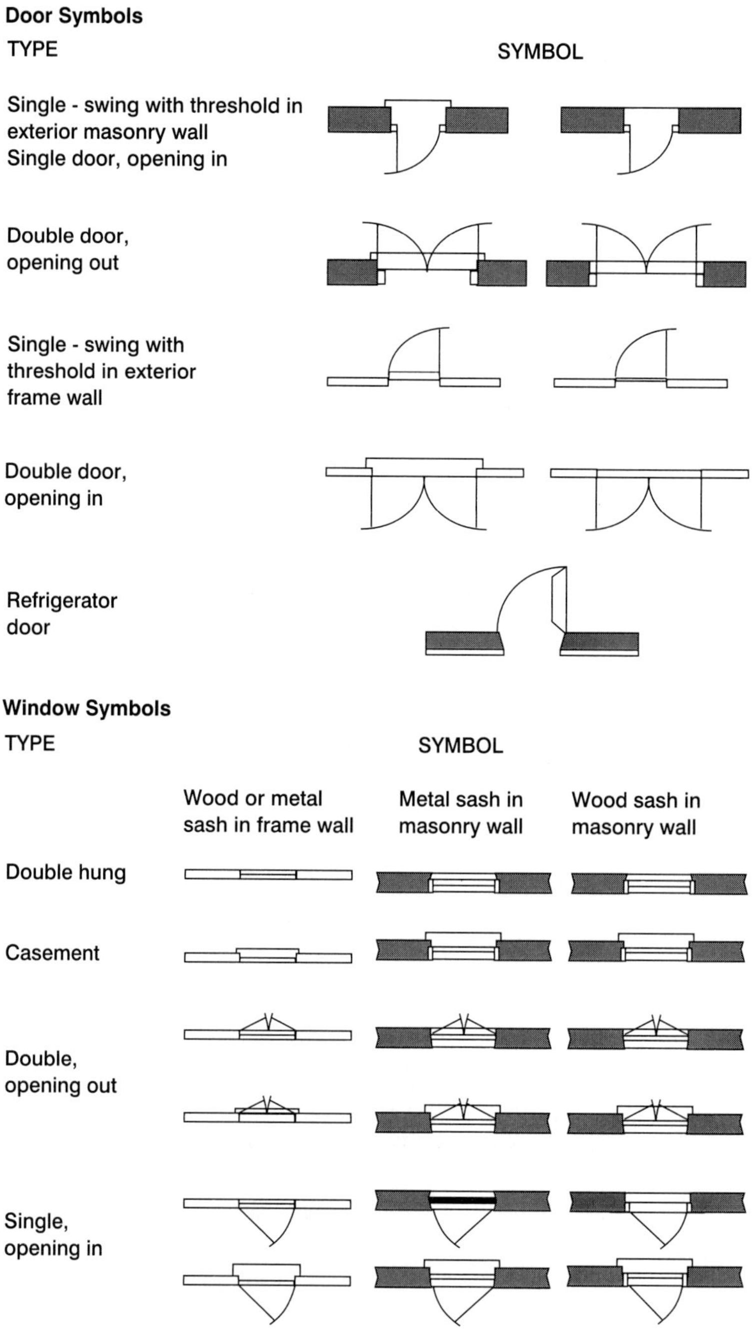

Door Symbols

TYPE | SYMBOL

Single - swing with threshold in exterior masonry wall
Single door, opening in

Double door, opening out

Single - swing with threshold in exterior frame wall

Double door, opening in

Refrigerator door

Window Symbols

TYPE | SYMBOL

Wood or metal sash in frame wall | Metal sash in masonry wall | Wood sash in masonry wall

Double hung

Casement

Double, opening out

Single, opening in

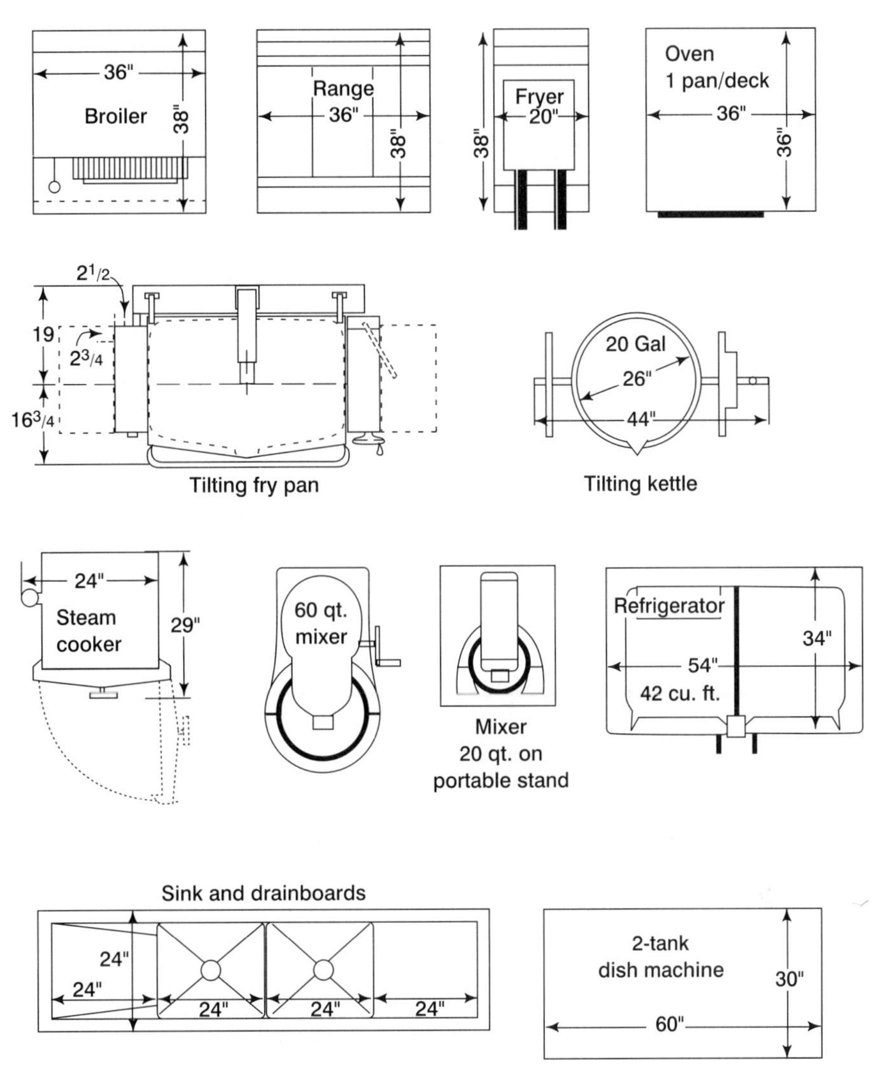

Figure 10.24 Templates of sample pieces of equipment drawn to ¼-inch scale. Cut-out templates may be moved about on a floor plan of the same scale to determine floor space needs, and to determine the most efficient arrangement of equipment.

A sheet of thin tracing paper is taped over the floor plan, and lines are drawn on it to show the route used in the preparation of several menu items. Drawing lines that trace the movements of food and workers from one key work point to the next within a unit, as well as from one work area or department to the next, is a good check on the efficiency of the arrangement. Actual measurement of the distances can be made by passing a string over pintacks at each key point as the preparation of a menu item progresses and then measuring the string. At this time, a check on width of aisles, work area space, location of hand-washing sinks for employees, storage space for carts and trucks, and similar details is made.

The above procedures afford good checks of the adequacy of the tentative floor plan, for necessary equipment, and of working areas before the final plan is made. The space allowances for passageways between working areas, between tables, between ranges and cook's table, and between other major pieces of equipment should also be checked for adequacy. Changes and adjustments should be made on paper instead of after construction has begun, because it is costly to make revisions at that time.

Separate drawings are made by the architect for plumbing, electric, and gas installations in addition to those for the building construction. All must be coordinated and checked carefully to ensure that gas, water, and waste outlets and vents will be in the correct positions for the equipment planned. Also, the electrical wiring with convenient switch control boxes, power and regular outlets and turn-on switches, and locations and kinds of light fixtures must be noted. Telephone conduits and outlets, wiring for computers, and intercom, public address, or TV system as decided on are indicated.

Designing by Computer

Computer-assisted design (CAD) planning began in the early 1960s and has grown and expanded rapidly during the past three decades. Computer capability for graphically designing a floor plan and equipment layout can replace the method just described in the "Mechanics of Drawing" section. However, all of the preliminary studies, analyses, and team input remain as necessary steps to obtain the data needed to create the design on a computer.

CAD for foodservices is based on an interactive graphic concept; that is, the software programs are developed to assist with schematic planning. Software programs use variables that must be identified by the foodservice manager and the planning team.

Today, programs are sophisticated and technologies that have been developed led to the use of the terms, "before CAD" and "after CAD." *Before CAD* is computer-modeling software meant to replace the use of "sketching on napkins and tissues" as the starting point for conceptual design. *After CAD* refers to computer-aided facilities management (CAFM), which provides a greater range of services for managing the building project beyond the designing function.

CAD software systems have been adapted and expanded by many companies. An update of the most recent systems is best obtained by reviewing trade journals and contacting companies. Many software programs can be used on personal computers, as well as on more powerful, networked workstations. CAD systems run on a variety of platforms, such as UNIX, DOS, MAC, and VAX.

Many add-on packages are available today to make floor plan design faster and easier. They have features such as instant viewing and zooming, display of several views simultaneously, cutting and pasting of drawings to create new ones, and marking drawings for modifications and easier version control. The output devices may include high-end digital plotters, interactive video displays, or virtual reality programs.

Foodservice managers or planning teams wishing to design facilities by computer will want to search the market carefully for appropriate software and add-ons. New developments appear almost daily and any listing of components today may soon be outmoded or obsolete. Some resources for keeping abreast of developments are listed at the end of this chapter under "Resources for CAD Programs."

Architect's Blueprints

After the dietitian, foodservice manager, and others thoroughly check the preliminary plans, the architect prepares a complete set of drawings that are reproduced as blueprints. Blueprints always include the name and address of the facility, the scale used, and the date the plan was prepared. Details of construction, material, plumbing, and electrical wiring, connections, and fixtures are indicated and coded. Side elevation drawings are included for door and window finishings, stairways, and built-in or attached equipment.

When reading and checking blueprints, one must constantly consider the scale to which they are drawn. The scale should be sufficiently large to permit detailed study. The heavy, solid lines indicate walls; the space between lines indicates the wall thickness; and the markings in between denote the kind of materials, such as stone, brick, and concrete blocks. Three or four parallel lines at a break in the wall denote the position and size of windows. The direction in which doors will open appears in blueprints as an arc extending from the door hinge to the door's fully open position (see Figure 10.23). Steps are shown as parallel lines with an arrow and the words "up" or "down." Dimensions of all spaces are indicated, and rooms and equipment are labeled. Architects use a variety of symbols to identify special features; for example, some electrical symbols are shown in Figure 10.25. All the symbols the architect uses are explained in a legend on the drawing.

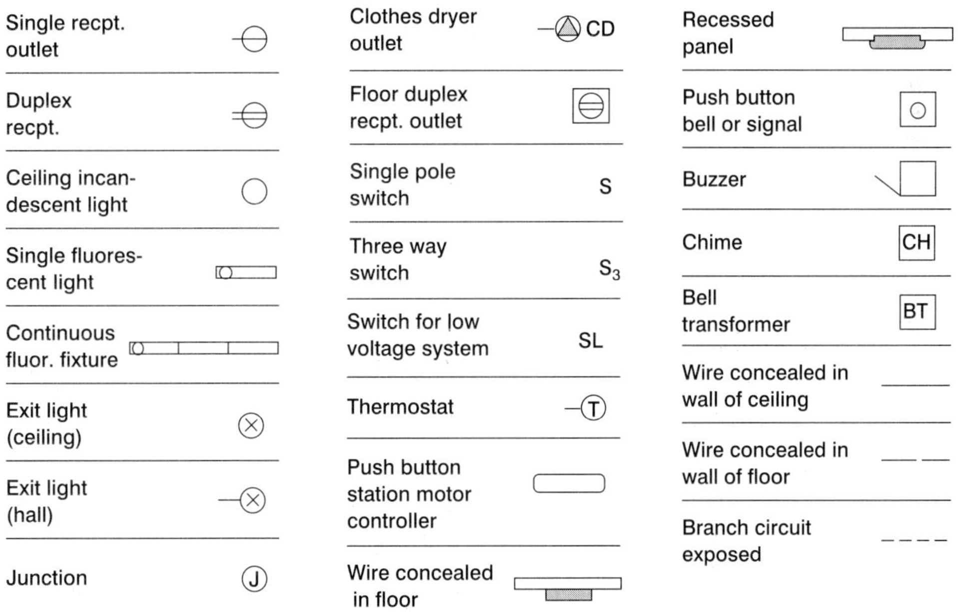

Single recpt. outlet

Duplex recpt.

Ceiling incandescent light

Single fluorescent light

Continuous fluor. fixture

Exit light (ceiling)

Exit light (hall)

Junction — J

Clothes dryer outlet — CD

Floor duplex recpt. outlet

Single pole switch — S

Three way switch — S_3

Switch for low voltage system — SL

Thermostat — T

Push button station motor controller

Wire concealed in floor

Recessed panel

Push button bell or signal

Buzzer

Chime — CH

Bell transformer — BT

Wire concealed in wall of ceiling

Wire concealed in wall of floor

Branch circuit exposed

Figure 10.25 Electrical symbols used on blueprints to indicate type and location of wiring and outlets.

Specifications and Contract Documents

The architect must also prepare a set of written documents to accompany the blueprints when presented to contractors for bid. These documents include a statement of general conditions and scope of the work to be done; a schedule of operation, which includes a timetable for contractors to complete their work and detail of penalties resulting from failure to meet deadlines; who has the responsibility for installations and inspections; and specifications for all aspects of the work and for the equipment required.

Specifications include details such as the location of the building; type of base construction; mix of cement; size and kinds of conduits, drains, and vents; type and installation of roofing and flooring; wall finishes and colors; hardware; doors and windows; and all other construction features. Equipment specifications generally include the brand name and model number, material to be used, size or capacity, and the number required (see Chapter 11 for details). In large installations, separate contract documents may be prepared for bids on the electrical or HVAC system. All specifications must meet applicable building and installation codes, and all the documents must be clearly worded to avoid misinterpretations.

Bids, Contracts, Construction, and Inspection

When the contract documents are completed, they are advertised and made available to interested bidders. Certain reputable contractors and equipment dealers may be notified that the plans are complete and be invited to bid on the project.

The contract is generally awarded to the low bidder, who then works closely with the architect until construction is complete. The foodservice manager closely monitors developments during the construction phase of the project, checking frequently with the architect. Conditions of the contract, as well as the individuals concerned, will determine what adjustments can be made after the contract is signed.

The actual construction time will vary, depending on the type and size of the building and the availability of labor, materials, and equipment. During construction, the architect will frequently check the progress and quality of work to be sure that both meet contract specifications. In addition, the architect must inspect and approve all construction, equipment and installations before the sponsoring organization accepts the facility. At least 2 to 3 weeks before the scheduled opening, a punch list should be prepared. A *punch list* is a detailed checklist that would reveal any defective, substitute, or inferior equipment so that corrections could be made prior to the opening or training dates. A qualified professional who is neither supplying nor installing the kitchen should prepare the punch list. Each item of equipment is performance tested to see that it meets specifications and claims and that it has been installed correctly.

In addition, performance tests, usually conducted by the equipment vendor's representative to demonstrate proper operation, care, and maintenance of the equipment, should be attended by the dietitian, foodservice manager and assistants, the kitchen supervisor, maintenance personnel, and the architect. The demonstrations may be videotaped, too, for use in training future employees and for later review sessions for present personnel.

Usually, the various contractors guarantee necessary adjustments and some service for a specified period following the project's completion. After some predetermined date, all repairs and full maintenance become the foodservice management's responsibility. Any warranty contract forms supplied should be completed and returned promptly to the manufacturer.

SUMMARY

The principles and guidelines for facility design planning presented in this chapter apply to all types of foodservice building projects. In fact, the general considerations for making and checking floor plans are similar for different kinds of institutions, regardless of the type of service, menu, clientele, and other governing conditions. Parts of a project that were originally eliminated can possibly be included at lower cost in the future if basic plans for them are incorporated during the construction period. For example, if a monorail system for transporting supplies and food is anticipated in the future, the necessary overhead rails and other requirements could be incorporated into the original construction.

During the project's planning phase, foodservice managers would have been collecting a list of items that should be included in the proposed plans. These could range from a telephone jack in the dining room to storage space for banquet tables, high chairs, reserve china, and utensils. A written list of such details is an excellent way to ensure that these items are included in the final plans.

A balance of beauty and utility in the structure, furnishings, and equipment is helpful for successful foodservice planning. Colorful walls and floor coverings, modern lighting, streamlined modular kitchen equipment made of well-finished metals, machines with mechanical parts and motors enclosed, and the use of attractive woods and metals in dining room furniture are but a few of the many features contributing to the functionality of modern foodservice areas. Sanitation, ease of maintenance, noise reduction, and controlled environmental temperatures for comfort are built-in features that contribute to making a facility successful and help to achieve the objectives outlined in the prospectus for the foodservice operation.

REVIEW QUESTIONS

1. What does facilities planning and design encompass?
2. What preliminary studies and data collection are essential to prepare for a facilities planning project and why?
3. How can a foodservice manager keep abreast of new developments in foodservice design equipment?
4. What are some of the sources of information on the ADA regulatory considerations that must be observed in planning a new foodservice facility?

5. Generally, who are the members of the team that cooperatively plan a foodservice facility? What contributions does each make? What information must the foodservice manager be prepared to provide for the other team members?
6. Why is a prospectus an important document in a planning project? What are the three parts of a prospectus?
7. Point out some ways that energy conservation, sanitation, safety, and noise control can be built in to a facility plan.
8. For greatest efficiency, what is the recommended flow of work and people, and the space relationships for a foodservice facility?
9. What determines the number and kind of work units that are to be included in any given floor plan design?
10. What are the mechanics of drawing a floor plan and arranging the equipment layout by hand? by computer-aided facility design?
11. What contract agreements and specifications must be prepared and included in the documents sent out for bid for the facility construction?
12. After construction is complete, what inspections and performance tests should be made before the sponsoring organization formally and finally accepts the facility?

SELECTED REFERENCES

Accommodating Disabilities—Business Management Guide—1992. (*Contact:* Commerce Clearing House, Inc., P.O. Box 5490, Chicago, IL 60680-9882)

Bean, N. R.: Planning for catastrophe: Fast track to recovery. Cornell HRA Quarterly. 1992; 32(2):64.

Borsenik, F. D., and Stutts, A. T.: *The Management of Maintenance and Engineering Systems in the Hospitality Industry,* 2nd ed. New York: John Wiley & Sons, 1987.

Breit, A.: Satisfying the health department. Food Mgmt. 1991; 26(8): 60.

Frable, F., Jr.: Timing is key to new renovated kitchens. Nations' Rest. News. 1994; 28(41):24.

Haasis, J., Jr.: Design and chlorofluorocarbons (CFCs). The Consultant. 1990; 23(3):31.

Hall, L.: Design considerations for correctional facilities. The Consultant. 1989; 22(3):45.

Kaye, R. E.: The food storage area—Just another closet? The Consultant. 1990; 23(2):54.

Kazarian, E. A.: *Food Service Facilities Planning,* 3rd ed. New York: Van Nostrand Reinhold, 1989.

Kohl, J. P., and Greenlaw, P. S.: Here comes the Americans with Disabilities Act. Part II. Implications for managers. Cornell HRA Quarterly. 1992; 33(1):33.

Kotschevar, L., and Terrell, M. E.: *Foodservice Planning: Layout and Equipment,* 3rd ed. New York: John Wiley & Sons, 1985.

Lundberg, D. E.: *The Restaurant: From Concept to Operation.* New York: John Wiley & Sons, 1985.

Markham, J. E.: The blueprint for success. (About feasibility studies). The Consultant. 1990; 23(3):34.

Norton, L. C.: Planning for the '90s. Food Mgmt. 1990; 25(10):75.

Norwich, R.: Safe and hygienic floors in food preparation areas. The Consultant. 1991; 24(3):59.

Rethmeyer, A.: Evaluating a planned renovation. Food Mgmt. 1989; 24(4):48.

Richardson, J. K., Jr.: CFCs and the foodservice industry. The Consultant. 1990; 23(2):27.

Riggs, J. L.: *Production Systems: Planning, Analysis and Control.* New York: John Wiley & Sons, 1981.

Unklesbay, N., and Unklesbay, K.: *Energy Management in Foodservice.* Westport, Conn.: AVI Publishing Company, 1981.

Von Egmond-Pannell, D. V.: *School Foodservice,* 3rd ed. Westport, Conn.: AVI Publishing Company, 1985.

Woods, R. H., and Kavanaugh, R. R.: Here comes the Americans with Disabilities Act. Part I. Are you ready? Cornell HRA Quarterly. 1992; 33(1):24.

RESOURCES FOR CAD PROGRAMS

Architectural Record (professional magazine; each issue has a section on computers and product reviews). Subscription address: P.O. Box 564, Highstown, NJ 08520-9885.

Gen CADD Architecular (a general drafting program for laying out space, drawing walls, footings, doors, windows, stairs, and elevations. IBM PC compatible). *Contact:* DCA Softdesk, Henniker, NH.

Stover, R. N.: *An Analysis of CAD/CAM Applications.* Englewood Cliffs, N.J.: Prentice Hall, 1984.

Versa CAD (for Macintosh 4.0; compatible with Systems and Support; one of the fastest project microcomputer managers available for large projects). *Contact:* Computervision, Inc., 100 Crosby Drive, Bedford, MA 01730.

Zawacki, G.: A review of CAD—Past, present, and future in the foodservice industry. The Consultant. 1989; 22(3):42.

11

Equipment and Furnishings

OUTLINE

Complete coverage of this broad subject area is impossible in a general textbook, but an effort is made here to include pertinent basic information that can be supplemented by current literature from the manufacturers and from observations of equipment in use.

The selection and purchase of furnishings and equipment for any foodservice are major responsibilities of the director and the staff, and the wisdom with which selection is made determines in large measure whether lasting satisfaction will be attained. Employee and customer safety, the efficiency of work units, and the beauty of the environment may be marred by poor selection and placement of furnishings and equipment, and the quality of service that an organization may render is influenced, if not limited, by these features.

The wise selection of equipment for any foodservice can be made only after a thorough study of all factors affecting the particular situation has been made. Items are available in many designs, materials, sizes, and within a wide cost range, but only those items that will help to meet the specific needs of the foodservice and contribute to its efficient operation should be purchased.

The problem of selection is so important and errors so costly that major characteristics to consider in the selection of certain basic pieces of equipment are included in Appendix B.

A section devoted to dining room furnishings concludes this chapter. Basic information needed for the wise selection of dinnerware, tableware, glassware, and table covers is presented in Appendix B.

KEY CONCEPTS

1. Specific characteristics of the foodservice operation must be carefully considered before making any equipment selection decisions.
2. The first consideration for any equipment decision is the menu.
3. Equipment features such as design and function, size or capacity, material and construction, and initial and operating costs must be thoroughly studied and considered before choosing each piece of equipment.
4. Successful maintenance of equipment requires definite preventive maintenance plans to prolong its life and maintain its usefulness.
5. Well-written (specific and definite) specifications are an absolute necessity of any good equipment purchasing program.
6. Maintenance of high standards of sanitation in foodservice is aided by selecting equipment that meets the standards set by the National Sanitation Foundation International (NSFI).
7. Stainless steel is widely used in foodservice equipment construction because of its permanence, resistance to stains and corrosion, lack of reaction with food, appearance, ease of cleaning and fabrication, and price.
8. Stainless steel may be chosen by gauge and finish. The gauge number is a measure of weight (pounds per square foot), which in turn determines the thickness of the steel.

9. A record of maintenance and repair performed on each piece of equipment should be maintained in order to provide data for appraising upkeep costs and depreciation of equipment.
10. Dining room furnishing and tabletop items should be pleasing, durable, serviceable, and easy to maintain.

FACTORS AFFECTING SELECTION OF EQUIPMENT

Equipment for any foodservice should be selected on the basis of a thorough study of all major considerations. Important among these are the following:

1. The menu,
2. Number and type of patrons to be served,
3. Form in which the food will be purchased,
4. Style of service and length of serving period,
5. Number of labor hours available,
6. Ability of employees to do the work,
7. Accessibility and cost of utilities,
8. Budget and amount of money allotted for equipment, and
9. Floor plan and space allotments.

Most foodservices include one or more of each of the following: oven, range, tilting frypan, fryer, broiler, steam-jacketed kettle, pressure steam cooker, coffee maker, refrigerator, freezer, ice maker, mixer with attachments, food cutter, sinks, tables, and carts. A wide variety of additional equipment may be purchased as necessity demands and money permits.

Before final decisions are made, individual pieces of equipment should be considered as to design, materials in relation to suitability for the purpose, durability and cleanability, construction and safety, size and capacity, installation, operation, and performance, maintenance, and replacement of parts. Cost and method of purchase are also major considerations in the selection of equipment.

Sound generalizations concerning equipment needs are difficult to formulate because each foodservice presents an individual problem with an interplay of factors not exactly duplicated elsewhere. The determination of these needs, therefore, should be one of the first and most important considerations of the foodservice manager as a basis for deciding what equipment should be purchased. Each item selected must accomplish those definite tasks peculiar to the specific situation. If the installation is new, information concerning the demands to be made of the facility and the ways in which the furnishings and equipment may help to meet these demands is of primary importance in planning the layout and selecting the equipment. If the installation is already in operation and has been found to be inefficient, an analysis should be made of the layout and equipment as it exists. This study can be used as a basis to rearrange the floor plan and include any additional furnishings and equipment needed.

The Menu

The menu pattern and typical foods to be served must be known in order to determine the extent and complexity of the required food preparation. Detailed analysis of the preparation requirements of several typical menus provides the best basis for estimating foodservice equipment needs for a particular situation.

Standardized recipes that include AP and EP (as purchased and edible portion; see Chapter 6) weights of ingredients, yields, pan sizes, and portion size are invaluable aids to planning for efficient equipment. Batch size and how often a procedure is repeated are important considerations for determining equipment needs. A large mixer and both large-capacity and duplicate steam-jacketed kettles or tilting frypans might be advisable, since they are used in the preparation of many menu items. An increase in the amount of time needed to prepare 500 portions over that needed for 100 portions would be necessary but not always proportional to the increase in quantities. In general, little difference in time is required for chopping various amounts of food in less than machine-capacity quantities or for mixing or cooking an increased amount of food in larger equipment. Repetitive processes such as hand rolling of pastry or batch cooking of vegetables in a small pressure steamer require almost proportional quantity, time, and space increases.

Once the equipment has been installed, care must be taken that menus are planned with consideration for its balanced use. This means that the person responsible for planning menus must be familiar with the facilities at hand and know the capacities of the equipment and timing of processes for the amounts of food to be prepared. Demands for oven cooking beyond the capacity load may lead to much unhappiness between manager and cook and may also encourage the production of inferior food or too-early preparation. Preparation timetables, equipment capacity charts, and standardized recipes that indicate AP and EP weights of ingredients, yield, and pan size for the particular setup can contribute much to effective planning for the efficient use of equipment.

Number and Type of Patrons

The number and type of patrons are important factors in selecting the appropriate amount and kind of equipment for a foodservice. The equipment needs for the preparation and serving of a plate lunch to 500 children in a school dining room are quite different from those of a service restaurant offering a diversified menu to approximately the same number of people three times daily. In the school foodservice there probably would not be more than two hot entrées on the menu for any one day, but all food would have to be ready to serve within a short period of time. In the restaurant, a variety of items would be ready for final preparation over extended serving periods; also, some items would be cooked in small quantities at spaced intervals according to the peak hours of service. Obviously, smaller and more varied types of equipment would be needed in the restaurant than in the school dining room. Production schedules in a short-order operation would require duplicates of such items as griddles, broilers, and fryers, whereas a residence hall foodservice would need steam-jacketed kettles, steamers, and ovens to produce a large volume of food within a specified time period.

The number of people to be fed determines to a great extent the total volume of food that must be prepared, but numbers in themselves cannot be used to evaluate equipment needs. Estimates of numbers of persons to be served during each 15-minute interval of the serving period will provide a guide to food and equipment needs. Amount and capacity of equipment to select are based on the number served at the interval of greatest demand in relation to cooking time required for specific items.

Form of Food Purchased and Styles of Service

The form in which the food is to be purchased will greatly influence equipment needs. The selection of fabricated meats and poultry, frozen portioned fish, frozen juices and vegetables, juice concentrates, ready-to-bake pies, and some cooked entrées, chilled citrus fruit sections, washed spinach and other greens, and processed potatoes, carrots, and apples eliminates the need for space and equipment usually required for preparation and disposal of waste. Adequate facilities for short and long storage at the proper temperatures must be provided, but other equipment needs would be limited primarily to those pieces required in the final stages of production and the serving of the finished products.

Various styles of service, such as self-service in a cafeteria, table or buffet service in a public dining room, or vended service, require particular kinds of equipment for their efficient functioning. Length of serving period is another factor.

Labor Hours and Worker Abilities

The labor hours available and the skill of the workers cannot be overlooked in considering the equipment needs of any foodservice. If the labor budget or local labor market is limited, usually the selection of as much labor-saving equipment as possible is warranted. Judgment must be exercised in deciding what equipment will provide for the smooth functioning of the organization and also give the best return on the investment. Will the increased productivity of employees with automated equipment compensate for the possible increased payroll costs, initial costs, and maintenance costs? With the rising pay rates for employees at all levels, managers must weigh values carefully when selecting equipment they can operate successfully, efficiently, and economically to accomplish the job to be done.

Utilities

The adequacy of utilities for the successful installation and performance of commercial cooking and warming or power-driven equipment must be checked before the final decision is made on selections. Often the choice between gas, electric, or steam-heated cooking equipment demands considerable investigation of the continuing supply of the source of heat, replaceability of parts, relative costs of operation and maintenance, and the probable satisfaction received from use in the particular situation. High-pressure steam is not always available, thus, self-generating steam units would be a necessary choice, Power-driven equipment is equipped with motors of the proper size for the capacity of the machine, but cycle and current would have to be designated so that the machine would operate properly for the wiring and power in the building.

The Budget

The budgetary allowance must cover not only the initial cost of the equipment but often the additional cost of installation. Available funds determine to a great extent the possible amount and quality of equipment that can be purchased at any given time. If the initial equipment budget is adequate, the choice among various pieces becomes mere determination of the superior and preferred qualities for each article desired. Sometimes the equipment budget is so limited that the food director is forced to decide between certain desirable articles and to weigh with serious thought the relative points in quality grades of the pieces believed to be essential. It is advisable then to list all the needed equipment so that unbalanced expenditure will not result. Lack of such thought or insistence on the best may lead to disastrous spending.

Consensus is that equipment of good quality is the most economical. Generally, if the amount of money is limited, it is better to buy a few well-chosen pieces of equipment that will meet basic needs and make additions as funds are available than to purchase many pieces of inferior quality that will need to be replaced in a short time. In contrast, some consultants warn that because of the rapid change in the trend toward the use of prepared foods, it may be preferable in some installations to plan equipment for a short life span and early replacement until such developments are stabilized. The initial cost of equipment is influenced by the size; materials used; quality of workmanship; construction, including special mechanical features; and finish of the article. The limitation of funds may lead to the necessity of a choice as to which one or more of these points may be sacrificed with least jeopardy to the permanence of the article and satisfaction in its use.

Estimates of cost for foodservice equipment are difficult to ascertain because each operation must be considered individually. It is advisable to learn the costs of comparable situations before making tentative estimates for a new or remodeled setup.

The Floor Plan

Space allocation for the foodservice may restrict the amount and type of equipment and its placement, especially in old buildings where architectural changes are limited and in new ones where the original planning may have been ill advised regarding the functions and needs. The size and shape of the space allotted to food preparation and its relation to receiving, storage, and dining areas greatly influence the efficiency of operation and, ultimately, customer satisfaction, as discussed in the preceding chapter. Floor space either too small or too large to accommodate the equipment suitable and desirable for the volume of food production anticipated creates an unsatisfactory situation. In the first instance, the overcrowding of work makes for confusion and frustration, limits the amount and type of preparation that can be done, and slows production. When the space is too large, much time and effort can be wasted by workers in transporting food long distances. Also, there can be a tendency to overequip with needless items simply because ample space is available. In any case, a complete analysis of the real needs is necessary before an equipment investment is made.

FEATURES OF EQUIPMENT

General objectives and trends in current equipment developments include an increase in the number and kind of specialized items, many of which are adaptable to multiple use; function and attractiveness in appearance; compactness and efficient utilization of space to reduce labor hours and time requirements to a minimum; speed output of quality products; modular planning of matched units as shown in Figure 11.1; mobility and flexibility of arrangement; exact engineering tolerances, effective insulation; computerized and solid-state controls (Figure 11.2) for even temperatures and operation; built-in sanitation; and fuel efficiency. With the change in the type and amount of food preparation in the individual units has come a corresponding change in equipment to meet the particular production needs.

Design and Function

The design of equipment and furnishings for the foodservice should be in close harmony with the general plan of the building, especially in the decorative features and items such as table appointments. This is particularly noticeable in summer resorts, children's hospitals, and certain types of restaurants, where not only the modern trend of foodservice planning and interior decoration has been followed, but also some specialized idea or theme has been expressed through the design and type of

Figure 11.1 A compact arrangement of modular cooking units can be fitted into a continuous framework for beauty, space saving, operational efficiency, and maintenance of sanitation. This one includes cabinet-mounted steamers, kettles, and a tilting braising pan.
Courtesy of Groen Division, Dover Corporation, Elk Grove Village, Illinois.

Figure 11.2 This computerized convection oven utilizes solid-state temperature devices accurate to within one degree (±1°F). A digital control panel replaces conventional control knobs and allows entry of exact cooking temperatures and times.
Courtesy of Lang Manufacturing Company, Redmond, Washington.

furnishings selected. Sensitivity to the artistic design of foodservice furnishings and equipment is often more acute than to the design of similar items for the home, because of the larger size of items required and duplication in number, as in dining room tables and chairs. Generally speaking, heavy-duty equipment is designed to give a streamlined effect.

Beauty and utility may be combined in foodservice equipment through the application of art principles and consideration of the functions of various items by the designer. The gadget or piece of equipment may be beautiful in line and design, but it is of little value if it serves no real purpose or if an unreasonable amount of time is required for its operation or care. The design of cutlery such as a chef's knife with a heavy wide blade shaped for cutting on a board and a long-handled cook's fork are examples of how closely design is related to the use of an article. Also, the design may influence the timing, efficiency, and comfort of operation, as is the case with the utensils shown in Figure 11.3. These *spoodles,* or combination spoon and ladle, are not only color coded for portion control, but feature a handle that is designed for control and comfort.

Simplicity of design is pleasing and restful and usually results in a minimum amount of care. The maintenance of high sanitation standards in a foodservice is aided if the equipment that is selected is designed so that sharp corners, cracks, and crevices are eliminated and all surfaces are within easy access for cleaning. The Joint Committee on

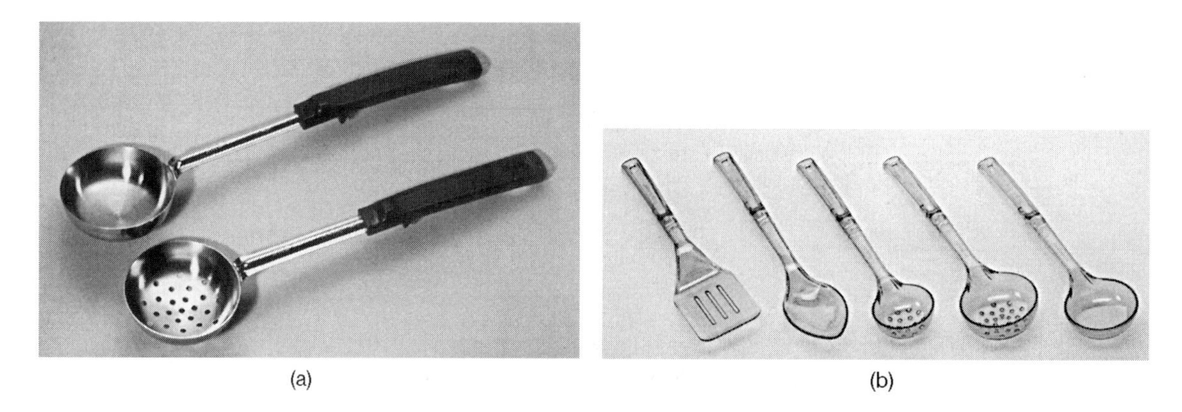

(a) (b)

Figure 11.3 Spoodles are designed to give the portion control of a ladle with the ease and balance of a spoon. Made of either (a) stainless steel with plastic handles or (b) entirely of high-impact plastic in a variety of capacities with perforated or solid bowls.
Courtesy of Vollrath Company, Sheboygan, Wisconsin.

Food Equipment Standards of NSFI has stressed the sanitation aspect of kitchen equipment design and construction as exemplified in the following statement:

> Foodservice equipment and appurtenances shall be fabricated to exclude vermin, dust, dirt, splash, or spillage as may be encountered under normal use, and shall be easily cleaned, maintained, and serviced.

All equipment mounted on legs or casters, such as the tray unit shown in Figure 11.4, should be designed to have a minimum clearance of 6 inches, but preferably 8 inches, between the floor and bottom surfaces of equipment, shelves, pipes, drains, or traps, to permit ease of cleaning. Heavy stationary equipment such as ranges and cabinets can be mounted successfully on a raised masonry, tile, or metal platform at least 2 inches high, sealed to the floor at all edges. Usually, this type of island base is recessed to allow for toe space beneath the equipment.

Specially designed mountings on wheels for specific purposes have become an important feature of foodservice planning for convenience, sanitation, and economical use of space and labor. Portable back-of-the-counter breakfast service units, including toasters, waffle irons, and egg cookers, can be transported out of the way during the remainder of the day. Dispenser units can be filled with clean trays in the dishwashing room and wheeled into position at the counter with minimum handling (see Figure 11.4). Portable bins for flour and sugar are more convenient to use and easier to keep clean than built-in bins. Sections of shelves in walk-in refrigerators and dry storage rooms mounted on wheels make for convenience in cleaning and rearrangement of storage. The importance of designing general utility trucks and dollies to fit into the places in which they are to be used cannot be overestimated.

Heavy-duty wheeled equipment, such as range sections, tilting frypans, fryers, ovens, reach-in refrigerators, and the many mobile work and serving units, make rearrangement possible in order to adapt to changing needs at minimum cost. Often the conversion of certain spaces from limited- to multiple-use areas can be effected through the inclusion of mobile equipment. Also, thorough cleaning in back of and underneath equipment is made easier when it is movable and accessible from all sides.

One of the outstanding improvements in serving equipment has been effected through a change in the design and construction of heated serving counters. This change from the old pattern of a given number of rectangular and round openings, far apart, in an elongated steam-table arrangement with limited fixed storage, to a condensed type with fractional size containers, has been estimated to permit up to 50 percent greater food capacity in the same amount of space. This arrangement also makes possible almost unlimited flexibility in service through the close arrangement of a few regular 18- × 12-inch rectangular top openings into which full-sized or combinations of fractional-size pans of different depths may be fitted with or without the aid of adaptor bars. Hot food serving counters may be designed and constructed for two or more openings, moist or dry heat, gas or electricity, separate heat controls for individual sections or for the unit, and space below enclosed or fitted for dish storage.

The selection of inserts for this type of counter should be made to meet the demands at peak times for the best service of all the usual types of hot foods included on a menu. The number of each size and depth of pans to purchase can be determined easily by careful analysis of several sample menus, the quantities of each type of food required, and the most satisfactory size and depth of pans for their preparation and service. In most instances, this will mean a relatively small number of sizes with ample duplication of those that will be used the most.

Common depths of the counter pans are 2½, 4, and 6 inches with some sizes available 1 and 8 inches deep. Capacities are listed for each size, for example, as shown in

Figure 11.4 Mobile self-leveling tray unit, filled at dish machine and transported to the point of use.
Courtesy of Precision Metal Products, Inc., Miami, Florida.

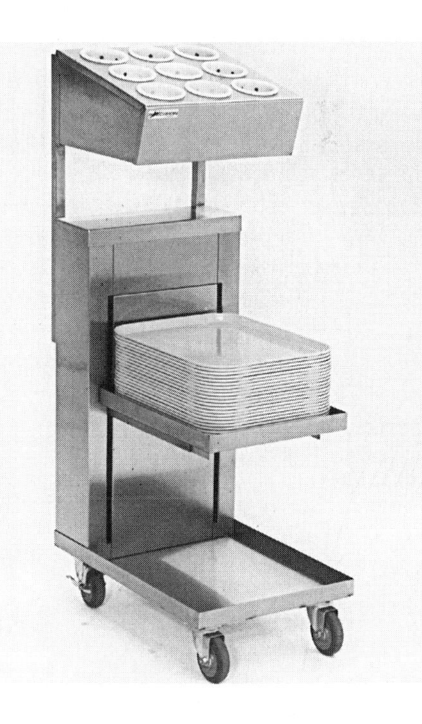

Table 11.1. All inserts fit flush with top openings, except the 8-inch-deep pans, which have a 2-inch shoulder extending above the opening. Pans of one size and depth are designed to nest together for convenient storage. Since these pans are made of non-corrosive well-finished metal, certain types of menu items may be cooked in and served directly from them, whereas other foods will need to be transferred to them for serving. Recipes can be standardized for a specific number of pans of suitable size and depth for a product and with the exact number of portions predetermined. The pan shown in Figure 11.5 is designed with reinforced corners to add strength and prevent vacuum-sticking of stacked pans. As with most hotel pans, they are available in full, half, one-third, one-fourth, and one-sixth sizes.

Size or Capacity

The size or capacity of equipment to select for a given situation is determined largely by the type of menu and service offered and the quantities of different types of foods to be prepared at one time. More pieces of heavy-duty equipment of larger capacities are required for the preparation of food for a college residence hall serving a nonselective menu at a set hour than for the preparation and service for a short-order lunch counter serving comparable or even greater numbers throughout an extended meal hour. Batch cooking, the cooking of vegetables in not more than 5-pound lots, timed at intervals to provide for a continuous supply to meet the demands of the service, is far preferable to cooking the entire amount at one time and holding the cooked product through the serving period. The latter would require one or two large steam-jacketed kettles instead of a battery of small ones and would mean less effort and time for the cook, but at the sacrifice of eye appeal, flavor, crispness, nutritive value of the food served, and the satisfaction of the guests.

Large equipment, such as ranges, ovens, tilting frypans, mixers, and dishwashers, may be obtained in more or less standard sizes, with slight variations in the articles produced by different manufacturers. For example, range sections may vary a few inches in the overall measurements and the inside dimensions of ovens may differ, whereas the capacities of mixers made by most firms are comparable.

Charts are available from most manufacturers that show the capacity or output per hour for each size of machine. For example, the capacity of a dishwasher is measured by the number of dishes that can be washed in an hour. The size of mixer to purchase would be determined by the volume of a product to be prepared each mixing, the time required for mixing or mashing each batch, and the total quantity of the produce

Table 11.1 Sample capacities of hotel pans.

One-Half Size		One-Fourth Size	
Depth (in.)	Capacity (qt)	Depth (in.)	Capacity (qt)
1	1¾		
2½	4½	2½	2⅛
4	7⅛	4	3⅜
6	10⅞	6	4¾
8	15		

Figure 11.5 A hotel pan designed to last longer, not stick together when stacked, be more comfortable to carry, provide a better steam table seal, and be easier to clean.
Courtesy of Vollrath Company, Sheboygan, Wisconsin.

needed within a given period of time. Obviously, the size and number of pieces of each item of equipment required will depend on the needs of the particular institution.

The articles most often fabricated or built to individual specifications are those that must conform to a given size or are desired because of special material. Special orders make the equipment more expensive and often delay delivery; however, to most people, the satisfaction of having a piece of equipment that exactly fits usually more than compensates for the disadvantages.

Standards of uniformity in size of both small and large equipment have become fairly well established through the experience of users and their work with designers, manufacturers, and consultants. Many kitchens of the past have had a multiplicity of sizes of cooking utensils, baking pans, and trays that may or may not have made economical use of range, oven, refrigerator, cabinet, or truck spaces in the particular situation. An example is the large oval serving tray that would never fit on a rack, shelf, or truck. Alert foodservice directors and planning experts have come to recognize some of these problems and to note the advantages that are gained by simplification of the whole setup through improved planning for the efficient and interrelated use of the items selected.

The selection of certain modular items of equipment, or those of uniform size, has proven advantageous in quantity food operations. When a specified size pan, tray, or rack fits easily in the refrigerator, storage cabinet, serving counter, or on racks or carts, great adaptability in and economical utilization of space are made possible. Also, manpower efficiency is increased and labor hours are reduced; less floor area is required with improved use of vertical space; the use of pans and trays of the same size or in their multiple units reduces the total number and kind to buy, their cost, and the storage space needed; the number of shelves in refrigerators, cabinets, and carts can be reduced when trays and pans can be inserted at close intervals on angle runners or glides; the rehandling or transfer of foods or dishes is reduced, since the tray rack fits into any unit, either on a shelf, on glides, or in the counter; sanitation is improved through reduced handling of food or dishes, low spillage, and machine washing of trays and pans.

Common modules are the 12- × 18-inch and 18- × 26-inch trays, which are easily accessible in several materials and convenient to use. The 12- × 18-inch trays fit into the standard dishwashing racks of conveyor-type machines. Cabinets, shelves, refrigerators, and carts are readily available to accommodate one or a combination of such trays. Some spaces could be sized so that either one 18- × 26-inch bun pan or two 12- × 18-inch trays could be used. Another common module is space into which 20- × 20-inch dishracks would fit, for storage of cups and glasses in the racks in which they were washed.

This system merits careful consideration in planning equipment for simplified operation with maximum efficiency and economy. Each unit will continue to need a certain amount of its equipment custom built according to specification, but certainly there should be uniformity within each operation.

Materials

Materials for the various pieces of foodservice equipment should be suitable for the purpose and give the best satisfaction possible. The materials used in the equipment influence price, wearing qualities, sanitation, satisfaction, and usefulness. The weight, finish, and workmanship of the materials are important factors in determining their suitability and wearing qualities.

The Joint Committee on Food Equipment Standards has established minimum requirements for materials and construction of certain foodservice equipment items as follows:

> Materials shall withstand normal wear, penetration of vermin, corrosive action of refrigerants, foods, cleaning and sanitizing compounds, and other elements in the intended end use environment.

The committee further specifies that surface materials in the food zone

> shall not impart toxic substances, odor, color, or taste to food. Exposed surfaces shall be smooth and easily cleanable.
>
> Nonfood zone materials shall be smooth and corrosion resistant or rendered corrosion resistant. Coatings, if used, shall be noncracking and nonchipping.
>
> Solder in food zones shall be formulated to be nontoxic and corrosion resistant under use conditions. Lead based solder shall not be used.

Metal. Metals have become increasingly important in foodservice planning. Today, we depend on them for nearly everything, from structural features such as doors, flooring under steam units, and walk-in refrigerators to tables, sinks, dishwashers, and cooking equipment. A wide variety of old and well-known metals and alloys, such as copper, tin, chromium, iron, steel, and aluminum, were used in the foodservices of the past, but have been outmoded by the chromium and chromium-nickel stainless steels. At one time, copper cooking utensils and dishwashers were commonly found in institutional foodservices. Their care and upkeep were high because they required frequent polishing and replacement of nickel or tin linings to prevent the reaction of foodstuffs with the copper. Such utensils were heavy to handle and were used mostly in hotels and the military where male cooks were employed. Nickel was used considerably as a plating for equipment trim, rails of cafeteria counters, and inexpensive tableware.

Aluminum lends itself to fabrication of numerous kinds and will take a satin, frosted, or chrome-plated finish. It can be painted, etched, or engraved. It is relatively light in weight, has high thermal and electrical conductivity, does not corrode readily, and if cold rolled, is relatively hard and durable. It is capable of withstanding pressure at high temperature, which makes it particularly well suited for cooking and baking utensils and steam-jacketed kettles. Aluminum cooking utensils often become discolored by food or water containing alkali, certain acids, and iron. Many items are manufactured from anodized aluminum that has been subjected to electrolytic action to coat and harden the surface and increase its resistance to oxidation, discoloration, marring, and scratching. Anodized aluminum is often used for items such as dry storage cabinets and service carts and trays. Its strength and light weight are factors in its favor for mobile equipment. Aluminum may be combined with other metals to produce alloys of higher tensile strength than aluminum alone.

Cast iron is used in institutional equipment as braces and castings for stands and supports, for pipes, and for large pieces of equipment such as ranges. Its use in small equipment is restricted to skillets, Dutch ovens, and griddles.

Galvanized steel and iron were long used for such equipment as sinks, dishwashers, and tables. In the process of galvanizing, a coating of zinc, deposited on the base metal, protects it to a certain extent from corrosion. The initial cost of equipment made of galvanized material is comparatively low, but the length of life is short, repair and replacement expenses are high, sanitation is low, contamination is likely, and the general appearance is undesirable and unattractive in comparison to equipment made of noncorrosive metal.

The use of *noncorrosive metals*, mainly the alloys of iron, nickel, and chromium, for equipment at food-processing plants such as bakeries, dairies, canneries, and in-home and institution-size kitchens has increased tremendously within recent years until at present all such units are planned with widespread usage of this material. These materials are available in forms suitable for fabrication into any desired type of equipment. If the sheets are too small for the particular item, they may be joined and welded most satisfactorily. The price is not prohibitive, so that this type of material functions in many and varied instances from decorative effects in or on public buildings to heavy-duty equipment, cooking utensils, and tableware. Improved methods of fabrication and the unprecedented emphasis on sanitation have been important factors in the high utilization of noncorrosive metal in items of equipment.

The outstanding characteristics of noncorrosive metals for foodservice equipment include permanence, resistance to ordinary stains and corrosion, lack of chemical reaction with food, attractive appearance, ease of cleaning and fabrication, and nonprohibitive price. Tests show that with proper construction and care noncorrosive metals wear indefinitely, and equipment made from them may be considered permanent investments. The strength and toughness are so high that even a comparatively lightweight metal may be used for heavy-duty items. These metals do not chip or crack. High ductility and weldability also make for permanence of the equipment made from them; thus, the upkeep costs are reduced to a minimum.

Resistance to stains and corrosion is a major feature in foodservice equipment where cleanliness, appearance, and sanitation are of utmost importance. The freedom from chemical reactions of the noncorrosive metals with foodstuffs at any temperature makes their use safe in food preparation. Tests show few or no traces of metals or metallic salts present after different foods have been heated and chilled for varying periods of time in containers made of these metals.

The appearance of noncorrosive metal equipment when well made and carefully finished is satisfying and conducive to the maintenance of excellent standards of cleanliness and order. The smooth, hard surface is not easily scratched or marred, and the cleaning methods are simple. Special metal cleaners are available, but a good cleaner and water and the usual polishing should be enough to keep the equipment in good condition. Common steel wool, scouring pads, scrapers, or wire brushes may mar the surface or leave small particles of iron imbedded in the stainless steel, which can cause rust stains. Darkened areas are caused usually by heat applied either in fabrication or in use and may be removed by vigorous rubbing with stainless steel wool, a stainless steel pad and powder, or a commercial heat-tint remover. To avoid heat tinting of cooking utensils, they should be subjected to no more heat than required to do the job effectively and should never be heated empty or with heat concentrated on a small area.

The noncorrosive alloys manufactured most often into institutional equipment are nickel-copper and the stainless steels. Monel metal is a natural alloy that contains approximately two-thirds nickel and one-third copper, with a small amount of iron. The supply is fairly limited so it is seldom selected for fabrication into foodservice equipment.

By far the greatest amount of foodservice equipment is made of some type of stainless steel. Each company producing stainless steel under its own trade name may use a slightly different formula, but the important elements are practically the same. A relatively low carbon content in the stainless steels gives high resistance to attack by corrosive agents. A chromium-nickel stainless steel alloy commonly called "18-8" (number 302) is a favorite material for foodservice equipment. As its name indicates, it contains approximately 18% chromium and 8% nickel with no copper present. Heavy-duty equipment made of the noncorrosive alloys retains its appearance and sanitary qualities over long-term use.

Standard Gauge. The gauge of thickness of metals is an important consideration in selecting materials for equipment. The adoption of the micrometer caliper scale to indicate the thickness of sheet metal in decimal parts of an inch and the abolition of gauge numbers are strongly recommended. However, the U.S. standard gauge is used by most manufacturers of iron and steel sheets. This system is a weight, not a thickness, gauge. For instance, number 20 U.S. gauge weighs 1.5 pounds per square foot, subject to the standard allowable variation. Weight always is the determining factor. That this gauge is 0.037 inch thick is secondary in the system. Numbers 10 to 14 gauge galvanized steel or 12 to 16 noncorrosive metals are most generally used for foodservice equipment. Metal lighter than 16 gauge is commonly used for sides or parts where the wear is light (see Figure 11.6).

Finish of Metals. The surface or finish of metals may be dull or bright; the higher the polish, the more susceptible the surface is to scratches. The degree of metal finish is indicated by a gradation in number. The larger numbers indicate a finer finish and a higher degree of polish. Standard finishes for the steels in sheet form are listed In Table 11.2. Numbers 4, 6, and 7 are produced by grinding and polishing the sheets of metal with different grades of abrasives. These original finishes are capable of being retained in the usual fabrication of equipment, which requires only local forming. Materials with a number 4 grind surface are more often selected for such items as tabletops, sinks, and counters than are those with shiny or mirror-like finishes.

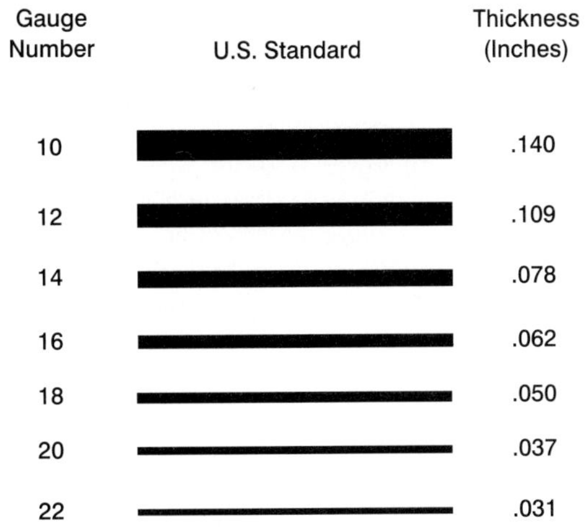

Figure 11.6 A diagram showing actual thickness of commonly used gauges of metals.

Gauge Number	U.S. Standard	Thickness (Inches)
10		.140
12		.109
14		.078
16		.062
18		.050
20		.037
22		.031

Glass. Glass and ceramic-lined equipment, such as drip coffee pots, are most satisfactory for certain purposes. They protect against metallic contamination, corrosion, and absorption. Glass-lined equipment is highly acid resistant and will withstand heat shock. This last quality is due to the fact that the coefficient of expansion of the glass enamel is similar to that of the steel shell. Most ceramics will break readily when exposed to extreme heat or mechanical shock.

Other Materials. Items such as counter fronts and ends and food tray delivery carts made of mirror-finish *fiberglass* with stainless steel structural trim are available in many beautiful colors. The interior and exterior walls of the food delivery carts are molded in one piece, then insulated with polyurethane foam. The surfaces are strong, dent and scratch resistant, and light in weight. *Porcelain* (glass on steel) or *vinyl* covered galvanized steel may be used satisfactorily on outside walls of refrigerators and on counter fronts at less cost than stainless steel. The materials just mentioned contribute to a colorful and pleasing decor, reduce reflected glare of light, and are easily maintained. Detached well-laminated and sealed *hardwood* cutting boards are permissible in some cities and states, although for purposes of sanitation,

Table 11.2 Stainless steel finishes.

Finish	Description
No. 1	Hot-rolled, annealed, and pickled
No. 2B	Full finish—bright cold-rolled
No. 2D	Full finish—dull cold-rolled
No. 4	Standard polish, one or both sides
No. 6	Standard polish, tampico brushed one or both sides
No. 7	High-luster polish on one or both sides

an increasing number of operators are choosing to use cutting boards made of reversible nontoxic, nonabsorbent polyethylene or hard rubber.

Carts, racks, stands, and dollies made of *polycarbonate* are light in weight, but capable of carrying heavy loads; they resist stains, dents, and scratches, will not rust or crack, and are easily disassembled for cleaning in a conveyor-type dishwashing machine. Side panels may be of a solid color or transparent, and most models are designed to accommodate 18- × 26-inch food boxes with fitted lids, trays, and bun pans (see Figure 11.7). All items can be fitted with nonmarking neoprene brake wheels and ball bearings.

Construction

The construction and workmanship of equipment determine whether or not it is durable, attractive, and sanitary. High-quality material and a perfect design for the purpose do not ensure good construction, although they contribute to it. Accurate dimensions, careful and well-finished joinings, solidarity, pleasing appearance, and ease of cleaning are important factors. Sinks, drainboards, and dishtables sloped to drain; tables and chairs properly braced; hinges and fasteners of heavy-duty materials and drawers constructed to function properly; adequate insulation where needed; and safety features are a few of the points to consider under construction. In addition, all parts must be easily cleanable.

Welding has replaced riveting, bolting, and soldering of both surface and understructure joinings in metal foodservice equipment. Great emphasis is placed on the impor-

(a) (b)

Figure 11.7 A selection of food storage containers that stack or nest easily and are available with snap-tight covers facilitates sanitary and efficient storage and transportation of food. Boxes made from (a) clear polycarbonate and (b) white high-density polyethylene are shown. Courtesy of Cambro Manufacturing Company, Huntington Beach, California.

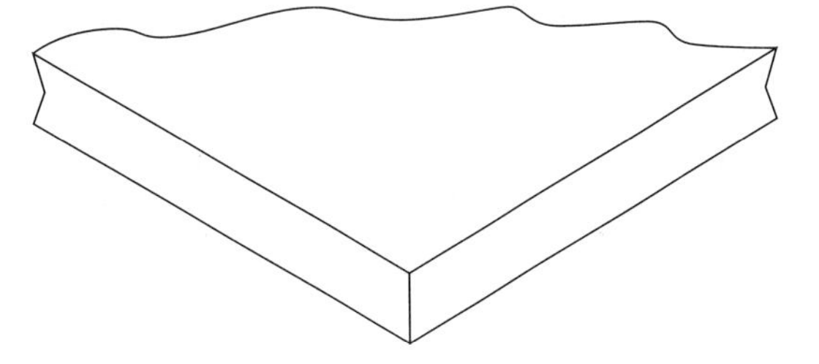

Figure 11.8 A mitered corner, welded and finished smooth.
Courtesy of National Sanitation Foundation International.

tance of grinding, polishing, and finishing of the surfaces and welded joints for smoothness and to ensure against possible progressive corrosion. Mitered corners (shown in Figure 11.8) that are properly welded and finished smooth, in items such as dishtables and sinks, are superior to deep square corners or those filled with solder. The construction recommended for items of equipment used for unpackaged food is for rounded internal angles with a minimum continuous and smooth radius of ⅛ inch and internal rounded corners with a minimum continuous and a smooth radius of ¼ inch for vertical and horizontal intersections and ⅛-inch radius for the alternate intersection.

The bull-nosed corner construction is used most often in finishing off the corners of horizontal surfaces such as worktables. The corner section of the top material is rounded off and made smooth both horizontally and vertically as an integral part of the horizontal surface. If the edge is flanged down and turned back, a minimum of ¾ inch should be allowed between the top and the flange, and the same distance should be allowed between the sheared edge and the frame angle or cabinet body to provide easy access for cleaning (see Figure 11.9).

To simplify construction and eliminate some of the hazards to good sanitation, fittings and parts have been combined into single forgings and castings wherever possible, and tubular supports sealed off smooth or fitted with adjustable screw-in solid pear-shaped feet have replaced open angular bracings with flange bases. In many instances, mobile, self-supporting, or wall-hung structures have replaced external framing. Several items welded or fitted together into a continuous unit may need to be brought into the facility and positioned before construction of the building is complete and while there is ample space for transporting the unit into the area.

The Joint Committee on Food Equipment Standards of NSFI outlines in detail permissible methods for construction of such general parts as angles, seams, finishes of joinings, openings, rims, framing and reinforcement, and body construction. Specifically, they give construction features for special items such as hoods, water-cooling units, counter guards, doors, hardware, sinks, refrigerators, power-driven machines and their installation. Many health departments use the recommended standards as

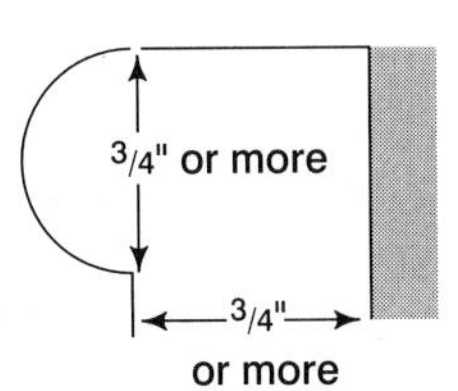

Figure 11.9 A bull-nosed corner, rounded off and finished smooth.
Courtesy of National Sanitation Foundation International.

a basis for approving equipment and its installation. An example of such a standard follows. Figure 11.10 shows the diagram.

Food Shields. Display stands for unpackaged foods are to be effectively shielded so as to intercept the direct line between the average customer's mouth and the food being displayed and shall be designed to minimize contamination by the customer.

Shields shall be mounted so as to intercept a direct line between the customer's mouth and the food display area at the customer-use position. The vertical distance from the average customer's mouth to the floor shall be considered to be 4 feet 6 inches (1.4 meters) to 5 feet (1.5 meters) for public eating establishments. Special consideration must be given to use location conditions such as tray rails and average customer's mouth height in educational institutions and other installations. Such shields are to be fabricated of easy-to-clean, sanitary materials conforming to materials specifications.

Safety Features. Safety features for the protection of workers in the use and care of equipment and for the production of safe food are important factors in the design, choice of materials, and construction of kitchen equipment. There is also a close relationship between these and the standards and controls for sanitation in a foodservice operation. Smooth, rounded corners on work surfaces, table drawers with stops and recessed pulls, automatic steam shut-off when cooker doors are opened, temperature controls, guards on slicers and chopping machines, brakes on mixers, recessed manifold control knobs on ranges and ovens, smooth, polished, welded seams, rounded corners, and knee-lever drain controls on sinks are a few examples of built-in safety in heavy-duty kitchen equipment.

Installation, Operation, and Performance

Proper installation is a necessity for the successful operation of all equipment. The best design and construction would be worthless if electrical, gas, or water connections were inadequate or poorly done. The dealer from whom the equipment was purchased may not be responsible for its installation by contract but will usually deliver, uncrate, assemble, and position the item ready for steam fitting or electrical and plumbing connections. In many cases, the dealer will supervise the installation and test it out to be certain that the equipment will function properly and instruct personnel in its operation and maintenance.

Architects, contractors, and engineers are responsible for providing proper and adequate plumbing, electrical wiring, and venting facilities for the satisfactory installation of kitchen equipment according to the standards of the local building, plumbing, electrical, and sanitation codes. Water, steam, gas, and waste pipe lines, and electrical conduits must be planned for each piece of equipment so that proper joinings can be made at the time of installation to avoid the necessity of extra pipe or wiring that might interfere with cleaning or placement of other equipment items.

The sanitation and safety aspects of equipment installation are important to the convenience and safety of its use and care. Sinks that drain well, wall-hung or mobile equipment that permits easy cleaning under and around it, equipment sealed to the

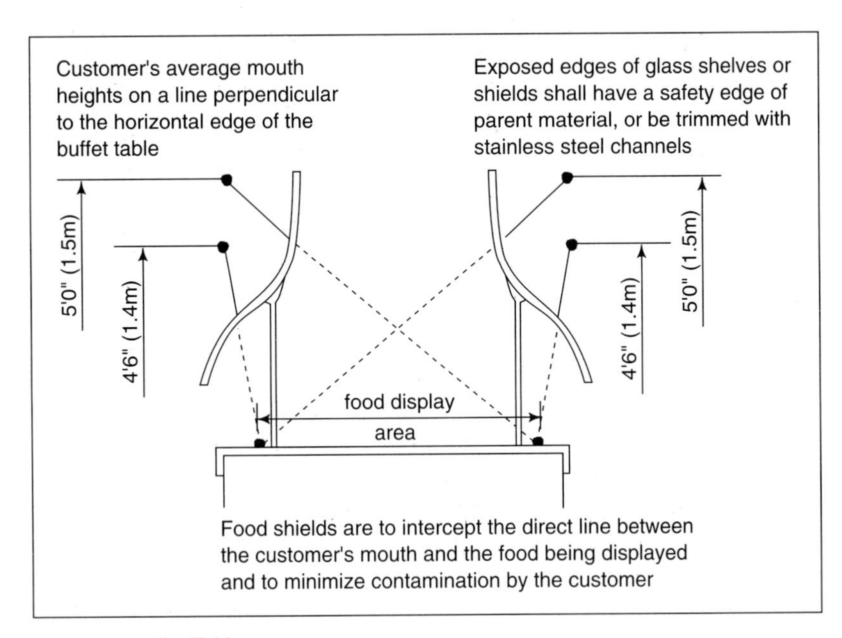
Customer's average mouth heights on a line perpendicular to the horizontal edge of the buffet table
Exposed edges of glass shelves or shields shall have a safety edge of parent material, or be trimmed with stainless steel channels
5'0" (1.5m)
4'6" (1.4m)
4'6" (1.4m)
5'0" (1.5m)
food display area
Food shields are to intercept the direct line between the customer's mouth and the food being displayed and to minimize contamination by the customer

(a) Typical Buffet Table

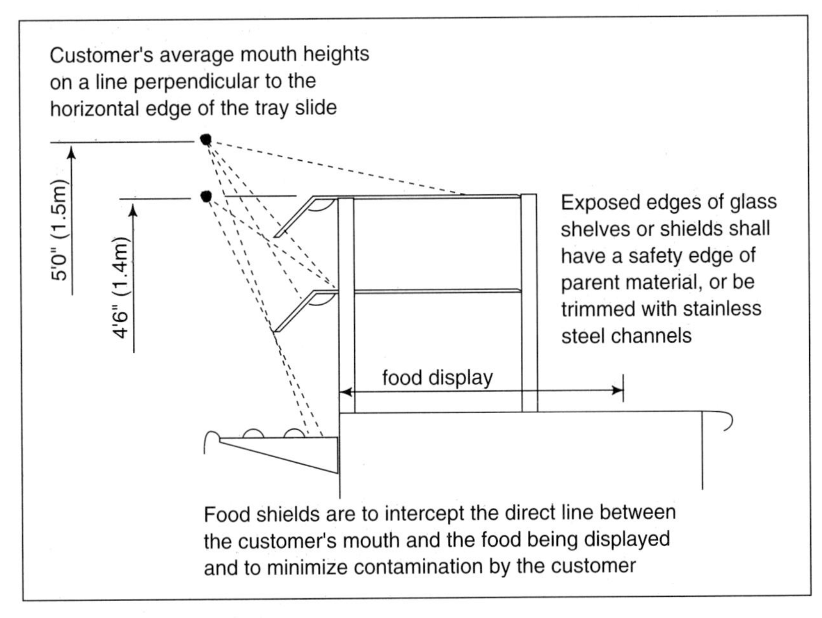
Customer's average mouth heights on a line perpendicular to the horizontal edge of the tray slide
5'0" (1.5m)
4'6" (1.4m)
Exposed edges of glass shelves or shields shall have a safety edge of parent material, or be trimmed with stainless steel channels
food display
Food shields are to intercept the direct line between the customer's mouth and the food being displayed and to minimize contamination by the customer

(b) Typical Cafeteria Counter

Figure 11.10 Standards for food shields.
Courtesy of National Sanitation Foundation International.

wall, and adequate aisle clearance so that food and supplies can be transported easily and safely on carts are but a few of the considerations to make in planning installations. Refer to Figure 11.1 for an example of an arrangement that successfully combines related pieces of equipment into a single continuous unit.

The operation of each piece of equipment must be checked many times by both the contractors and service engineers before it is ready for actual use. Full instruction for the proper operation and satisfactory performance of each piece of equipment should be given to all persons who work with it. They must know the danger signals, such as the sound of a defective motor, so that preventive measures can be taken early.

Maintenance and Replacement

The cost of care and upkeep on a piece of equipment may determine whether or not its purchase and use are justified. The annual repair and replacement of equipment should be made with consideration of the unit as a whole, and labor and operating costs should be checked constantly. If these are too high, they limit other expenditures that might promote greater efficiency in the organization. The dispersion of outlay between care and repair is important in more ways than one. Money, attention, and effort spent on care assume the continuance of the necessary equipment in use; money and the effort spent on repair are often attended by a disrupted work schedule, unpleasant stresses and strains on personnel, and sometimes definite fire hazards.

Many questions arise in regard to care and upkeep costs when equipment is selected. Are parts readily available, easily replaced, and relatively inexpensive? Does the replacement require the services of a specialist, or can a regular employee be trained to do the work? Should some piece of equipment fail to operate when needed, has provision been made so that operations may continue? Are special cleaning materials needed in caring for the equipment?

The care and repair of electrical equipment represents a major item in the maintenance cost of many foodservices. The adequate care of electric motors requires expert attention by technically trained and responsible engineers. Arrangements for such care are commonly made with the maintenance department on a contract basis, covering weekly inspection and other checkups necessary for good maintenance. Competent maintenance personnel will have a record card for every motor in the plant. All repair work, with its cost, and every inspection can be entered on the record. If this system is used, excessive amounts of attention or expense will show up, and the causes can be determined and corrected. Inspection records will also serve as a guide to indicate when motors should be replaced because of the high cost of keeping them in operating condition.

To evaluate a piece of equipment in use, an analysis of the expenditures for care and upkeep is made, and the condition of the equipment is checked to determine if the deterioration has been more rapid than it should have been under normal usage, exposure, cleaning operation, and contacts with food and heat. A factual basis for appraising upkeep costs and depreciation of equipment can be obtained by keeping careful records on each major piece. Figure 11.11 is a suggested method for keeping such records.

Successful maintenance of equipment requires definite plans to prolong its life and usefulness. Such plans place emphasis on a few simple procedures: keep the

Name of Institution:

Equipment or appliance item: Purchase Date:

Motor serial number	Motor make model	Equipment number	Location
Original cost	Estimated period of use: Months ☐ Years ☐	Make of equipment item:	Description: Type _____ Size _____ Capacity _____ Design _____
Appraisal	Motor specification: W __ V __ Amp. __ H.P. __	Estimated depreciation per Month __ Year __	Date fully depreciated _____

Appraisal		Repairs and replacements				
Date	Value	Date	Nature	By whom	Cost	Remarks

(A) Large equipment record

Name of Institution:

Name of item:	Purchase date:	Purchased from:	Location
Style ___ Size ___	Amount of original purchase _____	Quality or grade _____	Uses

Appraisal		Repairs or replacement				
Date	Value	Date	Nature	By whom	Cost	Amount on hand

(B) Small equipment record

Figure 11.11 Suggested form for recording information on each piece of (a) large equipment and (b) small equipment.

equipment clean; follow the manufacturer's printed directions for care and operation, including lubrication; keep the instruction card for each piece of equipment posted near it; stress careful handling as essential to continued use; and make needed repairs promptly. Some pertinent suggestions for the care of machines and instructions for their use include these: the assignment of the care of each machine to a responsible person; daily inspection for cleanliness and constant supervision by the manager when in use; immediate completion of even minor repairs; thorough knowledge of operating directions; regular oiling and inspections; and repairs by a competent person. Printed instructions should be easily available; directions for operation with a simple diagram should be posted by the machine; and any special warning should be printed in large or colored letters. When explaining its operation, the function and relationship of each part should be described in detail so that it is understood by the operator. There should also be a demonstration of proper use of the machine and an explanation of its value and a cost of repairs. (See Figure 8.13, Chapter 8, for the procedure for cleaning a food slicer.) Similar directions should be formulated for each piece of equipment and made into a manual for use by employees responsible for the care and cleanliness of the various items.

The operating cost is an important feature often overlooked in purchasing equipment. In some localities, electricity may be available for cooking purposes at a lower operating cost than gas, or vice versa. When all factors are considered, an electric range may be more economical in this particular instance, even though the initial cost may be more. Due consideration and investigation of the relative efficiency of various models and types are also necessary in selecting any piece of equipment.

According to the National Restaurant Association (NRA), energy management programs and the wise selection of equipment can result in savings of up to 20 percent on utility costs for the average foodservice establishment. Further information on energy conservation may be found in Chapter 9.

METHOD OF PURCHASE

The method of purchase of equipment varies somewhat with the institution. However, regardless of whether the actual placing of the order is done by the director of the foodservice, the purchasing agent, or the superintendent of the hospital, the preliminary procedures are much the same. All available data as to the needs and requirements of the institution are collected by the director, who is responsible for the smooth operation of the service and the satisfaction of its guests. Usually representatives of different firms are willing to demonstrate equipment and to give the prospective buyer information concerning the particular piece of equipment needed. Visits may be made to various institutions to see similar models in operation. After such investigations are made, and a definite idea of what is wanted is established, specifications are written and submitted to reliable firms. Written bids are then received and tabulated and a comparison is made, after which the order is placed.

The reliability of the firm from which the equipment is purchased means much to any institution. A reputable company with a record of successful operation usually strives to sell dependable merchandise of good quality. The company can be counted on to honor the guarantee and to do everything possible to keep the goodwill and confidence of the customer. In their planning and engineering departments, equipment dealers employ experts whose services are always available to the prospective customer. Years of experience and constant contact with both the manufacturing and operating units in the field enable them to be of valuable assistance. Most companies keep records of the sale, service calls, and repairs of the various pieces of equipment. In return, they deserve fair treatment and consideration from the director of the foodservice or the purchasing agent for the institution.

To be of value, a specification for equipment must be specific and definite. It covers every detail in relation to material, construction, size, color, finish, and cost, eliminating any question in the mind of either the buyer or the manufacturer as to what the finished product will be. When delivered, if the equipment does not measure up to the specified order, it need not be accepted. If the buyer is disappointed but has permitted loopholes in the specification, it must be accepted. However, most firms are so desirous of selling satisfaction that they check orders carefully with the buyer to see that everything is included before the equipment is made or delivered.

The vague and the definite specifications that follow for a particular piece of equipment illustrate the difference between the two types. Specifications may be indefinite, and yet to the casual observer all points may seem to be included. After reading the second example, one can readily see the weak spots in the first.

Vague Specifications

- Item number: xx
- Name of item: Cook's table with sink
- Dimensions: 8 ft long, 2 ft 6 in. wide, 3 ft high
- Material and construction: Top of this table to be made of heavy-gauge stainless steel with semi-rolled edge and to be furnished with one sink, 18 in. long, 24 in. wide, 12 in. deep, fitted with drain. Sink to be located 3 in. from left end of table. The under side of this table to be reinforced with channel braces. Table to be supported by four stainless steel tubular standards with adjustable feet. Stainless slatted shelf to rest on cross rails 10 in. above floor. Table to be equipped with one drawer, 24 in. long, 22 in. wide, and 5 in. deep. Drawer to be made of heavy stainless steel, reinforced on front facing. All joints of this drawer to be welded, and drawer equipped with ball-bearing drawer slides. This drawer to be fitted with a white metal handle.
- Price: $. . . . . . .

Definite Specifications

- Item number: xx
- Name of item: Cook's table with sink
- Dimensions: 8 ft long, 2 ft 6 in. wide, 3 ft high
- Material and construction: Top of this table to be made of No. 14 gauge. No. 4 grind, No. 302 stainless steel with all edges turned down 1½ in., semi-rolled edge. All corners to be fully rounded bull-nose construction and integral with top. Top of this table to be fitted with one sink, 18 in. long, 24 in. wide, and 12 in. deep, with all corners and intersections fully rounded to a 1-in. radius. All

joints to be welded, ground smooth, and polished. Bottom sloped to drain in center. Sink to be located 3 in. from left end of table, 3 in. from each side. Sink to be equipped with 2-in. white metal drain with plug and chain complete.
The under side of this table top to be properly reinforced and braced with 4-in. No. 14-gauge stainless steel channel braces welded on. Four tubular leg standards to be welded to these channel cross braces. Standards to be made of seamless stainless steel tubing 1⅝ in. outside diameter, cross rails and braces of the same material, fitted and welded together. Resting on these cross rails and braces will be a slatted bar shelf elevated 10 in. above floor. Slats to be made of No. 16 stainless steel, No. 4-grind, welded to 2-in. No. 16 stainless steel supports. Slats 2 in. wide and bent down at ends and formed to fit over cross rails. Slatted shelf to be built in two removable parts of equal length. Leg standards to be fitted with adjustable inside threaded, stainless-steel, tubular, closed, smooth-finish feet.
Table to be equipped with one drawer, 24 in. long, 22 in. wide, and 5 in. deep. Drawer to be made of No. 16-gauge, No. 4-grind stainless steel throughout, reinforced on front facing with No. 14-gauge, No. 4-grind stainless steel. All joints of this drawer to be welded, ground, and polished. Each drawer to be equipped with nontilting, easy-glide roller-bearing drawer slides, and all metal tracks welded to under side of table top. This drawer to be fitted with a polished white metal pull handle.

- Price: F.O.B $. . . . .
- Delivery date: Not later than. . . . .

When purchasing electrically operated equipment, it is essential that exact electrical specifications be given to the manufacturer at the time the order is placed. A motor is wound to operate on a certain voltage current, and when set up to operate on another, it may run more slowly or more rapidly than was intended, causing its output to be greater or less than its rated horsepower. There is danger of overheating and a breakdown of insulation, which will result in short circuits and the necessity for motor repairs or replacements. A three-phase motor is desirable because the absence of brushes lessens the maintenance problems. Motors of less than 1 horsepower may be used equally well on 110- or 220-volt currents but motors of larger horsepower should be operated on a 220-volt current. Manufacturers now use ball-bearing motors, fully enclosed and ventilated, which eliminate the need for frequent oiling. Most motors are built especially for the machines they operate. They must be adequate in power to easily carry the capacity loads of the machines.

SELECTION OF SOME BASIC ITEMS

An analysis of the basic considerations discussed thus far helps to determine whether the selection of certain items of kitchen equipment is justified and gives attention to the mechanics of buying. Standards for various types of equipment have been mentioned. The problem of selection is so important and errors are so costly that major characteristics to consider in the selection of certain types of items are given in Appendix B. No attempt is made to evaluate or identify equipment by trade name. The buyer may need to make a selection between the products of several competitive manufacturers or jobbers, each of whom may have quality products but with a wide variance in some details. All equipment should be a sound investment for the operator, be easily cleaned, safe to operate, and accomplish the work for which it was designed. Wise selection can be made only after an exhaustive study of all available data and observation of similarly installed equipment have been accomplished.

Manufacturers' specification sheets, brochures, and catalogs, trade shows, current professional and trade journals and magazines, and the representatives of the manufacturing companies are the best sources of up-to-date information on specific items. Special features may be changed fairly often so that detailed information on certain models is soon outdated in a publication like this one.

Some points for consideration when selecting foodservice equipment, other than price, cost of operation, and maintenance, are included in Appendix B to help acquaint the reader with possible features and variations of certain items. The availability of utilities and other factors might predetermine some decisions; for example, the choice between an electric or gas-heated range presents no problem if the advantages of one source of heat over the other are evident in the particular situation. Instead, the problem becomes one of a choice between various models manufactured by several different firms. Space permits only a limited amount of basic information on certain fundamental items. It is expected that supplementary material will be kept up to date and made available in library or office files for students and foodservice operators.

Cooking Equipment

This equipment must conform to requirements for material, construction, safety, and sanitation established by groups such as the American Standards Association, American Gas Association, National Board of Fire Underwriters, Underwriters Laboratories, Inc., American Society of Mechanical Engineers, and NSFI. One should be sure that parts are replaceable and service is available for all items selected, and should also give consideration to original and operating costs, effectiveness in accomplishing the task to be done, and the time and skill required for ordinary maintenance. The life expectancy requirement depends somewhat on the situation, but the selection of durable high-quality equipment is usually economical.

Electric, Gas, and Steam Equipment. *Electrically heated* cooking equipment designed for alternating or direct current of specified voltage; rating required expressed in watts or kilowatts (1,000 watts = 1 kilowatt) per hour; wiring concealed and protected from moisture; switches plainly identified; thermostatic heat controls; flues not required for electric cooking equipment but the usual hood or built-in ventilating system necessary to remove cooking vapors and odors.

Gas-fired cooking equipment designed for natural, manufactured, mixed, or liquefied petroleum fuel; adapted to given pressures; rating requirement expressed in

British thermal units (Btus) per hour; individual shut-off valve for each piece of gas equipment; manifolds and cocks accessible but concealed; removable burners; automatic lighting with pilot light for each burner; thermostatic heat controls; gas equipment vented through hood or built-in ventilator instead of kitchen flue to exhaust combustible gases.

The most commonly used pieces of gas- and electric-heated cooking equipment are ranges, griddles, broilers, fryers, tilting frypans, and ovens.

Steam-heated cooking equipment includes steam-jacketed kettles, cabinet steamers, steam tables, and combination ovens, which combine steam with gas or electric heat. In a steam cabinet, steam is injected into a closed cavity where it comes into direct contact with the food. Under pressure steam has a higher temperature than nonpressurized steam, thereby allowing quicker cooking times. A *low-pressure steamer* utilizes steam at 5 psi (pounds per square inch), which converts to approximately 227°F. The standard pressure in a *high-pressure steamer* is 15 psi or 250°F. A batch of peas cooked in the low-pressure steamer will take 8 minutes and in the high-pressure steamer 1 minute. A third type of steam cabinet is the *pressureless convection steamer* in which steam enters at atmospheric pressure (0 psi) or 212°F but is convected or circulated continuously over the food. This constant movement of steam shortens cooking time to less than that of low-pressure static steamers. Specifics about these types of cooking equipment are included in Appendix B.

Noncooking Equipment

Power-Operated Equipment. Modern foodservices depend on motor-driven machines for rapid and efficient performance of many tasks. Safety precautions are necessary. Capacity charts for all types of machines are available from manufacturers and distributors. Motors, built according to capacity of machine, must carry peak load easily; specify voltage, cycle, and phase; a three-phase motor is usually used for ¾ horsepower or larger; sealed-in motors and removable parts for ease of cleaning.

Power-operated noncooking equipment includes mixers, choppers, cutters, slicers, vertical cutter mixers, refrigerators, freezers, dish and utensil cleaning equipment, waste disposals, and transport equipment. Among the more common pieces of nonmechanical kitchen equipment are tables, sinks, storage cabinets, racks, carts, scales, cooking utensils, and cutlery. More detailed information about each of these types of noncooking equipment is included in Appendix B.

The most commonly used types of serving equipment are counters, utensils, dispensers, coffee makers, and mobile serving carts. These are also discussed in Appendix B.

DINING ROOM FURNISHINGS

A dining area that is attractive and appealing does much to make patrons feel comfortable and adds to the enjoyment of the food they are served. The foodservice director may be responsible for the selection of some of the furnishings, especially for the dinnerware, tableware, and glassware. The services of an interior designer or decorator may be employed, however, to help create the desired atmosphere through selection of the appropriate style and type of tables and chairs, window treatment, and a color scheme that will coordinate all furnishings into a harmonious effect.

Basic information needed for the wise selection of dinnerware (dishes), tableware (knives, forks, spoons), glassware, and table covers is presented in Appendix B. Specialized assistance likely will be sought for the purchase of furniture, drapes, curtains, and other furnishings. All furnishings should be pleasing, durable, serviceable, and easy to maintain.

Dinnerware

Many types of material are used in making dishes for today's foodservice market, including china, glass, and melamine or other plastic ware or combinations of other materials kept secret by their manufacturers. Dinnerware suitable for foodservices varies with the type of service given. A club or luxury restaurant may wish to use fine china such as that found in the homes of its patrons. In contrast, a school foodservice needs more durable ware to withstand the hazards attendant on its use. Fast-food establishments usually find disposable dinnerware best fills their needs.

Tableware

The most satisfactory type of eating utensils for institutions is that which has been designed and made especially for heavy-duty use. Such ware falls into two classes: *flatware* of the usual array of knives, forks, and spoons; and items such as teapots, sugar bowls, pitcher, and platters, known as *hollow ware* if made of silver. All must be durable and serviceable and, at the same time, attractive in line and design. Silver and stainless steel tableware are the two types used, and the decision to select one or the other will depend largely on the type of foodservice, the tastes of the clientele served, and the amount of money available for this expenditure and upkeep.

Glassware

Glassware is a major item of purchase for dining room furnishings for food services, because it is easily broken and replacement is frequent. It is usually more economical to purchase good quality glassware than inexpensive types.

Table Covers

One other furnishing to consider is the type of covering, if any, to be used on the dining tables, or trays if that type of service is used. Many tabletop surfaces are attractive, durable, and suitable for use without a cover. Simplicity and informality in dining have made this custom popular, and it does reduce laundry costs.

For many people, much of the charm of a foodservice is conditioned, if not determined, by the use of a clean tablecloth of good quality, freshly and carefully placed. Paper napkins and placemats and plastics have replaced cloth in many foodservices for convenience and economy. Whatever the choice, the cover should be of a type and color appropriate for the facility, contribute to the total atmosphere of the room, and be harmonious with the dishes to be used.

SUMMARY

Prospective foodservice managers, as well as those already employed in the field, should have a "working" knowledge of equipment and furnishings—construction, materials suitable for various uses, something of the sizes or capacities available, and how to relate that information to meeting the needs of the individual foodservice. Wise selection and proper care of the many items that must be provided for efficient operation of the foodservice should result in economies for the organization and a satisfied working crew because they have been supplied with the correct tools to accomplish their task.

REVIEW QUESTIONS

1. List the many factors that affect decisions regarding equipment selection.
2. If labor cost is high in a particular area, what type of equipment should be given strong consideration?
3. In an assembly/serve foodservice system, what pieces of equipment are most warranted?
4. Discuss the advantages of modular equipment.
5. Discuss the factors that determine the size and capacity of equipment chosen.
6. What are the most common materials used in the manufacture of foodservice equipment and for what are they usually used?
7. List and briefly explain the desired characteristics of a metal chosen for foodservice equipment.
8. Describe how the gauge of metals is determined in the United States.
9. Identify what should be included in a specification for a piece of foodservice equipment.
10. What is the role of the National Sanitation Foundation International regarding foodservice equipment and furnishings?
11. List and briefly explain the various types of ovens now in use in foodservice operations.

SELECTED REFERENCES

A guide to purchasing food, disposables, tabletop items, supplies, and equipment. Voice of Foodservice Distr. 1993; 29(14; special issue):125.

Cheney, K.: Set a sensational tabletop. Restaurants and Institutions. 1993; 103(24):105.

Durocher, J.: Keeping your cook with properly maintained refrigeration. Restaurant Business. 1994; 93(15);172.

Durocher, J.: Steam esteem. Restaurant Business. 1994; 93(12):158.

Frable, F.: A checklist: What to look for when evaluating equipment. Nation's Rest. News. 1994; 28(27):94.

Frable, F.: Selecting right wall finish solves maintenance problems. Nation's Rest. News. 1994; 28(45):54.

Frable, F.: Storage equipment helps manage shrinking space. Nation's Rest. News. 1994; 28(44):34.

Frable, F.: Strange business: How kitchen equipment is sold. Nation's Rest. News. 1994; 28(37):183.

Friedland, A.: Warewashers. Food Mgmt. 1995; 30(4):42.

Koss-Feder, L.: Equipment heads menu for f & b profits. Hotel and Motel Mgmt. 1994; 209(8):21.

Lorenzini, B.: Equipment innovations. Restaurants and Institutions. 1994; 104(6):131.

Lorenzini, B.: New from abroad: Multienergy cooking tunnel. Restaurants and Institutions. 1994; 104(21):46

MacDonald, J.: New & improved: The latest advances in foodservice technology. Hotel and Motel Mgmt. 1993; 208(11):40.

Patterson, P.: New products waiting in the wings. Nation's Rest. News. 1993; 27(36):86.

Patterson, P.: Unique equipment surfaces at NAFEM. Nation's Rest. News. 1993; 27(39):100.

Standard 2 Food Service Equipment. Ann Arbor, Mich.: National Sanitation Foundation International, 1988.

Varga, J.: Gas tilting skillets. Cooking for Profit. 1995; 526:14.

4

Organization and Administration

12

Designing and Managing the Organization

OUTLINE

Society is composed of a variety of interdependent, overlapping organizations. To function effectively, organizations must be properly designed and managed by competent managers. Organizations need competent managers to be able to meet their objectives both efficiently and effectively. The return that organizations realize from human resources is determined, in large part, by the competence of their managers.

This chapter examines theories of management, the functions performed by a manager, the requisite managerial skills, management activities and roles, the tools of management, and the principles of organizational structure.

Historical theories of management have contributed a great deal to modern management theories. The growing complexity of organizations today results in a greater need to examine them as a whole. Using the systems approach, a manager recognizes that the organization as a whole is greater than the sum of its parts. The systems manager sees the contributions of each part to the whole system and that a change in one part will have an impact on other parts of the system. This approach allows a manager to diagnose and identify reasons for the occurrence of a situation. The contingency approach leads management to apply different basic guidelines to leading, motivating, and organizing depending on the particular situation.

To accomplish the common managerial functions of planning, organizing, staffing, directing, coordinating, reporting, and budgeting, a manager engages in a variety of activities that can be grouped into three basic categories: interpersonal relationships, information processing, and decision making. The three categories can be further divided into 10 observable working roles. In this chapter, the functions and roles of managerial work are explored. The skills required to perform these various functions and roles are also described.

The first two functions of management, planning and organizing, are discussed in greater detail in this chapter. (The remaining functions are explored further in subsequent chapters.) The role of strategic planning, the development of a mission statement, and the steps necessary to develop the framework of the organization are included in the discussion. The various types of organizational structures found in foodservice operations are described.

The chapter concludes with a discussion of some important management tools. The organization chart is a map of the organization. The organization manual with its job descriptions, job specifications, and job schedules goes even further as a model of the organization.

KEY CONCEPTS

1. The four important and predominant theories of management are classical or traditional, human relations, management science or operations research, and the modern or systems approach.
2. Classical management theory contributed a number of principles for the successful division, coordination, and administration of work activities: scalar principle, delegation, unity of command, functional principle, and the line and staff principle, among others.

3. Important systems theory concepts are feedback, hierarchy of systems, interdependency, and wholism.
4. The basic functions performed by managers are planning, organizing, staffing, directing, coordinating, reporting, and budgeting.
5. The planning function involves a sequence of steps including the writing of a vision, philosophy, slogan, mission, strategic plans, intermediate plans, policies, procedures, schedules, and rules and then implementing, following up, and controlling the plans.
6. The four basic steps necessary to develop the framework of an organization's structure are to determine and define objectives, analyze and classify work to be done, describe in detail work to be done, and determine and specify the relationship between and among workers and management.
7. Organizations may be structured on a line, line and staff, or functional basis.
8. Managers need varying degrees of three skills (technical, human, and conceptual skills) depending on the level in the hierarchy at which the manager is working.
9. Managerial roles may be classified as interpersonal (figurehead, leader, and liaison), informational (monitor, disseminator, and spokesperson), and decisional (entrepreneur or initiator, disturbance handler, resource allocator, and negotiator).
10. Useful mechanical or visual tools of management are organization charts, job descriptions, job specifications, and work schedules.

THEORIES OF MANAGEMENT

Many of the challenges that managers faced many years ago are the same as those faced by today's managers. For example, increasing worker productivity, decreasing production costs, maintaining employee motivation and morale, and meeting the challenges of stiff competition are just some of the issues that have persisted through the years.

People have always been of prime importance in all thinking about management. What has changed through the years is views on why people work and how they are best managed. In some cases these changes have occurred because the historical conditions dictated the need for such changes in thinking.

Modern management theories are both a reflection and a reaction to past management theories. There has never been nor is there now, one best theory of management. Each has its own particular worthwhile applications as well as some limitations.

Classical

Classical theory is a grouping of several similar ideas that evolved in the late 1800s and early 1900s. Pioneers in this theory were Frederick W. Taylor, who was known as the "father of scientific management," Max Weber, Frank and Lillian Gilbreth, and Henri Fayol. The basic tenets of classical theory were that (1) there is one best way to do each job; (2) there is one best way to put an organization together; and (3) the organization should be arranged in a rational and impersonal manner. Fayol's (1949) principles encompass these tenets:

1. *Division of work:* This is essential for efficiency, and specialization is the most efficient way to use human effort.
2. *Authority and responsibility:* Authority is the right way to give orders and obtain obedience, and responsibility is the natural result of authority.
3. *Discipline:* The judicious use of sanctions and penalties is the best way to obtain obedience to rules and work agreements;
4. *Unity of command*: This specifies that each person should be accountable to only one superior.
5. *Unity of direction:* This specifies that all units should be moving toward the same objectives through coordinated and focused effort.
6. *Subordination of individual interest to general interest:* The interests of the organization should take priority over the interest of individuals.
7. *Remuneration of employees:* Pay and compensation should be fair for both employee and organization.
8. *Centralization:* Subordinates' involvement through decentralization should be balanced with managers' final authority through centralization.
9. *Scalar chain:* In a scalar chain, authority and responsibility flow in a direct line vertically from the highest level of the organization to the lowest.
10. *Order:* People and materials must be in the appropriate places at the proper time for maximum efficiency.
11. *Equity:* All employees should be treated equally to ensure fairness.
12. *Stability of personnel:* Employee turnover should be minimized to maintain organizational efficiency.
13. *Initiative:* Workers should be encouraged to develop and carry out plans for improvements.
14. *Esprit de corps:* Managers should promote a team spirit of unity and harmony among employees.

Classical theory continues to have great relevance and application to basic managerial problems, but it has been criticized as being too mechanistic and not recognizing the differences in people and organizations. Not all people are motivated by economic rewards and not all organizations can take the same approach to managing their employees.

Human Relations

The **human relations theory** evolved during the 1920s through the 1950s from the effort to compensate for some of the deficiencies of the classical theory. Where classical organization advocated focus on tasks, structure, and authority, human relations theorists have introduced the behavioral sciences as an integral part of organization theory. They view the organization as a social system and recognize the existence of the informal organization, in which workers align themselves into social groups within the framework of the formal organization. Many human relations theorists hold that employee participation in management planning and decision making yields positive effects in terms of morale and productivity. This theory is discussed in more detail in Chapter 14.

Management Science/Operations Research

The *management science theory* of management combines some of the ideas from the classical and human relations theories. It emphasizes research on operations and the use of quantitative techniques to help managers make decisions. Advances in computer technology have made possible the wide variety of mathematical models and quantitative tools that are integral to this approach to management.

One extension of this theory has been the development of **management information systems** (MIS). MIS include such tools as linear programming, queuing models, and simulation models to facilitate decision making. The **program evaluation and review technique** (PERT) is another tool for the effective planning and control functions of management.

Modern Management Theories

Modern management theories have evolved because of the complex nature of today's organizations. Ideas from classical, human relations, and management science have been integrated into the modern theories. The understanding that organizations and people are complex entities with differing motives, needs, aspirations, and potentials has led to the widely held belief that there can be few static and universal management principles. This complex view is evident in the two modern management theories: the systems theory and contingency theory.

Systems Approach. The systems approach, introduced in Chapter 2, is defined as a set of interdependent parts that work together to achieve a common goal. Several important concepts from the systems approach apply to organizations. Being predominantly open systems, human organizations interact with various elements of their environment. (For example, a hospital dietary department interacts with many external groups such as patients, customers, medical staff, hospital administration, and some regulatory agencies. The department, in turn, affects the external groups with which it interacts.)

Organizations tend toward a dynamic or moving equilibrium. Members seek to maintain the organization and to have it survive. They react to changes and forces, both internal and external, in ways that often create a new state of equilibrium and balance. *Feedback* of information from a point of operation and from the environment to a control center or centers can provide the data necessary to initiate corrective measures to restore equilibrium.

Organizations and the world of which they are a part consist of a *hierarchy of systems.* Thus, a corporation is composed of divisions, departments, sections, and groups of individual employees. Also, the corporation is part of larger systems, such as all the firms in its industry, firms in its metropolitan area, and perhaps an association of many industries such as the National Restaurant Association (NRA) or the American Hospital Association.

Interdependency is a key concept in systems theory. The elements of a system interact with one another and are interdependent. Generally, a change in one part of an organization affects other parts of that organization. Sometimes the interdependencies are not fully appreciated when changes are made. A change in organizational structure and work flow in one department may unexpectedly induce changes in departments that relate to the first department.

Systems theory contains the doctrine that the whole of a structure or entity is more than the sum of its parts. This is called *wholism.* The cooperative, synergistic working together of members of a department or team often yields a total product that exceeds the sum of their individual contributions.

Systems theory helps organize a large body of information that might otherwise make little sense. It has made major contributions to the study of organization and management in recent years. Systems theory aids in diagnosing the interactive relationships among task, technology, environment, and organization members.

In contrast to the classical models of organization, the systems approach has shown that managers operate in fluid, dynamic, and often ambiguous situations. The manager generally is not in full control of these situations. Managers must learn to shape actions and to make progress toward goals, but they know that results are affected by many factors and forces.

Systems theory has become popular because of its apparent ability to serve as a universal model of systems, including physical, biological, social, and behavioral phenomena.

Contingency Approach. The **contingency approach** holds that managerial activities should be adjusted to suit the situations. Factors within the situation, such as characteristics of the workforce, size and type of organization, and its goals, should determine the managerial approach that is used. The contingency approach is dependent on seeing the organization as a system and emphasizes the need for managers to strategize based on the relevant facts. Important principles of the contingency approach are as follows: Individual motivation may be influenced by factors in the environment; managers must adjust their leadership behavior to fit thc particular situation; and the structure of the organization must be designed to fit the organization environment and the technology it uses.

FUNCTIONS OF MANAGEMENT

The basic purpose of management has been recognized as the leadership of individuals and groups in order to accomplish the goals of the organization. Henri Fayol, a French mining engineer/ manager, recognized that managerial undertakings require planning, organization, command, coordination, and control. Luther Gulick (1937) developed the following seven major functions of management: planning, organizing, staffing, directing, coordinating, reporting, and budgeting. Various combinations of these functions with some modifications, deletions, and additions are found in modern management texts.

POSDCORB, the acronym created from the names of the seven functions described by Gulick, is still widely accepted as describing the basic framework of the manager's job. Some disagreement does exist as to whether these functions are common to all levels of management. Others believe that even more functions should be included. Leading, actuating, activating, motivating, and communicating are concepts often fitted into the POSDCORB framework. That there is a degree of overlapping is evident in the functions themselves and in the efforts to classify them.

Planning

The **planning** function, described by Gulick in 1937 and still relevant today, involves developing in broad outline the activities required to accomplish the objectives of the organization and the most effective ways of doing so. Planning is a basic function and all others are dependent on it. The objective of planning is to think ahead, clearly determine objectives and policies, and select a course of action toward the accomplishment of the goals. Day-to-day planning of operational activities and short- and long-range planning toward department and institution goals are part of this function. Overall planning is the responsibility of top management, but participation at all levels in goal setting and development of new plans and procedures increases their effectiveness.

The first steps in the planning process are to develop a *vision,* a *philosophy* or *core values statement,* a *slogan,* and a *mission statement.* Each of these should be simple, easily understood, attainable, measurable, desirable, and energizing. Many commercial foodservices print one or more of these statements on their menus or other printed materials to communicate their business philosophy to their customers.

It is desirable to have each of these developed cooperatively by all members of the organization, not just by top management. The vision is the organization's view of the future; the philosophy contains the organization's set of core values for the attainment of that vision; the slogan is a short, memorable statement of "who we are"; and the mission statement is the summary of the organization's purpose, customers, product, and service. An example of a hospital foodservice's mission statement might be:

> The Department of Hospitality Services at Malibu Hospital is a multifaceted, service-oriented department that provides comprehensive nutrition care, foodservices, and educational programs for the patients, employees, visitors, and members of the community. All programs will be conducted with highest standards of quality and service within budgetary limitations.

Within multilevel organizations, department mission statements must be written based on the mission statement of the organization.

Once these basic planning statements have been written, *strategic planning* can take place. Usually accomplished at the top levels of the management hierarchy, strategic planning involves making some decisions based on environmental conditions, competition, forecasts of the future, and the current and anticipated resources available (Figure 12.1). Strategic plans are usually done every 3 to 15 years depending on the changing nature of the existing conditions. Regardless of how often they are developed, they are reviewed on a regular basis. The strategic planning docu-

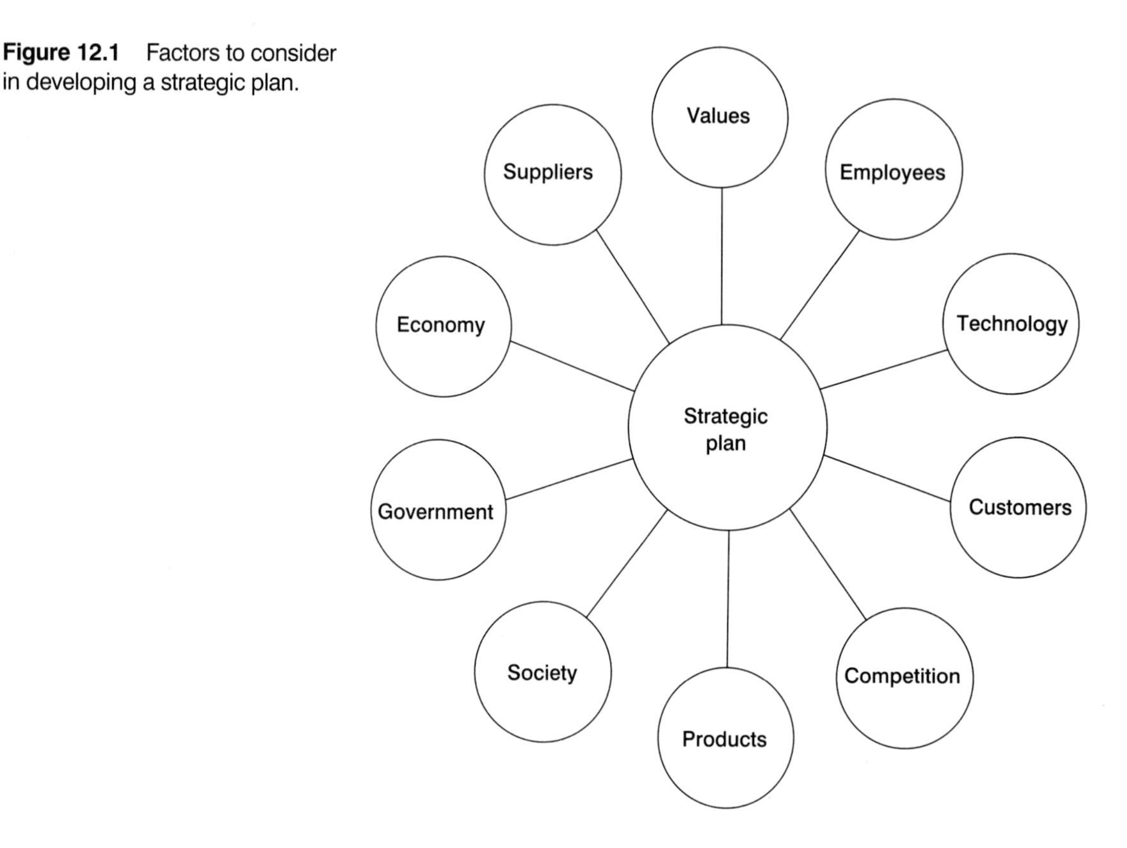

Figure 12.1 Factors to consider in developing a strategic plan.

ment serves as the foundation for *intermediate plans,* which cover a period of from 1 to 3 years. Intermediate planning is based on the strategic plan and begins with the development of *policies.*

Policies are decision-making guides that are written to ensure that all actions taken by organizational members are consistent with the organization's strategy and objectives. Once strategic and intermediate plans have been written, corresponding *operating plans* and budgets may be developed. Operating plans lay out the plans for the current period of time, usually 1 year or less. They are designed to provide the framework for implementing the strategic and intermediate plans at the departmental level. Operating plans specify the *procedures* to be used, the schedule to be followed, and the budget to stay within.

Procedures are detailed guidelines for planned activities that occur regularly. These are sometimes called *standard operating procedures* (SOPs). Many organizations have manuals containing their policies and procedures, which serve as helpful guides for managers and new employees. Schedules are guides for the actual timing of activities. In addition to the types of schedules shown later in this chapter, a number of formal techniques can be used to develop activity schedules. PERT, mentioned earlier, is one such management tool and is discussed in Chapter 15.

If policies and procedures do not change over time, they are referred to as *standing plans.* A plan to be used only once or infrequently is termed a *single-use plan.* Special functions or catered events often call for a single-use plan in a foodservice operation.

Rules are simply written statements of what must be done. For example, a foodservice will have a written procedure that covers correct hand-washing techniques. One of the rules in this area would be that all employees will thoroughly wash their hands when returning to the production area after a break.

To be effective, plans must be implemented, followed up on, and controlled. To implement a plan, a manager makes decisions that initiate the actions called for in the plans. In the follow-up stage the manager compares actual outcomes with those that were planned. Corrective action may be required when actual does not match planned outcome. The most common form of control is feedback, in which the manager monitors performance and takes any corrective action required.

The sequence of the planning process is diagrammed in Figure 12.2. Planning is a continuous process and requires that management be diligent in conducting periodic reviews of all plans to ensure that they are still appropriate as conditions change.

Organizing

Organizing includes the activities necessary to develop the formal structure of authority through which work is subdivided, defined, and coordinated to accomplish the organization's objectives. The organizing function identifies activities and tasks, divides tasks into positions, and puts like tasks together to take advantage of special abilities and skills of the workers and to use their talents effectively. Perhaps the chief function of the organizing process is the establishment of relationships among all other functions of management.

Organizational Structure. An organization is a system, having an established structure and conscious planning, in which people work and deal with one another in a coordinated and cooperative manner for the accomplishment of common goals.

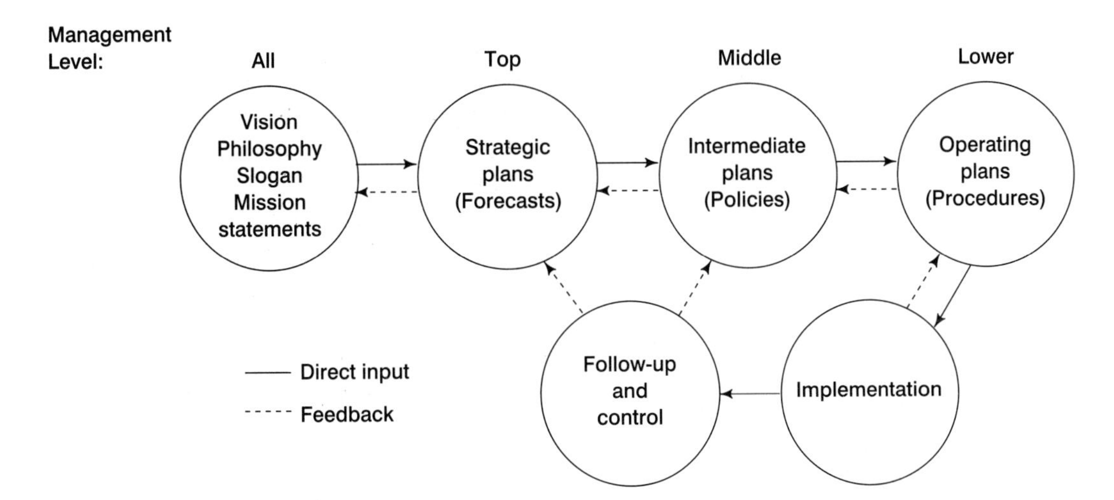

Figure 12.2 The planning process.

The formal organization is the planned structure that establishes a pattern of relationships among the various components of the organization. The informal organization refers to those aspects of the system that arise spontaneously from the activities and interactions of participants.

Whenever several people work together for a common goal, there must be some form of organization; that is, the tasks must be divided among them and the work of the group must be coordinated. Otherwise, there may be duplication of effort or even work at cross purposes. Dividing the work and arranging for coordination make up the process of organization, and once that is completed the group may be described as "an organization."

Certain steps are necessary in developing the framework of an organization's structure if goals of an enterprise are to be accomplished and the workers' talents developed to their fullest potential. These steps may be summarized as follows:

1. *Determine and define objectives:* The purpose of every organization dealing with personnel is to accomplish, with the efforts of people, some basic purpose or objective with the greatest efficiency, maximum economy, and minimum effort, and to provide for the personal development of the people working in the organization. Specifically, a foodservice has as its goal the production and service of the best food possible within its financial resources. It is important that these objectives and the plans and policies for their achievement be presented in writing and understood by all responsible.

2. *Analyze and classify work to be done:* This is accomplished by dividing the total work necessary for the accomplishment of overall goals into its major parts and grouping each into like, or similar, activities. Examination of the work to be done will reveal tasks that are similar or are logically related. Such classification may be made by grouping activities that require similar skills, the same equipment, or duties performed in the same areas. There are no arbitrary rules for grouping. In a foodservice, the activities could be grouped as purchasing and storage preparation and processing, housekeeping and maintenance, and service and dishwashing. Each of these groupings might be broken into smaller classifications, depending on the type and size of the enterprise. With the increasing complexity of foodservice organizations and the trend toward centralization of certain functions, the organizational structure takes on new dimensions and must consider the total management structure as well as the organization of its individual units.

3. *Describe in some detail the work or activity in terms of the employee:* This step is discussed in more detail later in this chapter under "Job Description."

4. *Determine and specify the relationship of the workers to each other and to management:* The work should be grouped into departments or other organizational units, with responsibility and authority defined for each level. It is generally understood that each person assigned to a job will be expected to assume the responsibility for performing the tasks given him or her and that each person will be held accountable for the results. However, persons can be accountable only to the degree that they have been given responsibility and authority. Responsibility without authority is meaningless. An assignment should be specific and in writing. For an

organizational structure to become operational, of course, requires the selection of qualified personnel, provision of adequate financing and equipment, and a suitable physical environment. No successful organization structure remains static. It must be a continuing process that moves with changing concepts within the system and with changing conditions in its environment.

Application of the principles of organization and administration to a specific situation should precede any attempt at the operation of a foodservice unit. A detailed plan may be outlined for use as a guide in initiating a new foodservice of any type or for reorganizing one previously in operation.

Types of Organizations. Two types of authority relationships most often found in foodservice operating systems are line and line and staff. Large, complex operations may be organized on a functional basis.

Line Authority. In the *line organization,* lines of authority are clearly drawn, and each individual is responsible to the person ranking above him or her on the organizational chart. Thus, authority and responsibility pass from the top-ranking member down to the lowest in rank. In such an organizational structure, each person knows to whom he or she is responsible and, in turn, who is responsible to him or her.

The organizational line structure can grow in two directions, vertically or horizontally. Vertical growth occurs through the delegation of authority, in which the individual at the top delegates work to her immediate subordinates, who redelegate part or all of their work to their subordinates, and so on down the line. For example, the director of a growing cafeteria operation may add an assistant manager, thus creating another level in the chain of command. When the distance from the top to bottom becomes too great for effective coordination, the responsibilities may be redistributed horizontally through departmentalization. In establishing departments, activities are grouped into natural units, with a manager given authority and accepting responsibility for that area of activities. There are several ways of dividing the work, but, in foodservices, the usual way to do it is by function, product, or location. The work may be divided in a restaurant into production, service, and sanitation; in dietetics, by administrative, clinical, and education; in a central commissary, by meat, vegetable, salad, and bakery departments; or by individual schools in a multiunit school foodservice system.

Advantages of the line organization include expediency in decision making, direct placement of responsibility, and clear understanding of authority relationships. A major disadvantage is that the person at the top tends to become overloaded with too much detail, thus limiting the time that he or she can devote to the planning and research necessary for development and growth of the organization. There is no specialist to whom one can turn for help in the various areas of operation.

Staff Authority. As an enterprise grows, the line organization may no longer be adequate to cope with the many diversified responsibilities demanded of the person at the top. Staff specialists, such as personnel director, research and development specialist, and data processing coordinator, are added to assist the lines in an advisory capacity. The line positions and personnel are involved directly in accomplishing the work for which the organization was created and the staff advises and supports the line in a **line and staff** organization. A staff position also may be an assistant who serves as an extension of a line officer. The potential for conflict exists between line and staff personnel if the lines of authority are not clearly understood. For example, if a staff specialist recommends a change in procedure, the order for the change would come from the line personnel. Friction may arise if a strong staff person tries to overrule the manager or if the manager does not make full use of the abilities of the staff.

Functional Authority. Some researchers include *functional authority* under the staff-type organization, but others consider it to be a distinct type in itself. Functional authority exists when an individual delegates limited authority over a specified segment of activities to another person. In a multiunit foodservice company, for example, the responsibility for purchasing or for menu planning and quality control may be vested in a vice president who then has authority over that function in all units.

Staffing

Staffing is the personnel function of employing and training people and maintaining favorable work conditions. The basic purpose of the staffing function is to obtain the best available people for the organization and to foster development of their skills and abilities. Chapter 13 discusses this important management function in more detail.

Directing

Directing requires the continuous process of making decisions, conveying them to subordinates, and ensuring appropriate action. Delegation of responsibility is essential to distribute workloads to qualified individuals at various levels. Those delegating responsibility should not do so without detailed instructions as to what is expected of the subordinate and the necessary authority to carry out the responsibilities. If a subordinate is not given sufficient authority, the job is merely assigned, not delegated.

A very important part of the directing function is the concern with employees as human beings. Studies have shown that most people work at only 50% to 60% efficiency, and some investigators place this figure as low as 45%. The alert manager is aware that through careful, intelligent guidance and counseling and by effective supervision, the worker's productivity may be increased as much as 20%. This may mean the difference between financial success and failure of an enterprise.

Coordinating

Coordinating is the functional activity of interrelating the various parts of work so they flow smoothly. To function effectively, organizations must be properly designed. Division of work is usually accomplished through departmentalization, or specialization by function, product, client, geographical area, number of persons, or time. Different methods of coordination are required for different types of departmentalization. As stated in Chapter 2, management's role in the systems approach is one of coordinating. The manager must recognize the need of all the parties and make decisions based on the overall effect on the organization as a whole and its objectives.

Reporting

Reporting involves keeping supervisors, managers, and subordinates informed concerning responsibility through records, research, reports, inspection, and other methods. Records and evaluations of the results of work done are kept as the work progresses in order to compare performance with the yardstick of acceptability.

Budgeting

Budgeting includes fiscal planning, accounting, and controlling. Control tends to ensure performance in accordance with plans and is a necessary function of all areas of foodservice. This necessitates measuring quantity of output, quality of the finished product, food and labor costs, and the efficient use of workers' time. Through control, standards of acceptability and accountability are set for performance. A good control system prevents present and future deviation from plans and does much to stimulate an employee to maintain the standards of the foodservice director. The budgeting function should be one of guidance, not command. It is concerned with employees as human beings with interests to be stimulated, aptitudes and abilities to be directed and developed, and comprehension and understanding of their responsibilities to be increased (see Chapter 16).

SKILLS OF MANAGERS

The most widely accepted method of classifying managerial skills is in terms of the three-skill approach initially proposed by Robert L. Katz (1974). He identified *technical, human,* and *conceptual* skills as those that every successful manager must have in varying degrees, according to the level of the hierarchy at which the manager is operating. Katz contended that managers need all three skills to fulfill their role requirements, but the relative importance and the specific types within each category depend on the leadership situation.

Based on the concept of skill as an ability to translate knowledge into action, the three interrelated skill categories may be briefly summarized as follows: (1) **technical skills**—performing specialized activities, (2) **human skills**—understanding and motivating individuals and groups, and (3) **conceptual skills**—understanding and integrating all the activities and interests of the organization toward a common objective. Technical skills are usually more important than conceptual ones for lower level managers. Human skills are needed at all levels of management, but are relatively less important for managers at the top level than the low level. Conceptual skill becomes more important with the need for policy decisions and broad-scale action at upper levels of management.

MANAGERIAL ACTIVITIES AND ROLES

Whereas a number of studies have investigated the personal styles and characteristics of managers, relatively few have researched what managers actually do in fulfilling job requirements. After reviewing and synthesizing the available research on how various managers spend their time, Harold Koontz (1980) reported that Henry Mintzberg designed a study to produce a more supportable and useful description of managerial work. Mintzberg's resulting role theory of management has received attention as a useful way of describing the duties and responsibilities of managers. Mintzberg (1973) defined a role as an organized set of behaviors belonging to an identifiable position. He identified 10 roles common to the work of all managers and divided them into *interpersonal* (three roles), *informational* (three roles), and *decisional* (four roles). Although the roles are described here individually, in reality they constitute an integrated whole. In essence, the manager's formal authority and status educe interpersonal relationships leading to information roles and these, in turn, to decisional roles.

The three interpersonal roles of figurehead, leader, and liaison devolve from the manager's formal authority and status. As *figureheads*, managers perform duties of a symbolic, legal, or social nature because of their position in the organization. As *leaders,* managers establish the work atmosphere within the organization and activate subordinates to achieve organizational goals. As *liaisons,* managers establish and maintain contacts outside the organization to obtain information and cooperation.

The three informational roles of monitor, disseminator, and spokesperson characterize the manager as the central focus for receiving and sending of nonroutine information. In the *monitor* role, managers collect all information relevant to the organization. The *disseminator* role involves the manager in transmitting information gathered outside the organization to members inside. In the *spokesperson* role, managers transmit information from inside the organization to outsiders.

The four decisional roles are entrepreneur or initiator, disturbance handler, resource allocator, and negotiator. Managers adopt the role of an *entrepreneur* when they initiate controlled change in the organization to adapt and keep pace with changing conditions in the environment. Unexpected changes require the manager to perform as a *disturbance handler.* As *resource allocators,* managers make decisions concerning priorities for utilization of organizational resources. Finally, managers must be *negotiators* in their dealings with individuals or other organizations. These interlocking and interrelated roles are shown in Figure 12.3.

TOOLS OF MANAGEMENT

Directors of foodservice commonly use the organization chart as a means of explaining and clarifying the structure of an organization. They also use job descriptions, job specifications, and work schedules as devices for the clear presentation of personnel and their responsibilities to top management and to employees. These mechanical or visual means are indispensable in the able direction and supervision of a foodser-

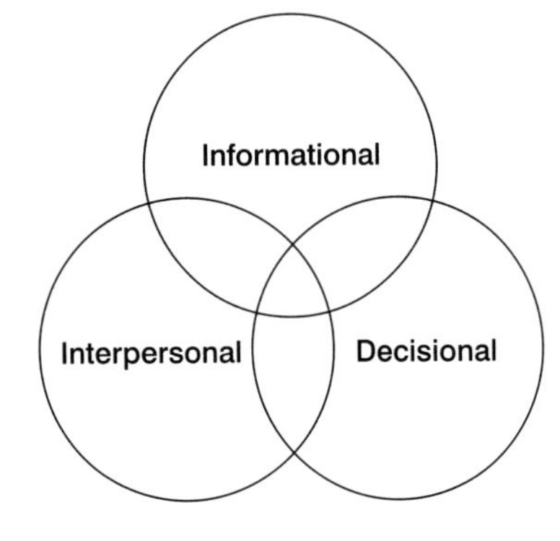

Figure 12.3 Interlocking and interrelated roles of managers.

vice, and for convenience they may be called *tools of organization and management.* Performance appraisal as a tool of management is discussed in Chapter 13.

Organization Chart

The **organization chart** may be considered the first tool of management. It graphically presents the basic groupings and relationships of positions and functions. The chart presents a picture of the formal organizational structure and serves many useful purposes, but it does have some limitations. Whereas lines of authority are depicted on the chart, the degree of authority and responsibility at each level is not shown. Informal relationships between equals or between people in different parts of the organization are not evident. For this reason, job descriptions and organization manuals are valuable supplements to the organization chart.

The organization chart is usually constructed on the basis of the line of authority, but it may be based on functional activity or a combination of the two. Functions and positions are graphically presented by the use of blocks or circles. Solid lines connecting the various blocks indicate the channels of authority. Those persons with the greatest authority are shown at the top of the chart and those with the least at the bottom. Advisory responsibility and lines of communication often are shown by use of dotted lines. Organization charts for two hospitals are shown in Figures 12.4 and 12.5 and for college foodservices in Figures 12.6 and 12.7.

Figure 12.5 is an example of the organizational structure of a hospital after extensive restructuring to eliminate multiple levels of middle management and to broaden the span of responsibility for all managers. Downsizing and streamlining operations such as this have resulted in previously tall, narrow organization charts becoming much broader and flatter. The "upside-down" organization chart shown in Figure 12.7 is an example of a creative approach to depicting the relative importance of various positions in the organization.

Job Description

A **job description** is an organized list of duties that reflects required skills and responsibilities in a specific position. It may be thought of as an extension of the organization chart in that it shows activities and job relationships for the positions identified on the

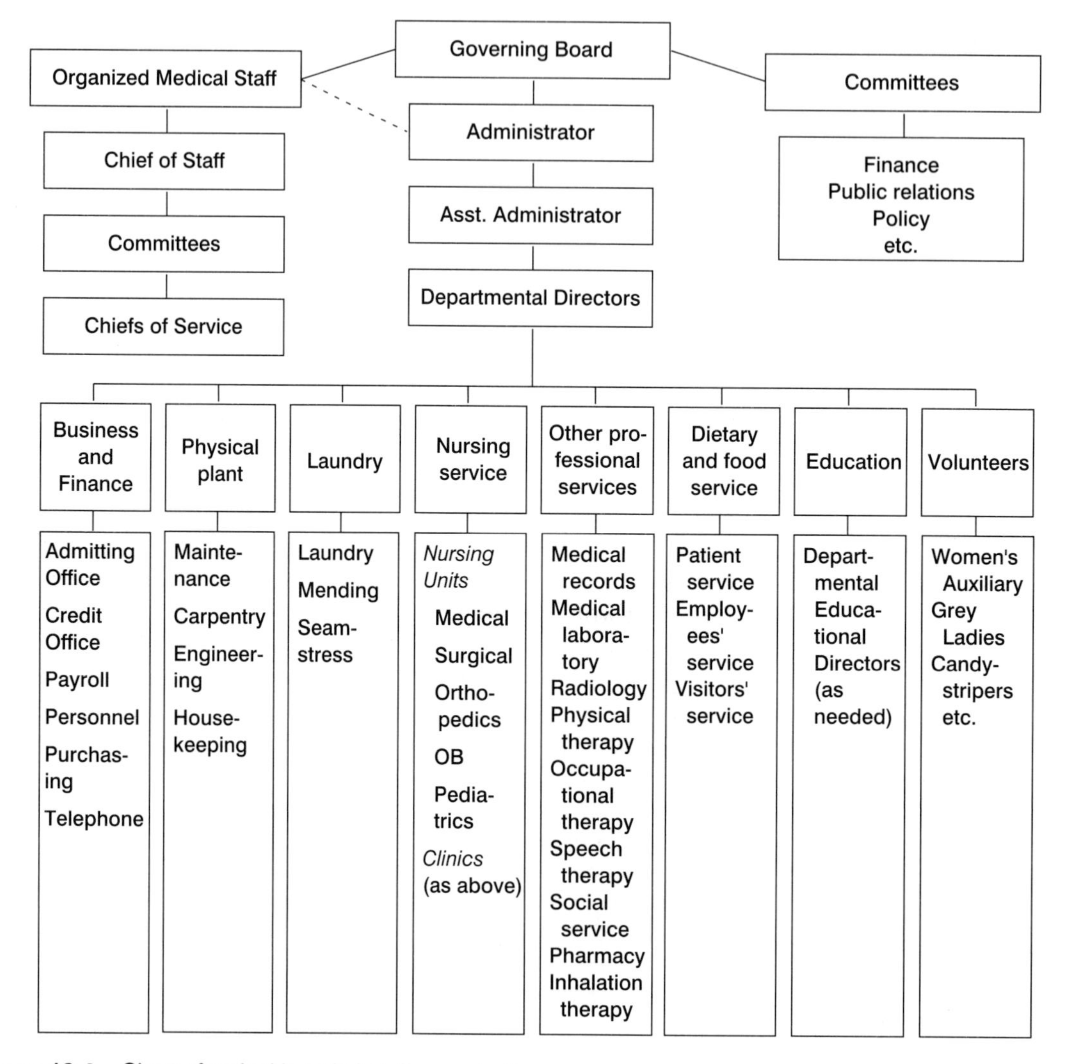

Figure 12.4 Chart of typical hospital staff organization.

organization chart. Job descriptions are valuable for matching qualified applicants to the job, for orientation and training of employees, for performance appraisal, for establishing rates of pay, and for defining limits of authority and responsibility. They should be written for every position in the foodservice and should be reviewed and updated periodically. In many organizations, the job descriptions are incorporated into a procedure manual or kept in a loose-leaf notebook for easy access.

Job descriptions may be written in either narrative or outline form or a combination of the two. The format probably will vary according to the job classification; for example, the work of the foodservice employee is described in terms of specific duties and skill requirements, but the job description for the professional position is more likely to be written in terms of broad areas of responsibilities. Most job descriptions include identifying information, a job summary, and specific duties and requirements. The initial job descriptions for a new facility would reflect the responsibilities delegated to each position on a trial basis and subject to early revision. In the case of an established unit, they are developed from information obtained from interviews with employees and supervisors and from observations by the person responsible for writing the job description. A *job analysis,* in which all aspects of a job are studied and analyzed, may be conducted first to collect information for the job description.

The job description shown in Figure 12.8 may be useful as a guide. The exact content and format, however, vary according to the position being described and the needs and complexities of the institution.

To empower employees, job descriptions have been replaced by a skills matrix system in some progressive companies. Each skills matrix describes steps in the career ladder along a vertical axis, as well as skills and competencies that are required for each step across the horizontal axis. The skill matrices specify roles and levels of performance for a "family of jobs," rather than a description of a specific job.

Job Specification

A **job specification** is a written statement of the minimum standards that must be met by an applicant for a particular job. It covers duties involved in a job, the work-

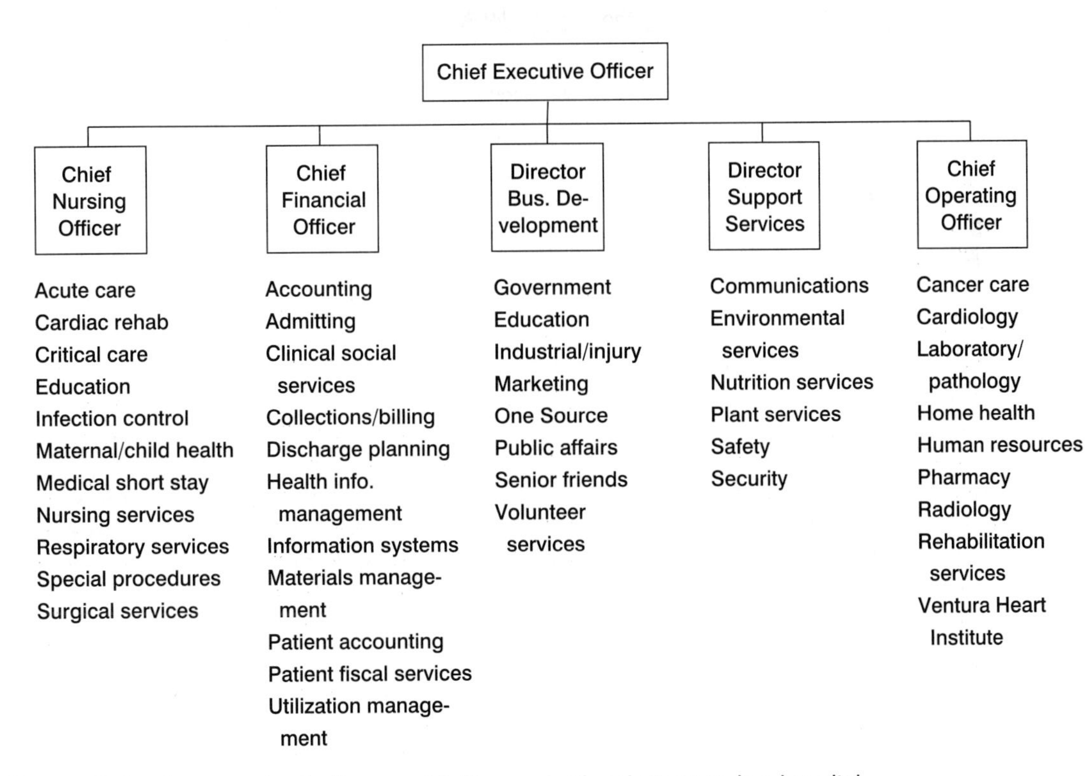

Figure 12.5 An example of a "lean and flat" organizational structure in a hospital.
Courtesy of Linda Dahl, RD, Los Robles Regional Medical Center.

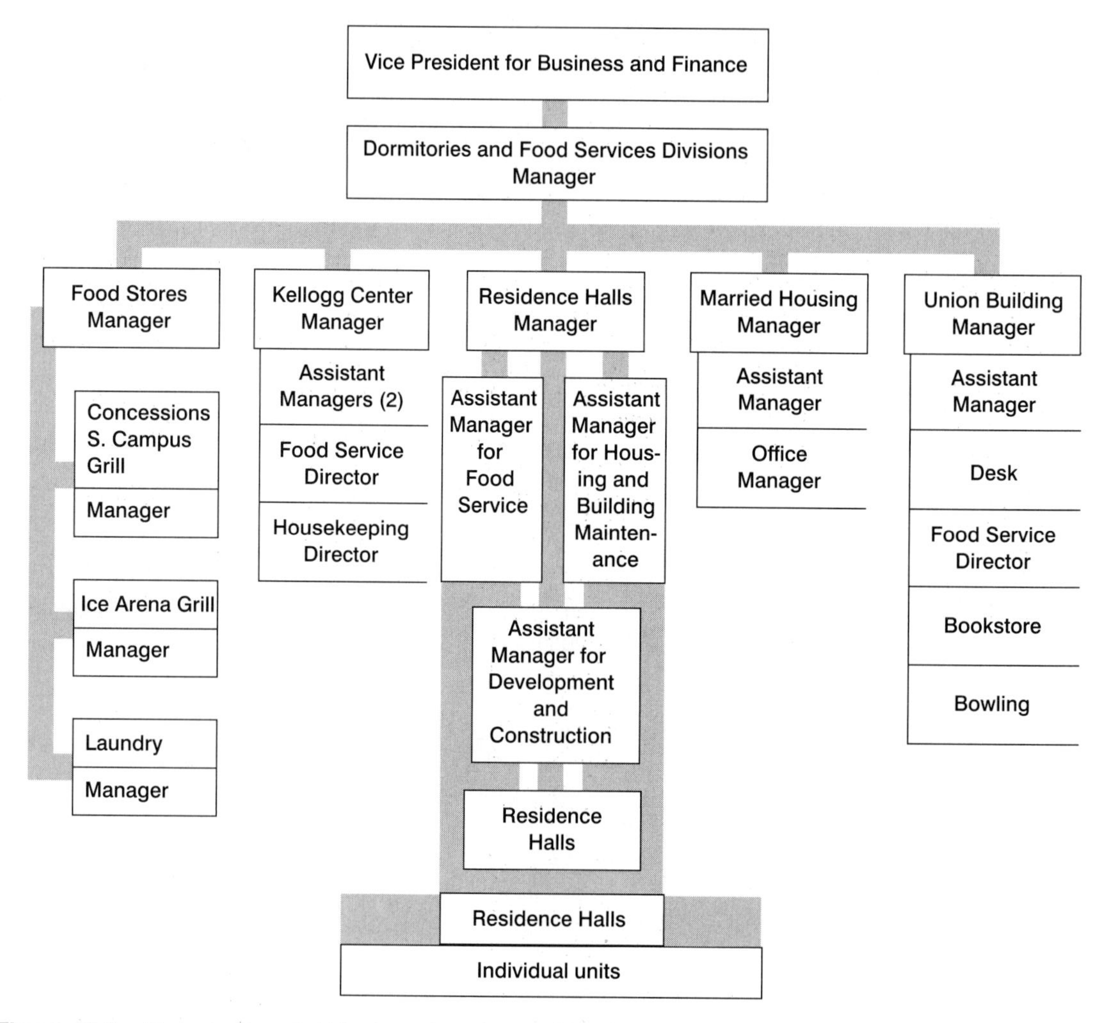

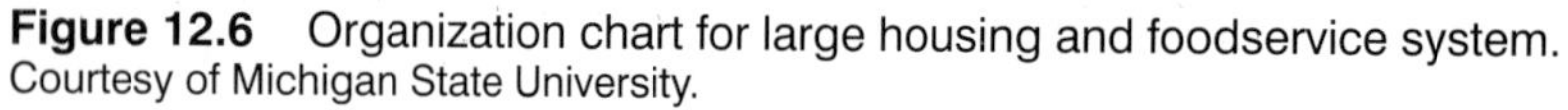

Figure 12.6 Organization chart for large housing and foodservice system.
Courtesy of Michigan State University.

ing conditions peculiar to the job, and personal qualifications required of the worker to carry out the assigned responsibilities successfully. This tool is used primarily by the employing officer in the selection and placement of the right person for the specific position. Many small institutions use the job description as a job specification also (see Figure 12.9).

Work Schedule

A **work schedule** is an outline of work to be performed by an individual with stated procedures and time requirements for his or her duties. It is important to break down the tasks into an organized plan with careful consideration given to timing and sequence of operations. Work schedules are especially helpful in training new employees and are given to the employee after the person has been hired and training has begun. This is one means of communication between the employer and employee. Work schedules should be reviewed periodically and adjustments made as needed to adapt to changes in procedures.

An example of a work schedule for a cafeteria worker is given in Figure 12.10. For food production employees, the individual work schedule would outline in general terms the day's work routine, but would need to be supplemented by a daily production schedule giving specific assignments for preparation of the day's menu items and pre-preparation for the next day. A more detailed discussion of production scheduling is included in Chapter 6.

VanEgmond-Pannell (1987) suggests three basic types of work schedules: individual, daily unit, and organization. Because the individual schedule on a daily basis

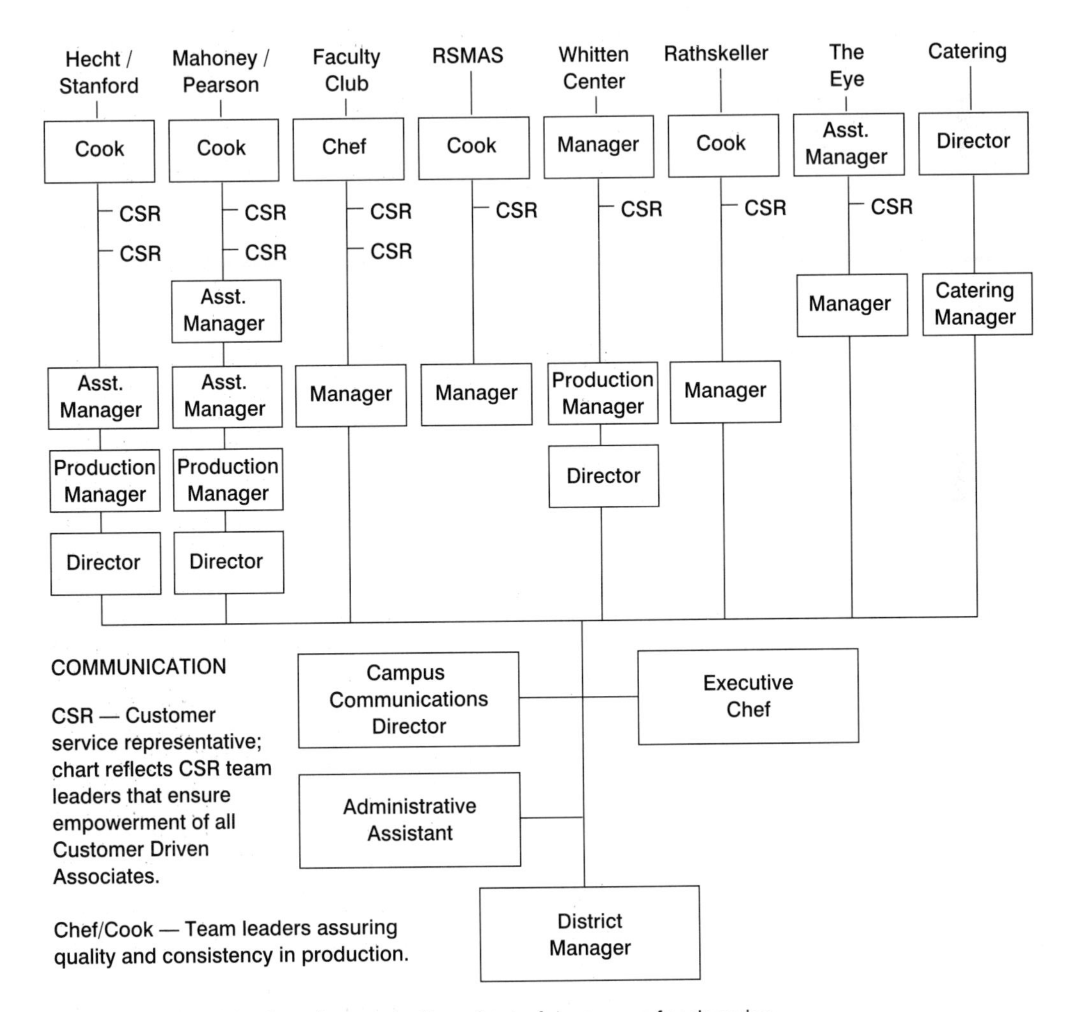

Figure 12.7 An "upside down" organization chart of a campus foodservice.
Courtesy of Gene Perkins, Marriott Corporation.

JOB DESCRIPTION

Job title: First Cook *Date:* September 2, 19____

Job code: 2–26.32 Dept 10 *Location:* Kitchen of University Cafeteria

Job summary

Prepares meats and main dishes, soups and gravies for noon meal.
Cleans and washes small equipment used in cooking.
Keeps own working area clean.

Performance requirements

Responsibilities: responsible for the preparation of meat and main dishes, soups and gravies to be served at a stated time.
Job knowledge: plan own work schedule, know basic principles of quantity food cookery and how to use certain equipment.
Mental application: mentally alert.
Dexterity and accuracy: accurate in weighing and measuring of food ingredients and portions.
Equipment used: food chopper, mixer, ovens, ranges, steam cooker, steam-jacketed kettle, fryer, broiler, meat slicer.
Standards of production: preparation of foods of high quality in specified quantities.

Supervision

Under general supervision of dietitian.
Gives some supervision to assistant cooks.

Relation to other jobs

Promotion from: Salad maker or vegetable preparation worker.
Promotion to: Foodservice supervisor (if education and ability warrant).

Qualifications

Experience desirable but not required.
Education and training.
Technical or vocational training: none.
Formal education: grammar school.
Ability to read, write, and understand English.

Figure 12.8 Job description; write-up for a cook's position. Job description should be available for all positions in the department.

would be too time consuming for most managers, she recommends the daily unit schedule shown in Figure 12.11.

The organization work schedule gives the standing assignments by half-hour periods for all employees in chart form. It does not relate specifically to the day's menu. This type of schedule shows graphically the total workload and its division among employees, but would not be effective unless accompanied by daily assignments or a production schedule.

Scheduling of Employees

Workers may be scheduled successfully only after thorough analysis and study of the jobs to be done, the working conditions, and the probable efficiency of the employees. The menu pattern, the form in which food is purchased, the method of preparation, and the total quantity needed are important factors in determining the amount of preparation time and labor required to produce and serve meals in a given situation. Good menu planning provides for variation in meal items and combinations from day to day, with a fairly uniform production schedule. Workers cannot be expected to maintain high interest and to work efficiently if they have little to do one day and are overworked the next.

Analysis of several sample menus in terms of total labor hours and the time of day required for the amounts and types of preparations is a basic consideration in determining the number of employees necessary in any foodservice. The total estimated work hours required to cover all activities in the organization divided by the number of working hours in the day would give an indication of the number of full-time employees needed. However, careful attention must be given to time schedules so that each employee will be occupied during his or her hours on duty. Certain preparations or service duties may require a reduction in the estimated number of full-time workers and the addition of some part-time workers during peak periods in order to maintain the desired standards at an even tempo. A graphic presentation of the estimated work hours needed for each job, as shown in Figure 12.12, helps to clarify the problems of scheduling and the distribution of the workload.

JOB SPECIFICATION

Payroll title: First Cook

Department: Preparation Department *Occupational code:* 2–26.32

Supervised by: Dietitian

Job summary: Prepares meat, main dishes, soups, and gravies for noon meal.

Educational status: Speak, read, write English. Grammar school graduate or higher.

Experience required: Cooking in a cafeteria or restaurant 6 months desirable but not required.

Knowledge and skills: Knowledge of basic principles of quantity food preparation; ability to adjust recipes and follow directions; ability to plan work.

Physical requirements: Standard physical examination.

Personal requirements: Neat, clean; male or female.

References required: Two work and personal references.

Hours: 6:30 a.m. to 3:00 p.m., 5 days a week; days off to be arranged; 30-minute lunch period.

Wage code: Grade 3.

Promotional opportunities: To foodservice supervisor.

Advantages and disadvantages of the job: Location, environment, security.

Tests: None.

Figure 12.9 Job specification. Example of a typical format used for each job in the department.

WORK SCHEDULE FOR CAFETERIA COUNTER WORKER

Name: ______________________

Hours: 5:30 to 2:00 p.m.
30 min for breakfast
15 min for coffee break

Position—Cafeteria Counter Worker—No. 1
Days off: ______________________

Supervised by: ______________________
Relieved by: ______________________

5:30 to 7:15 A.M.:

1. Read breakfast menu
2. Ready equipment for breakfast meal
   a. Turn on heat in cafeteria counter units for hot foods, grill, dish warmers, etc.
   b. Prepare counter units for cold food
   c. Obtain required serving utensils and put in position for use
   d. Place dishes where needed, those required for hot food in dish warmer
3. Make coffee (consult supervisor for instructions and amount to be made)
4. Fill milk dispenser
5. Obtain food items to be served cold: fruit, fruit juice, dry cereals, butter, cream, etc. Place in proper location on cafeteria counter
6. Obtain hot food and put in hot section of counter
7. Check with supervisor for correct portion sizes if this has not been decided previously

6:30 to 8:00 A.M.:

1. Open cafeteria doors for breakfast service
2. Check meal tickets, volunteer lists, guest tickets, and collect cash as directed by supervisor
3. Replenish cold food items, dishes, and silver
4. Notify cook before hot items are depleted
5. Make additional coffee as needed
6. Keep counters clean; wipe up spilled food

8:00 to 8:30 A.M.:

Eat breakfast

8:30 to 10:30 A.M.:

1. Break down serving line and return leftover foods to refrigerators and cook's area as directed by supervisor
2. Clean equipment, serving counters, and tables in dining area
3. Prepare serving counters for coffee break period
   a. Get a supply of cups, saucers, and tableware
   b. Make coffee
   c. Fill cream dispensers
   d. Keep counter supplied during coffee break period (9:30–10:30)
4. Fill salad dressing, relish, and condiment containers for noon meal

10:30 to 11:30 A.M.:

1. Confer with supervisor regarding menu items and portion sizes for noon meal
2. Clean equipment, counters, and tables in dining area
3. Prepare counters for lunch:
   a. Turn on heat in hot counter and dish warmers
   b. Set out tea bags, cream, ice cups, glasses
   c. Place serving utensils and dishes in position for use
4. Make coffee
5. Fill milk and clean dispensers
6. Set portioned cold foods on cold counter

11:30 A.M. to 1:30 P.M.:

1. Open cafeteria doors for noon meal service
2. Replenish cold food items, dishes, and silver as needed
3. Keep counters clean; wipe up spilled food
4. Make additional coffee as needed

1:30 to 2:00 P.M.:

1. Turn off heating and cooling elements in serving counters
2. Help break down serving line
3. Return leftover foods to proper places
4. Serve late lunches to doctors and nurses
5. Clean equipment and serving counter as directed by supervisor

2:00 P.M.

Off duty

Figure 12.10 Work schedule for a counter worker in any type of cafeteria.

WORK SCHEDULE

MENU:

Lasagna casserole | Buttered french bread | Chilled peach halves

Tossed salad | Milk

Time	*Manager 7½ hr*	*6–Hr Assistant*	*5–Hr Assistant*	*4–Hr Assistant*
7:30–8:00	Make coffee or tea for teachers			
8:00–8:30	Help with lasagna sauce	Prepare lasagna		
8:30–9:00	Lunch count—Tickets		Dip up fruit and refrigerate	
9:00–9:30				
9:30–10:00	Teachers' salads		Wash vegetables for salad	
10:00–10:30	Cut bread and butter	Prepare bread crumbs for fried chicken tomorrow		Cut up vegetables for salad
10:30–11:00			Put out desserts	Set up line—napkins, straws, dishes
11:00–11:30	Eat lunch—20 min Put food on steam table	Eat lunch—20 min Put food on steam table	Eat lunch—20 min Wash pots and pans	Mix salad for first lunch
11:30–12:00	Cashier	Serving Set up for next line	Serving Help in dishroom	Back up line Dishroom
12:00–12:30	Serving	Serving Set up for next line	Cashier Help in dishroom	Back up line Dishroom
12:30–1:00	Serving	Serving Put away food	Cashier 10 min break	Back up line Dishroom
1:00–1:30	Count money—10 min break Help to clean tables	10 min break Clean tables	Clean steamtable	Eat lunch (eat on own time)
1:30–2:00	Prepare reports	Clean up		Clean dishroom
2:00–2:30	Place orders			Help with kitchen cleanup
2:30–3:00	Take topping out of freezer and put in refrigerator for tomorrow			

Figure 12.11 Sample work schedule for school foodservice employees.
Courtesy of Dorothy VanEgmond-Pannell, *School Foodservice,* 3rd ed., Westport, Conn.: AVI Publishing Company, 1987.

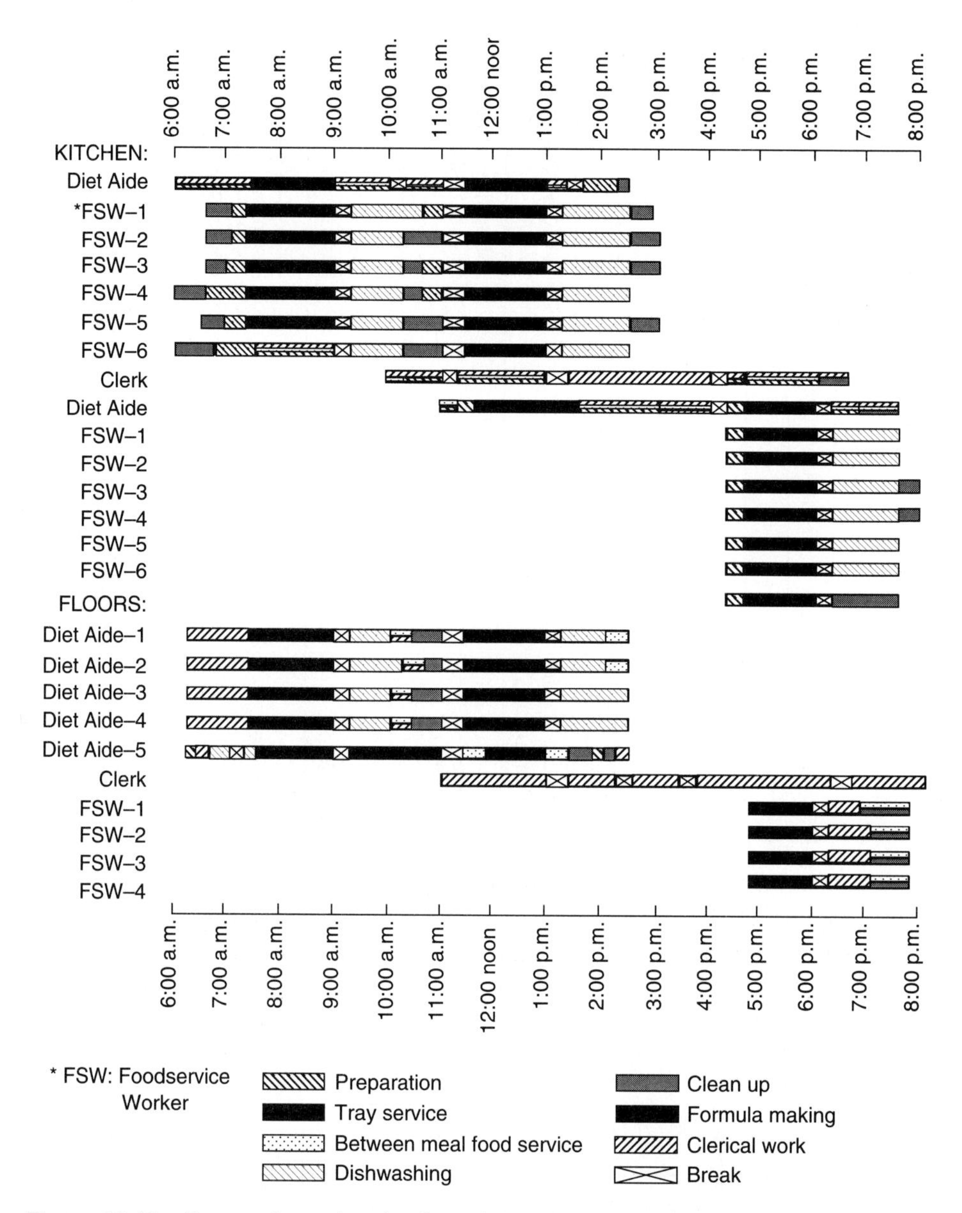

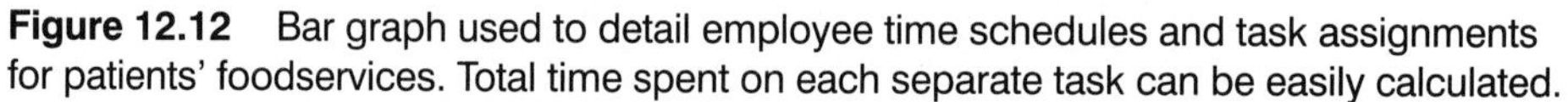

Figure 12.12 Bar graph used to detail employee time schedules and task assignments for patients' foodservices. Total time spent on each separate task can be easily calculated.

Working conditions such as the physical factors of temperature, humidity, lighting, and safety influence the scheduling of personnel and affect workers' performance. Of particular importance is the amount and arrangement of equipment. The distance each employee must travel within his or her work area should be kept at a minimum in order to conserve the individual's energy and time. Use of mechanical devices in the processing and service of food may decrease the total labor hours needed and increase the degree of skill and responsibility of employees. Arrangement of work areas for efficient operation cannot be overemphasized.

A work distribution analysis chart of the total activities within a department will show where tasks may be eliminated, combined, or modified in the overall picture. One must be sure that the activities are so organized and combined that efficient use is made of the labor hours of each individual worker. Studies may be made to determine a good standard for each procedure, such as the time required for the average worker to combine and shape 25 pounds of ham loaf mixture into loaves in pans for baking. The standards for each procedure should be such that the workers in a particular organization will be able to maintain them. The standard time should be set at a level that the average employee could do 20% to 30% more work without undue fatigue.

Written schedules clarify the responsibilities of workers and give them a feeling of security. It is wise to include a statement indicating that additional duties may have to be assigned from time to time. However, work schedules must be kept flexible and adjustments made as needed to adapt to the daily menu. Also, the introduction of new food products may decrease the amount of time needed for pre-preparation as well as the time of cookery; likewise, additional processes may become necessary.

SUMMARY

The contributions of classical, human relations, and management science theories to current-day management thought have been numerous. Systems and contingency theories of management incorporate many concepts from early management history. The division of management work into functions is one such concept. Each of the functions of management is described in this chapter.

Managers utilize three skills in proportion to their place in the organizational hierarchy: technical, human, and conceptual skills. Managerial work may also be classified into activities and roles. For some, this classification appears to be more useful than functions and skills.

Various tools are utilized by managers in the fulfillment of organizational responsibilities. These are organization charts, job descriptions, job specifications, and work schedules.

Information pertaining to the broad subjects of organization and management is voluminous. Only basic concepts with limited application are included in this chapter. The following chapters discuss some special areas of concern to persons in the management of foodservices. Supplementary reading of current literature is advised to become acquainted with newer developments as they evolve.

REVIEW QUESTIONS

1. Discuss how concepts from the classical, human relations, and management science theories have been incorporated into the modern management theories today.

2. Give several examples of the application of human relations theory used today.
3. Describe the systems approach as it pertains to the management of a foodservice organization. Why is interdependency considered such a key concept in systems theory?
4. Discuss the statement: "A foodservice operation is an interdependent, open system."
5. Explain how contingency theory is really a combination of many other theories.
6. Explain the difference between a line and a line and staff organization. What are the advantages and disadvantages of each?
7. What are the seven widely accepted functions of management? Compare and contrast the division of managerial work into functions versus roles.
8. List and briefly explain (or diagram) the relationship between Katz's three managerial skills and the manager's level in the organizational hierarchy.
9. List the steps required to establish an organizational structure.
10. Outline the sequence of steps required in an effective planning process. Where do you find some concepts from historical management theory applied to planning?
11. How do job descriptions, job specifications, and work schedules differ, and for what purpose is each used?

SELECTED REFERENCES

Armstrong, L., and Symonds, W. C.: Beyond "May I help you?": Services finally are trying to put their house in order. Business Week. 1991; Special issue 3237:100.

Axline, L. L.: TQM: A look in the mirror. Mgmt. Rev. 1991; 80(7):64.

Drucker, P. F.: *Managing in Turbulent Times.* New York: Harper & Row, 1980.

Fayol, H.: *General and Industrial Management.* London: Sir Isaac Pittman and Sons, 1949.

Gift, B.: On the road to TQM. Food Mgmt. 1992; 27(4):88.

Gulick, L.: Notes on the theory of organization. In L. Gulick and L. Urwick (Eds.), *Papers on the Science of Administration.* New York: Columbia University Press, 1937.

Hitt, M. A., Middlemist, R. D., and Mathis, R. L.: *Management Concepts and Effective Practice.* Saint Paul, Minn.: West Publishing Company, 1989.

Katz, R. L.: Skills of an effective administrator. Harvard Bus. Rev. 1974; 52:90.

King, P.: A total quality makeover. Food Mgmt. 1992; 27(4):96.

Koontz, H.: The management theory jungle revisited. Academy of Mgmt. J. 1980; 5:175.

Mintzberg, H.: *The Nature of Managerial Work.* New York: Harper & Row, 1973.

Moravec, M., and Tucker, R.: Job descriptions for the 21st century. Pers. J. 1992; 71(6):37.

Rinke, W. J.: Total quality management: Just another management fad? Dietary Manager. 1995; 4(1):10–13.

Schecter, M.: The quest for quality. Food Mgmt. 1992; 27(4):82.

———: Restaurant makes the CQE commitment. Food Mgmt. 1992; 27(4):112.

VanEgmond-Pannell, D.: *School Foodservice,* 3rd ed. Westport, Conn.: AVI Publishing Company, 1987.

———: *School Foodservice Management,* 4th ed. New York: Van Nostrand Reinhold, 1990.

13

Staffing and Managing Human Resources

OUTLINE

Staffing
Skill Standards
The Employment Process
Recruitment
Selection
The Worker on the Job
Personnel Records
Orientation
Training
Performance Appraisal
Promotions and Transfers
Discipline
Dismissals
Supervision
Decision Making
Handling Grievances
Staff Conferences
Labor Policies and Legislation
Labor-Management Relations
Legislation

Staffing and managing human resources involve all the methods of matching tasks to be performed with the people available to do the work. The *acquisition* of human resources (e.g., job analysis human resource planning, recruitment, and selection); the *development* of human resources (e.g., placement, orientation, training, performance appraisal, and personnel development); the *rewards* of human resources (e.g., compensation and promotion); and the *maintenance* of human resources (e.g., health and safety, transfer, discipline, dismissal, supervision, decision making, grievance handling, and labor relations) are all intertwined in this process. Each of these subjects is discussed in this chapter.

An important aspect of the human resource management function is the establishment, acceptance, and enforcement of fair labor policies within an organization. We discuss four areas in which policies are generally established: wages and income maintenance, hours and schedules of work, security in employment, and employee services and benefits. Major federal legislation that has an impact on these organizational policies is included in each section. The chapter concludes with a discussion of labor-management relations including relevant legislation and the impact of unionization on foodservices.

KEY CONCEPTS

1. Staffing is the managerial function of matching requirements of tasks to be performed with skills available through the use of effective hiring, placement, promotion, transfer, job design, training, supervision, decision making, performance appraisal, and discipline.
2. Employment needs are outlined in various tools of management such as the organization chart, job description, and job specification.
3. Internal recruitment sources include promotions-from-within and referrals by present employees. The most commonly used external sources are newspaper advertising, employment agencies, schools, and labor unions.
4. An effective hiring process includes a carefully prepared application form and an equally carefully prepared interview.
5. A new employee should be introduced to the job and the company first in a formal orientation program and then in a well-organized training program.
6. Behaviorally anchored rating scales (BARS) are generally considered to be the best and most effective means of performance appraisal.
7. Any disciplinary action taken should be immediate, consistent, impersonal, based on known expectations, and legally defensible.
8. Supervision encompasses the coordination and direction of the work of employees through personal contact, which is reinforced through checking by observation or checking records and charts.
9. The five-step decision-making process provides a framework for making better decisions and problem solving.
10. A number of pieces of legislation establish fair labor policies and attempt to balance the power between labor and management.

STAFFING

Foodservice has been called the ultimate people business. As the single most important resource in any enterprise, the human factor is the key to success. The ability of foodservice managers to understand people, recognize their potential, and provide for their growth and development on the job is of inestimable worth in helping to create good human relations. Realization by workers that they are useful and important to efficient functioning of the business contributes to their sense of responsibility, ownership, and pride in the organization. An increase in pay alone does not buy goodwill, loyalty, or confidence in self and others. Often, just simple changes or considerations such as beautification of the work area, elimination of safety hazards, rearrangement of equipment, modification of work schedules, or even cheerful words of appreciation and encouragement produce incentives that result in increased and improved quality output. Mutually understood and accepted objectives and policies of the foodservice and well-defined channels of communication also contribute significantly to high-level employer-employee relationships.

Beyond good pay and benefits, these characteristics are common to those companies considered the best to work for in the United States: encouraging open communication, flowing up and down; promoting from within; stressing quality to foster a sense of employee pride in output; allowing employees to share in profits; reducing distinctions between ranks; creating as pleasant a workplace environment as possible; encouraging employees to be active in community service; helping employees save by matching funds that they save; and making people feel part of a team. The presence of these characteristics results in a good workplace climate and low absenteeism and turnover of good employees.

Staffing is not simply a synonym for employment. **Staffing** includes all the methods of matching requirements of tasks to be performed with skills available. Hiring, placement, promotion, transfer, job design, and training are all intertwined in this process. People must be hired and promoted who can be trained to perform the necessary tasks. And training must be designed around the needs of the employees and the organization. Staffing may be thought of as an integrated system for moving people into, through, and eventually out of the organization. The components of an **integrated staffing** system are shown in Figure 13.1.

A detailed plan of organization for a foodservice indicates the number and types of human resources needed, presents their distribution among the various work areas of the service, and shows their work schedules, the provision made for their training, and the responsibilities assigned to each. Far more difficult than the formulation of such a plan on paper is its actual inauguration. Then, all the neat little blocks on the chart designating individuals assigned certain responsibilities become persons with diverse energies and loyalties, egocentric ideas, and unclarified codes of values, some skillful, others not, some with acceptable food standards, and others apparently totally lacking in this regard. Left to chance, the introduction of the human element into an orderly plan is likely to plunge it into chaos. With wise selection, intelligent and adequate direction, and careful supervision, the human element vitalizes and enriches the plan.

Foodservices in many large companies have personnel departments responsible for the staffing function. In such organizations, the foodservice director works closely with the personnel department. However, in many small foodservices, personnel management responsibilities are assumed by the director of that department. Thus, the director may be responsible for the organization plans and the procurement, placement, induction, on-the-job training, and supervision of all employees in the department.

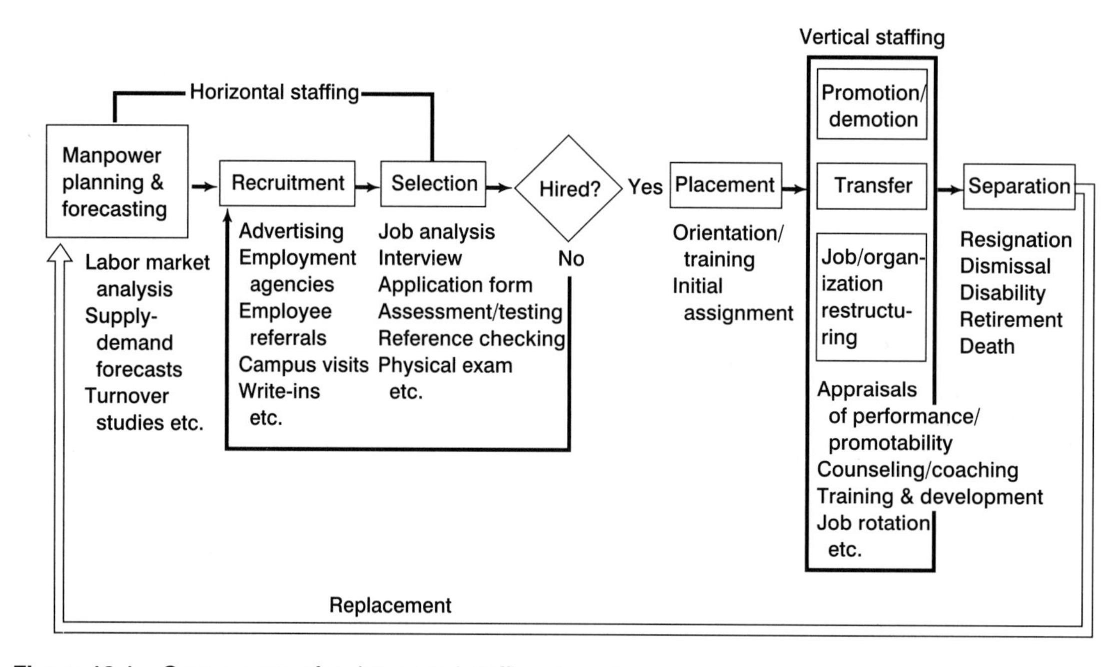

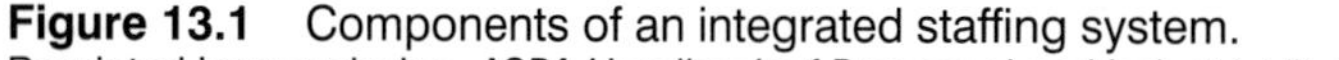

Figure 13.1 Components of an integrated staffing system.
Reprinted by permission. *ASPA Handbook of Personnel and Industrial Relations,* by Dale Yoder and H. G. Heneman Jr., p. 4–3, copyright © 1979, by Bureau of National Affairs, Inc., Washington, DC.

Management of human resources presents unique problems that can be solved only by persons with an understanding of human nature, a respect for the personalities of others, and an appreciation of the labor requirements and employment opportunities of the company. Insight into and respect for the rights of all individuals in an organization are the responsibility of the person in charge. These time-honored rights include (1) the right to be treated as an individual and respected as a person, (2) the right to have a voice in one's own affairs, which includes the right to contribute to the best of one's own abilities to the solution of common problems, (3) the right to recognition for one's contribution to the common good, (4) the right to develop and make use of one's highest capabilities, and (5) the right to fairness and justice in all dealings with one's supervision.

As soon as he or she is employed, a worker becomes a member of the group and begins to share in forming that intangible but all-important element termed group morale or group spirit. An understanding, cooperative, and helpful worker contributes to group morale; an irritable, carping, complaining, and obstructive worker destroys it. Many organizations have learned through sad experience how very great a destructive force one malcontent member can exert on group spirit. Because the disciplining of workers who are a disturbing force or their elimination from the work group is neither easy nor pleasant, the selection of those who will build morale rather than destroy it is of great importance in the choice of staff.

The skill, craftsmanship, dependability, and regularity of workers and their contribution to group morale in previous positions can determine whether they are chosen as new employees. Certain other things indicate probable contributions in the future. For example, capacity for growth, desire for self-improvement in order to render greater service, ambition for promotion, and identification with the firm are all important in the selection of a workforce for tomorrow. However, not every person is equally desirous of assuming responsibility and carrying out a project to conclusion. This is also true of workers who are unwilling to face problem solving. Some people are overdependent and eager to avoid directing themselves or others.

After the foodservice director has considered personnel needs, what the foodservice has to offer in return should be considered. Everybody works smarter when there's something in it for them. Part of the reward is money. Adequate compensation and steady employment are basic to any satisfactory employer-employee relationship.

Another part of the compensation may be intangible; that is, just as employees contribute to the morale of the group spirit of the service, so the administration will contribute to their sense of personal satisfaction. The provision of meaningful work and the recognition of achievements are important motivational elements.

A third part of the compensation will be the opportunity to do a good job. Full instructions as to accepted procedures and standards and adequate on-the-job supervision are vital to satisfactory performance by workers. Only then will they experience pride in their accomplishments and attain and maintain a high level of performance on the job.

The job should provide opportunity for growth and a reasonable chance for promotion. Workers should have an opportunity to make their service a creative experience. They should be encouraged to regard improvement in techniques as possible and welcome and to feel that suggestions toward this end will be cordially received. They have a right to expect fairness in dealing with management, freedom from misrepresentation and misinformation about the employing organization, a reasonable opportunity for continued education, promotion when earned, and provision for satisfying recreation.

The foodservice director should synthesize the two points of view—the employer's and the employee's—into an adequate, functioning personnel program. Such a program should be characterized by wise selection, careful placement, adequate supervision, and education for the present job and for the future; fair employment policies; services desirable for the comfort and welfare of employees; and the keeping of records that will facilitate the evaluation and revision, if needed, of the management program.

Skill Standards

The changing nature of today's business environment and labor markets has given impetus to the development of national *skill standards* by a number of professional organizations. These skill standards define the level of performance necessary to be successful on the job. More specifically, they include the steps involved in completing critical tasks; tools and equipment used; descriptions of possible problems and their responses; and the knowledge, skills, and abilities basic to completing these tasks. Everyone benefits from the use of skills standards:

- Employers can recruit, screen, place, train, and appraise employees more effectively, efficiently, and fairly.

- Workers can know what to expect on the job and be better prepared, thereby increasing their mobility and opportunity for advancement.
- Labor organizations can increase employment security through portable skills and credentials.
- Students can have clear direction to help them set goals and train for future employment.
- Educators can design quality curriculum and instruction consistent with the needs of the industry.
- Consumers can expect high-quality, efficient service from well-trained employees.

THE EMPLOYMENT PROCESS

Organization charts indicate the number of workers needed in each department of a foodservice, and job descriptions and job specifications outline the specific conditions under which each employee will work, the job requirements, and the training and other personal qualifications deemed desirable. Such information affords the foodservice director charged with human resource management an inventory of employment needs.

Recruitment

The next step is to survey the sources of labor supply and determine which one or ones are to be used to bring the open positions to the attention of the best qualified prospective employees. Minority group members should be actively recruited if the organization is to stay in line with public policy. Labor sources are many and varied, dependent somewhat on which sources are available locally and on the general labor market. Most sources may be classified as either internal or external.

Internal Sources. Promotion of employees to a higher level position, transfer from a related department or unit, and the rehire of a person formerly on the payroll are examples of internal sources. Promotions or transfers within an organization help to stimulate interest and build morale of employees when they know that on the basis of measured merit they will be given preference over an outsider in the event of a good vacancy. Caution must be taken to ensure that the individual has the necessary personal attributes as well as training and experience for the position open and that equal opportunity employment regulations have not been violated.

An indirect internal source of labor is a present employee who notifies friends or relatives of vacancies and arranges for an interview with the employer. Recruiting labor in this way has advantages and disadvantages. Present employees generally prefer working with those who are congenial, and a pleasant spirit within the group may be built by utilizing this source of labor supply. In contrast, personal ties may be stronger than business loyalties, and inept, unskilled workers may be highly recommended by relatives and friends. Furthermore, a strongly clannish feeling among the workers may lead to an unfortunate generalized reaction against any disciplinary measure, no matter how well justified. The many phases of this situation should be considered before extensive use is made of this source of labor.

External Sources. Some foodservice organizations may plan to fill vacancies by promoting from within; however, replacements will eventually be needed to fill the depleted ranks. The most common external sources are the press, employment agencies, schools, and labor unions.

Advertising. Newspaper advertising is a means of reaching a large group of potential applicants. Such advertisements should cite the qualifications desired; otherwise, many unqualified applicants may respond. Definite statements as to desired training and experience in the foodservice field tend to limit the applicants to those who are truly qualified. Details concerning salary, sick leave, time schedule, and vacations are much better left until the personal interview. The advertisement should state whether application is to be made first in person or by letter.

Employment Agencies. Private employment agencies have long served as a means of locating labor. Usually, these agencies are supported by a registration fee charged to persons who are seeking employment. They generally provide a preliminary "weeding out" of would-be applicants, eliminating the unfit. Often these agencies tend to deal with specialized groups in the professional or technical areas and are of most value to those seeking employees on the managerial level.

Federal, state, and local employment agencies are a significantly important labor source. The value of these agencies lies in the fact that they have studied the employer's needs and have set up the machinery needed to test the aptitudes and skills of the workers. Such procedures benefit foodservice managers who are endeavoring to reduce turnover to a minimum and develop a stable workforce.

School. In some localities, vocational and technical schools offer training for the food industry. These graduates are excellent candidates for available foodservice positions. The adequacy of their specific preparation for this work may greatly shorten the period of preliminary training necessary.

Another source, important in the foodservices of colleges and universities, is the student employment office of the college. Utilizing this type of labor offers financial assistance to worthy students and often provides experience to those majoring in food systems management. Perhaps the greatest advantage of student employees to the college foodservice manager is their availability for short work periods during the peak of the service load. However, the labor cost is high because the workers are inexperienced, and the labor turnover is great, thus much energy is expended in introducing new workers to the jobs. The short work periods necessitated by student classroom assignments make the planning of work far more complicated than when full-time employees are used. The immaturity and inexperience of the worker can also result in a waste of food supplies and labor hours unless constant and thorough supervision is provided. The maintenance of high food standards and acceptable service is often much more difficult with student employees than with carefully chosen, well-trained employees of long-time service.

Labor Unions. In organizations where employees are unionized, the labor union may be an important source for workers.

Selection

After the prospective workers have been recruited, the next step is for the employer to select the most capable person available for the particular opening. The cost of hiring, training, and discharging or transferring a worker is too great to allow many mistakes in the employee procurement process. Failure at this point is far more expensive than is commonly recognized.

Recognition of the heavy initial cost of employment means, when the labor market permits, a trend toward careful selection of each appointee.

Application Form. The application form plays an important role in the employment of any worker. The information requested should be phrased in direct simple statements pertinent to the particular job in which the applicant shows interest, and questions raised should be easily answered. Obviously, quite different information would be required of the person applying for a management position than for one who expects to be a counter worker. However, both application forms, when completed, must contain biographical data that will provide the employer with all the facts necessary not only to determine the fitness of the applicant for the job, but also to compare the qualifications of all applicants.

The Fair Employment Practice laws adopted by many states make it illegal to ask questions that would be discriminatory because of race, religion, sex, age, marital status, or national origin. After the employee has been hired, such information can be obtained for the individual's personnel records. The manager should check with the personnel department or other authoritative source regarding restrictions in the application form and the interview. References of former employment are usually requested and should be checked.

The Interview. The purposes of the selection interview are (1) to get information—not only all the facts, but attitudes, feelings, and personality traits that determine "will-do" qualifications; (2) to give information—just as it is essential that the interviewer know all about the applicant, it is also essential that the applicant know all about the establishment and the job; and (3) to make a friend—treat an applicant with the same courtesy that you would give to a customer, because every applicant is a potential customer.

The direct personal interview is advantageous in that the interviewer has the opportunity to become acquainted with the applicant and to observe personal characteristics and reactions that would be impossible to learn from an application form or letter. Also, the great majority of employees of a foodservice are relatively untrained persons, whose qualifications cannot be ascertained in any other way than by a personal interview and by possible communications with previous employers. Documents that could be termed credentials are rarely available; therefore, the personal interview becomes of great importance in making a wise choice. In filling administrative positions, the personal interview serves as a final check of the fitness of a person whose credentials have been considered carefully.

The applicant should be treated as a person whose concern with the decision is as real and vital as that of the employing agency. The job should be discussed in relation to other positions in the foodservice to which the job might lead. Reasonable hopes for promotion should be discussed and fringe benefits should be presented. Appraisal of the job specifications in terms of the applicant's own fitness should motivate the applicant toward either self-placement or self-elimination.

The development of a successful technique in interviewing requires thought, study, and experience. Some interviewing suggestions include these:

The "Do's" of Interviewing

1. Have a purpose and a plan for the interview—a guided interview pattern.
2. Have on hand during the interview, and study carefully, an analysis of the job, a job description, and a job specification.
3. Provide a private place for the interview, free from interruptions and distractions.
4. Put the applicant at ease; establish confidence and a free and easy talking situation.
5. Use the pronoun *I* very, very sparingly; *we* is much better.
6. Listen with sincere and intensive interest.
7. Do ask questions beginning with *what, why,* and *how.* Useful phrases to keep in mind are:
   Would you give me an example?
   For instance . . .
   In what way . . .
   Suppose . . .
8. Do safeguard personal confidences.
9. Do strive to learn not only what the applicant thinks and feels, but *why the applicant thinks and feels that way.*
10. Do be pleasant and courteous.
11. Do strive to be a good sounding board or mirror for the applicant's expressions of attitudes, feelings, and ideas.
12. Do ask questions that encourage self-analysis.
13. Do prod, search, and dig courteously for all the facts.
14. Maintain an attitude of friendly interest in the applicant. Make a friend and a customer even though you do not hire the applicant.
15. Do make notes for record purposes either during or after the interview.
16. Do, immediately after the interview, write a summary on the interview form in the space provided.

Suggestions of things to avoid in interviewing include these:

The "Don'ts" of Interviewing

1. Do not interrupt the applicant.
2. Do not talk too much. Talkative interviews usually are failures.

3. Do not rush the interview. This is not only discourteous, but it results in failure of the interview.
4. Do not ask leading questions. If the question is so worded that the answer you want is apparent to the applicant, you are actually interviewing yourself.
5. Do not ask questions that demand only a yes or no response.
6. Do not merely talk when the applicant has finished a statement. Use such responses as "I see," "I think I understand," or "What else can you add to that statement?"
7. Do not agree and do not disagree. Be interested but noncommittal.
8. Do not argue, or else the interview is finished.
9. Do not lose control of self or the interview.
10. Do not get in a rut. Do not leave any impression with the applicant that the interview is routine and perfunctory.
11. Do not "talk down."
12. Do not express or imply authority. The good employment interview is a free and easy exchange of attitudes and ideas between equals.
13. Do not jump to conclusions. The purpose of the interview is to get information. Appraisal and conclusions will come *after* the interview.
14. Do not preach or moralize. This is not the purpose of the interview.
15. Do not interview when either you or the applicant is upset.

Tests. Impressions of the prospective employee gained in the interview and from the follow-up of references are admittedly incomplete. They may be checked or replaced by tests of various types, the most common being intelligence, trade, and aptitude. A number of companies, including foodservices, have improved the results of their selection decisions by the use of psychological tests. These companies have found that the benefits derived from psychological testing far exceed the costs. An applicant's probable tenure, customer relations, work values, and safety record may be predicted with such tests. To be considered legal, all psychological test questions must be job related and legal to ask. In addition, all applicants must be asked the same questions, and scoring methods must be the same for all applicants.

The physical fitness of an applicant for a foodservice appointment is highly important. A health examination should be required of all foodservice workers. Only physically fit persons can do their best work. Quite as important is the need for assurance that the individual presents no health hazard to the foodservice. Managers are well aware of the devastation that might result from the inadvertent employment of a person with a communicable disease.

THE WORKER ON THE JOB

Personnel Records

After an agreement on employment terms has been reached, a record of appointment is made. This becomes the nucleus of the records of the activity and progress of the worker within the organization. Records may be kept on computer, in card files, or in loose-leaf form. Included among the items listed on the forms are name, address, name of spouse, number of children, other dependents, educational background, former employment (including company and length of time), date of hiring, job assigned, wage rate, whether or not meals are included, absences with reasons, adjustments in work and wages, promotions, demotions, or transfers with reasons, and information concerning insurance and health benefits. Such complete records are useful in indicating the sense of responsibility and the serious intent of employees, and as a basis for merit ratings, salary adjustments, or other benefits.

Orientation

The induction of the newly employed worker to the job is a most important phase of staffing. Smith (1984) outlines 10 steps to be included in an orientation program that is designed to challenge the new employee's interest and elicit support for the goals and objectives of the company.

1. *Introduction to the company:* Introductions are simply a matter of identifying the company, where it has been, and where it is going. the key is to make the new employee feel good about the company and begin to instill the pride of belonging, being a part of the company.
2. *Review of important policies and practices:* Policy review will vary from company to company, but certainly must include standards of conduct and performance, and an introductory (probation) period of employment, a discipline policy, and a safety policy.
3. *Review of benefits and services:* A review of benefits is crucial. It is not as important to sell a benefits program and all its virtues as it is to communicate what is provided and at what expense. Employees need to appreciate the cost of benefits, and the employer should be able to relate the percentage of payroll spent on their behalf. In addition, discuss services that the employee might not construe as benefits, such as a credit union, parking, food, medical care, discounts, and social-recreational services.
4. *Benefit plan enrollment:* Complete necessary benefit enrollment forms with the assurance that the employee understands his or her options. Provisions should be made to allow the employee to discuss plan options with a spouse before making a commitment.
5. *Completion of employment documents:* Payroll withholding, emergency information, picture releases, employment agreements, equal employment opportunity data, and other relevant and appropriate documents must be completed.
6. *Review of employer expectations:* This deals with employer-employee relationships. Discuss teamwork, working relationships, attitude, and loyalty. A performance appraisal form makes a good topical outline for a discussion of employer expectations.
7. *Setting of employee expectations:* If employees meet company expectations, what can they expect from the company? Training and development, scheduled wage and salary reviews, security, recognition, working conditions, opportunity for advancement, educational assistance programs, counseling, grievance procedures, and other relevant and appropriate expectations should be detailed.

8. *Introduction to fellow workers:* Introduce people a few at a time to let the names be assimilated. Use of name tags is helpful and so is the buddy system. Assign someone to be mentor to the new employees, to introduce them, take them to break periods, have lunch with them. A few days is usually enough introduction time.
9. *Introduction to facilities:* Take a standard tour of the facility. This is more effective if you break it into several tours. On the first day, tour the immediate work area and then expand the tour on subsequent days until the facility is covered.
10. *Introduction to the job:* Have your training program in place. Be prepared and ready to get the employee immediately involved in the work flow. (p. 48)

Training

After the individual worker has been properly introduced to the job, the employee still needs to be thoroughly trained, especially in the initial period of employment. Familiarity with established operational policies and procedures, presented by management in a well-organized manner, can do much to encourage the new worker and help in gaining self-confidence. Generally, advantages of a good training program include reduction in labor turnover, absenteeism, accidents, and production costs, and an increase in the maintenance of morale, job satisfaction, and efficient production at high levels.

The first step in establishing a training program is to decide when training is needed. Next, determine exactly what needs to be taught and who should receive that training. Goals should be established for the program and an outline developed containing the steps required to help meet those goals.

Adult Training. The unique characteristics of adults as learners must be considered when planning for on-the-job training. Children learn for the future and in order to advance to the next level of learning. Adults, however, learn for immediate application or to solve a present problem. For this reason, they require practical results from the learning experience. Other distinguishing characteristics of the adult learner are a reduced tolerance for disrespectful treatment, the preference for helping to plan and conduct one's own learning experiences, and a broader base of life experiences to bring to the learning activity.

Group Training. Often training can be given efficiently and economically through group instruction. This type of teaching saves time for the instructor and the worker, and also has the advantage of affording the stimulus that comes as the result of group participation. In a foodservice, basic group instruction concerning the policies of management is practical and valuable. Among the areas that might be included are the history and objectives of the organization, relationships of departments and key persons within the particular department, the operational budget as it affects the workers, the preparation and service of food, the sanitation and safety program, and the principles and values of work improvement programs.

Perhaps the most important psychological principle of group training is the use of well-prepared teachers instead of a fellow worker who may have had successful experience in a limited area. Often the stimulation and the inspiration given to the employee by an able instructor are highly motivating and more important in the development of the individual worker than the immediate mastery of routine skills. Tools found to be of value in such an instructional program are audio and visual aids, including films and television, illustrative material, such as posters, charts, and cartoons, and demonstrations in which both the instructor and the employees participate. Spending time and money merely showing films in group training classes is wasteful unless the workers have been alerted to the points of emphasis, time is allowed for discussion after the presentation, and follow-up occurs through application on the job. Other psychological principles of group education are not considered here, although they should be understood by those in charge of such programs.

On-the-Job Training. Some large foodservice organizations have inaugurated rather extensive programs to provide on-the-job training of employees, with highly satisfactory results. Important objectives of such programs are (1) to reduce time spent in perfecting skills for the production and service of attractive, wholesome food of high quality at reasonable cost; (2) to avoid accidents and damage to property and equipment; and (3) to promote good understanding and close working relationships among employees and supervisors. In these programs, emphasis is given to certain requirements common to all good job instruction, such as job knowledge, manipulative skills, human relations, adaptability, and ability to express oneself. These requirements are necessary for the instructor to be an effective teacher.

Teacher preparation for instruction to be given on the job includes the following tasks:

1. *Break down the job:* List principal steps, pick out the key points.
2. *Have a timetable:* How much skill do you expect your pupil to have and how soon?
3. *Have everything ready:* Make sure the right tools, equipment, and materials are at hand.
4. *Have the workplace properly arranged:* Arrange it in the way in which the worker will be expected to keep it.

After the preparation, the teacher sets about with the actual instruction:

1. *Prepare the worker:* Put the worker at ease. A frightened or embarrassed person cannot learn. Find out what is already known about the job. Begin where knowledge ends. Interest the worker in learning the job. Place the worker in the correct position so that the job won't be viewed from the wrong direction.
2. *Present the operation:* Tell, show, illustrate, and question carefully and patiently. Stress key points. Make them clear. Instruct slowly, clearly, and completely, taking up one point at a time, but no more than the trainee can master. Work first for accuracy, then for speed.
3. *Try out the worker's performance:* Test by having the worker perform the job under observation. Have the worker tell and show you, and explain key points. Ask questions and correct errors patiently. Continue until you know the worker knows.
4. *Follow up the worker's performance:* Let the worker perform alone. Check frequently, but do not take over if you can give the help needed. Designate to

whom the worker goes for help. Encourage questions. Get the worker to look for key points as progress is made. Taper off extra coaching and close follow-up until the worker is able to work under usual supervision. Give credit where credit is due.

A job breakdown is the analysis of a job to be taught and a listing of the elemental steps of what to do and the key points of how to do them. This serves as a guide in giving instruction so that none of the necessary points will be omitted. Figure 13.2 is an example of a job breakdown for making change. There should be a job breakdown for every task and/or job to be performed in the organization.

Slide-tape programs for individual instruction in work methods and procedures have proven to be satisfactory, and although it is time consuming to prepare such a program, the results appear to justify their use. Slides showing correct techniques are accompanied by oral explanations on tape. For techniques involving motion or rhythm, videotaping may be helpful.

Encouragement of the worker by the supervisor during the first days on the job and during the training period is important in stabilizing interest and sustaining a sense of adequacy. Informal interviews may serve as a means of determining areas in which help is needed, as well as those in which ability is most marked. Every expres-

Job: Making Change
Equipment and Supplies: Money and Cash Register

Important Steps	*Key Points*
REGISTER FIRST—WRAP AFTERWARD	
1. Accept money from customer.	1. State amount of sale* "out of" amount received from customer.
2. Place customer's money on plate.	2. Stand in front of cash register. Do not put bill in drawer until after change has been counted.
3. Record the sale on cash register.	3. Check amount of change recorded on viewer.
4. Count change from till.	4. Begin with amount of sale picking up smallest change first, up to amount received from customer.
5. Count change carefully to customer.	5. Start with amount of sale—stop counting when amount is the same as the customer gave.
6. Place customer's money in till.	6. Close the drawer immediately.
7. Deliver change, receipt or sales slip, and merchandise to customer.	7. Say *Thank You.* Let customer know you mean it.

*Including tax (state and federal).

Figure 13.2 Job breakdown for making change at the cash register. Important steps are "what to do"; key points are "how to do."

sion of friendly, courteous interest is appreciated by the worker and aids in a successful adjustment to the new environment.

In addition to the satisfaction attained by establishing pleasant employer-employee relations, the right induction of the new worker has a dollar-and-cents value that cannot be overlooked. An employee who is unhappy, disinterested, and discontented will tend to look for placement elsewhere after a short experience with the company. Then all the money, time, and effort spent in obtaining and introducing the employee to the job will have been lost, and a similar expenditure must be made before another worker can be assigned the task.

Training budgets have been steadily increasing in restaurants, hotels, and fast-food operations because industry experts see employee training as a solution not only to increased turnover, but also as a way to solve other current problems. Among employees and in business, decreased productivity and intense competition have been the stimulants for training programs that are both intensive and progressive.

Performance Appraisal

For maximum effectiveness from the workforce, every employee should know what is expected and how he or she is performing on the job. Workers are entitled to commendation for work well done and to the opportunity to earn greater responsibility, either with or without increased remuneration. One of the responsibilities of management and supervision is the performance appraisal. Management then has an obligation to communicate this information to each worker regarding individual progress. The personal development of and efficient production by each worker are of concern to management, but an individual worker cannot be expected to improve if evaluations are not made known or counsel is not made available for assistance. Performance appraisals are used to determine job competence, need for additional training or counseling, and to review the employee's progress within the organization. Ratings made objectively and without prejudice furnish valuable information that can be used in job placement, training, supervision, promotion, replacement, and future recommendations. Careful selection and placement and proper training of employees are prerequisites to a successful evaluation program. The performance appraisal may be accomplished by several methods, including rating scales, checklists, narrative evaluation, personal conferences, and management by objectives.

There are few, if any, objective standards that can be used for measuring subjective personal characteristics such as character, reliability, and initiative. Yet these traits, as they relate to the capabilities, efficiency, and development of each employee, are important to an organization. Such characteristics must be appraised in some way if management is to have an intelligent basis for classifying workers according to rank or grade and, thus, help to provide a standard for salary increases, promotions, transfers, or placement into a job for which the worker is well suited.

Rating procedures have been developed that provide a measurement of the degree to which certain intangible personality traits are present in workers and in their performance on the job. Care should be taken to design the scale to meet the objective desired. Will this estimate of the relative worth of employees be used as a basis for rewards or recognitions or as a tool for explaining to workers why they may or may not be making progress on the job? In the hands of competent administrators, the rating form can be designed to obtain information to accomplish both purposes.

Distinguishable personal traits most likely to affect performance are honesty, initiative, judgment, and ability to get along with other workers. Examples of qualities on a rating chart are quality of work, quantity of work, adaptability, job knowledge, and dependability.

These so-called rating scales, from which the variously known merit, progress, development, or service ratings are derived, are not new in industrial management, although few are directly applicable to foodservices. Some administrators prefer a system of gradation checking where each quality, factor, or characteristic may be marked on a scale ranging from poor to superior, or the reverse, with two or three possible levels within each grade. For example:

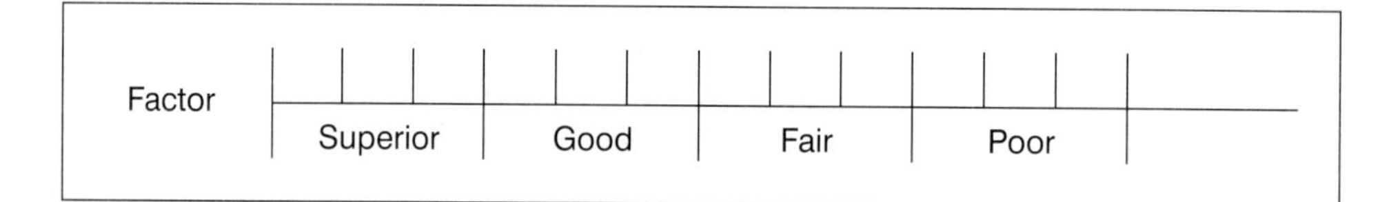

Another format might describe the grade for each factor listed.

Figure 13.3 is an example of a rating scale with definitions of the various factors attached for use of the rater. Figure 13.4 shows one with a different format.

The critical-incident appraisal method requires that supervisors identify behaviors that are indicators of excellent and poor performance. Throughout the rating period, records are kept of such **critical incidents** for each employee. These records are then compared to the previously identified indicators of performance appraisal.

Combining both the rating scale and critical-incident methods led to the **behaviorally anchored rating scale** (BARS) method of performance appraisal. BARS is generally considered the best and most effective means of appraisal. The disadvantage of this method is that the scales are difficult and time consuming to develop. An example of a BARS rating scale is shown in Figure 13.5. Specific behaviors are identified for each performance level for each job category. Employees are then rated according to the scale developed.

One of the original purposes of management by objectives (MBO) was to simplify and overcome the limitations of the more traditional performance appraisal. This approach emphasizes the setting of measurable performance goals that are mutually agreed on by the employee and the immediate supervisor. At stated intervals, the employee's progress toward the goals is assessed by the employee and the superior. Participation by employees in the performance appraisal process has resulted in favorable perceptions regarding the performance appraisal interview as well as positive performance outcomes.

Regardless of the rating systems selected, the person making the ratings should be well qualified for the responsibility of evaluating people. Usually the immediate supervisor is in a position to do the best job, since this person can observe activities continuously. However, adequate instructions are needed as to the purpose and values of the program so that follow-through with assistance is provided when needed. Also, a thorough explanation and understanding of the factors to be rated is neces-

EMPLOYEE PERFORMANCE REVIEW RATING

Name ________________________________ Date ____________________

Present job ____________________________ Department ________________

Pay range ______________ To ______________

Current rate______________ New rate ____________

Review period from ____________ to ______________
Date Date

Reason
for review: ☐ Probationary ☐ Annual ☐ Special ____________________
Explain

Date to be returned to Personnel Department ________________________

PART I

The purpose of the Employee Performance Review is twofold.

1. To identify the areas in which the employee is proficient, to encourage more effective utilization of known and demonstrated strengths, and to apply these proficiencies to the best advantage in accomplishing the objectives of the job and the department.
2. To identify areas in which improvement is desirable and to assist the employee to plan and execute a program of study, practice, and/or discipline to develop total competency in the job.

It is the responsibility of the employee's supervisor to do the utmost to see that these purposes are accomplished. With this in mind, make every effort to state your comments so that they are constructive and positive.

It is important to remember that a rating of "Successful Performance" represents the performance level of a fully competent employee who is successfully performing all job duties and responsibilities and is doing a good job. Performance levels either above or below "Successful Performance" should be fully documented in Part II with regard to what was done to merit the rating given.

Indicate your evaluation of the employee's performance on each of the factors below by placing a (✓) at the proper place on each scale.

For your convenience the midpoint on the rating scale has been indicated *but* it is important to remember that the evaluations criteria have been weighted. Therefore, the relationship between the score checked on the rating scale and the amount of the merit increase is minimal (a score of 5 on the rating scale does not mean a 5 percent increase).

Knowledge of Work. Consider the knowledge and understanding the employee has regarding the job. Does the employee know the methods or techniques to be used? Does the employee know the reasons for the procedure to be followed? Does the employee know the purpose of the job and what is required to be accomplished and how it contributes to the objectives of the department?

Evaluate:

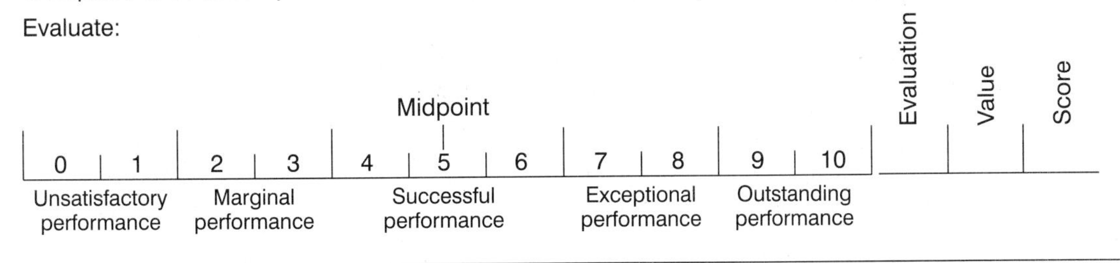

Figure 13.3 Scale for review of employee performance.

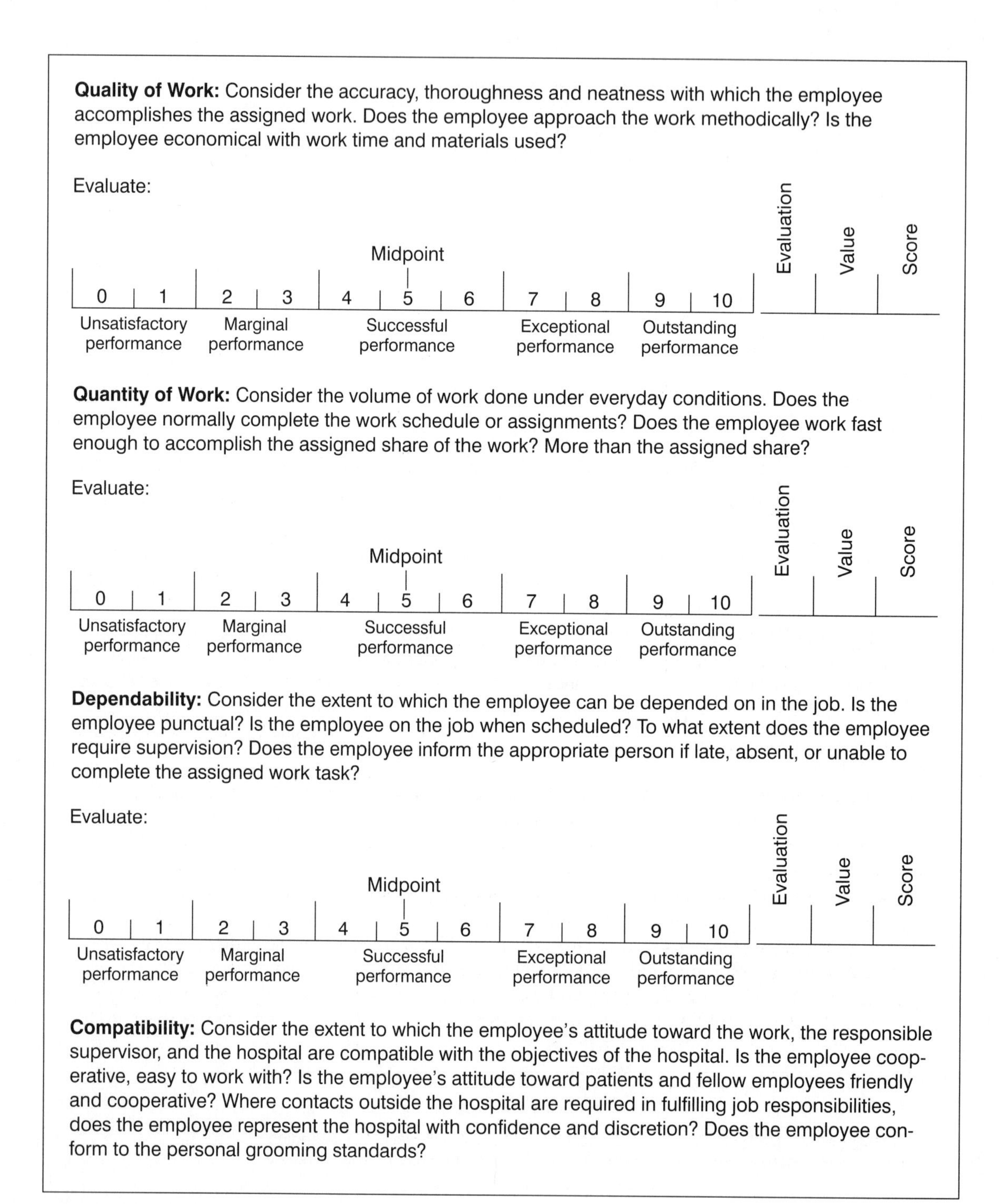

Quality of Work: Consider the accuracy, thoroughness and neatness with which the employee accomplishes the assigned work. Does the employee approach the work methodically? Is the employee economical with work time and materials used?

Evaluate:

Quantity of Work: Consider the volume of work done under everyday conditions. Does the employee normally complete the work schedule or assignments? Does the employee work fast enough to accomplish the assigned share of the work? More than the assigned share?

Evaluate:

Dependability: Consider the extent to which the employee can be depended on in the job. Is the employee punctual? Is the employee on the job when scheduled? To what extent does the employee require supervision? Does the employee inform the appropriate person if late, absent, or unable to complete the assigned work task?

Evaluate:

Compatibility: Consider the extent to which the employee's attitude toward the work, the responsible supervisor, and the hospital are compatible with the objectives of the hospital. Is the employee cooperative, easy to work with? Is the employee's attitude toward patients and fellow employees friendly and cooperative? Where contacts outside the hospital are required in fulfilling job responsibilities, does the employee represent the hospital with confidence and discretion? Does the employee conform to the personal grooming standards?

Figure 13.3 *Continued*

Initiative: Consider how well the employee applies the knowledge of work. Is the employee a self-starter who makes frequent practical suggestions? Does the employee proceed on assigned work voluntarily and readily accept suggestions? How much does the employee rely on others in getting started on assigned work? How effectively does the employee share acquired skills with others?

Evaluate:

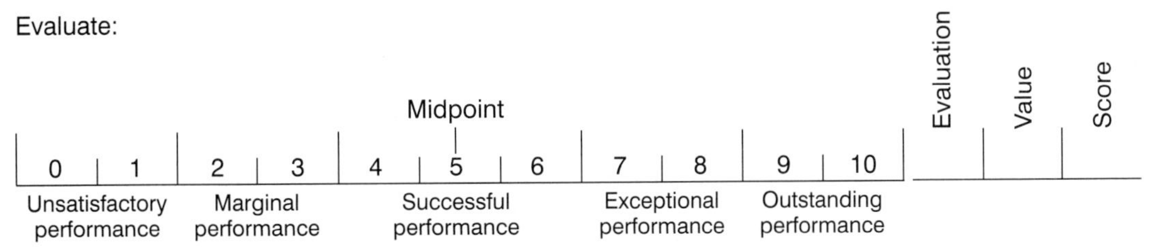

0	1	2	3	4	5 (Midpoint)	6	7	8	9	10	Evaluation	Value	Score
Unsatisfactory performance		Marginal performance		Successful performance			Exceptional performance		Outstanding performance				

Safety: Consider the extent to which the employee is aware of unsafe practices and conditions in the work setting. Is the employee quick to sense possible hazards and then to take the appropriate steps to get them corrected? How careful is the employee insofar as safe work practices are concerned with regard to regular work assignments? Does the employee always follow the established work procedure for assigned tasks?

Evaluate:

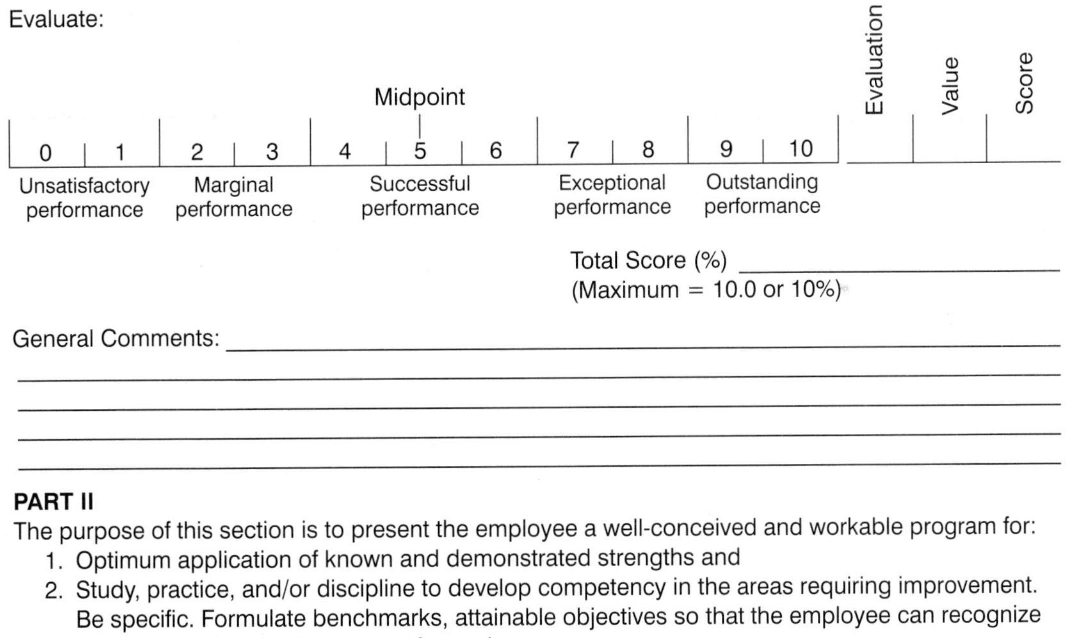

0	1	2	3	4	5 (Midpoint)	6	7	8	9	10	Evaluation	Value	Score
Unsatisfactory performance		Marginal performance		Successful performance			Exceptional performance		Outstanding performance				

Total Score (%) ____________________
(Maximum = 10.0 or 10%)

General Comments: ____

____

____

____

____

PART II

The purpose of this section is to present the employee a well-conceived and workable program for:

1. Optimum application of known and demonstrated strengths and
2. Study, practice, and/or discipline to develop competency in the areas requiring improvement. Be specific. Formulate benchmarks, attainable objectives so that the employee can recognize accomplishment and be aware of growth.

Knowledge of work: ____

____

____

____

Figure 13.3 *Continued*

Quality of work: ______________________________

____

____

____

____

Quantity of work: ______________________________

____

____

____

____

Dependability: ______________________________

____

____

____

____

Compatibility: ______________________________

____

____

____

____

Initiative: ______________________________

____

____

____

____

Safety: ______________________________

____

____

____

____

Prepared by ____________ ____________ ____________

Name Title Date

Department head ______________________ ____________

Name Date

FOR EMPLOYEE: My signature on this form indicates that I have reviewed my performance evaluation as presented on this form and a copy has been given to me.

For Probationary Rating ONLY

Recommended for:

☐ Permanent Status

☐ Extension of Probation

☐ Termination

____

Signature Date

Figure 13.3 *Continued*

Courtesy of Lincoln General Hospital, Lincoln, Nebraska.

Associate Rating Form: Kitchen Worker

1. *Job Skill* (Max. points: 25) Consider job performance and skill. Does the worker keep up with the work and keep the station clean; make all products uniformly, waste conscious, and economical; work quietly and reasonably fast; refrain from visiting with fellow associates while on duty?	Excellent.....25 Good...........20 Average ......15 Fair..............10 Poor 5
2. *Cooperation* (Max. points: 25) Consider attitude. Does the worker respond quickly to a call for assistance from a fellow associate? Have a spirit of willingness? Receptive to change and new ideas? Accept new suggestions regarding his/her work?	Excellent.....25 Good...........20 Average ......15 Fair..............10 Poor 5
3. *Sanitation* (Max. points: 10) Consider health regulations: "No Smoking—Wash Hands When Leaving Rest Rooms." Does worker keep paper, trash, liquids, vegetable leaves, and other foreign materials off floor; keep hot foods hot, and other food under refrigeration?	Excellent.....10 Good............8 Average ....6 Fair................4 Poor 2
4. *Care of equipment* (Max. points: 10) Does worker keep equipment clean and everything returned to proper place, know correct way to operate ovens, steamers, mixers, and other appliances?	Excellent.....10 Good............8 Average ....6 Fair................4 Poor 2
5. *Safety* (Max. points: 10) Does worker work safely and is worker safety conscious? Correct or report all hazards that may cause an accident? Know whereabouts of fire extinguisher and how to use it?	Excellent.....10 Good............8 Average ....6 Fair................4 Poor 2
6. *Appearance* (Max. points: 10) Consider personal cleanliness and neatness. Does worker seem to enjoy the work? Is he/she clean of body? Are clothes clean and appropriate?	Excellent.....10 Good............8 Average ....6 Fair................4 Poor 2
7. *Attendance* (Max. points: 10) Consider regular daily attendance and promptness. Does worker return from 10 minute breaks and meal periods on time?	Excellent.....10 Good............8 Average ....6 Fair................4 Poor 2

Figure 13.4 An example of a rating form designed for evaluating a kitchen worker. Comparable forms could be made applicable for each classification of worker.

Performance Factors	Superior	Very good
1. Quality of work Evaluate accuracy, thoroughness, and neatness of completed work. Disregard the quantity of work.	Rarely commits an error. Defect or reject rate is less than 0.003%. Quality of work is consistently exceptional. 20	Makes only an occasional error. Defect or reject rate is consistently less than 0.010%. Quality of work is high grade, but not exceptional. 16
2. Quantity of work Evaluate amount of work performed and/or number of assignments completed, sales calls made, and so on. Disregard quality of work.	Regularly exceeds specified output, volume of work, or number of assignments by more than 15%. Consistently completes an extraordinary amount of work. 20	Regularly exceeds specified output, volume of work, or number of assignments. Consistently turns out a good volume of work. 16
3. Dependability Evaluate ability to meet commitments and deadlines and the extent of required supervision.	Consistently meets commitments and deadlines. Needs no supervision on routine tasks. 20	Meets commitments and deadlines 95% or more of the time. Needs minimum supervision on routine tasks. 16
4. Attitude Evaluate general demeanor toward job, co-workers, supervisor, and company.	Displays enthusiasm about the work and the company. Cooperates freely with associates. Always accepts suggestions and criticism in a constructive manner. 10	Appears to be happy at his or her work. Cooperates freely with associates. Usually accepts suggestions or criticism in a constructive manner. 8
5. Initiative Evaluate ability to recognize problems and take corrective action, make suggestions for improvements, and accept responsibility for accomplishing unassigned tasks.	Regularly recognizes job-related problems and initiates corrective action. Has made at least three well-thought-out suggestions for work improvement (formally or informally) during the past evaluation period. 5	Usually recognizes job-related problems and initiates corrective action. Has made at least one well-thought-out suggestion for work improvement during the past evaluation period. 4
6. Housekeeping Evaluate cleanliness and orderliness of workplace and in-process storage areas and close-of-shift cleanup.	Maintains an exceptionally neat and orderly workplace, file drawers, and storage areas. Work surfaces are clean and uncluttered. Tools are always arranged neatly and/or put away at close of shift. 10	Maintains a neat and orderly workplace, file drawers, and storage areas. Work surfaces are usually clean and uncluttered. Tools are put away at close of shift. 8
7. Attendance Evaluate attendance and tardiness.	No days lost. No times late. 10	1 to 2 days out sick or 1 day absent of own accord, or 1 time late. 8
8. Potential for growth and advancement Evaluate potential for increasing job knowledge and for advancing to other jobs in the department and other jobs in the organization.	Given an opportunity, can assimilate more knowledge of company operations. Has demonstrated great potential for advancement in the entire organization. 5	Given an opportunity, can assimilate more knowledge of company operations. Has demonstrated good potential for advancement in the department. 4

Figure 13.5 Example of the behaviorally anchored rating scale (BARS).

At expected level	Below expected level	Unsatisfactory
Errors are only occasionally troublesome. Defects or reject rate rarely exceeds standard of 0.010%. Quality of work is average. 12	Errors are frequently troublesome. Defect or reject rate often exceeds standard of 0.010%. Quality of work is below average. 8	Errors are frequently troublesome. Defect or reject rate regularly exceeds 0.010%. Quality of work is unsatisfactory 4
Usually meets the specified output, volume of work, or number of assignments. Amount of work completed is about average for this job. 12	Often fails to meet specified output, volume of work, or number of assignments. Amount of work completed is about 10% less than average for this job. 8	Regularly fails to meet specified output, volume of work, or number of assignments. Amount of work is almost always greater than 11% less than average for this job. 4
Meets commitments and deadlines 90% or more of the time. Needs occasional supervision on routine tasks. 12	Meets commitments and deadlines less than 85 to 90% of the time. Needs constant checking, even on routine tasks. 8	Meets commitments and deadlines less than 95% of the time. Work and progress must be checked all the time. 4
Accepts most assignments without complaint. Cooperates with associates when requested to do so. Follows instructions. 6	Frequently questions suitability of assignments. Complains regularly about the nature of the work. Cooperates with associates when requested to do so. Often rejects suggestions or criticism. 4	Constantly complains about the work and the company. Regularly voices objection to assignments. Does not cooperate with co-workers, is always negative toward suggestions and criticism. 2
Occasionally recognizes and acts upon job-related problems. Occasionally makes well-thought-out suggestions for work improvement. 3	Fails to recognize job-related problems or to initiate action to correct them. Usually waits to be told what to do. 2	Fails to recognize job-related problems or to initiate action to correct them. Always needs to be told what to do. Never displays any kind of initiative. 1
Maintains a reasonably neat and clean workplace. Work surfaces are acceptably free from soil or debris that would interfere with work. Only occasionally does not put tools away. 6	Often fails to maintain a reasonably neat and clean workplace. Work surfaces are often cluttered and not conducive to quality craft. Often fails to put tools away. 4	Consistently fails to maintain a neat and clean workplace. More often than not, fails to clean up or put tools away at close of shift. 2
2 to 3 days out sick, or 2 days absent of own accord, or 2 times late. 6	3 to 5 days out sick, or 3 days absent of own accord, or 3 times late. 4	More than 5 days out sick, or more than 3 days absent of own accord, or more than 3 times late. 2
Has pretty much acquired as much knowledge here as he or she can assimilate. Has demonstrated some potential for advancement in the department. 3	Has difficulty in acquiring knowledge and skills here. Has demonstrated only limited potential for advancement. 2	Has great difficulty in acquiring knowledge and skills here. Has demonstrated no potential for advancement here. 1

Figure 13.5 *Continued*
From Bittel, L. R., and Newstrom, J. W.: *What Every Supervisor Should Know,* 6th ed. New York: McGraw-Hill, 1990, pp. 189–190. Reproduced with permission of McGraw-Hill.

sary to avoid misinterpretation of the forms or to avoid failure to meet the intended standards. The person who is charged with the responsibility of rating employees must be objective and able to evaluate individuals in terms of the factors to be rated, to be guided by the pattern of performance instead of isolated happenings, to communicate fairly and accurately what is observed, and to be consistent from one time to the next. These are prime requisites of the rater.

An interview is a vitally important part of the performance appraisal process. The purpose of the interview is to provide information and set goals. It should be scheduled in advance and both employee and supervisor should be prepared. The proper atmosphere for two-way communication must be established. The supervisor should begin with a statement of purpose and then encourage the employee to participate in the dialogue. Total performance, both positive and negative, should be discussed. The evaluation stage of the interview should conclude with a summary and documentation of the interview for the employee's file. In the second stage of the interview, the emphasis should be on setting mutual goals, including personal growth and formulating follow-up procedures. The employee should never be left in doubt about the rating given, or about what must be done to change or improve if there is a need to do so.

A relatively new trend that has been gathering momentum is the implementation of upward feedback performance appraisals, in which employees rate their bosses' people management skills. Upward feedback performance appraisals have proven to be valuable tools for the improvement of management performance and subordinates' morale. They provide employees with an opportunity to have a say in how they are supervised.

Promotions and Transfers

On the basis of sound ratings by members of staff, the foodservice manager is fairly able to predict the probable future development of various members of the organization. In the application of a rating scale, one group may rate high. This scale helps to identify the people deserving the stimulus and encouragement toward promotion. The term *promotion* commonly implies an increase in responsibility and salary. Sometimes promotion carries only the opportunity for experience in a desired field. It may mean shorter hours and greater assurance of security. Regardless of the nature of the promotion, it is an expression of appreciation of an individual's worth.

Often a worker found unfit for one job may do well in another. The apparent lack of fitness may arise in the supervisory relationship or in contacts with coworkers. Personal prejudice against a particular type of work or physical inability to do the job may be the cause. In some cases, a minor shift may enable the worker to become a contented and valuable employee. Transfer of an employee who is not finding satisfaction in the current job to another opening within the organization offering a different challenge or opportunity has salvaged many workers. Different jobs may present wide variation in skill requirements, which makes possible the transfer of workers if necessary. Relative levels of difficulty should be considered in placement and in replacement. Continued training of the employee in this new position is critical to the success of a relocated worker.

Discipline

Discipline is required when other measures have failed to make sure that workers perform according to accepted standards. Leadership must first ensure that work rules are clear, reasonable, fair, reviewed regularly, and consistent with the collective bargaining agreement. Rules must be disseminated to employees orally and posted in a visible location. And they must be enforced promptly, consistently, and without discrimination. Leadership must set a good example by complying with all rules and requirements.

Any disciplinary action must be undertaken with sensitivity and sound judgment. The supervisor should first thoroughly investigate what happened and why. As a general rule, disciplinary action should be taken in private. Personnel policies usually fit the severity of the penalty to the severity of the infraction, with the steps in progressive discipline ranging anywhere from an informal talk, an oral unrecorded warning or reprimand, a written or official warning, a disciplinary suspension, a demotion or transfer, to a discharge.

As an aid to supervisors, Metzger (1982) proposed the "hot stove" analogy to disciplinary action. Experiencing discipline should be like touching a hot stove. The burn gives advance warning, is immediate, consistent, and impersonal:

- *Advance warning:* Everyone knows what will happen if you touch a hot stove. Employees should know what is expected of them.
- *Immediate:* The burn is immediate. Discipline should not be hasty, but should be taken as soon as possible after the infraction.
- *Consistent:* The hot stove burns every time. Disciplinary action should be taken every time an infraction occurs.
- *Impersonal:* Whoever touches the stove is burned. The act not the person should be disciplined.

After disciplinary action, the employee should be treated as before.

Dismissals

If an employee is terminated without the consent of the employee, the act is termed *dismissal.* An individual may be discharged because of failure to perform assigned duties, but this should be the final step and should follow counseling, warning, or possibly disciplinary layoff. Each person discharged from a foodservice should be given a terminal or "exit" interview in which strong points are recognized and the reasons for dismissal are dispassionately reviewed. If the situation merits that the employee be recommended for another position, aid should be given in the placement problem. In any event, the discharged employee should not leave the service without having had a chance to speak regarding the dismissal and without being made aware, if possible, of the fair deal given by the supervisor.

Opinions differ regarding the discussion of a dismissal with other employees. If there is a possibility that the incident may foster a sense of insecurity among the group, a presentation of the facts, not necessarily full and complete, may be desirable from the standpoint of group morale. Often, employees understand far more of such situations than the director believes.

Supervision

Supervision encompasses coordinating and directing the work of employees to accomplish the organization's goals. In small foodservice systems, the total supervisory function may be the responsibility of the manager. In larger systems, the supervision of the day-to-day technical operations may be delegated to foodservice supervisors, dietetic technicians, or cook-managers. The manager is thus able to concentrate on planning, policy and goal setting, and interdepartmental relationships, and on solving overall problems of the department. In a large department, the director, chief dietitian, or other administrator may delegate these management functions in part to other professionally trained staff.

When responsibility and authority are delegated, management must provide guidance so that the supervisor understands the limits of authority; that is, what decisions can be made without consultation and what actions can be taken on one's own. Management has a responsibility for training supervisors so that they can solve problems and meet emergencies.

The supervisor represents both management and employees. In a foodservice unit, as with industry in general, the supervisor is one of the key persons in the organization. The supervisor is the one to whom the employees look as a representative of management, whereas to management the supervisor represents the working force. Both groups, therefore, are interested in the quality of the supervision as represented by this staff member. The supervisor must be able to (1) interpret the objectives and policies of the company to the employees in such a way as to encourage their cooperation and elicit their confidence, and (2) inspire and lead employees as evidenced through fair and intelligent dealings with them and through the personnel program.

Throughout an employee's term of service, supervision should play a large part in relating the employee to the task and to coworkers. When the probation period is past and the employee is regarded as a member of the permanent staff, familiar with the task and able in its performance, supervision is still necessary to maintain interest and provide for personal growth. To a large extent, recognition and approbation by superiors remain potent incentives to the average worker. The supervisor must accept the responsibility of finding and using incentives that lead to sound development. Adjustments in work assignments to meet changes in the individual's abilities and interests are wisely made only when supervision is adequate, both in kind and amount.

Routine Supervision. Routine supervision varies with the situation, but it is, for the most part, a matter of personal contact reinforced through checking by observations, records, and charts. Routine supervision may consist of greeting employees each day by name; checking for cleanliness, appearance, and state of health; checking menus and work schedules; making work assignments; explaining to employees any instruction they seem not to understand; checking continuously for quality and quantity of production and service; inspecting for sanitation of work areas and equipment; and, in general, maintaining good working conditions. The supervision of personnel is too often left to chance or to the "free time" that never seems a part of the foodservice manager's busy day. To avoid the hit-and-miss contact with employees, the wise supervisor sets aside a certain time each day for checking on the work in progress and for stimulating interest and cooperation in the individual and in the group. Schedules are needed for checking daily, weekly, and periodic jobs. Checking at the end of the day to see that the work as scheduled has been carried out completes the "routine" supervision.

Decision Making

Much of the supervisor's time is spent making decisions and solving problems. Decision making can be thought of as the generic process, whereas problem solving is one type of decision making that applies to a specific situation.

The ability to make decisions in a timely and logical manner is an important skill for supervisors to possess. Better decisions are likely to be made when a number of steps are followed in the proper sequence. These steps are to (1) define the problem—nothing is as useless as the right answer to the wrong question, (2) analyze the problem, (3) develop many alternatives—brainstorm and then think them through considering the consequences of each, (4) choose the alternative with the most positive consequences and the least negative consequences, and (5) follow-up and appraise the decision.

Handling Grievances

The wise supervisor gives active supervision; that is, the wise supervisor does not sit at a desk waiting for employees to come with problems. The wise supervisor foresees and is prepared to meet possible difficulties instead of merely waiting for something unpleasant to happen. Grievances are not always expressed in verbal or written form. Supervisors should be alert for symptoms of unexpressed dissatisfaction such as excessive tardiness or absenteeism, decline in quantity or quality of work, change in attitude, or indifference.

Many grievances can be settled by the supervisor and employee on an informal basis. If the employees are unionized, the contract includes formal grievance procedures, which usually include presentation of the grievance in writing (see Figure 13.6) and an attempt to settle the dispute at the first-line supervisory level. If this is not possible, the grievance moves through higher levels of authority until settled.

Staff Conferences

Regular staff conferences, department meetings, and the use of rating scales are all valuable in personnel direction. Continued effort to relate workers to their tasks and to the organization as a whole is often expressed in conferences scheduled at regular intervals by the supervisor. At these conferences, points of general interest are presented and suggestions for improvement of the foodservice are exchanged. Knotty problems, such as waste, breakage, and low productivity, that have not been mastered by a direct supervisory approach can be resolved focusing the interest and awareness of the whole group on them. Never should a staff conference be used for disciplinary action for certain members of the group. As previously stated, the adult worker, like the school-child, rarely benefits from public reprimand and unkind ridicule.

Lincoln General Hospital
Lincoln, Nebraska

GRIEVANCE CONTROL FORM

1. ____
(Name) (Position Title) (Dept.) (Employment Date)

______________________ ______________________
(Supervisor) (Department Head)

2. *Nature of Dissatisfaction:* (Be specific in writing up the dissatisfaction. Be sure to state who is involved, what happened, where did it occur, what were the circumstances, how did it happen, and what is the solution being sought.)______________________

3. *The Final Settlement Was:* (Resolution at what step; what was the recommended solution and all other pertinent information significant to the solution.)______________________

4. *Action Taken:*

*Work Days	Step	Date	Action Taken By:	Schedules Met (Yes/No)	**Work Days	Resolved (Yes/No)
2	I	______	______ (Supervisor, Verbal)	______	2	______
2	II	______	______ (Department Head, Verbal)	______	2	______
5	III	______	______ (Grievance Committee, Written)	______	2	______
4	IV	______	______ (Administrator, Written)	______	2	______

Comments: (Regarding any exceptions made at any step)______________________

* Any of the time limits specified in the procedure may be extended by mutual agreement of the parties involved.

** In the event that the employee fails to appeal the decision to the next step in the grievance procedure within two (2) working days, the employee will be deemed to have accepted the decision and waived the right to further appeal.

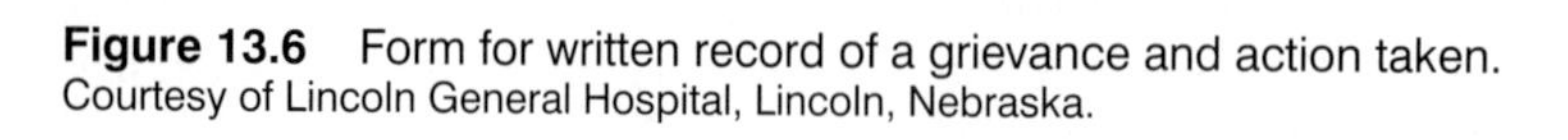

Figure 13.6 Form for written record of a grievance and action taken.
Courtesy of Lincoln General Hospital, Lincoln, Nebraska.

In addition to group contact, time should be taken by the supervisor for a talk with each individual worker at least once a week. All employees want to feel that someone is interested in them as an individual and recognizes their present and potential worth to the organization.

Labor Policies and Legislation

Policies are guides for future action. They should be broad enough to allow some variation in management decisions at all levels, yet offer guidelines for consistency in interpretation, and to commit personnel to certain predictable actions. Policies should not be confused with directives or rules. Policies are adopted to provide meaning or understanding related to a course of action; directives and rules are aimed at compliance.

An important aspect of personnel management is the labor policies accepted and put in force. This is true regardless of the size of the organization. There is an old saying that when two men (or women) meet, there is a social problem; when one undertakes a task at the other's behest, there is a labor problem; and when wages are paid for this labor, there is an economic problem. The policies controlling the approach to these problems have slowly developed as civilization has grown and as the number of workers has increased. They have been formed, reformed, and revised, particularly in recent years, due to legislation enacted at federal, state, and local levels.

Policies relating to personnel are known as labor policies. Procurement policies may be related to preferred sources to be used for obtaining applicants, instruments such as tests to be used in selection, or a ratio of employees, such as women to men or minority to major racial groups to be hired. Policies for development of personnel may concern the type of training programs the company offers, whether or not fees or tuition for continuing education are paid, the amount of time to be allowed from work for personnel to attend classes or meetings, and the bases for promotions and transfers.

Those policies regarding compensation have to do with wage scales to be followed; vacation, sick leave, and holiday pay to be given; bonus or profit-sharing plans to be offered; and group insurance or other benefits that are available to the personnel.

Integration policies refer to whether labor unions are recognized, the way in which grievances and appeals are handled, or the degree of employee participation to be permitted in decision making.

Maintenance policies are about the services to be provided for employees' physical, mental, and emotional health. They may be related to safety measures, compensation for accidents, retirement systems, recreational programs, or other services, all of which are a part of the organizational plan.

Once policies have been developed and accepted, they should be written. The wise employer of today makes available to every worker a copy of the labor-management policies presented in a company handbook. This publication may be an impressive volume of many pages and elaborate illustrations or a few mimeographed sheets, but whatever its format, the contents should include information that the worker wants to know about the organization and that the employer wants him or her to know. Employees are not interested in cooperating as members of the team without understanding the policies, especially as these affect them. They want to know what is expected of them and to be kept informed of their accomplishments, the basis for promotion and/or wage increases, fringe benefits, opportunities for steady work, and the possibilities of any seasonal layoffs.

From the standpoint of the employee, labor policies should be explicit in their provisions for a fair rate of pay, for promotions and transfers, for stabilization of employment, and for ways of keeping jobs interesting so that life does not become mere dull routine. They should offer provisions for fair disciplinary action among employees, recognition of industrial health hazards and provisions for their control, participation in the formulation of future plans and policies of the company, usually expressed by demands for collective bargaining, and certain fringe benefits.

Managers wish to have employees informed of policies about the goals and objectives for which the organization is in business, the goods and services offered, the effect of high productivity as a benefit to both the employee and the company, cost-expenditure ratios and how they affect profits and resulting benefits, and the relationships desired with the public and with other departments of the organization.

There is general agreement on the list of topics that the employer has found must be covered in labor policies conducive to productive management and those desired by employees as vital to satisfactory working conditions. The ones cited by both—wages and income maintenance; hours of work; schedules and overtime provisions; security in employment, including transfers and promotions; safe and otherwise satisfactory working environment; insurance, retirement, or pension plans; equal employment opportunities; and fair employment practices and civil rights—may be regarded as the major issues in labor policies for most foodservice operations.

These topics can be grouped under four headings: wages and income maintenance, hours and schedules of work, security in employment, and employee services and benefits. Major federal legislation applicable to employment in the private sector is included as appropriate under each of the following discussions.

Wages and Income Maintenance. From the point of view of the worker, the most important characteristic of the wage, the take-home pay received for labor performed, is its purchasing power. This represents the measure of the wants that the worker is able to satisfy and largely determines the adequacy of his or her standard of living, sense of financial security, and identification of self as a worthy and responsible member of the community. In the past, foodservices, like other service organizations, have tended to offer an annual wage rate below that necessary for a fair standard of living. This situation has improved as desirable policies on wages have been adopted and as state and federal laws have been enacted.

The formulation of satisfactory policies regarding wages and other income maintenance is contingent on many factors, among them (1) the desire and intent of the company to pay fair wages to all employees and at the same time to maintain just control over labor costs; (2) recognition of the relationship between the duties and responsibilities of various jobs within the organization and the wages paid; and (3) acknowledgment of individual differences in experience, ability, and willingness to take responsibility. Management has the obligation to reflect such differences in the wage scale established for a particular job and to communicate freely with the workers on these points. Policies based on such considerations will lead to a systematic classification of jobs and wages that could be developed jointly by the employer and the employees. It would then be possible to express the value or worth of each job in terms of wages.

The application of the wage policy to kitchen and dining room personnel would lead to certain groupings, such as

1. Busers, pot and pan washers, dishwashers;
2. Workers in preliminary or pre-preparation;
3. Foodservice groups, including counter workers and waiters or waitresses;
4. Cook's assistants and second cooks, dining room host or hostess, cashiers;
5. Cooks, including meat, vegetable, salad, and pastry cooks; and
6. Supervisors on the nonprofessional level.

A wage differential will exist between groups. Civil service and labor unions, as well as many other organizations, have established steps within each wage level or grade so that employees who merit wage increases may be given such recognition for superior service, although not qualified for advancement to a higher grade or job category.

The *Fair Labor Standards Act of 1938,* also known as the Federal Wage and Hour Law, was first enacted to help eliminate poverty, to create purchasing power, and to establish a wage floor that would help prevent another depression. The minimum wage set at that time was $.40 per hour! The base has gradually increased over the years. The act was amended in 1966, and, under new provisions, most foodservice employees were included for the first time. The minimum wage that year was $1.60, and the law included provisions for gradual increases that would continue as cost of living increased. The act applies equally to all covered workers regardless of sex, number of employees, and whether they are full-time or part-time workers.

The *Equal Pay Act,* a 1963 amendment to the Fair Labor Standards Act, prohibits employers from discriminating on the basis of sex in the payment of wages for equal work for employees covered by the act. It requires employers to pay equal wages to men and women in their employ doing equal work on jobs requiring equal skill, effort, and responsibility that are performed under similar working conditions.

Another provision of the Fair Labor Standards Act of special interest to restaurant foodservice managers relates to wages for tipped employees. Tips received by an employee may be considered by the employer as part of the wages of the employee, but cannot exceed 50% of the applicable minimum rate. A "tipped" employee is a worker engaged in an occupation in which the worker customarily and regularly receives more than $30 a month in tips.

Many foodservice operations employ student workers; this is especially true in colleges and universities, schools, retirement homes, and other homes for congregate living. Minimum-wage laws adopted by various states may make provision for compensation at an adjusted rate below the federal standard. Usually students who work fewer than 20 hours per week are not affected by provisions of such laws.

Unless specifically exempt by this law, all employees must be paid at least one and one-half times the employee's regular rate of pay for all hours worked in excess of 40 hours in a work week of 7 days. Extra pay is not required for Saturday, Sunday, holiday, or vacation work.

All foodservice managers should become familiar with the state and federal laws regulating minimum wages for their various classifications of employees. Information may be obtained from the nearest office of the Wage and Hour Division of the U.S. Department of Labor.

Unemployment compensation is another piece of federal legislation that, in addition to regular pay for work on the job, partially assures income maintenance. This nationwide system of insurance to protect wage earners and their families against loss of income because of unemployment was first established under the Social Security Act of 1935. The purpose of this insurance is to provide workers with a weekly income to tide them over during periods of unemployment between jobs. Persons covered must have been employed for a specific period of time on a job covered by the law, be able and willing to work, and be unemployed through no fault of their own.

Unemployment insurance is a joint federal-state program, operated by the states with the assistance of the U.S. Department of Labor. Each state has its own specific requirements and benefits. Basically, employers pay a tax based on their payrolls. Benefits to unemployed workers are paid out of the fund built up from these taxes. In most states, firms employing three or four or more workers for 20 weeks throughout the year must participate. Each state law specifies conditions under which workers may receive benefits, the amount they receive, and the number of weeks they may draw benefits. In most states, the employer alone contributes to this fund; in only a few do employees make payment to it. Thus, unemployment compensation is an added payroll cost for many foodservice managers and an added benefit to the employees.

Hours and Schedules of Work. The 40-hour work week established under the Minimum Wage and Hour Law is generally in use throughout the United States. Some organizations have adopted a 37½- or a 35-hour week. Time worked beyond 40 hours in a 7-day or 80 hours in a 14-day period (in hospitals or other facilities that care for the sick, elderly, or persons with mental illness) as specified under the law requires extra compensation, as previously noted.

The schedule of specific hours of the day when each employee is to be on duty should be carefully considered by all foodservice managers. As discussed, many different factors enter into the planning of satisfactory schedules. Employers have a responsibility for scheduling their employees so that their time at work is as needed and is used to their best advantage in order to help control labor costs. Split shifts are almost a thing of the past; straight shifts are usually preferred. An 8-hour day, 5 days a week is common practice also. However, some organizations have experimented with variations, notably a 10-hour day and 4-day week to allow a 3-day off-duty period for the employees and a 12-hour day, 3-day week. Most foodservice organizations have not found this scheduling practical because of the nature of the work to be done.

In addition to the needs of the employer and the organization, consideration is given to the employee and to stipulations in union contracts, if in effect, when planning scheduled time on and off duty for each member of the staff. Most state labor laws require break times for meals and between-meal rest periods for employees, which is a further consideration when planning schedules to cover work that must be done. Familiarity with these regulations is a necessity for the manager.

Security in Employment. One of the major concerns of the working world in recent years is equal opportunity for employment for all persons who desire employment and who are qualified. **Equal employment opportunity** (EEO) is the umbrella term that encompasses all laws and regulations prohibiting discrimination or requiring affirmative action. The **Equal Employment Opportunity Commission** (EEOC) regulations and interpretive guidelines provide guidance to management for compliance with Title VII of the Civil Rights Act of 1964, the Age Discrimination in Employment Act, the Pregnancy Discrimination Act, and the Americans with Disabilities Act (ADA) of 1990, all of which are federal EEO statutes.

Most states also have legislation that prohibits discrimination. In some cases, these statutes are broader than the federal laws. Marital status, sexual preference and orientation, race, color, religion, national origin, or ancestry are demographics protected from discrimination in various states. These are designated as fair employment practice laws. A public accommodation law, when in effect, requires that service be given in an equal manner to all persons.

The *Civil Rights Act of 1964* stipulates that "No person in the United States shall, on ground of race, color, or national origin be excluded from participating in, be denied the benefits of, or be subjected to any program or activity receiving Federal Financial assistance." Title VII under this act extended the provision to include prohibition of discrimination "by employers, employment agencies and labor unions."

Thus, employees who are in covered positions are entitled to be free of unlawful discrimination with regard to recruitment, classified advertising, job classification, hiring, utilization of physical facilities, transfer, promotion, discharge wages and salaries, seniority lines, testing, insurance coverage, pension and retirement benefits, referral to jobs, union membership, and the like. All potential employees have equal opportunity, regardless of background.

In 1974, Title VII of the Civil Rights Act was amended to include prohibition against discrimination based on religion and sex. Then, in 1978, sexual discrimination was further broadened with the passage of the Pregnancy Discrimination Act, which prohibited discrimination due to pregnancy, childbirth, or related medical conditions.

The *Civil Rights Act of 1991* increases the likelihood that employees will sue because discrimination cases will be easier to win and the damages that are awarded would be more substantial. This act does not make anything illegal that wasn't already illegal, but it does relax the burden of proof and make possible recovery for pain, suffering, and punitive damages.

The *Age Discrimination in Employment Act of 1967* promotes the employment of the older worker, based on ability instead of age. It prohibits arbitrary age discrimination in employment and helps employers and employees find ways to meet problems arising from the impact of age on employment. The act protects most individuals who are at least 40 but less than 70 years of age from "discrimination in employment based on age in matters of hiring, discharge compensation or other terms, conditions, or privileges of employment."

Sexual harassment takes two forms and is a form of sexual discrimination that violates federal, state, and most local laws. The first form is *quid pro quo* and occurs when a supervisor either rewards or punishes a subordinate for providing and or not providing sexual favors. The second form is the *hostile work environment,* which occurs when an employee's ability to work is undermined by an atmosphere infused with unwelcome sexually oriented or otherwise hostile conduct created by supervisors or coworkers.

The **Americans with Disabilities Act** of 1990 (ADA) prohibits discrimination against qualified persons with disabilities in all aspects of employment from application to termination. No job may be denied an individual with a disability if the individual is qualified and able to perform the essential functions of the job, with or without reasonable accommodation. If needed, an employer must make the reasonable accommodation unless it would result in undue hardship for the employer. Existing performance standards do not have to be lowered, but such standards must be job related and uniformly applied to all employees and applicants for that particular job. Equal opportunity must be provided to individuals with disabilities to apply and be considered for a job. The applicant cannot be asked preemployment questions regarding his or her disability, but the applicant can be asked about his or her ability to perform specific job functions. He or she may also be asked to describe or demonstrate how these job functions could be accomplished. Medical histories and preemployment physical exams are not allowed under the ADA legislation; however, the job offer may be conditional depending on the results of a post-offer physical exam. This exam must be required of all applicants in the same job category. Tests for the use of illegal drugs are not considered to be physical exams under the ADA and are, therefore, still legal.

Quotas are fixed, inflexible percentages or numbers of positions that an employer decides can be filled only by members of a certain minority group. This is a form of reverse discrimination and is almost always illegal.

Affirmative action does not involve the setting of specific quotas but rather the desire to reach general goals to increase the numbers of women and minorities in specific positions. Required for certain federal contractors, affirmative action is legal if ordered by a court to remedy past discrimination or if limited in time and scope. Currently, the continued existence of affirmative action policies is being hotly debated based on the assertion by some that it results in reverse discrimination.

In an effort to stem the tide of illegal immigrants coming into the country, Congress passed the Immigration Reform and Control Act in 1986. This act makes it illegal to recruit or hire persons not legally eligible for employment in the United States. Employers must complete an I-9 form for each employee to verify eligibility to work in the United States. Further, any employer who has four or more employees is prohibited from discriminating against employees or job applicants on the basis of national origin or citizenship status.

As can be seen, our economic society is characterized by many areas of friction in industry. Students of labor quite generally agree that in no area is there an economic problem more important to human beings than security of job tenure, which means assurance of the satisfaction of physical needs, a place in the esteem and affection of others, an opportunity for self-expression, and a chance to enjoy leisure. The three risks that more than any others tend to make the position of most wage earners in industry insecure are unemployment, physical impairment, and old age. The definition for unemployment used by the Bureau of Census in making its enumeration is: "Unemployment may be described as involuntary idleness on the part of those who have lost their latest jobs, are able to work, and are looking for work." This definition is obviously narrow, because it excludes all those persons who are unwilling to work, are unemployable because of physical or mental disabilities, or are temporarily idle for seasonal reasons. However, the definition covers the group whose unemployment usually arises from conditions inherent in the organization and management of industry.

Problems of tenure must concern all persons charged with the direction of the foodservice industry. Fortunately, foodservices on the whole lend themselves to steady employment, and many managers take pride in the long tenure of large numbers of their workers. Sometimes, however, the workers' acceptance of tenure as a matter of course brings definite problems such as laxity and inefficiency in the performance of assigned tasks and lack of interest in improved practices. Standards of performance in some instances have been lowered as security of employment has been assured. Personnel policies should cover such contingencies.

Employee Services and Benefits. Benefits that employees receive often represent as much as 39% of wages earned. Some of these are so taken for granted that they are scarcely realized or appreciated by those who receive them. Yet, if such services were not provided, the lack would be acutely noticed. Managers recognize the humanistic desirability of making available certain programs and services in addition to a fair wage for their employees' comfort and well-being. A less altruistic point of view may cause managers to offer those same benefits in order to compete in the job market and attract desirable applicants.

Extra benefits, sometimes called "fringe" benefits, fall into three general groups: health and safety, economic, and convenience and comfort. The first, *health and safety,* is an important basic factor in all personnel matters. This factor affects social and economic life, being of interest not only to the employee but to the employer and the public as well. Time lost because of illness and accidents is expensive for both management and labor, results in lowered productivity and increased losses for the employer, and directly affects the income of the employee. Maintaining the good physical condition of employees is economically desirable as well as necessary for achievement of the many goals of the department. Also, managers of any foodservice recognize that the health of the worker may affect the health of the public through both direct and indirect contact. Additional discussions regarding the importance of good health for the foodservice employee are given in Chapters 3 and 6.

Safe working conditions are of first importance to employer and employee alike. A foodservice does not present the identical hazards found in any other industry, but duplicates some of those found in several industries. Falls, burns, shocks, and cuts are possible, as they are in any other place where mechanical equipment is used. It is the responsibility of the manager to see that safeguards are maintained, that the equipment is kept in safe condition, and that all working conditions are safe and clean.

The *Occupational Safety and Health Act of 1970* (OSHA) has forced managers to look critically at working conditions and to bring any that are undesirable up to a standard demanded by law. Every employer covered by the law is required to furnish employment and places of employment that are free from recognized hazards that are causing or are likely to cause death or serious physical harm and must comply with all safety regulations promulgated by the Secretary of Labor in accordance with the provisions of the act.

Another benefit for employees is provided for in the **workmen's compensation insurance** program. This legislation is administered by the states, and the liability insurance premiums are paid for by employers. Workmen's compensation laws are based on the theory that the cost of accidents should be a part of production costs, the same as wages, taxes, insurance, and raw materials.

This insurance covers employers' liability for the costs of any accident incurred by an employee on or in connection with the job. The workers must show that they were injured on the job and the extent of their injuries. Compensation laws state the specific amount of payment allowed for each type of injury in addition to hospital, surgical, and, in case of death, funeral expenses. All foodservice directors will need to determine, through their state department of labor, who can be covered by workmen's compensation, the methods of payments, and the amount of benefits to which the worker is entitled.

The *Family and Medical Leave Act of 1993* gives employees a maximum of 12 weeks of unpaid and job-protected leave per year for themselves, or a spouse, parent, or child with a serious health condition.

Health and accident insurance plans provide some assistance to employees who may become ill or who are injured off the job. Fear of injury or illness is the cause of much worry, even when an insurance plan is available to employees. Without it, many workers would be in financial straits if they were compelled to pay medical and hospital bills on their own.

Many forms of health and accident insurance are available for groups. In some cases, the company alone pays for the employees; in others, it is jointly borne by the company and those who participate in it. Through labor union efforts and the efforts of concerned managers, more and more health services are being made available to employees, many at employer expense. Flexible benefit plans, which allow the employee to choose from a wide array of benefit options, are gaining in popularity. Whereas early benefit plans included only health and retirement, employees can now often choose profit sharing; stock ownership benefits; legal, educational, and child care assistance; dental and vision insurance; and life insurance, depending on their own particular needs and wants.

The extent to which foodservices provide these benefits to employees usually depends on the size of the organization and the facilities it has available, for example, the emergency room of a hospital and the concern of those at the decision-making level.

The second group of employee services and benefits are those labeled *economic*. Most of the programs discussed so far provide some economic benefit to workers, even if indirectly. All insurance plans undoubtedly could be put under this classification instead of putting some under the classification of health and safety. However, benefits to be discussed in this economic group have a direct monetary value in returns to the employee; the employer carries the cost of some, and others are shared by the employee.

Social Security benefits are provided by the *Social Security Act,* a nationwide program of insurance designed to protect wage earners and their families against loss of income due to old age, disability, and death. A designated percentage of the salary of each employed person must be withheld from his or her wages and the same amount from the business added to the Social Security Fund, or to a comparable retirement-system fund if a nonprofit type of organization is involved. Provisions and benefits of Social Security change from time to time, so details soon become outdated. Managers must keep in touch with their local Social Security office to stay informed of current changes.

Other economic benefits offered by some organizations to their employees may include group life insurance programs, profit-sharing plans, and pensions or retirement plans. All of these add to the economic security of those who continue in the service of a particular organization long enough to build a fund that is significant for them after regular employment ceases, either because of retirement or death. Vacations, holidays, and sick leave, all with pay, are other forms of fringe benefits for personnel. Properly administered, they are of advantage to the organization as well.

Employee convenience and comfort benefits make up the third group of fringe benefits. Services provided for the comfort or convenience of employees compile a long list and include, among others, adequate rest and locker rooms, meal service available to employees often at reduced or at-cost levels, free medical service on an emergency basis, credit unions, and recreational facilities. Educational tuition or fees for personnel to attend workshops or classes for self-development and skill development is also included among these benefits. These types of benefits help to build a loyal, contented working group with high morale.

Although many of the labor laws enacted are directed toward the protection of specific groups, the regulations applicable to all workers are well established. Familiarity with federal, state, and local laws applying to foodservice employees is obligatory for every foodservice administrator and manager. Only then can labor policies be of benefit to both the worker and his or her organization and be put into action for implementation.

LABOR-MANAGEMENT RELATIONS

Foodservice managers are concerned with problems arising from directing employees' activities; that is, in handling the people who must translate the policies, procedures, and plans into action. When groups of people work together, the potential for conflict always exists. Some people must manage and some must carry out the technical operations. Everyone wants more of whatever improves his or her position. The closer the relationship between the employee and manager, with open and free discussion on both sides, the less danger there is for grievances to arise.

Many foodservices are so small that the relationship between employer and employee is immediate and direct. Under such circumstances, discussion of mutual concerns is possible right in the workplace. Direct face-to-face contact tends to develop a sense of real association and mutual interest. Employees with a somewhat complete picture of a relatively small business may see their job in relation to the whole. Many services, however, are so large that there is limited personal contact between employer and employee. Workers may feel there is little chance for the individual to be recognized as an important person in the organization. Also, they may not have an overall view of the business that would make possible self-evaluation of their own jobs in terms of the whole. Workers engaged in a limited phase of total large-scale production may find that they lack the direct contact that tends to humanize employer-employee relationships in a small foodservice.

Legislation

Managers who are not attuned to the concerns of employees, who do not recognize that a small complaint or conflict that arises is probably a symptom of a deeper problem, and who fail to investigate and correct the situation are opening the door for labor unions to come in to represent the employees better.

During the years, much legislation has been enacted to attempt to balance the power between labor and management. In 1935, the passage of the *Wagner Act* [or *National Labor Relations Act* (NLRA)] was the beginning of positive support of unionization and collective bargaining by the federal government. Prior to this legislation, workers had been exploited by management. This exploitation is documented in court cases as early as 1806. The terms of the Wagner Act regulate employees' rights to join a union, prohibit unfair management practices, prohibit management from interfering with their employees who wish to join a union and from discriminating against those who do join, and require employers and union members to bargain collectively (an obligation to meet and discuss terms with an open mind, but without being required to come to an agreement). The majority of today's collective bargaining agreements provide for an impartial arbitrator to hear and decide grievances under the bargaining agreement. The NLRA also established the National Labor Relations Board (NLRB) to administer and interpret the provisions of the act.

These procedures are followed in unfair labor practice cases: (1) A private party files a charge that an unfair labor practice has been committed; (2) the regional office at which the charge is filed investigates the case and decides whether to proceed with the complaint; (3) if the regional director issues a complaint, an attorney from the regional office will prosecute the case; (4) if the case is not settled at this level, a hearing is necessary with a staff attorney representing the NLRB controlling the case; and (5) finally, an administrative law judge hands down a recommended decision and order.

One of the major responsibilities of the NLRB is to determine whether employees should be represented by a union. Employees usually initiate a union campaign if they desire union representation. At least 30% of the employees in the bargaining unit must support their petition to the NLRB; this support is shown by employees' signatures on authorization cards. These cards are investigated and authenticated by the NLRB before ordering that an election be held. The secret ballot election is conducted, and the results are tabulated by an NLRB representative. If the union wins the election, a contract is then drawn up, and the union designates a bargaining unit employee as its union "steward." This person handles the union business in the workplace.

Employees of hospitals operated entirely on a nonprofit basis were exempt from the original NLRA. However, an amendment to the NLRA brought nonprofit hospitals under the provisions of the act. In such situations, dietitians may be called on to defend their positions as "management" instead of as "labor."

The *Taft-Hartley Act* (or *Labor-Management Relations Act*), passed in 1947, was enacted to offset some of the power and unfair practices that labor unions seemed to acquire since 1935. Among other provisions, it prevents unions from coercing employees to join, outlaws the **union shop** (which requires an employee to become a member of the union in order to retain a job) and the **closed shop** (which obligates an employer to hire only union members and to discharge any employee who drops union membership), and makes it illegal for unions to refuse to join in **collective bargaining.** This statute marked a shift away from encouragement of unionization to a more neutral position on the part of the federal government.

Further legislation, passed in 1959, was the *Landrum-Griffin Act* (or *Labor-Management Reporting and Disclosure Act*), which is in the interests of both labor and management but is especially pro-individual labor union member. It contains a bill of rights for union members, requires certain financial disclosures by unions through a specified reporting system, prescribes procedures for the election of union officers, and provides civil and criminal remedies for financial abuses by union officers.

The labor legislation discussed and the areas of human resource management on which it makes an impact are summarized in Table 13.1.

Table 13.1 Selected labor legislation classified by the relevant area of human resource management.

Acquisition of Human Resources	Development of Human Resources	Rewarding of Human Resources	Maintenance of Human Resources
Equal Pay Act, 1963	Equal Pay Act, 1963	Fair Labor Standards Act, 1938	State Fair Labor Practices Act, 1913
Civil Rights Act, 1964	Civil Rights Act, 1964	Equal Pay Act, 1963	Wagner Act, 1935
Age Discrimination in Employment Act, 1967	Age Discrimination in Employment Act, 1967	Civil Rights Act, 1964	Social Security Act, 1935
Civil Rights Act, 1974	Civil Rights Act, 1974	Age Discrimination in Employment Act, 1967	Taft-Hartley Act, 1947
Pregnancy Discrimination Act, 1978	Pregnancy Discrimination Act, 1978	Civil Rights Act, 1974	Landrum-Griffin Act, 1959
Immigration Reform and Control Act, 1986	Immigration Reform and Control Act, 1986	Pregnancy Discrimination Act, 1978	Civil Rights Act, 1964
Americans with Disabilities Act, 1990	Americans with Disabilities Act, 1990	Immigration Reform and Control Act, 1986	Age Discrimination in Employment Act, 1967
Civil Rights Act, 1991	Civil Rights Act, 1991	Americans with Disabilities Act, 1990	Occupational Safety & Health Act, 1970
		Civil Rights Act, 1991	Civil Rights Act, 1974
			Pregnancy Discrimination Act, 1978
			Immigration Reform and Control Act, 1986
			Americans with Disabilities Act, 1990
			Civil Rights Act, 1991
			Family and Medical Leave Act, 1993

There are numerous reasons for employers to become the target of union organizing attempts or for employees to turn to a union. Chief among these reasons are poorly developed or administered personnel policies and practices, or a breakdown in some facet of employer-employee relations. A number of steps should be taken by managers long before organizational attempts begin. Most important among them is a review of personnel policies and employee relations, making every effort to maintain good personnel practices, put policies into writing, and communicate them to employees with frequent reviews and discussions.

The impact of unionization on foodservices may be great for those who are naive in the ways of collective bargaining. Legal counsel to assist in negotiating a fair, workable contract for both labor and management is to be encouraged. If unionization is to become a reality, it is important to create a favorable climate for cooperation, to make sure that the negotiator understands the economic as well as the administrative problems of a foodservice operation, such as scheduling required to cover meal hours, the services necessary, especially to patients in health care facilities, the equipment to be used, and the prices charged in relation to the labor costs.

Certain rights of management may be lost when unionization takes place, since some of the authority but little of the responsibility will be shared with the union. Some of the freedoms lost are the right to hire, discharge, change work assignments and time schedules, set wages and fringe benefits, change policies without appeal, discipline workers without being subject to appeal to the union, and receive and act on grievances directly. The loss of the right to use volunteer workers in the department may also be realized.

It is imperative, therefore, that the collective bargaining agreement contain a management rights clause. There are two major categories of management rights clauses. One is a brief, general clause not dealing with specific rights, but with the principle of management rights in general. The other is a detailed clause, which clearly lists areas of authority that are specific to management.

Certain cost increases should also be noted: for time loss from the job by the person selected to be the union steward, and the cost of management support to the union based on a given sum per member per month in contributions.

The nature of labor organizations and the methods they use differ according to the understanding and goals of the leaders and members, their convictions as to remedies needed, and by legal and other forms of social control. Ordinarily, management and organized labor have different approaches to solving their problems. This often leads to long hours of negotiations before a satisfactory mutual agreement can be reached. It is important that each group try to see the other's viewpoint with fairness and with an honest belief in the good faith of the other.

REVIEW QUESTIONS

1. Name the characteristics beyond good pay and benefits that are common to the companies considered the best to work for in the United States today.
2. What are the rights of individuals in an organization?

3. Identify the tasks that are included in the staffing function of management.
4. Graph an integrated staffing system.
5. List and briefly describe the sources of potential employees.
6. Discuss the purposes of the employment interview.
7. Describe what should be included in an employee orientation program.
8. List the five teaching steps that should be included in an on-the-job training program.
9. What are the purposes of the performance appraisal interview?
10. Discuss when and how disciplinary action should be taken.
11. What are the five steps recommended for good decision making?
12. Identify legislation that has had an impact on foodservice management operations.
13. Discuss the impact of unionization on a foodservice management operation.
14. Define the term "policy."

SELECTED REFERENCES

Barber, A. E., Dunham, R. B., and Formisano, R. A.: The impact of flexible benefits on employee satisfaction: A field study. Pers. Psych. 1992; 45(1):55.

Barlow, W. E., and Hane, E. Z.: A practical guide to the Americans with Disabilities Act. Pers. J. 1992; 71(6):53.

Beasley, M. A.: Developing and applying effective personnel policies. Food Mgmt. 1995; 30(3):42.

Bittel, L. R., and Newstrom, J. W.: *What Every Supervisor Should Know,* 6th ed. New York: McGraw-Hill, 1990.

Burnett, D.: Exercising better management skills. Pers. Mgmt. 1994; 26(1):42.

Christine, B.: Steps involved in ergonomic training. Risk Mgmt. 1994; 41(8):72.

Clark, R. C.: The causes and cures of learner overload. Training. 1994; 31(7):41.

Daily Labor Report: Statements on bill (H.R. 1435) to amend age discrimination in employment act with respect to public health. Washington, D.C.: The Bureau of National Affairs, Inc. March 13, 1986.

Dolliver, S. K.: The missing link: Evaluating training programs. Supervision. 1994; 55(11):10.

Far West Laboratory for Educational Research and Development: *Health Care Skill Standards.* San Francisco: Far West Laboratory, 1995.

Heerwagen, P.: Management training. North Valley Bus. J. 1993; 4(10):10.

Hobbs, B.: Safety is the top priority of training programs: Employee training in concessions companies. Amusement Bus. 1994;106(21):40.

Hospitality and Tourism Skills Board and CHRIE: *Building Skills by Building Alliances.* Washington, D.C.: CHRIE, 1995.

Leeds, D.: Show-stopping training. Training and Development. 1995; 49(3):34.

Leonard, B.: Creating opportunities to excel: Skill-based pay-for-performance plan. Human Resource Mgmt. Mag. 1995; 40(2):47.

Leslie, D. L.: *Labor Law in a Nutshell.* St. Paul, Minn.: West Publishing Company, 1992.

Levering, R., Moskowitz, M., and Katz, M.: *The 100 Best Companies to Work for in America.* Reading, Mass.: Addison-Wesley, 1984.

Lloyd, C.: HTM—The formula for reducing turnover: Hiring, training, motivation. Telemarketing. 1994; 13(2):60.

Martin, S. L., and Lehnen, L. P.: Select the right employees through testing. Pers. J. 1992; 71(6):46.

Metzger, N.: *The Health Care Supervisor's Handbook,* 2nd ed. Rockville, Md.: Aspen Systems Corporation, 1982.

Moravec, M., and Tucker, R.: Job descriptions for the 21st century. Pers. J. 1992; 71(6):37.

Morgan, C.: Employee training that works. Supervision. 1994; 55(6):11.

Plavner, J. T.: Employment law lingo. Human Resource Mgmt. Mag. 1992; 37(5):48.

Rogers, B.: The making of a highly skilled worker. Human Resource Mgmt. Mag. 1994; 39(7):62.

Schuster, K.: How to target training. Food Mgmt. 1994; 29(7):77.

Smith, R. E.: Employee orientation. Pers. J. 1984; 63(12):43.

Stephenson, S.: Hiring the right stuff: Personnel management at institutional foodservice departments. Restaurants and Institutions. 1994; 104(26):128.

The manager as trainer: Who trains the trainers? Supervisory Mgmt. 1995; 40(3):1.

Total quality training. American Printer. 1994; 213(3):86.

United States Code, Volume 12: Title 29 (Labor). Washington, D.C.: U.S. Government Printing Office, 1989.

14

Administrative Leadership

OUTLINE

Organizational effectiveness depends not only on the financial and physical resources of a company, but also on the skills and abilities of its employees. And, regardless of how carefully those employees have been selected and trained, it is difficult to ensure that they will apply their full energies to the job. One of the greatest challenges facing a manager is understanding the differing needs of individuals and thus the forces that will motivate him or her to be a productive employee. Balancing the roles of manager and leader is yet another challenge. It is possible to be one and not the other. In today's business environment, however, the ability to be both is essential.

As administrative leaders, those individuals who assume the management of foodservice organizations will be successful to the degree that they are willing to assume responsibility and are able to maintain good human relations. The goals and objectives of the department cannot be attained by the administrator alone; working satisfactorily through other people constitutes the major part of the job.

Most people would assume that without administrative leadership no organization could achieve its goals and plans. This assumption is generally valid, but what is meant by administrative leadership? In this chapter, the difference between leadership and management is discussed, and the characteristics displayed by managers and leaders in administrative positions are compared.

The topic of leadership effectiveness is of special interest. An historical view of leadership is presented that traces the evolution of effective leadership theories from the era of scientific management to the present-day systems concept and contingency approach. The major contributions of each period are summarized.

The judicious use of power is an important factor in leadership success. Therefore, an understanding of how a leader acquires power is essential.

Communication is another key factor for effective leadership. Some of the barriers to successful communication are described as well as some techniques to improve in this area.

Although profit and productivity are still major goals of administrative leadership, managerial ethics and social responsibility have assumed equal importance. This chapter concludes with a discussion of the ethical and social responsibilities required by administrative leaders in today's foodservice industry.

KEY CONCEPTS

1. An individual's motivations stem from energizing forces within the individual (needs, attitudes, interests, perceptions) and within the organization (rewards, tasks, coworkers, supervisors, communication, feedback).
2. Leadership is the activity of influencing other people's behavior toward the achievement of desired objectives. Management is the function of running an organization by effectively and efficiently integrating and coordinating resources in order to achieve desired objectives.
3. As a leader, the foodservice manager must empower employees by clearly communicating the organization's mission, accepting the responsibility for leading the group, and earning employees' trust.
4. Early theories of leadership include scientific management, in which a leader's role was to motivate employees with rewards of money, and human relations theory, in which a leader improved productivity by showing an interest in the employee as an individual.
5. McGregor's Theory X and Theory Y are based on the idea that a leader's attitude toward employees has an impact on job performance and may lead to different management strategies.
6. Situational management theory holds that effectiveness as a leader depends on the characteristics of the leader and the subordinates as well as the situational variables involved.
7. The contingency theory of leadership holds that there is no one "best" style of leadership but that style must be adjusted to fit the situation.
8. Leaders acquire power from their ability to reward and punish, position in the organization, expertise, and/or personal characteristics.
9. Communication, or the constant development of understanding among people, is central to leadership effectiveness.
10. Effective communication means that there is successful transfer of information, meaning, and understanding from a sender to a receiver.

11. Types of communication include oral, written, visual aids, body language, facial expressions, gestures, and actions. The effectiveness of communication can be improved by using multiple forms of communication.
12. Barriers to effective communication can be overcome by being aware of their existence and employing some of the suggested techniques to improve communication.

MOTIVATION

Motivation is the sum of energizing forces internal (individual) and external (organizational) to an individual that results in behavior. It is not possible to motivate another person to do anything he or she does not want to do. Motivation must come from within the person. It is only possible to create an environment in which one can motivate one's self. To do this, a leader must understand the concept of human motivation. Abraham Maslow's (1954) classic research on motivational theory has provided the foundation of most current thinking in this area.

According to Maslow's *need hierarchy theory* a person is motivated by his or her desire to satisfy specific needs. These needs are arranged in a hierarchical order (Figure 14.1). Maslow theorized that only an unsatisfied need motivates behavior; when a need is satisfied, it is no longer a primary motivator; higher order needs cannot become motivating forces until preceding lower order needs have been satisfied; and, finally, people want to move up the hierarchy.

A second theory of motivation based on needs was put forth by David McClelland (1961). *Achievement motivation theory* holds that an organization offers an individual the opportunity to satisfy three needs: the need for power, the need for achieve-

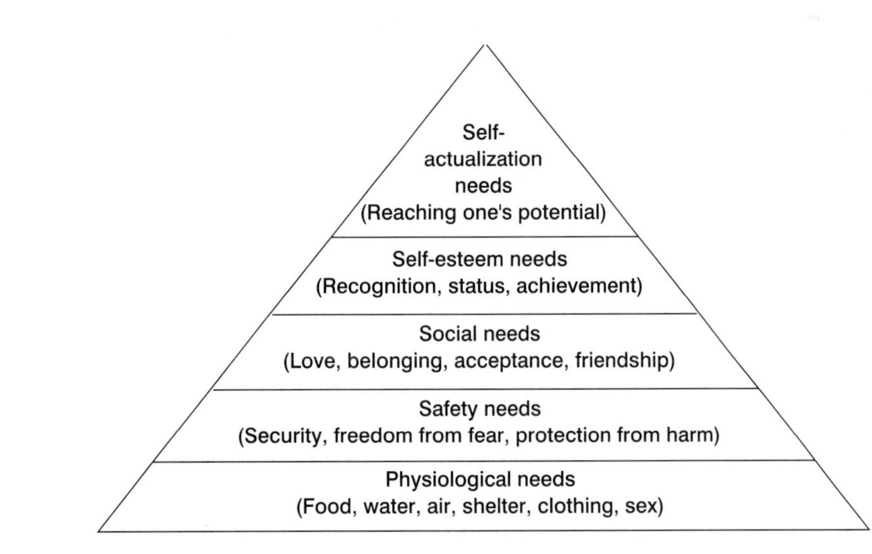

Figure 14.1 Hierarchy of needs.

ment, and the need for affiliation. Depending on the individuals' particular needs they will be motivated by tasks that provide an opportunity to attain that need.

Both Maslow and McClelland based their theories on differences among people. In contrast, organization theories of motivation emphasize task elements with less consideration of individual differences. Frederick Herzberg's (1959) *dual-factor theory* or *motivation-hygiene theory* purports that factors such as achievement, recognition, responsibility, opportunity for advancement, and the work itself are all motivators, whereas factors such as the company policies, supervision, salary, working conditions, and interpersonal relations are hygiene factors. Hygiene factors do not motivate but simply prevent dissatisfaction and act as a precondition for motivation by motivators.

Operant conditioning is a second theory of motivation based on organizational factors. The basic idea behind operant conditioning is that people will perform in order to receive rewards and avoid punishment.

Expectancy theory combines individual factors and organizational factors into a theory of motivation based on the interaction of the two. Expectancy theory states that people make decisions about their behavior on the expectation that the choice they make is more likely to lead to a needed or desired outcome. The relationship between behavior and outcome is affected in complex ways by individual and organizational factors.

LEADERSHIP

Leadership is one of the most observed and least understood phenomena on earth according to Burns (1978); in fact, he cites more than 130 definitions of the term. After reviewing more than 5000 research studies and monographs on leadership in *Stogdill's Handbook of Leadership* (Bass, 1981), the editor of the book concluded that there is no common set of factors, traits, or processes that identifies the qualities of effective leadership.

Leadership, like the concept of management, means different things to different people—ranging from being the first to initiate a change to inspiring bravery on the battlefield. This fact has caused many to use other, more definitive terms, such as *activating* or *influencing.* At times, leadership and management have been used synonymously. However, *leadership* is essentially the business or activity of trying to *influence* people to strive willingly to attain the goals and plans of the organization. *Management* is the function of running an organization from a conceptual or policy standpoint. Leadership may then be defined as working with people to get them to produce willingly the results the leader wants or needs to accomplish.

Hickman (1990) suggests that although managers and leaders both have minds and souls, they tend to emphasize the use of one over the other as they function in the organization. That is, the mind represents the analytical, calculating, structuring, and ordering side of tasks, while the soul represents the visionary, passionate, creative, and flexible side.

Abraham Zaleznik (1989), a professor of social psychology of management, writes that managers and leaders are very different kinds of people who differ in motivation, personal history, and how they think and act. Management and leadership require different responses to different demands, and there are situations when each is required. To understand how the roles of leaders and managers differ, some of the characteristics of each are tabulated in Table 14.1.

Koontz and O'Donnell (1968) suggest that subordinates will respond to authority alone to do the bare minimum to maintain their jobs. But "to raise effort toward total

Table 14.1 Some characteristics of managers and leaders.

Manager	Leader
How They View Themselves:	
Managers have a strong sense of belonging to their organizations.	Leaders see themselves as separate from their organizations and the people of their organizations.
Managers see themselves as protectors of existing order with which they identify.	Leaders have strong personal mastery that impels them to struggle for change in existing order.
How They View Their Function:	
Managers work through other people within established organizational policies and practice to reach an organizational goal. They limit their choices to pre-established organizational goals, policies, and practices. Managers are concerned with process.	Leaders question established procedures and create new concepts. They inspire people to look at options. They are concerned with results.
Personality:	
Managers have a strong instinct for survival.	Leaders seek out risks, especially if rewards seem high.
Managers can tolerate mundane, practical work.	Leaders dislike mundane tasks.
Relationships:	
Managers relate to people according to their role—boss, employee, peer, and so forth.	Leaders relate to people in an intuitive and empathetic way.
Primary Concern:	
Managers are concerned with achievement of organizational goals.	Leaders are concerned with achievement of personal goals.
Place in Organization:	
Managers are supervisors, department heads, and administrators. They are usually considered higher echelon.	Leaders may be found at any level in the plan of organization from technician to highest echelon.
Power:	
Managers derive power from their positions.	Leaders derive power through personal relationships.
Goals:	
Managers are concerned with pre-established, organizational goals. Their personal and subgoals arise out of necessity to conform to organizational structure, rather than a desire to change.	Leaders are concerned with personal goals. They are not comfortable with the status quo of established organizational goals and policies. They enjoy innovating.

Source: Tamel, Mary E., and Reynolds, Helen: *Executive Leadership.* Englewood Cliffs, N.J.: Prentice Hall, Inc., 1981, p. 59. Used with permission.

capability, the manager must induce devoted response on the part of subordinates by exercising leadership."

Leadership has been viewed as a special form of power involving relationships with people. These relationships are developed when leaders successfully fuse organizational and personal needs in a way that allows people and organizations to reach levels of mutual achievement and satisfaction. This can be an exceedingly difficult task. Each employee is different with different motivations, ambitions, interests, and personalities. As a result, each must be treated differently. Work situations differ. How managers can handle these divergent factors effectively has been the subject of study for many years, and such research, both past and present, can be used to improve managerial leadership effectiveness.

Drucker (1992) contends that the essence of leadership is performance—not charisma or a set of personality traits. He states that there are three basic requirements for effective leadership: (1) The leader must think through the organization's mission, defining it and establishing it, clearly and visibly. Any necessary compromises made are compatible with the leader's mission and goals, and standards are maintained. (2) The leader sees leadership as a responsibility, not rank and privilege. The effective leader accepts responsibility for subordinates' mistakes but sees their triumphs as triumphs. For this reason, effective leaders do everything possible to surround themselves with able, independent, and self-assured people. (3) Last, the leader must earn trust. This means that a leader's actions and professed beliefs must be congruent. Drucker states that being a good leader is not based on being clever but on being consistent, and that these are the same characteristics required of a good manager.

Bennis and Nanus (1985) agree with Drucker's idea and, based on his research, find that managers must grow to become leader-managers. They put forth four essential traits of effective leadership: (1) the capacity to engage people and draw them to a compelling vision of what is possible; (2) the ability to communicate their vision in a way that allows people to make it their own and give it personal meaning; (3) trust, total reliability, and integrity, as well as the performance of actions that are congruent with their vision; and (4) the possession of high regard for self and others. The combined effect of these personal characteristics empowers people by (1) making them feel significant, (2) focusing on their developing competence rather than their failures, (3) creating a shared sense of community, and (4) making work exciting and worthy of dedicated commitment.

As a leader of people, the foodservice manager has the task then to empower employees by clearly establishing and communicating the organization's mission, accepting the responsibility of leading the group, and earning employees' trust by showing a high regard for self and others. An empowered team is necessary to create an effective, smoothly operating work unit.

The Traditional Leadership Role

Scientific Management. The era of **scientific management** was founded on the belief that the main common interest of both the organization and the employee was money, and only money. The leader-manager's role consisted of issuing orders and handing out rewards and punishment. The founders of the scientific management theory, such as Frederick W. Taylor and Frank and Lillian Gilbreth, were primarily concerned with the best method and "right wage" for doing the job. The employee was viewed as a machine or tool. This type of thinking met the needs of the day. But times change. The practices of the scientific management movement began to be questioned in the late 1920s.

Human Relations Approach. The turning point came as a result of the Hawthorne studies. Western Electric Company conducted some experiments at their Hawthorne plant outside Chicago to determine the relationship between the physical working environment and productivity. Lighting was one variable that was tested. Researchers were surprised to find that no matter how they varied the intensity of the lighting, productivity increased. They concluded that the level of performance had nothing to do with the lighting intensity but rather was a result of the interest shown in the worker as a person rather than as a machine. Thus, the human relations theory era was born. Human relationists such as Mayo, Maslow, Roethlisberger, and Dickson brought a more tolerant approach to the leadership of people—consideration of the individual and an understanding of why people work. The theory was good. The implementation, in many cases, was poor.

Newer Approaches to Leadership

Theory X and Theory Y. The human relations movement began to lose favor in the early 1950s. McGregor introduced his Theory X and Theory Y analysis of leadership strategies, suggesting that the basic attitude of a manager toward employees has an impact on job performance. He divided these supervisory attitudes into two categories—Theory X and Theory Y. **Theory X** attitude was held by the traditional and "old-line" managers and is pessimistic about employees' abilities and skills. **Theory Y** was the attitude held by the emerging manager of the 1960s and 1970s and is optimistic. However, again implementation was the problem. Managers trained in Theory Y management found that, in many cases, the resulting job performance did not yield the desired level of quality (Figure 14.2).

Situational Management. The work done at The Ohio State University and by Blake and Mouton (1986), as well as others, culminated in the theory that effectiveness as a leader depends on a multiplicity of factors, not only human behavior and motivation. The **situational management** approach concentrates on the theory that leadership effectiveness is a function of the individual leader (including traits and personalities), of that leader's subordinate (including attitude toward working, socioeconomic interests, and personality), and of the situational variables involved.

Because followers are the ones who determine whether a person possesses leadership qualities, the expectations of followers have been studied. Table 14.2 shows the characteristics that followers admire in superiors. From this study, it may be concluded that honesty, competence, a forward-looking attitude, and inspiration are important leader attributes.

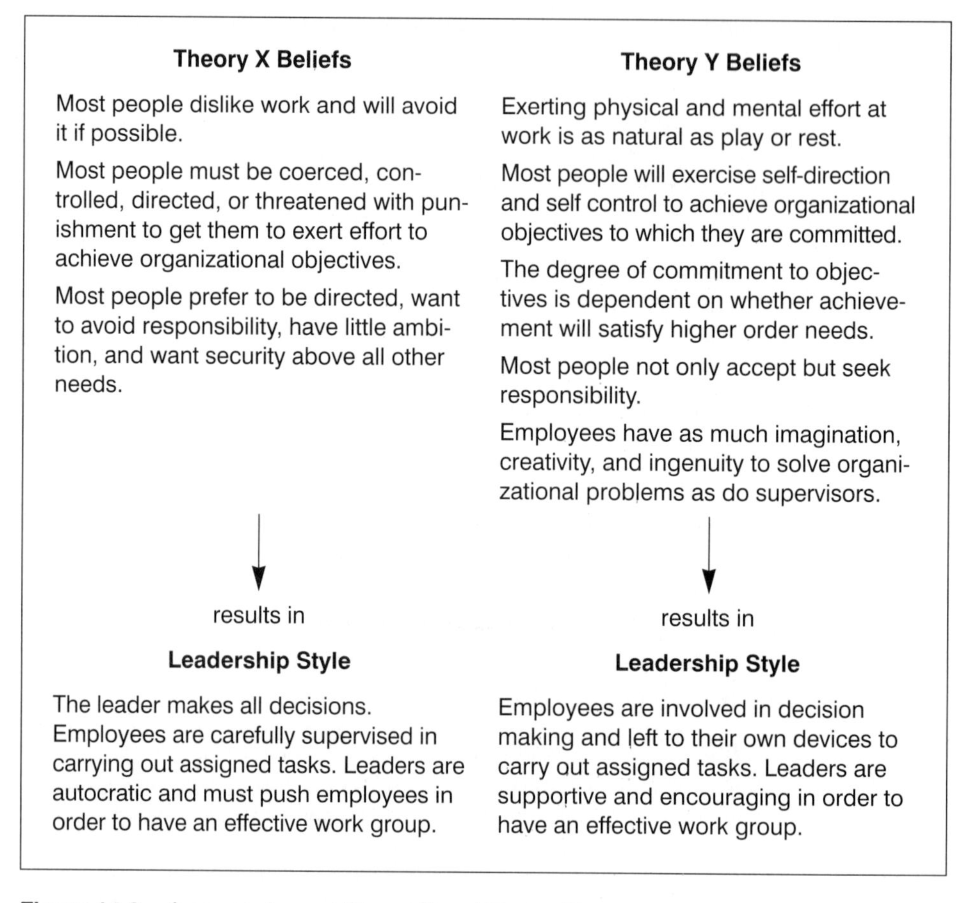

Figure 14.2 A comparison of Theory X and Theory Y.

Early studies by Stogdill at The Ohio State University attempted to define more global leader-type behaviors. Two separate and distinct dimensions of leader behavior were identified—initiating structure and consideration. *Initiating structure* refers to the relationship between the leader and the members of the work group in seeking to establish well-defined patterns of organization, channels of communication, and procedures. **Consideration** refers to behavior that indicates friendship, mutual trust, respect, and warmth in the relationship between the leader and the work group. This was the first study to plot leader behavior on two axes. Leadership quadrants drawn from the Ohio State studies are illustrated in Figures 14.3, 14.4, and 14.5. The effects of initiating structure on employee satisfaction and performance have been found to depend entirely on the situation. Research has shown that consideration behavior is positively related to employee satisfaction, but its effect on performance is unclear.

Building on the work done at Ohio State, Blake and Mouton developed the managerial grid, shown in Figure 14.6. Originally, the team leadership style (9,9) was believed ideal—high concern for people and production. However, it was soon found that some of the other leadership styles were equally effective and that some managers using a team leadership style were not effective. The managerial grid labels were modified as shown in Figures 14.7 and 14.8.

One researcher, House (1971), proposed a theory that helps to explain the situational nature of the initiating structure dimension of leader behaviors. Called the **path-goal theory** of leadership, it states that the functions of a leader should consist of increasing personal rewards for subordinates for goal attainment and making the path to these rewards easier to follow by clarifying it, removing and reducing roadblocks, and increasing opportunities for satisfaction along the way. House based his theory on the expectancy theory of motivation proposed by Vroom (1964), which states that motivation is a function of both the person's ability to accomplish the task and his or her desire to do so. In the late 1960s, the theory that leadership effectiveness is contingent on not only leadership style but the attitude and outlook of the follower and the situational constraints came to be accepted.

Contingency Theories of Leadership. Basically, **contingency theory** holds that there is no one "best" style of leadership but that style must be adjusted to fit

Table 14.2 Characteristics of superior leaders.

	U.S. Managers (N = 2,615)	
Characteristic	Ranking	Percentage of Managers Selecting
Honest	1	83
Competent	2	67
Forward-looking	3	62
Inspiring	4	58
Intelligent	5	43
Fair-minded	6	40
Broad-minded	7	37
Straightforward	8	34
Imaginative	9	34
Dependable	10	33
Supportive	11	32
Courageous	12	27
Caring	13	26
Cooperative	14	25
Mature	15	23
Ambitious	16	21
Determined	17	20
Self-controlled	18	13
Loyal	19	11
Independent	20	10

Source: Kouzes, James M., and Posner, Barry Z.: *The Leadership Challenge: How to Get Extraordinary Things Done in Organizations,* Table 1, p. 17. Copyright © 1987 by Jossey-Bass Inc., Publishers. Used with permission.

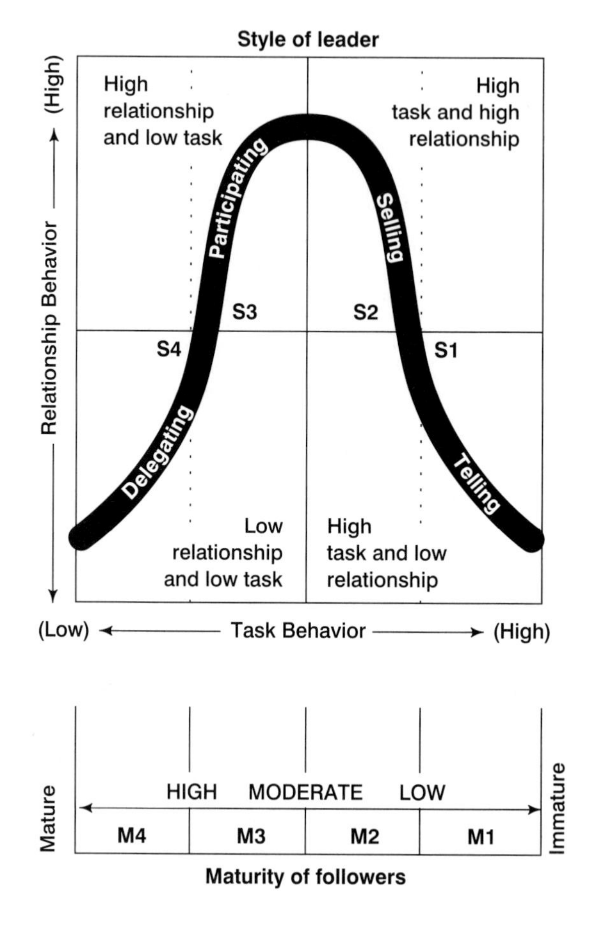

Figure 14.3 Style of leader versus maturity of follower(s). From Hersey, Paul, and Blanchard, Kenneth: *Management of Organizational Behavior: Utilizing Human Resources,* 4th ed., © 1982, p. 152. Reprinted by permission of Prentice Hall, Englewood Cliffs, New Jersey.

the situation. Effective leadership in any given situation is dependent on a number of circumstances: The situation—how structured is the task involved, whether or not the leader has any power as perceived by subordinates, and how well the leader gets along with subordinates. Fiedler and Garcia (1987) developed a continuum identifying possible states of each of the three factors. He concluded that in very "favorable" or in very "unfavorable" situations for getting a task accomplished by group effort, the task-oriented management style works best. In intermediate situations, the human relations style is more successful.

In 1958, Tannenbaum and Schmidt wrote a now classic article, "How to Choose a Leadership Pattern," in which they described how a manager should successfully lead his organization. Fifteen years later, they reconsidered and updated their original statements to reflect new management concepts and societal changes. The revised continuum of manager-nonmanager behavior is shown in Figure 14.9.

The total area of freedom shared by managers and nonmanagers is constantly redefined by interactions between them and the forces in the environment. The points on the continuum designate types of manager and nonmanager behavior that are possible with the amount of freedom available to each. This continuum allows

High Relationship — Low Task: **Leadership through Participation** Use when followers are "able" but "unwilling" or "insecure."	High Relationship — High Task: **Leadership through Selling** Use when followers are "unable" but "willing" or "motivated."
Low Relationship — Low Task: **Leadership through Delegation** Use when followers are "able" and "willing" or "motivated."	Low Relationship — High Task: **Leadership through Telling** Use when followers are "unable" and "unwilling" or "insecure."

Figure 14.4 Interpretation of the situational leadership model.
From Bolman, Lee G., and Deal, Terrance E.: *Reframing Organizations: Artistry, Choice, and Leadership,* p. 418. Copyright © 1991 by Jossey-Bass, Inc., Publishers. Used with permission.

managers to review and analyze their own behavior within the context of alternatives available. It is important to recognize that there is no implication that either end of the continuum is inherently more effective than the other. The appropriate balance is determined by forces in the manager, in the nonmanager, and in the particular situation. The model also suggests that neither manager nor nonmanager has complete control. The nonmanager always has the option of noncompliance, and managers can never relieve themselves of all responsibility for the actions and decisions of the organization.

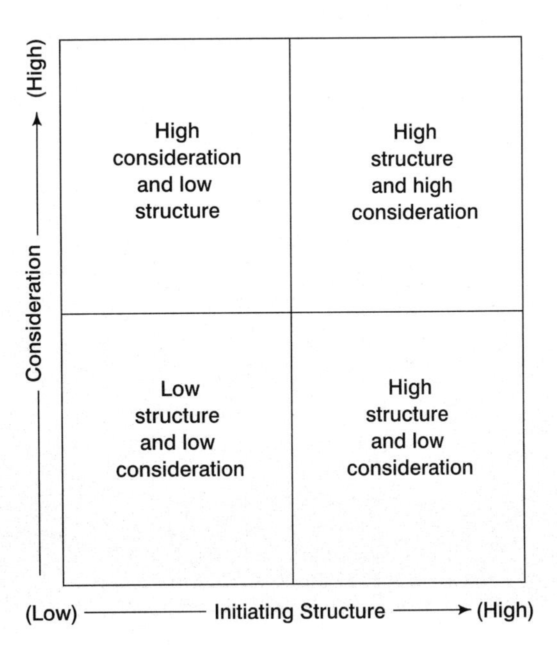

Figure 14.5 The Ohio State leadership quadrants.
From Hersey, Paul, and Blanchard, Kenneth: *Management of Organizational Behavior: Utilizing Human Resources,* 4th ed., © 1982, p. 96. Reprinted by permission of Prentice Hall, Englewood Cliffs, New Jersey.

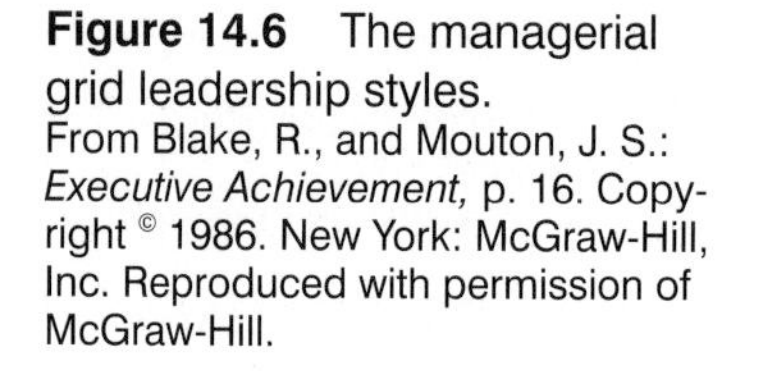

Figure 14.6 The managerial grid leadership styles.
From Blake, R., and Mouton, J. S.: *Executive Achievement,* p. 16. Copyright © 1986. New York: McGraw-Hill, Inc. Reproduced with permission of McGraw-Hill.

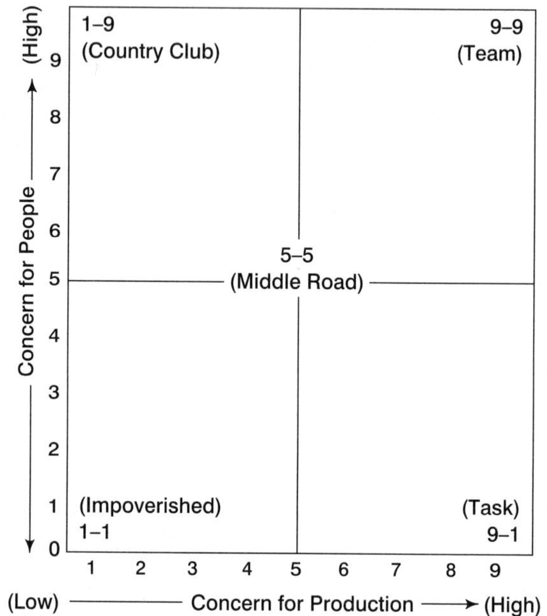

Also based on the premise that there is no one best way to influence people, Hersey and Blanchard's situational leadership model, shown in Figure 14.3, bases the recommended style of leadership on (1) the amount of guidance and direction (task behavior) a leader gives; (2) the amount of socioemotional support (relationship behavior) a leader provides; and (3) the readiness (maturity) level that followers exhibit in performing a specific task, function, or objective. Each of the leadership styles shown (delegating, participating, selling, and telling) is a combination of task and relationship behavior (the two dimensions identified in the Ohio State studies).

It is probable that most people are able to operate within a narrow band of preferred ways of leading and tend to use these styles over and over. Self-development and training should be directed to a wider range of styles for use in the appropriate situations. Ideally, persons in foodservice management positions should accept as a personal philosophy that their human resources are their greatest assets and that to improve their value is not only a material advantage but a moral obligation as well.

The historical view of leadership and the contributions of each of the periods is summarized in Figure 14.10.

Leadership Power

Because of his or her position in the organization, a leader possesses *position power* and because of personal characteristics or expertise she or he may also possess *personal power.* Power is used to influence the behavior of others, an important part of a leader's job. Some of the specific ways that leaders acquire power were identified by French and Raven (1959) in a now classic study. They are:

Coercive power: Followers believe that the leader has the authority to punish them and the punishment will be unpleasant such as a salary reduction, a demotion or termination, or assignment to unpleasant tasks.
Reward power: Followers believe that the leader has the authority to reward them and the rewards will be pleasant such as an increase in salary, a promotion, or assignment to preferred tasks.
Legitimate power: Followers believe that the leader has the right to give directions because of his or her position in the organization.
Expert power: Followers believe that the leader has expertise or knowledge that will be of help to them.
Referent power: Followers believe that the leader has charisma or personal characteristics that result in admiration and respect and therefore want to follow that leader.

Expert and referent power evolve from the traits, skills, and beliefs of the leader, whereas coercive, reward, and legitimate power are all based on the organization's support of the leader. Any type of power, when properly used, is of value to a leader. To be most effective, leaders should develop as many sources of power and influence as possible. In the end, what separates the effective leader from the ineffective one is how the power that one possesses is used.

Effective Communication

One leadership model shows the central role of communication for leadership effectiveness. In this model, communication is the glue that binds the behavior between

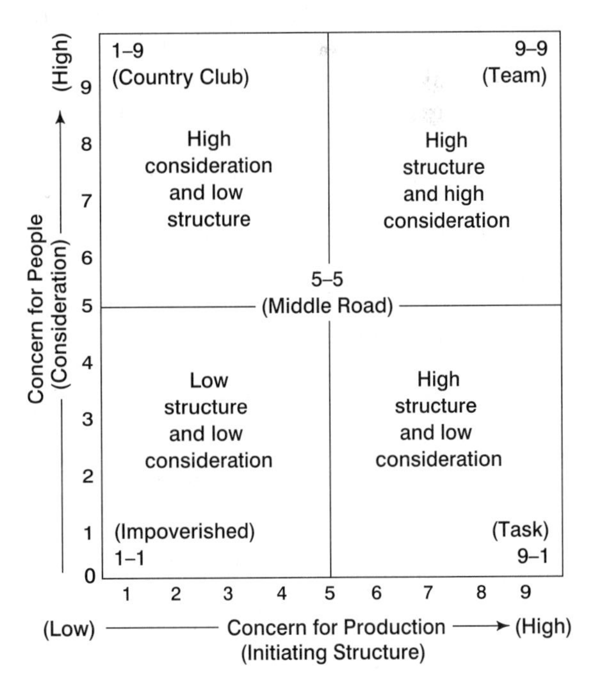

Figure 14.7 Merging of the Ohio State and the managerial grid theories of leadership.
From Blake, R., and Mouton, J. S.: *Executive Achievement,* p. 16. Copyright © 1986. New York: McGraw-Hill, Inc. Reproduced with permission of McGraw-Hill.

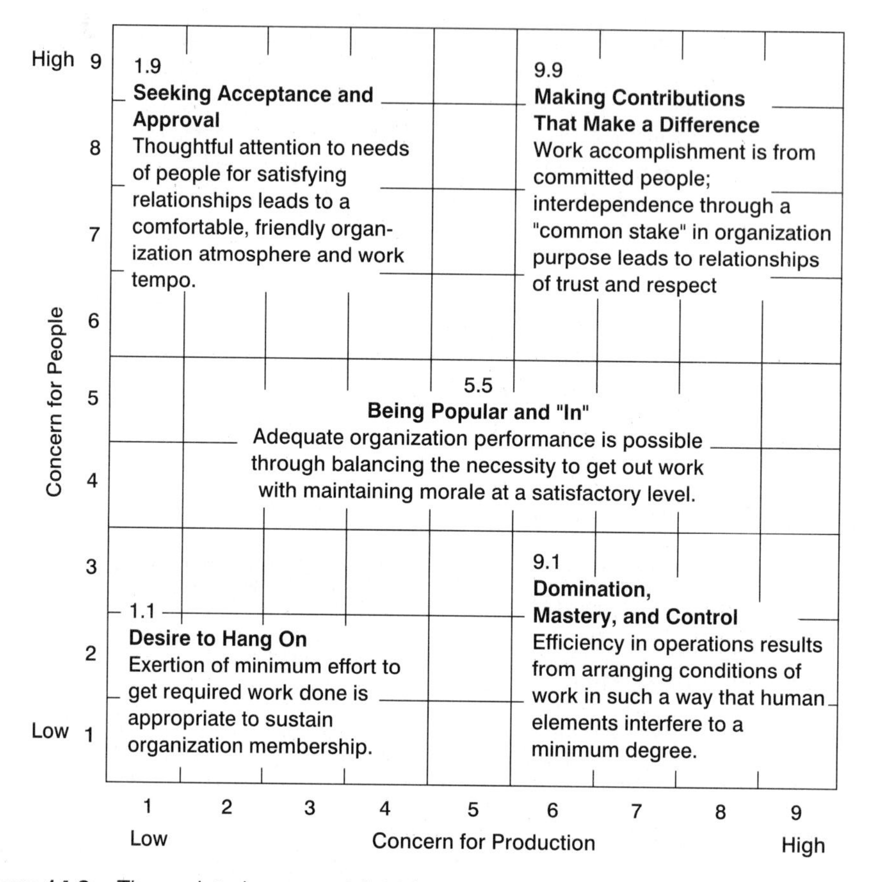

Figure 14.8 The updated managerial grid.
From Blake, R., and Mouton, J. S.: *Executive Achievement,* p. 16. Copyright © 1986. New York: McGraw-Hill, Inc. Reproduced with permission of McGraw-Hill.

leader and follower. The messages transmitted between them present the styles, attitudes, values, motives, skills, and personality variables that are possessed by the leader. The amount of control exerted will vary depending on the situation, task, personnel, and the interrelationships of these components. Good communication is a critical component of effective leadership. It is the process that links all of the management functions. In fact, estimates indicate that between 70% and 90% of a manager's time is spent communicating.

Definitions. *Communication* can be defined as the constant development of understanding among people. *Effective communication* means that there is successful transfer of information, meaning, and understanding from a sender to a receiver. It is not necessary to have agreement, but there must be mutual understanding for the exchange to be considered successful. For a leader to lead, directions must be followed. For directions to be followed, they must be understood. The best plans will fail if communication is not comprehended. It is almost certain that no message will be transmitted or received with 100% accuracy. The average employee remembers

10% to 15% of what (s)he hears
15% to 30% of what (s)he hears and sees
30% to 50% of what (s)he says
50% to 75% of what (s)he does

but (s)he remembers

75% of what (s)he does with proper instruction.

Proper instruction includes hearing, seeing, saying and doing, and then repeating it all again.

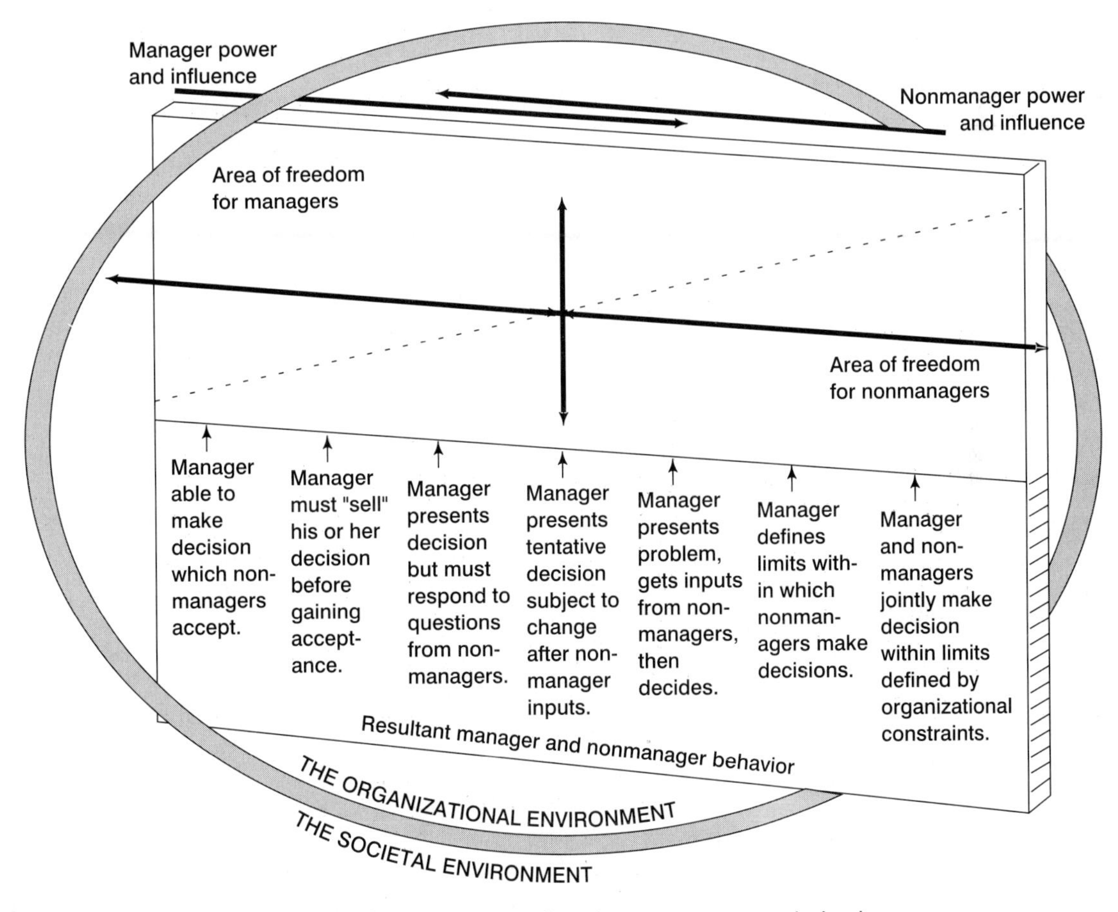

Figure 14.9 Tannenbaum-Schmidt's continuum of manager-nonmanager behavior.
Reprinted by permission of *Harvard Business Review.* An exhibit from "How to choose a leadership pattern" by Robert Tannenbaum and Warren H. Schmidt (May/June 1973).

Scientific Management—1910 to 1926	
Taylor and Gantt	The one best way
F. and L. Gilbreth	"Efficiency" and work simplification
Human Relations—1926 to 1947	
Mayo	Employees must be treated like people, not machines
Roethlisberger and Dickson	Satisfaction in work
Maslow	
The Behavioral Scientists—1947 to 1967	
McGregor	Under the right conditions, people will manage themselves
Argyris	People work best when their social and psychological needs are met
Likert	
Drucker	
Herzberg	
The New Thinkers—1967 to 2000	
Reddin	The situation, the type of work, and the type of manager all determine the most appropriate leadership style
Tannenbaum and Schmidt	There is no "one best way"
Hersey and Blanchard	
Blake and Mouton	
House	

Figure 14.10 A summary of the historical view of leadership.

Channels of Communications. An organization's channels of communication can be divided into the formal channel (that established by the organizational structure) and the informal channel. In the formal channel, communication can be downward, upward, or horizontal. Communication from the top down is the most frequently used channel. Equally important is the upward flow of information. Management should encourage the free flow of suggestions, complaints, and facts.

The informal channel of communication includes the "grapevine." News acquired through the grapevine contains some factual information but, most of the time, it carries inaccurate information, half truths, rumors, private interpretations, suspicions, and other kinds of distorted information. The grapevine is constantly active and spreads information with amazing speed, often faster than formal channels.

An alert manager acknowledges the grapevine's presence and tries to take advantage of it, if possible. The grapevine can carry a certain amount of useful information. It can help to clarify and disseminate formal channel information. Rumors and inaccuracies should be dispelled by stating the facts to as many people as possible.

Types of Communication. Oral, or spoken, communication is the most common form of communication and is generally superior. Oral communication takes less time and is more effective in achieving understanding. Face-to-face communication has the advantage of also providing information through body language, personal mannerisms, and facial expressions. Oral communication should be used when (1) instruction is simple, (2) quick action is required, (3) a method to be followed needs to be demonstrated, (4) privacy is required, and (5) employees have proven they are capable and meet their commitments.

Written communication should be used in some circumstances, particularly when (1) a policy or some other authority is being quoted, (2) employees are to be held strictly accountable, (3) a record is needed, (4) employees are inexperienced, and (5) distance makes oral communication impossible. A well-balanced use of both oral and written communication is often very effective.

Other types of communication include visual aids, gestures, and actions. Visual aids, such as pictures, charts, cartoons, symbols, and videos, can be effective particularly when used with good oral communication. "Actions speak louder than words" is sage advice to any manager. Gestures, handshakes, a shrug of the shoulders, a smile, and silence all have meaning and are powerful forms of communication to subordinates.

Barriers to Good Communication. Some of the barriers to communication have to do with the language used, the differing backgrounds of the sender and receiver, and the circumstances in which communication takes place. The receiver hears what he or she expects to hear and may shut out or ignore what is not expected. There is a tendency to infer what is expected even when it has not been communicated. Senders and receivers have different perceptions based on their different backgrounds. It is important to consider where the other person "is coming from." Receivers evaluate the source and interpret or accept communication in light of that evaluation. A trusted and respected leader will have more open channels of communication than a leader who does not command trust and respect. Conflicting information is often ignored. Different people most often attach various meanings to certain words. The sender or communicator must not only choose words that convey the meaning to the receiver but also give attention to the message transmitted by nonverbal cues. Body language and facial expression often say more than the words they accompany. A receiver who is emotionally upset often stops listening in order to think about what he or she will say next. Noise and the environment often form a physical barrier to communication. There is a right place and a wrong place to conduct good communication. There is a right time and a wrong time.

A network breakdown occurs when there is a disruption or closure of a communication channel. This can be caused by a number of factors, both intentional and unintentional. Some reasons for the network to break down are forgetfulness, jealousy, fear of negative feedback, and to gain an edge over the competition. Information overload occurs when someone receives more information than they are able to process. Time pressures on the sender form a barrier to effective communication because of hastily developed messages, use of the most expedient (often, not most effective) channel, and allowing insufficient time for feedback. These barriers to successful communication are summarized in Figure 14.11.

Improving Communication. Communication is not a one-way process. One of the most important parts of effective communication is to listen to the reply, which

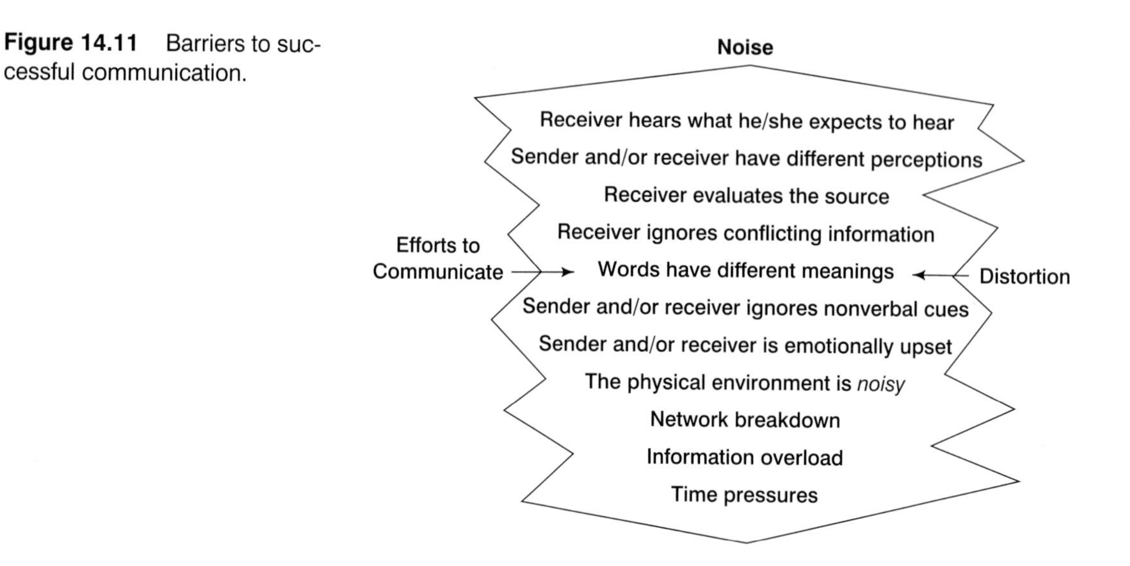

Figure 14.11 Barriers to successful communication.

may be either words, facial expression, body language, or even silence. The evaluation of feedback can tell much about how the message has been received. Empathy or the ability to put yourself in the receiver's shoes in a conversation can also be crucial to mutual understanding.

Face-to-face communication is advantageous to use, when possible, because of the ability to gain immediate feedback from multiple channels, such as oral, facial expression, and body language. Long, technical, and complicated words should be avoided as much as possible. To secure understanding, repeating information using slightly different words, phrases, or approaches is often effective. Being sensitive to the receiver or being able to put yourself in the receiver's place can improve communication. One should be aware that some words or phrases can have symbolic meaning to others and avoid using them. Proper timing is also important. The old maxim "criticize in private, praise in public" is an example of timing. Reinforcing words with congruent actions has already been discussed as essential to effective communication. Finally, an atmosphere of openness and trust, fostered by self-disclosure, builds healthy relationships that contribute to effective communication. These methods to improve communication are summarized in Figure 14.12.

Managerial Ethics and Social Responsibility

The major goals of administrative leadership in the scientific management era were profit and productivity. Today, leadership in organizations involves managerial **ethics** and social responsibility as well. Among the ethical challenges facing foodservice managers today are identifying and understanding different cultural values, dealing with unethical behavior in the organization, balancing the organization's need to know with the employees' rights to privacy, balancing management and employees' rights, and identifying and implementing programs to ensure that the organization is operating in a socially responsible manner. Some areas of each of these are controlled by government regulatory agencies and legal mandates, but many other areas are not.

Ethics can be defined as the rightness or wrongness of actions and as the goodness or badness of these actions' objectives. Many professional organizations have a code of ethics that provides guidelines for their members to use in their work with others. A bill of rights for employees also provides a valuable set of guidelines for managers to use in dealing with subordinates and, at the same time, assures employees of certain rights, such as the right to follow a grievance procedure or the right to a safe workplace.

Social responsibility is an ethical issue because it deals with the goodness or badness of organizational actions in terms of their impact on society as a whole. The classical view of social responsibility is that organizations have no obligations to society other than to achieve organizational objectives. The modern view holds that organizations must operate to achieve the greatest good for the greatest number of people. In other words, social responsibility is demonstrated when a company goes beyond profit maximization in order to benefit society in other ways. This may be demonstrated in a number of ways. On the lowest level, organizational "image building" occurs when managers support good causes in an effort to promote the company and its products. The middle-of-the-road approach includes "good citizenship," which is demonstrated by company support of charities or public interest issues, employee time off to work in problem areas, and employee wellness programs. At the highest level is "full corporate social responsibility," demonstrated by enthusiastic support for social problems.

The areas in which managers are expected to take a proactive, socially responsible stance include the environment, minority group relations, consumer responsibility, and employee rights. Foodservice has a major impact on the environment. Conservation, including the wise use of water and energy, and pollution prevention are real concerns

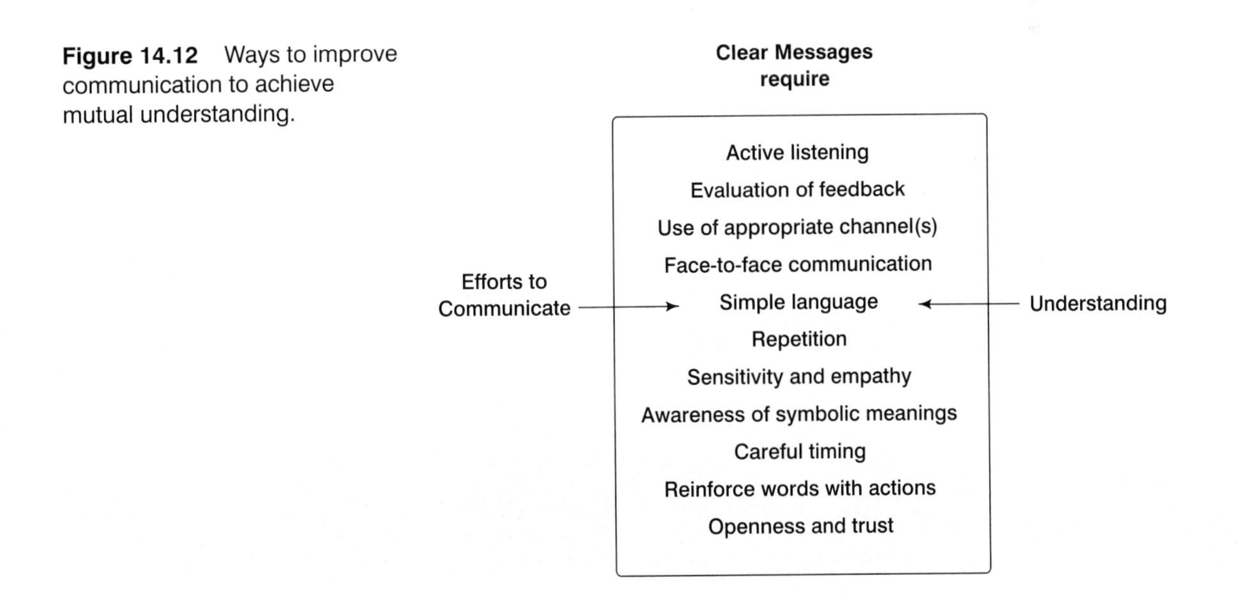

Clear Messages
require
Active listening
Evaluation of feedback
Use of appropriate channel(s)
Face-to-face communication
Simple language
Repetition
Sensitivity and empathy
Awareness of symbolic meanings
Careful timing
Reinforce words with actions
Openness and trust
Efforts to
Communicate
Understanding

Figure 14.12 Ways to improve communication to achieve mutual understanding.

for the industry, as well as being economically and socially mandated practices. The major pollution concern for foodservice is solid waste management. The National Restaurant Association (NRA) (the representative of 636,000 foodservice units) lends its full support to socially responsible, multitiered solutions to the solid waste crisis. To be successful, such a solution would require the integration of recycling, source reduction, incineration and landfilling, and the support of local and state government. Policies for solid waste management depend on (1) cooperation of customers to sort and separate solid waste, (2) interest and support of employees and management, (3) the ability to store waste, (4) the cost of transporting waste to recycling centers, (5) the availability of buy-back centers or reclaimers, and (6) a market demand for recycled products.

Social responsibility in minority group relations means going beyond the minimum EEOC requirements and actively recruiting and promoting members of minority groups, as well as training them if they lack necessary entry-level skills.

Responsibility to consumers in the foodservice field means ethical pricing and advertising practices. It also means serving good quality food that has been prepared under the most sanitary conditions humanly possible.

In the area of management and employee rights, several areas of social responsibility are important. Freedom of speech, assembly, due process of law, privacy, fair compensation, and safe working conditions are all mandated by law. As with EEOC regulations, OSHA sets minimum safety requirements. An employer must use his or her own sense of responsibility to determine how much additional cost to incur to ensure the safest of working conditions.

More controversial areas of employee rights are substance abuse and drug testing, smoking in the workplace, and surveillance of employees. Random drug testing is illegal unless certain criteria are met. For example, the employer must have employee written consent, the job consequences of abuse must be severe, and the results must be held strictly confidential. No-smoking facilities are becoming more widespread. This threatens the rights of the smoker; however, thus far, opposition has not been strong. Modern technology has made sophisticated employee surveillance techniques more effective, readily available, and affordable. If this practice occurs without employees' knowledge, a question of ethics could be raised.

Functional Responsibilities and Skills Required

Certain basic responsibilities are common to all foodservice administrators in whatever type of organization they are employed. Most of the responsibilities that are specific to foodservice are discussed in detail in other chapters of this book and include the following:

- Establishment of goals, objectives, and standards;
- Personnel selection, education, and maintenance of an effective staff;
- Overall planning and delegation of work to be done; scheduling of workers;
- Purchase of food, equipment, and supplies according to specification;
- Planning for physical facilities and equipment needs;
- Supervision of all technical operations: production, delivery and service of food, sanitation, safety, security; and

- Financial planning and control.

These responsibilities may be classified under the functions of management discussed in Chapter 12.

Effective administrative leadership is a professional responsibility. Some of the key characteristics of successful—and, thus, professionally responsible—leaders are given by Lester (1981):

> (1) Sense of responsibility—This may mean sometimes subordinating personal desires to the needs of the organization or the profession; (2) Technical and professional competence—The input of others to make decisions may provide guidance, but the final decision will require personal technical and professional knowledge and skills; (3) Enthusiasm—Honest, genuine enthusiasm for the goals and plans of the leader are vital to the generation of commitment and enthusiasm on the part of employees. New directions and unfamiliar areas should be sought without reluctance; (4) Communication skills—Communication is one of the vital linking processes that holds the organization together. It is a key variable in leadership effectiveness. Verbal, written, and nonverbal communications should be understood and utilized effectively. Active listening, avoiding distortion, is a key to good communication; (5) High ethical standards—Ethics are the basis of all group interaction and decision-making processes. Therefore, they play a key role in the leadership function. Professional ethics require leaders to maintain high standards of personal conduct in all situations so that employees may rely on their actions. Integrity is demonstrated when concern for company interests is greater than personal pride; (6) Flexibility—Leaders must have the ability to take whatever comes along and thrive on it. This requires an understanding and acceptance of the fact that no two people or two situations are ever exactly alike. Approaches must be adapted. Change and stress must be understood and managed; (7) Vision—An ability to see the organization as a whole made up of interdependent and interrelated parts; to see where it is going and how it can get there is necessary for effective leadership. Leaders with ideas and images that can excite people and develop timely and appropriate choices will inspire those they lead.

SUMMARY

Leadership is widely touted as a cure-all for organizational problems. It is also widely misunderstood—a fact that can result in oversimplified advice to managers. Leadership in foodservice organizations can be defined as the privilege of having the responsibility to direct the actions of others in carrying out the purposes of their organization. This can be accomplished at various levels of authority, but at all levels the leader is accountable for both successes and failures. Many researchers believe that they can systematize or develop models to capture the elusive nature of the leadership phenomenon; however, there are just as many researchers who believe that it cannot be systematized or modeled. This latter group contends that there are too many variables inherent in the environments in which the leader must function, and that the currently popular models of leadership, such as the managerial grid, the Tannenbaum-Schmidt continuum, and the Hersey-Blanchard situational model, neglect many of the most critical challenges that leaders must face.

Managers and leaders—are they different? Bennis (1985) makes this insightful distinction between managers and leaders: "Managers do things right" while "leaders do the right things." He contends that both functions are necessary in any organization, but that American businesses today are dangerously overmanaged and underled. Drucker (1992) agrees and suggests that leadership is made up of a few essential principles, the very same ones that constitute effective management. What needs to be done now is the integration of all the principles of management and leadership—a blending of innovation, stability, order, and flexibility.

Clearly, leadership has been the subject of exhaustive study for many years and will continue to be studied. Many researchers would agree that the knowledge that we already have is enough to improve the situation in American business. Managers now need to do a better job of applying what is known about motivation and communication in order to become leader-managers. To be competitive in today's world, one must not only make a profit but also develop competent and motivated people who can adapt rapidly to changing technology and markets, work together and create synergies, and interact with customers as though they were speaking for the company.

Many see the need for a new team-based style of leadership in which the power of ideas is recognized over the power of position. The differences inherent in this new style are highlighted in Table 14.3. This style requires that the leader:

1. Build trusting relationships within the team by being empathetic, providing honest and sincere appreciation for work, keeping confidences, and being a good listener.
2. Build a unified team by creating a shared sense of purpose, creating an environment where goals are considered team goals, recognizing and appreciating people for their individual differences, making each person responsible for the team product, building the confidence of each team member, becoming involved and staying involved, and becoming a mentor.

Table 14.3. Differences between the old and new styles of leadership.

Old Leadership Style	New Leadership Style
Power is concentrated with the leader.	Power is distributed throughout the team.
The leader is accountable and controls the organization.	The leader is accountable but surrenders control of the organization to teams.
The leader defines the vision, mission, and goals of the organization.	The leader defines the vision and mission of the organization. The team defines the goals of the organization.
The leader makes decisions. Employees implement decisions.	Each team member has input into the decision-making process. Decisions are agreed to by the whole team.
Individuals are recognized for achievement.	The team is recognized for achievement.
The leader takes credit for the end product of employees' work.	The team takes credit for the end product of the team's work.

3. Establish a clear communication style by fostering trust, accepting others' viewpoints, and being consistent in interactions with others.
4. Solve problems creatively by including all members of the team in the problem-solving process.
5. Develop an enthusiastic motivating style by including all members of the team in the decision-making process, giving all members of the team ownership in the end products, providing some form of recognition at the end of a project, and keeping the team focused on its goals and objectives.
6. Become a flexible risk taker and decision maker.

REVIEW QUESTIONS

1. Explain the need hierarchy, achievement motivation, motivation-hygiene, operant conditioning, and expectancy theories of motivation. Describe one management strategy using each of these theories.
2. Compare and contrast the concepts of management and leadership.
3. Describe the difference between McGregor's Theory X and Theory Y approaches to leadership.
4. Trace the history of leadership theories, and describe the contributions that each theory has made to the current approaches.
5. What is situational leadership?
6. Describe how the Tannenbaum-Schmidt continuum of manager behavior fits into a contingency theory of leadership.
7. Define the five types of leader power and give an example of the judicious use of each.
8. Define communication.
9. List the various modes of communication that are used in an organization and give an example of each.
10. List and briefly describe some of the barriers to good communication.
11. Describe some ways in which communication can be made a two-way process.
12. Identify some other techniques that may be used to improve communication.
13. Describe how social responsibility has been legally mandated.
14. Discuss a foodservice manager's professional responsibilities.

SELECTED REFERENCES

Barnes, L. B., and Kriger, M. P.: The hidden side of organizational leadership. Sloan Mgmt. Rev. 1986; 27:15.

Bass, B. M.: *Stogdill's Handbook of Leadership. A Survey of Theory and Research,* rev. ed. New York: The Free Press, 1981.

Bennis, W., and Nanus, B.: *Leaders: The Strategies for Taking Charge.* New York: Harper & Row, 1985.

Blake, R. R., and Mouton, J. S.: *Executive Achievement: Making It at the Top.* New York: McGraw-Hill, 1986.

Bolman, L. G., and Deal, T. E.: *Reframing Organizations.* San Francisco: Jossey-Bass, 1991.

Burns, J. M.: *Leadership.* New York: Harper & Row, 1978.

Dessler, G.: *Organization Theory.* Englewood Cliffs, N.J.: Prentice Hall, 1986.

Drucker, P. F.: *Managing for the Future.* New York: Truman Talley Books/E. P. Dutton, 1992.

Ewing, D. W.: *Freedom Inside the Organization.* New York: E. P. Dutton, 1977.

Fiedler, F. E., and Garcia, J. E.: *New Approaches to Effective Leadership.* New York: John Wiley & Sons, 1987.

French, J. R. P., and Raven, B. H.: The bases of social power. In D. Cartwright, ed., *Studies in Social Power.* Ann Arbor, Mich.: University of Michigan Press, 1959.

Hersey, P., and Blanchard, K. H.: *Management of Organizational Behavior: Utilizing Human Resources,* 4th ed. Englewood Cliffs, N.J.: Prentice Hall, 1982.

Herzberg, F., Mausner, V., and Snyderman, B.: *The Motivation to Work.* New York: Wiley, 1959.

Hickman, C. R.: *Mind of a Manager, Soul of a Leader.* New York: John Wiley & Sons, 1990.

Hinterhuber, H. H., and Popp, W.: Are you a strategist or just a manager? Harvard Bus. Rev. January-February 1992, pp 105–113.

House, R. J.: A path-goal theory of leader effectiveness. Admin. Science Quart. 1971; 16:321.

Koehler, J. W., Anatol, K. W. E., and Applbaum, R. L.: *Organizational Communication,* 2nd ed. New York: Holt, Rinehart and Winston, 1981.

Koontz, H., and O'Donnell, C.: The functions of the manager. In *Principles of Management.* New York: McGraw-Hill, 1968.

Kouzes, J. M., and Posner, B. Z.: *The Leadership Challenge.* San Francisco: Jossey-Bass, 1987.

Lester, R. I.: Leadership: Some principles and concepts. Pers. J. 1981; 50:868.

Levine, S. R., and Crom, M. A.: *The Leader in You.* New York: Simon & Schuster, 1993, pp. 31–97.

Maccoby, M.: *Why Work?* New York: Simon & Schuster, 1989.

Maslow, A. H.: *Motivation and Personality.* New York: Harper & Row, 1954.

McClelland, D. C.: *The Achieving Society.* Princeton, N.J.: Van Nostrand, 1961.

McGregor, D.: *The Human Side of Enterprise.* New York: McGraw-Hill, 1985.

Pace, R.D.: Dietetics leadership in the 21st century. J. Am. Diet. Assoc. 1995; 95(5):536.

Porter, L. W., and Lawler II, E. E.: *Managerial Attitudes and Performance.* Homewood, Ill.: Richard D. Irwin, 1968.

Roberts, W.: *Leadership Secrets of Attila the Hun.* New York: Warner Books, 1987.

Skinner, B. F.: *Contingencies of Reinforcement.* New York: Appleton-Century-Crofts, 1969.

Spence, H., and Duncan, P., eds.: Food service and the environment. In Proceedings of the Food Service and the Environment Conference, Victoria University of Wellington, New Zealand, April 6–7, 1992.

Stogdill, R. M., and Coons, A. E., eds.: *Leader Behavior: Its Description and Measurement,* Research Monograph No. 88. Columbus, Ohio: Bureau of Business Research, The Ohio State University, 1957.

Stone, F. M., ed.: *The AMA Handbook of Supervisory Management.* New York: AMACOM, 1989.

Tamel M. E., and Reynolds, H.: *Executive Leadership.* Englewood Cliffs, N.J.: Prentice Hall, 1981.

Tannenbaum, R., and Schmidt, W. H.: How to choose a leadership pattern. Harvard Bus. Rev. 1958; 36(2): 95–101.

———: How to choose a leadership pattern. Harvard Bus. Rev. 1973; 51(3):162.

Van Hooser, P.: The leadership challenge: Motivation and the "new breed" of employee. Dietary Manager. 1995; 4(4):6.

Vroom, V.: *Work and Motivation.* New York: Wiley, 1964.

Williamson, J. N., ed.: *The Leader-Manager.* New York: John Wiley & Sons, 1986.

Zaleznik, A.: *The Managerial Mystique.* New York: Harper & Row, 1989.

15

Work Improvement and Productivity

OUTLINE

Raising productivity has been called the first responsibility of management in a knowledge society. Predictions are that it will be the biggest and toughest challenge facing managers during the next few decades. Past experience has shown that an increase in productivity as a result of investment of capital and new technology depends entirely on how people use the new resources. However, people "working smarter" is believed to be the real key to productivity increases in foodservice operations.

Productivity, the ratio of input to output, can be used as a measure of work improvement. However, using the input/output ratio as a definition of productivity in the practical world of foodservice today is inadequate and somewhat irrelevant. A wider conception of productivity, which encompasses factors such as product quality and customer satisfaction, is necessary. In addition, any attempt at work improvement must take into consideration the people involved. An understanding of human nature on the part of management and improvement in the overall quality of work life are critical components of any productivity improvement program.

Increased production with less human effort has been another objective in the foodservice industry for years. Some methods for designing effective and efficient ways of accomplishing work are included in this chapter. This work design must consider improved job content, a safe and healthy work environment, and efficient and effective work methods.

Each of these improvements requires a careful study of existing conditions before any change is implemented. The study of work methods begins with the establishment of standards. The method employed can then be analyzed using one of several techniques. Work simplification is one such technique for identifying and eliminating the uneconomical use of time, equipment, materials, space, or human effort.

In foodservice operations, the focus on increased productivity has to center around performance. Work performance in some jobs means that quality, quantity, and/or both quality and quantity must be considered. Quality management philosophies, such as total quality management, are receiving increased attention in all segments of foodservice.

A step-by-step procedure for implementing a work improvement program is outlined in this chapter. One very important step requires that the job be broken down into its component parts in detail. This may be accomplished by work sampling, a pathway or flow diagram, operation and process charting, or micromotion studies, each of which is discussed briefly.

KEY CONCEPTS

1. *Productivity* is a measure of the output of goods or services in relation to the input of resources. People are the key factor to improving productivity.
2. The *QWL (quality of work life)* approach to improving productivity advocates a relationship between employees and management that incorporates the qualities of cooperation, trust, involvement, respect, rapport, and openness.
3. The goals of *work design* are to improve the content of the job, to provide a safe and healthy work environment, and to design a staff of fit people, an optimum work environment, and effective and efficient work methods.
4. *Work simplification* is the philosophy that there is always a better way of accomplishing a task that will eliminate the uneconomical use of time, equipment, materials, space, or human effort.
5. The fundamental *principles of motion economy* may be applied to foodservice operations in order to improve productivity.
6. Methods that can be used when conducting a work improvement study include *work sampling, pathway* or *flow diagrams, operation and process charting,* and *micromotion studies.*
7. *Total quality management* (TQM) is a management process and set of disciplines that are coordinated to ensure that the company consistently meets or exceeds quality standards as set by customers.
8. TQM involves all levels of the organization and requires high levels of employee participation and teamwork.
9. TQM companies employ measurement methods, tools, and programs to manage data in all processes in order to eliminate waste and pursue continuous improvement.
10. TQM techniques include baseline and benchmarking measurements, brainstorming, flowcharts, check sheets, cause-and-effect diagrams, Pareto charts, scatter

diagrams, histograms, sociotechnical systems, statistical process control, just-in-time inventory control, and ISO 9000.

WHAT IS PRODUCTIVITY?

The successful and efficient day-to-day operation of a foodservice is a constant challenge to its management group. No foodservice system can afford to remain static for long. Instead, it must keep pace with the socioeconomic changes and technological developments in food and equipment and their effects on the overall pattern of operation. Changes in consumer attitudes and behavior, labor and energy costs, regulatory considerations and the general business environment have created new and challenging problems. The present-day foodservice consumer shows a much greater awareness of economic value and sanitary requirements and demands quality food and efficient service at a reasonable price.

Increased production with less human effort has long been an objective in industry. Interest in the designing of work systems that could convert human work practices into those done by machines instead contributed to the Industrial Revolution. Since then the development has not been steady, but we do rely heavily on mechanization and automation to increase productivity and develop manpower effectiveness. Current high material and labor costs make it imperative that every effort be made to study the work design and to perfect efficient operation if high standards of production and quality of products are to be maintained at a reasonable cost.

The simplification of tasks and techniques designed to decrease worker fatigue is an effective aid to good management and is accorded wide recognition and attention by both managers and workers in the foodservice field.

Increased productivity and employee satisfaction are frequently considered to be the overall objective of work design. Many definitions have been given for productivity but essentially, in foodservice management, it is a measure or level of *output* of goods produced or services rendered in relation to *input* in terms of time (labor hours, minutes, or days), money spent, or other resources used.

To relate such diverse quantitative units of measurement as number of meals and amounts of service, pounds of materials, labor hours, BTUs, and capital equipment, we can express these units in dollar values. The resulting formula is a profitability ratio that must be greater than one to produce a profit.

A crucial problem facing some companies is their inability, due to competition (and cost containment), to recover increases in the cost of materials, labor, or other resources by raising prices. They are also unable to decrease the cost of the resources or substitute others. Therefore, if the profit margin is to be maintained or increased, productivity must be improved.

In foodservice organizations, productivity is measured using indicators such as meals per worked hour, meals per paid hour, meal equivalents per worked hour, meal equivalents per paid hour, transactions per worked hour, and transactions per paid hour. When measured for successive periods, these productivity indicators show a trend. Comparisons can also be made between similar institutions.

THE QWL APPROACH

Quality of work life (QWL) is a term that has been used to describe values that relate to the quality of human experiences in the workplace. QWL is affected by a composite of factors on the job, including factors that relate to work itself, to the work environment, and to the employee personally. People are the key factor in improving productivity. If productivity is to be improved, both the nature of people and the organizations in which they work must be understood. People are the highest order of resources and, as such, are responsible for controlling and utilizing all other resources.

If the source of improving the productivity position of an organization is directly traceable to people, then it follows that the achievement of a better bottom line of productivity must be *everybody's* business. Managers must be capable of utilizing the human resources of the organization and use a systems approach to productivity improvement in which all members of the organization are involved.

Increased productivity means motivation, dignity, and greater personal participation in the design and performance of work in the foodservice organization. It means developing individuals whose lives can be productive in the fullest sense.

Judson (1982) reported that a study of 195 U.S. companies found that management ineffectiveness was by far the single greatest cause of declining productivity and that the only successful effort to raise productivity was an integrated QWL approach.

QWL is a multifaceted concept. Incentive plans such as a contingent time-off plan under which the company agrees to award specific time off if the workers perform at an agreed-on level have been successful in improving productivity. Such factors as reducing worker fears, providing opportunity for advancement, implementing job enrichment by adding responsibility, budget, or staff to the job, allowing the exercise of professional skills, and improving communication skills also aid in increasing productivity.

A classic study conducted a number of years ago by Kahn and Katz (1960), two behavioral science researchers, found that a particular leadership style was more effective in increasing productivity and employee satisfaction. The characteristics of this style of leadership are (1) general supervision rather than close, detailed supervision of employees; (2) more time devoted to supervisory activities than in doing production work; (3) much attention to planning of work and special tasks; (4) a willingness to permit employees to participate in the decision-making process; and (5) an approach to the job situation that is described as being employee centered, that is, showing a sincere interest in the needs and problems of employees as individuals, as well as being interested in high production.

Increased involvement of workers in their organizations has received much attention in the last few years. Today's workers no longer want to be separated from responsibility. Productivity appears to be maximized when a unity of purpose and a feeling of ownership exist among employees. This unity is created when the greatest possible responsibility is given to the lowest possible levels of the organization; compensation systems are designed so that employees are salaried with incentive earnings tied to competence and performance; the greatest degree of involvement and consensus is sought from all levels; and management exhibits unity with the employee.

In addition to these characteristics, improvements in resources (supervision, methods, and technology) to facilitate greater effectiveness and reduce frustrations seem effective in improving productivity and employee satisfaction.

The QWL approach, in essence, attempts to replace the typical adversary relationship between management and employees with a cooperative one. The key words of QWL are cooperation, trust, involvement, respect, rapport, and openness.

Work Design

The overall objectives of work design are to increase productivity and employee satisfaction. The specific objectives are to improve the content of the job, to provide a safe and healthy work environment, and to design a staff of fit people, an optimum work environment, and effective and efficient work methods.

Job Content. Job content in foodservice systems is being improved through automation of the production and distribution systems. Food factories and commissary-type operations employ large-volume machinery in long, integrated production runs to prepare one specific product at a time. This system makes possible a more orderly pace and, usually, more desirable working hours.

Another approach to changing job content is to redelegate some parts of the job. A growing number of jobs are becoming encumbered with routine "busy work" that has little or no value and that could be delegated to less skilled employees. High levels of productivity, profitability, motivation, and morale are dependent on allowing employees to do what they have been trained and paid to do. The ingredient room where foodservice employees weigh, measure, and assemble all the ingredients for each production formula is an example of this downward shift of responsibility. Such work would normally be performed by a cook, chef, or cold prep person. The use of support personnel such as dietetic technicians and assistants is another example. Delegation of this type must be done carefully. It is a complex process requiring skill in planning, organizing, and controlling. The different needs and abilities of employees must be effectively managed.

Safety and Health. The provision of a safe and healthy work environment is both economically and sociologically important. From an economic standpoint, accidents and job-related illnesses are extremely costly in terms of productivity. The safety and health of the workforce is the responsibility of its management and is discussed further in the sections that follow.

Equipment Design. Kazarian (1983) stated that "Probably the greatest change that will be evident in the future planning of foodservice facilities is the physical arrangement of spaces and equipment to increase the productivity of workers." Avery (1985) said:

> In order for the equipment and workers to combine productively, one must call on the science variously known as ergonomics, human factors, or human engineering. This may be defined as adapting tasks, equipment and working environment to the sensory, perceptual, mental, and physical attributes of the human worker. The employee works best if his equipment is designed for the job to be done, his capabilities, and if it is well placed in pleasing surroundings. (p. 15)

To maximize productivity in foodservice, Avery (1985) set forth these principles of human engineering:

1. Design and arrangement of equipment should be such that the equipment's use requires a minimum application of human physical effort.
2. Only essential information should be provided for the equipment, and this should be presented when and where it is required with maximum clarity. It should be arranged in a step-by-step order.
3. Control devices on equipment should be easily identified, minimum in number, logical in placement, and in consonance with displays in operation. They should relate precisely to the functions they control.
4. Equipment should be designed to provide maximum productivity while utilizing the worker's physical and mental attributes most effectively. It should take into account the dimensions of the worker and his or her strengths.
5. Equipment should be selected on the basis of need in utilizing specific ingredients to prepare a selected menu, grouped in most used combinations, and arranged in order of most frequent interuse proceeding from left to right. Those tasks demanding the greatest skill should be grouped around the worker having these skills, and his or her movements to provide for his or her needs should be minimal.
6. The environment in which the foodservice worker operates should be designed and controlled to allow him or her to be most productive, comfortable, and happy with his or her job. This control involves consideration of lighting, facility and equipment coloration, temperature, humidity, noise, smells, facility design, floor conditions, and safety, among others. (p. 16)

Work Environment. One of the goals of human engineering is the prevention of fatigue. The manager of a foodservice may find that the fatigue or tiredness of some workers, with a resultant drop in their energy, enthusiasm, and production output, is due to external factors beyond his or her control, such as irregularities in the home situation, extraordinary physical exertion away from the job, or a nutritionally inadequate food intake. However, in the organization, while the workers are on the job, there are unlimited opportunities to study causes of fatigue and possibly to correct them.

Certain psychological factors such as attitude because of disinterest in and boredom with the job, dislike of the supervisor, or a low rate of pay may contribute to the fatigue and low output of some workers, but such situations often can be improved through changes in personnel policies and administration. Emphasis in this section is relative to the environmental and physical factors on the job that can affect fatigue and on work improvement methods.

With a given set of working conditions and equipment, the amount of work done in a day depends on the ability of the worker and the speed at which he or she works. The fatigue resulting from a given level of activity depends on such factors as (1) hours of work, that is, the length of the working day and the weekly working hours; (2) number, location, and length of rest periods; (3) working conditions such as lighting, heating, ventilation, and noise; and (4) the work itself.

The amount of reserve energy brought to the job varies with individuals. Some workers can maintain a fairly even tempo throughout the day, whereas others tire rather quickly and need to rest periodically to recoup nervous and physical energy. Short rest periods appropriately scheduled tend to reduce fatigue and lessen time taken by employees for personal needs.

Lighting, heating, ventilation, and noise are environmental factors that often contribute to worker fatigue. Satisfactory standards for the lighting of kitchen areas is 35 to 50 foot-candles on work surfaces with reflectance ratios of 80 for ceiling, 30 to 35 on equipment, to a minimum of 15 for floors. Temperature and humidity also influence worker productivity. A desirable climate for food preparation and service areas is around 68°F to 72°F in the winter and 74°F to 78°F in summer with relative humidity of 40% to 45%. Higher temperatures tend to increase the heart rate and fatigue of most workers. Air conditioning in hot and humid locales is considered a necessity, whereas in some parts of the country a good fan and duct system is satisfactory to change the air every 2 to 5 minutes. Hoods over cooking equipment provide for the disposal of much heat and odor originating from these units. Noise has a disturbing and tiring effect on most people. Effective control of the intensity of noise in a foodservice area is possible through precautionary measures such as installation of sound-absorbing ceiling materials, the use of rubber-tired mobile equipment and smooth-running motors, and training of the employees to work quietly.

Much has been written about the value of the study of physical facilities and the procedures followed in specific jobs. These studies are aimed at increasing efficiency in the operation of a foodservice. A thorough analysis of a floor plan, on paper or in actuality, would provide facts on which to base decisions regarding changes needed in order to make the most compact arrangement possible, yet provide adequate equipment in an efficient arrangement.

Work Methods. To design the most effective and efficient work methods, existing methods must first be studied. A very important part of the manager's job is the establishment and maintenance of standards by which performance can be judged. Standards are criteria against which results are judged.

Foodservice managers and administrative dietitians have responsibility for determining the standards of time, quantity, quality, and cost for their own department. How much work of what quality is to be accomplished in what length of time and at what cost? Managers must be able to answer these questions realistically and provide the information for the workers if any degree of competence and *high productivity* are to be achieved. Otherwise, employees develop their own standards and may never reach their potential or produce at an acceptable level. Managers who allow this to happen have lost control of operations; events then control the manager.

Standards are derived from objectives, the statements of what is to be achieved. Behavioral objectives relate specifically to what an individual should be able to do as a result of the learning process. Such objectives usually form the basis of a training program and give specific measures for trainee attainment. Although workers learn proper techniques and procedures through training programs, actual work experience is usually necessary for them to develop speed and accuracy in completing a given task. This is a situation in which preestablished standards are necessary so the employees will know exactly what is expected of them and so a goal is provided for them to attain. An example of a behavioral objective for a kitchen worker is "the employee will be able to make and wrap 90 sandwiches in 30 minutes."

Time and quantity studies must be made within a given foodservice department in order to establish desirable standards for that department with its own equipment, space arrangement, facilities, and procedures. These variables make it difficult if not impossible to have universal standards among foodservice organizations. The employee who knows she is expected to make 45 sandwiches in 15 minutes, or 3 sandwiches per minute, is much more apt to respond to meet the challenge of a goal to be reached than the employee who is told to "just work as fast as you can."

Each operation or task to be performed requires similar questions to be asked and answered. Actual time studies may be made in order to arrive at realistic time-quantity standards. Quality and cost standards come from the knowledge of the dietitian or foodservice manager, who must have acquired this from his or her basic education and preparation for the position. Standards for personnel to which applicants can be compared include minimum acceptable qualities necessary for adequate performance of the job duties. These standards or qualifications are stated in job specifications for use by those who select personnel. (See Chapter 12 for more on job specifications.)

Although no one can set standards for someone else's department, results of some research studies can prove helpful as a guide to foodservice managers undertaking a standards-setting project. A formula for determining how long it takes to accomplish a given task is

$$\frac{a + 4m + b}{6} = \text{Probable time required}$$

a = the "least" amount of time to do a given task
b = the "most" amount of time to do a given task
m = the most "likely" time to do a given task

Let us say, for instance, that we want to know how much time should be allowed to mop a heavily obstructed 1600-square-foot institutional kitchen. A sloppy, cursory job will take 20 minutes:

$a = 20$ (minutes)*

An extremely thorough job will take 120 minutes:

$b = 120$ (minutes)*

The job, done adequately, will most likely take 45 minutes:

$m = 45$ (minutes)*

Using the formula,

$$\frac{a + 4m + b}{6} = \frac{20 + (4 \times 45) + 120}{6} = \frac{20 + 180 + 120}{6} = \frac{320}{6} = 53\tfrac{1}{3} \text{ minutes}$$

Knowing that the most probable time to mop the floor is 53* minutes, we will allot that amount of time daily to that task.

Studies throughout the years have given suggested standards in terms of number of labor minutes required for one meal served. These ranged from a low of 4 or 5 minutes for some school foodservices, to 9 or 10 minutes for college and industrial cafeterias, up to 18 to 20 labor minutes per meal including supervisory time in hospitals. Greater efficiency (lower number of minutes required) is usually achieved when a larger number of meals are served as compared with a small organization serving few meals. Foodservice managers may wish to make a study of their own operations and develop a standard for themselves based on number of labor minutes needed for each meal served.

School foodservice standards for personnel staffing are stated in terms of number of labor hours needed for number of lunches served, as shown in Table 15.1.

Worker efficiency in foodservices is estimated to be from 40% to 55%. Although the nature of the work involved in preparing meals for specified periods of the day, causing uneven workloads, can be part of the reason for such low productivity as compared with that in other industries, foodservices can no longer afford the luxury of inefficiency. Labor costs are too high, and every means possible should be used to improve productivity.

Productivity is sometimes equated with efficiency. Kaud (1984) proposed this formula for expressing work efficiency:

$$\text{Efficiency \%} = \frac{\text{earned (standard) hours}}{\text{actual (worked) hours}} \times 100$$

The time during which the worker actually performs necessary work can be determined through the use of the work sampling (see the later section under "Methods for Work Improvement Study").

Work Simplification. Detailed studies of activities within an organization often reveal that cost and time requirements are high because of unnecessary operations and excess motions used by the workers in the performance of their jobs. When proper adjustments are made in both the physical setup and the work procedures, the conservation of energy of the workers, increased production, and a reduction in total person-hours should result. Such studies have proved highly effective in the simplification of effort in both repetitive and nonrepetitive activities and apply either in a new situation or where long-established procedures have become accepted practices.

* Derived from actual time study of cursory, thorough, and adequate sampling.

Table 15.1 School foodservice standards.

Number of Lunches Served	Meals Per Labor Hour	Total Hours
Up to 100	9½	9–11
101–150	10	10–15
151–200	11	15–17
201–250	12	17–20
251–300	13	20–22
301–350	14	22–25
351–400	14	25–29
401–450	14	29–32
451–500	14	32–35
501–550	15	35–36
551–600	15	36–40
601–700	16	40–43
701–800	16	43–50
801+	18	50+

What has become known as *work simplification* began in the late 1920s. An industrial engineer, Allan H. Mogensen, developed the philosophy as a result of his work at Eastman Kodak. He found that workers were creative at thwarting his attempts to prescribe more efficient methods and, when not under surveillance, would develop more productive methods that would enhance their rewards. He reasoned that this creativity could be harnessed in a way that would enable every employee to be his or her own industrial engineer. Thus, the slogan, "Work smarter, nor harder," emerged.

Work simplification is more than a technique or set of how-to-do-its. It is a way of thinking or a philosophy that there is always a better way. The emphasis is on the elimination of any uneconomical use of time, equipment, materials, space, or human effort. Conservative estimates show that through an effective work simplification program, foodservice worker productivity can be increased by 20% to 50%.

Employee interest, understanding, and cooperation are essential to the successful operation of a work simplification program. Thinking through and planning before starting any task are necessary if it is to be accomplished efficiently and in the simplest manner possible. The elimination of wasted effort is easy once the worker becomes "motion conscious," learns to apply the principles that may be involved, and sees objectively the benefits of changed procedures. Such benefits can be evidenced by lessened fatigue of workers, safer and better working conditions, better and more uniform quality production, and possibly higher wages through increased production. Agreement and understanding of the objectives and realization that benefits will be shared mutually by workers and management are factors for success. The solicitation and incorporation of suggestions for job improvement methods from the workers are conducive to enthusiastic interest and participation. Usually, any employee resistance to change in established work routines can be overcome by the proper approach of management before and after the inauguration of a work simplification program. The selection of personnel possessing the personality to be a leader and training for leadership in this work are of prime importance to its implementation.

Motion Economy. The same principles of motion economy adopted by engineers many years ago are applicable in a foodservice operation. Analysts and supervisors need to have an understanding of these principles and the ability to interpret them to workers effectively before job breakdown studies and revision in procedures, arrangement of work area, and equipment are inaugurated. A listing of these fundamental principles of motion economy is given in Table 15.2.

Practical application of most of these principles can be made easily in the foodservice field and lead to increased efficiency; that is, reduction in the motion and time required for the job. Also, the steady output of production with less fatigue on the part of the worker can be a result. A few specific examples of application follow:

- *Principles 1, 2, and 3:* To serve food onto a plate at the counter, pick up plate with left hand and bring to a center position while right hand grasps serving utensil, dips food, and carries it to the plate, both operations ending simultaneously; when panning rolls, pick up a roll in each hand and place on pan.
- *Principle 6:* Stir a mass of food easily and with minimum fatigue by grasping the handle of a wire whip (thumb up) and stirring in a circular motion instead of pushing the whip directly back and forth across the kettle. Principles 5 and 7 are also applied in this same example, since greater force may be gained easily at the beginning of the downward and upward parts of the cycle.
- *Principle 8:* Gain and maintain speed in dipping muffin or cupcake batter through the use of rhythmic motions; use regular and rhythmic motions in slicing or chopping certain vegetables and fruits with a French knife on a board.
- *Principles 8 and 10:* Equip and arrange each individual's work area so that body movements are confined to a minimum in his or her job performance.
- *Principle 11:* Store mixing-machine attachments and cooking utensils as close as possible to place of use; remove clean dishes from washing machine directly to carts or dispensing units that fit into serving counter; store glasses and cups in racks in which they were washed; cook certain foods in containers from which they will be served.
- *Principle 12:* Install vegetable peeler at end of preparation sink so that peeled potatoes can be dumped directly into the sink; install water outlets above range and jacketed kettles so utensils can be filled at point of use.
- *Principle 14:* For breading foods, arrange container of food to be breaded, flour, egg mixture, crumbs, and cooking pan in correct sequence so that no wasted motions are made.
- *Principle 16:* Provide some means of adjusting height of work surface to the tall and short worker; include one or two adjustable-height stools in list of kitchen equipment.
- *Principle 18:* Provide knee lever-controlled drain outlets on kitchen sinks; install electronic-eye controls on doors between dining room and kitchen.

Work Improvement Program. Improvement in any work program is contingent on a study of the environmental factors and the activities of the workers in meeting the objectives of the organization. Such a study can involve one or more jobs in an organization. The following steps explain the usual approach for analysis and revision of a job method:

Table 15.2 Principles of motion economy.

Use of the Human Body	Arrangement of the Workplace	Design of Tools and Equipment
1. The two hands should begin, as well as complete, their motions at the same time.	10. There should be a definite and fixed place for all tools and materials.	18. The hands should be relieved of all work that can be done more advantageously by a jig, fixture, or a foot-operated device.
2. The two hands should not be idle at the same time, except during rest periods.	11. Tools, materials, and controls should be located close to the point of use.	19. Two or more tools should be combined wherever possible.
3. Motions of the arms should be made in opposite and symmetrical directions, and should be made simultaneously.	12. Gravity feed bins and containers should be used to deliver material close to the point of use.	20. Tools and materials should be prepositioned whenever possible.
4. Hand motions should be confined to the lowest classification with which it is possible to perform the work satisfactorily.	13. "Drop deliveries" should be used wherever possible.	21. Where each finger performs some specific movement, such as in typewriting, the load should be distributed in accordance with the inherent capacities of the fingers.
5. Momentum should be employed to assist the worker wherever possible, and it should be reduced to a minimum if it must be overcome by muscular effort.	14. Materials and tools should be located to permit the best sequence of motions.	22. Levers, crossbars, and hand wheels should be located in such positions that the operator can manipulate them with the least change in body position and with the greatest mechanical advantage.
6. Smooth, continuous motions of the hands are preferable to zigzag motions or straight-line motions involving sudden and sharp changes in direction.	15. Provisions should be made for adequate conditions for seeing. Good illumination is the first requirement for satisfactory visual perception.	
7. Ballistic movements are faster, easier, and more accurate than restricted (fixation) or "controlled" movements.	16. The height of the workplace and the chair should preferably be arranged so that alternate sitting and standing at work are easily possible.	
8. Work should be arranged to permit easy and natural rhythm whenever possible.	17. A chair of the type and height to permit good posture should be provided for every worker.	
9. Eye fixations should be as few and as close together as possible.		

Source: Ralph M. Barnes, *Motion and Time Study,* 6th ed. John Wiley & Sons, New York, 1968, p. 220.

1. *Select the job to be improved:* What needs improvement most? A bottle-neck job is the best one to start thinking about. Jobs that require much time or that require much worker movement for materials, tools, and such are also good choices.

2. *Break down the job in detail:* Get in the habit of seeing every job as follows: make ready—the effort and time put into getting the equipment, tools, and materials

with which to work; do—the actual productive work; and put away—the clean up or disposal following the "do." Analyze each operation, noting procedure, equipment used, distance moved, and time required. Some methods for performing this step are discussed in the next section.

3. *Challenge every detail:* Ask these questions of the entire job and of every part of the job. *What* is done? *Why* is it done at all? Is it necessary? *Where* is it done? Why is it done there? Is this the right place for it? Where is the best place? *When* is it done? Why is it done then? Is there a better time? *Who* does it? Why does this person do it? Is he or she the right person? *How* is it done? Why is it done this way? Is there a better way? Is there an easier way? Also question the materials, tools, and equipment that are being used. A form such as that shown in Figure 15.1 is useful for this step.

4. *Develop a better method:* The process- or activity-analysis chart affords opportunity to study any job objectively and to evaluate the efficiency of performance. Checks can be made easily to determine the necessary and the excess operations, where and how delays occur, the distances either product or worker must travel, and where changes can be made.

5. *Put the new method into effect:* Teach the new method and follow up with proper supervision. Continue to seek new and better ways to do the job.

Much of the analysis or breakdown of jobs is accomplished through motion and time studies. Barnes (1980) defines motion and time study as

> The systematic study of work systems with the purposes of (1) developing the preferred system and method—usually the one with the lowest costs; (2) standardizing this system and method; (3) determining the time required by a qualified and properly trained person working at a normal pace to do a specific task or operation; and (4) assisting in training the worker in the preferred method.

Methods for Work Improvement Study. The breakdown of job activities can be made in various ways and recorded appropriately for analysis, study, and evaluation. Among the possibilities are work sampling, pathway or flow diagrams, operation and process charting, and micromotion studies. The objective is to be able to gain a complete and detailed picture of the process, regardless of the method of recording.

Work sampling is a tool for fact-finding and is often less costly in time and money than a continuous study. It is based on the laws of probability that random samples reflect the same pattern of distribution as a large group. The primary use of work sampling is to measure the activities and delays of people or machines and determine the percentage of the time they are working or idle instead of observing the detailed activities of a repetitive task. The shorter and intermittent observations are less tiring to both the worker and the observer than continuous time studies; several workers can be observed simultaneously; interruptions do not affect the results; and tabulations can be made quickly on data-processing equipment, although neither management nor workers may have the knowledge of statistics involved. This process is sometimes known as random ratio-delay sampling.

A **pathway chart** or flow diagram is a scale drawing of an area on which the path of the worker or movement of material during a given process can be indicated and

Situation Studied Lettuce Chopping			**Date** 9-17-81	
Why Studied To improve efficiency of the process			**Study by** Gregoire/Palacio	
Where Kramer Food Center Salad Area			31.03 **minutes/occurrence**	
How Often Repeated per Year 690			21410.7 **minutes/year**	

Purpose	What is achieved? Lettuce is chopped for salads	What would happen if it weren't done? There would be no tossed salads	What could be done and still meet requirements? Lettuce could be bought chopped	What should be done? Lettuce should be chopped at facility
Place	Where is it done? Salad area	Disadvantages of doing it there: Don't always have enough heads of lettuce to fill barrels of chopped lettuce	Where else could it be done? Veg. Prep (downstairs)	Where should it be done? Salad area for control and better use of existing space.
Sequence	When is it done? Before: Each meal After: Lettuce is cleaned	Disadvantages of doing it then: Requires set-up and clean-up three times a day	Advantages of doing it elsewhere: Barrels could be filled to capacity.	When should it be done? Once a day
Person	Who does it? Student salad employee	Why that person? Assigned to students because this job is routine and disliked by full-time employees	Advantages of doing it sooner: Clean-up and set-up only once	Who should do it? Salad foodservice worker
Means	What equipment and methods are used? Equipment: Cutting board French knife Method: Manual chopping	Disadvantages of equipment: Somewhat dangerous Method: Slow, tedious	Advantages of doing it later: Lettuce would be fresher List two others who could do it. Veg. Prep employee Salad foodservice worker How else could it be done? Purchase chopped Manual lettuce chopper VCM-Meat slicer Advantages: Faster, safer	Where should it be done? VCM

Figure 15.1 Critical examination of the process of lettuce chopping. Example of how such a study is recorded on a suitable form.

measured, but with no breakdown of time or details of the operation. Measurement of the distance traveled as the worker moves about in the performance of a task is made by computing the total length of lines drawn from one key point to another simultaneously with the worker's movements and multiplying by the scale of the drawing. A more convenient method is to set up pins or string supports at key points on a scale drawing of the worker's area and wind a measured length of string around the supports as the worker progresses from one position to another.

Operation charts can be used as simple devices to record, in sequence, the elemental movements of the hands of a worker at a given station, without consideration of time. A diagram of the work area might head the chart with the observed activities of both hands listed in two columns—left side for left hand, and right side for right hand. In such a chart, small circles are usually used to indicate transportation and large circles to denote action. Analysis of the chart gives a basis for reducing transportation to the lowest degree possible and for replanning the work area and procedures. It is important that both hands be used simultaneously and effectively. A chart showing the procedures used in making a pineapple and cottage cheese salad is given in Figure 15.2. The lettuce cup had been prearranged on the plate.

The **process chart** is a fairly simple technique for recording and analyzing the breakdown of a job. It graphically presents the separate steps or events by the use of symbols for a given process so that the entire picture of the job can be condensed into a compact yet easily interpreted form. A process chart can present either *product analysis,* which shows in sequence the steps that a product goes through, or *person analysis,* which is a study of what the person does. Many different symbols were used by the Gilbreths when they devised this method of recording job activity years ago. Today, for most practical purposes, only four or five symbols are used that simplify and list quickly the steps or activities in a process. The following symbols are often used:

○ = operation or main steps in the process

⇨ = transportation or movement

D = delay

▽ = storage or hold

□ = inspection such as examination for quality or quantity

Symbols can be arranged in a vertical line in sequence or, as in the activity-analysis charts (Figures 15.3 and 15.4), with lines drawn from one symbol to another in each succeeding step. This method of charting makes for ease in checking time and in determining the number of repeats in any process as a basis for their reduction through revision of the method.

Without benefit of an elaborate process chart form, a simple listing of the procedures and times used in the preparation of a menu item can be made and used to improve either the physical setup or the method, for example, the observance of the cook mixing and portioning meat balls, beginning at the worktable. A sample procedure chart is shown in Table 15.3.

A **micromotion study** is a technique whereby movements of the worker may be photographed and recorded permanently on film. This method affords a more accu-

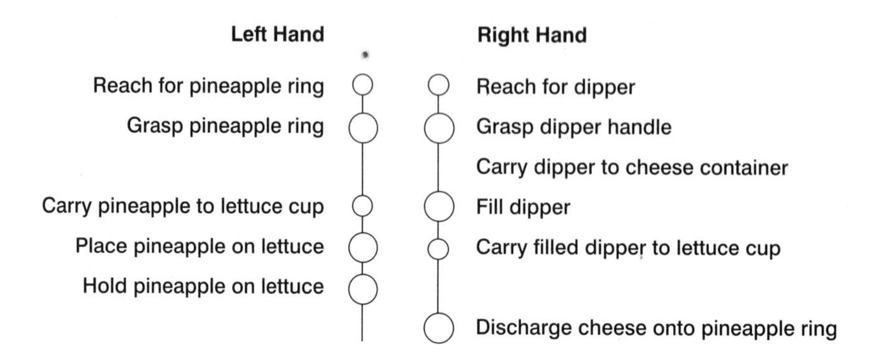

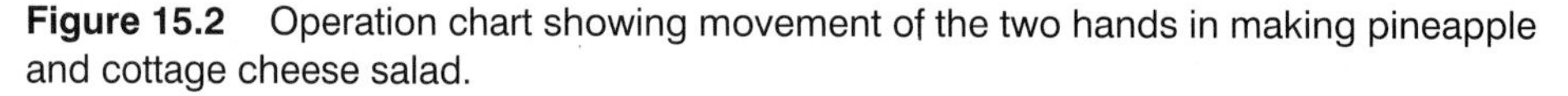

Figure 15.2 Operation chart showing movement of the two hands in making pineapple and cottage cheese salad.

rate presentation of detail than others, and projection for analysis can be made at different rates of speed. In addition, the time of each movement can be recorded.

A detailed motion breakdown of the activities portrayed on micromotion film is easily made and recorded in graphic form by use of **therbligs,** expressed through letter, line, or color symbols. The word *therblig,* formed by spelling Gilbreth backward but retaining the original order of the last two letters, was coined by Frank Gilbreth at the time he introduced the system of breaking down, into 17 subdivisions or elements, basic hand movements employed in job performance. The therbligs are search (Sh), select (St), inspect (I), transport empty (TE), grasp (G), hold (H), transport loaded (TL), release load (RL), position (P), preposition (PP), assemble (A), disassemble (DA), use (U), avoidable delay (AD), unavoidable delay (UD), plan (Pn), and rest (R).

Most often letter symbols are used in recording the breakdown of a procedure, for example, cutting a cake:

P	Place cake on table in position to cut
TE	Move right hand toward knife rack
Sh	Look over supply of knives
St	Select knife
G	Take knife in right hand
TL	Move knife to cutting position above cake
U	Cut cake

Any human activities can be analyzed by this system as a basis for eliminating the unnecessary and excess motions in the formulation of an improved procedure.

The **chronocyclegraph** is a photographic technique designed to show motion patterns of hands performing rapid repetitive operations. It is made by attaching lights to the hands, which show as dotted lines on the finished photograph. The entire workplace must be included in order to study the relationship of the worker and direction of his or her hand movements to the work setup. Complete calculation

PROCESS CHART

Present ☐ Proposed ☐

FILE NUMBER Page

SUMMARY		No.	TIME ( )
OPERATIONS	○	1,546	
INSPECTIONS	□	0	
MOVES	⇨	99	
DELAYS	D	70	
UNITS PRODUCED: 70 trays			1 1/2 hr.
TOTAL DISTANCE MOVED			

TASK or JOB	Dishwashing Procedure, Operations I Scrapping Trays
DEPT.	10th Floor Pantry
EQUIPMENT TOOLS etc.	Scrapping counter, prerinse counter with disposal unit, trash can, carts, cloth
OPERATOR	Pantry Worker A
ANALYST	DATE March 20

Descriptive Notes	Activity	Dist.	Time	Analysis notes
Rinses cloth at sink.	⊗□⇨D			Damp cloth is used to wipe trays.
Carries cloth to scrapping table.	○□⇨(x)D	8' 6"		
Brings loaded cart into pantry from hall.	○□⇨(x)D	5' 6"		Positions cart at left of operator. Each cart holds 6-9 trays.
Moves to side of table.	○□⇨(x)D			
Takes tray from cart and places on scrapping table.	⊗□⇨D			
Moves around in front of table.	○□⇨(x)D	1' 0"		Convenient position for working.
Changes position of tray.	○□⇨D(x)			
Places tray on stack of empty trays.	⊗□⇨D			
Pulls menu from tray.	⊗□⇨D			
Places name card on tray at extreme right of scrapping table.	⊗□⇨D			Name cards in stacks by sections.
Picks up salt and pepper with left hand.	⊗□⇨D			
Places salt and pepper on tray with name cards.	⊗□⇨D			
Empties coffee pots into disposal.	⊗□⇨D			
Places empty coffee pots on prerinse counter.	⊗□⇨D			
Picks up plate and scrapes waste into disposal.	⊗□⇨D			
Adds plate to stack on prerinse counter.	⊗□⇨D			
Picks up creamer and empties contents into disposal.	⊗□⇨D			
Places creamer on prerinse counter.	⊗□⇨D			
Places cup and saucer on prerinse counter.	⊗□⇨D			Saucers stacked.
Removes glasses from tray and empties contents into disposal.	⊗□⇨D			
Places empty glasses upside down in wash rack on prerinse counter.	⊗□⇨D			
Picks up bowl with right hand.	⊗□⇨D			
Transfers bowl to left hand.	○□⇨D(x)			Unnecessary handling. Rubber scraper better tool.
Scrapes waste food from bowl with spoon into disposal.	⊗□⇨D			
Stacks bowls on prerinse counter.	○□⇨D			

Figure 15.3 Process chart of tray scrapping as observed in a hospital floor-service pantry.

PROCESS CHART

Present ☐ Proposed ☒

FILE NUMBER Page

SUMMARY		No.	TIME ( )
OPERATIONS	○	1,024	
INSPECTIONS	□	2	
MOVES	⇨	13	
DELAYS	D	0	
UNITS PRODUCED: 70 trays			1 hr. total
TOTAL DISTANCE MOVED		117 feet	

TASK or JOB	Dishwashing Procedure, Operations I Scrapping Trays
DEPT.	10th Floor Pantry
EQUIPMENT TOOLS etc.	Scrapping counter, prerinse counter with disposal unit, trash can, carts, pan, rubber scraper
OPERATOR	Pantry Worker A
ANALYST	DATE March 20

Descriptive Notes	Activity	Dist.	Time	Analysis notes
Rinses cloth at sink and fills pan with water.	⊗ □ ⇨ D			Cloth is rinsed several times during operation.
Carries cloth and pan to scrapping table.	○ □ ⇨(x) D	7' 0"		Pan of water at counter reduces trips to sink.
Goes to silver storage unit.	○ □ ⇨(x) D	4' 9"		
Gets rubber spatula from drawer.	⊗ □ ⇨ D			
Returns and places spatula on scrapping table.	○ □ ⇨(x) D	5' 0"		
Brings loaded cart into pantry from hall.	○ □ ⇨(x) D	5' 6"		Pre-positioned tray on scrapping counter facilitates placement of name cards, and salts and peppers.
Takes tray from cart and places on scrapping table.	⊗ □ ⇨ D			
Picks up salt and pepper and places on tray at extreme right of worker.	⊗ □ ⇨ D			
Places name-card holder with menu on same tray.	⊗ □ ⇨ D			Menu pulled after name-card holder is on tray while hand is in position. Drops menu on table; all menus later put into trash can at one time.
Scrapes waste food from bowl with rubber scraper.	⊗ □ ⇨ D			
Stacks bowls on prerinse counter.	⊗ □ ⇨ D			All refuse from one tray scraped into one dish; then, into disposal.
Places silver in rack on prerinse counter.	⊗ □ ⇨ D			
Removes glasses from tray, empties contents into disposal.	⊗ □ ⇨ D			
Places glasses upside down in wash rack on prerinse counter.	⊗ □ ⇨ D			
Empties coffee pot and creamer into disposal, transfers directly to dish rack.	⊗ □ ⇨ D			Both hands used simultaneously. Movements combined or continuous wherever possible.
Removes paper tray-cover, folds once and places in trash can.	⊗ □ ⇨ D			
Wipes off bottom of tray, places on stack then wipes off top of tray.	⊗ □ ⇨ D			Handling of each tray reduced to a minimum.
Reaches to cart for next tray.	○ □ ⇨ D			
Operations repeated until all trays are scrapped.	○ □ ⇨ D			

Figure 15.4 Process chart of same operation as shown in Figure 15.3 after a study and revision of the original procedures had been made. By changing the sequence of operations and moving the tray cart near the working area, the operations, moves, and delays were reduced materially.

Table 15.3 Procedure chart.

Approximate Distance (ft.)	Time (min)	Description of Operation
6	9:00	Fasten bowl and beater in position on mixer
15	9:01	Go to refrigerator for milk, ground meat, and other weighed recipe ingredients (use cart)
15	9:035	Return to mixer
3	9:04	Place seasonings, eggs, milk, and cut-up bread in mixer bowl
	9:05	Mix slightly (observe)
	9:06	Add meat and mix to blend (observe)
3	9:075	Remove beater and take to wash sink
5	9:085	Lift bowl of meat mixture to low bench near worktable
50	9:09	Assemble portion tools and pans
	9:10	Portion onto pans with number 12 dipper
	9:25	Complete portioning

of velocity and acceleration of hand motions is limited by the two-dimensional factor of this technique.

Applications of Work Improvement

Analysis of the data accumulated in the study of the work situation and the methods used in a foodservice may show that certain changes could be made immediately, whereas others involve time, money, and an educational program for the workers. No one set of rules can be used to bring about the desired improvements, but through the cooperative effort of management and worker groups, many things can be made possible. A few suggestions for making improvements follow.

One of the first steps in a job improvement plan is to try to *eliminate unnecessary operations, delays, and moves* without producing deleterious effects on the product or worker. Habit plays an important part in the work routines of people, and it is easy for them to continue in the old pattern; for example, even though the improved methods of processing dehydrated fruits eliminate the necessity of soaking before cooking, some cooks might continue to soak them. A common example of good practice is to have one person fill and deliver storeroom requisitions once a day instead of each cook going to the storeroom for single items as needed.

Operations can be combined as in the making of certain types of sandwiches when the butter could be combined with the spread mixture and applied in one operation instead of two. Other examples of simplified practice are the one-bowl method for combining ingredients for cakes, and cutting a handful of celery stalks at one time on a board instead of singly in the hand.

A *change in sequence of operations* to make the most efficient use of time and equipment and to reduce distance is important. Instead of trying to pare and cut dry, hard squash, put it into the steamer for a short time until the hard cover softens; then it can be pared and cut quickly and easily.

The *selection of multiple-use equipment* reduces to a minimum the items needed. A mixing machine with all of the chopper, slicer, and grinder attachments might be more desirable for a given situation than the purchase of a chopping machine in addition to the mixer without attachments. Where and when the item will be used determines its best location. A mixing machine to be used in one department only should be located convenient to that center of activity, whereas a machine shared by two departments would be located between the two but nearest and most accessible to the department requiring the heaviest and most frequent usage. Duplication of some equipment can be compensated in reduced labor hours required for certain jobs.

Equipment can be relocated or removed entirely to facilitate a more direct flow of work in any area. To reduce time-wasting "searches," a definite place should be provided for every item and, in a well-regulated foodservice, everything is kept in the designated location except when in use. This storage location or prepositioning of the items should be convenient to the center of their first use; for example, the bowls, beaters, and attachments for the mixing machine should be stored next to or underneath the machine, the cook's cutlery stored in a drawer or on a rack at the cook's table, and clean water glasses returned to the water cooler in the wash racks for storage. Some kitchens may have retained a meat block, even though pan-ready meats are now used. Others may have more range space than is needed for modern cookery. In either case, the removal of certain equipment would provide space for more efficient utilization. Some kitchens may need additional equipment to provide adequate physical facilities for satisfactory operation. Most kitchen machines are designed as labor-saving devices and can do many times the amount of work that could be accomplished by hand in a comparable time and should be used whenever feasible.

Improvements in design and operation of kitchen machines influence the method of operation. Automatic timing and temperature-control devices release the worker for other duties more than was possible when frequent checking and manual control were necessary.

The *reduction of transportation or movement of materials and equipment* often can be made through rearrangement of equipment, mobile equipment, and the use of carts to transport many items at one time. The relation of the receiving, storage, and preparation areas requires careful planning to be sure that the flow of the raw product through preparation and service is kept direct and in as condensed an area as practicable. Some delays in operation can be avoided by the installation of additional equipment, by a change in the sequence of operation, such as the assembly-line technique in pie making, by the training of the workers to use both hands at one time and to practice certain shortcuts in preparation, or by a better understanding of the timing standard for various processes.

The *use of a different product* could become a deciding factor in changing the method of procedure. The present tendency is toward the increased use of prepared foods. Peeled carrots; processed potatoes; peeled and sectioned citrus fruit, frozen fruits, and vegetables; basic mixes for baked products, freeze-dried shrimp; and pan-ready poultry are only a few such items that definitely change the preliminary procedures necessary in many food production jobs.

Consideration of these suggestions and other factors peculiar to the situation provides a basis for outlining an improved method that can be tried and reevaluated for further streamlining. The advantages to be gained from such a revision are indicated by comparing the summaries at the tops of Figures 15.3 and 15.4.

Quality Management Approaches to Productivity

The discussions thus far on improving productivity have concentrated more on the quantitative aspects of work performance than the qualitative aspects. However, any attempt at productivity improvement must not leave out a consideration of such qualitative aspects. It is important for quality standards to be defined and built into the process. The difficulty has been in measuring quality. A number of attempts have been made in the past, but most, such as the quality assurance programs, were limited-time approaches. Currently, three quality management approaches are being implemented in foodservice operations—the quality improvement process (QIP), continuous quality improvement (CQI), and **total quality management** (TQM). Even though these approaches differ in title and purpose, they all have certain features in common. First, each encompasses a management strategy designed to improve the organization's quality of products and services continuously over time. Second, each promotes positive changes in the structure and culture of the organization to allow for employee participation in decision making and task shaping. This is often referred to as *empowerment* of employees. Finally, each provides means for assessment of the level of quality improvement and whether quality standards are being met.

As productivity concerns increased in the 1980s, management recognized the need to not only involve employees, but more importantly, to involve management in the integration of QWL groups, productivity teams, quality circles, and such into the organizational culture as part of a comprehensive improvement effort.

Total Quality Management. An understanding of what is referred to as *Japanese-style management* or *Theory Z* has led to the adoption of total quality management (TQM) methods in more than 3000 corporations and 40 governmental agencies in the United States. Eighty percent of the Fortune 1000 firms now have quality improvement programs in place and 76% of all companies see quality as a major goal.

TQM is based on the systems approach to management, namely, that the organization is viewed as a system of interrelated, interdependent parts. Fundamental to TQM is the fact that the organization is the focus of management, not the individual. The objective in TQM is to identify barriers to quality, satisfy internal and external customers, and create an atmosphere of continuous improvement.

TQM consists of five major subsystems: a customer focus, a strategic approach to operations, a commitment to human resource development, a long-term focus, and total employee involvement. All operations should center attention on satisfying customer needs by striving for continuous improvement in all areas. Personnel need to be trained in the TQM philosophies with strong encouragement to participate in operating decisions. A teamwork mentality is essential to TQM. And, finally, current decisions need to be evaluated based on their long-term rather than short-term consequences. As in the systems model, these five subsystems are interrelated and interdependent. No one of them can be left out and still have the system operate effectively.

At the heart of the TQM approach is the acceptance that variability is a natural and omnipresent condition. In the systems model inputs are "transformed" into outputs, outputs are evaluated, and adjustments made according to the feedback received. This is exactly where *quality assurance (QA)* is used. QA involves checking for adherence to quality standards or specification after the product has been produced. QA measurements include food temperatures, portion sizes, nutrient content, and so on. However, even though QA measures are important, they have been found to foster an inspection mentality that doesn't encourage empowerment or a sense of teamwork to improve quality when used alone.

When one considers the TQM concept of variation, the focus shifts from the outputs to the transformation process. Reducing variation in transformational activities within the organization is seen as the key to improving productivity and quality. Management moves from a policing role to that of coach, mentor, facilitator, and sponsor. This allows the management team to empower employees to work on quality improvement. Quality must first be assured within the system before it can be provided in the products or outputs.

W. Edwards Deming (1982) one of the most prominent pioneers in the quality movement, maintained that 90% of variation is due to systematic factors such as procedures, supplies, and equipment not under the employees' control. It is, therefore, management's responsibility to reduce variation and to involve employees in the continuous improvement of system processes.

TQM requires that management operate on the assumption that employees want to do their jobs well, are motivated, and have self-esteem, dignity, and an eagerness to learn. What has been called "a blinding flash of the obvious" because of its simplicity is that a well-managed organization takes advantage of all of its brain power. TQM requires a paradigm shift in the meaning of work and the system that supports it. One small part of the new paradigm is the requisite change in the way managers make decisions, allocate resources, and appraise employees.

To be an effective TQ manager five key competencies have been identified. They include the ability to:

- Develop relationships of openness and trust.
- Build collaboration and teamwork.
- Manage with statistical tools and quality processes based on collected facts.
- Support results through recognition and rewards.
- Create a learning and continuously improving organization.

The steps required to establish a TQM program may be outlined as follows:

1. Prepare for implementation:
    a. Analyze the processes (such as that shown in Figure 15.1).
    b. Be open-minded and flexible.
    c. Brainstorm.
    d. Be selective.
    e. Be persistent.
    f. Listen to employees' problems regarding their jobs.
    g. Learn from others, particularly from the best.

2. Train and develop employees and management on how to use these tools:
   a. The TQM approach to work including the new roles for managers and for employees and the fundamentals of teamwork.
   b. TQM tools for problem solving and measurement.
   c. TQM programs for improvement of work processes.
3. Develop and implement tools, programs, and work improvement strategies.
   a. Establish goals, timebound steps, and methods to implement an improvement.
4. Review, measure, and evaluate the results, then replan as needed.

Some tools for problem solving are *brainstorming, flow* or *process charting, check sheets, cause-and-effect diagrams, Pareto charts, scatter diagrams,* and *histograms.* Brainstorming is a useful technique for generating ideas about problems and opportunities for improvement. A process chart is shown in Figure 15.3 and again in Figure 15.4 after the process has been improved. Check sheets are used to show exactly what is happening and how often. It is a method of collecting data based on observations and may show a pattern of opportunities for improvement. A sample check sheet is shown in Figure 15.5. A cause-and-effect diagram, often referred to as Ishikawa's fish diagram, is used to focus on the different causes of a problem. This focus consequently allows for the grouping and organizing of efforts to improve a process. A fish diagram is shown in Figure 15.6. Pareto charts illustrate the relative importance of problems. They are essentially bar charts where the strategy is to work on the tallest bar or problem that occurs most frequently. A Pareto chart is shown in Figure 15.7. A scatter diagram is a tool used to determine the strength of a relationship between two variables and to determine the impact on one variable when the other is changed. A scatter diagram is shown in Figure 15.8. A histogram is a graphic means of depicting any frequency data that have been collected (Figures 15.9 and 15.10).

Baseline measurements provide the starting point in a TQM program against which progress and overall performance toward targets or goals may be assessed.

Problems: Customer Complaints	**Week one**					
	Mon	**Tue**	**Wed**	**Thu**	**Fri**	**Total**
food was cold	𝍸 |	||	||||	|	𝍸 | 𝍸	24
service was slow	||	|	|	|		5
prices too high	|||	||	||	𝍸	|	13
restrooms messy	|	𝍸	|			7
TOTAL	**12**	**10**	**8**	**7**	**12**	**49**

Figure 15.5 A check sheet for collating data.

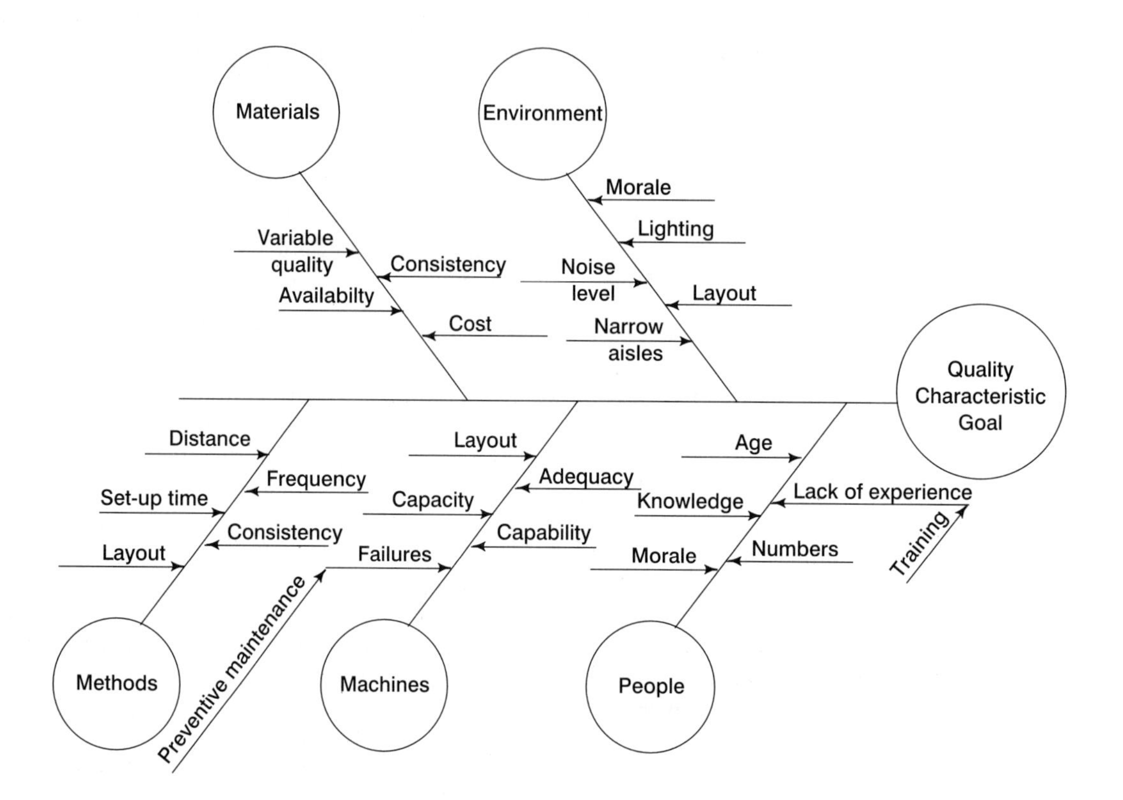

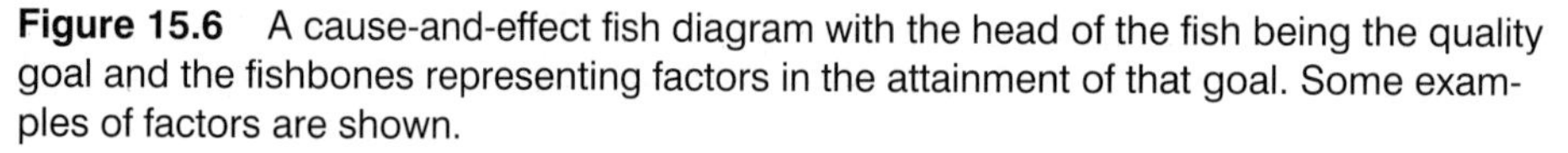

Figure 15.6 A cause-and-effect fish diagram with the head of the fish being the quality goal and the fishbones representing factors in the attainment of that goal. Some examples of factors are shown.

Benchmarking is the TQM measurement tool that provides an opportunity for a company to set attainable goals based on what other companies are achieving. Included in the benchmarking process is a profiling of the company and how it attained its results.

A number of programs have been used by companies to improve work processes. The primary ones include *sociotechnical systems* (STS), *statistical process control* (SPC), *just-in-time (JIT) inventory control,* and the *ISO 9000* program of the International Organization for Standardization. Briefly, STS begins with an analysis of the existing flow diagram, focusing on improvements in technical systems such as transportation, data capturing, and workstations. SPC uses such statistics as mean/average, range, and variation/standard deviation to establish control limits for a process. JIT is an inventory management system that links suppliers and customers to minimize total inventory-related costs. ISO 9000 is a series of five international standards that describes elements of an effective quality system.

Continuous improvement of quality and quantity of products and services is essential for the requisite productivity increases of the future. It is not enough to make a one-time attack. Continuous learning is required for increased productivity.

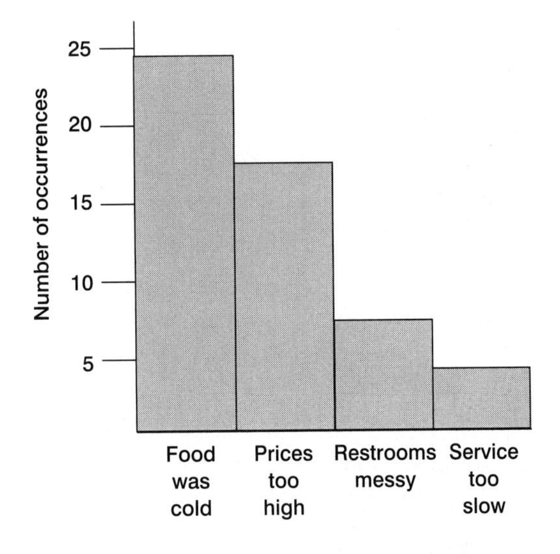

Figure 15.7 A Pareto chart.

The greatest benefit gained from training employees is not in learning new information, but in doing what is already done well even better. Continuous learning can also be promoted by utilizing the insight that people learn when they teach.

In seeking to empower employees, the goal should be to build responsibility for productivity and performance into every job regardless of level, difficulty, or skill.

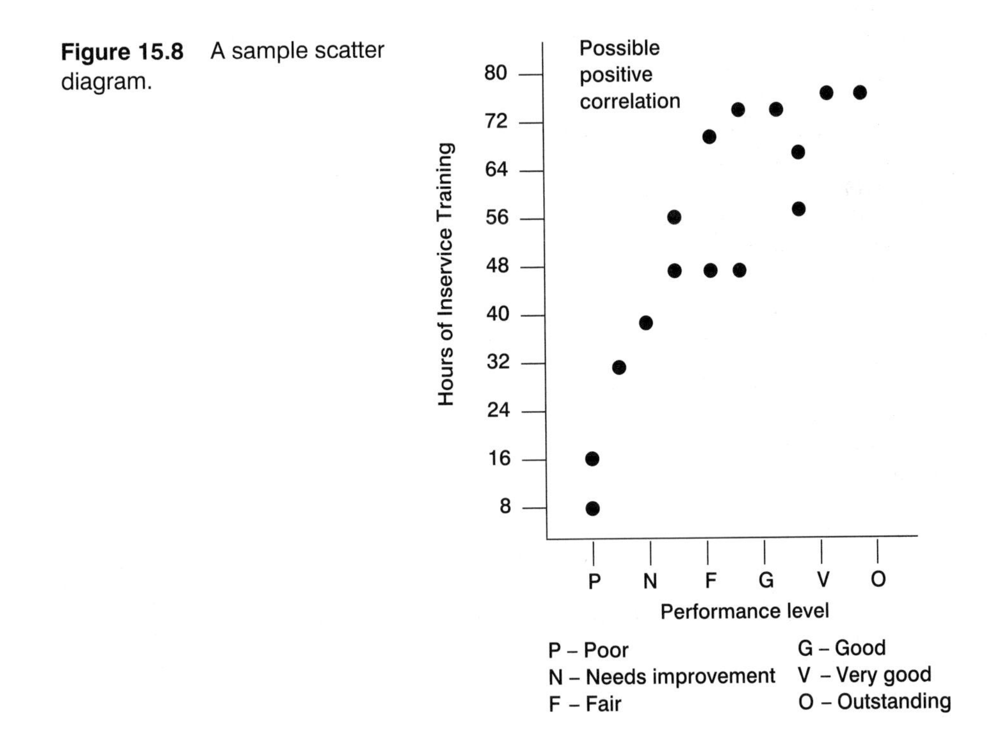

Figure 15.8 A sample scatter diagram.

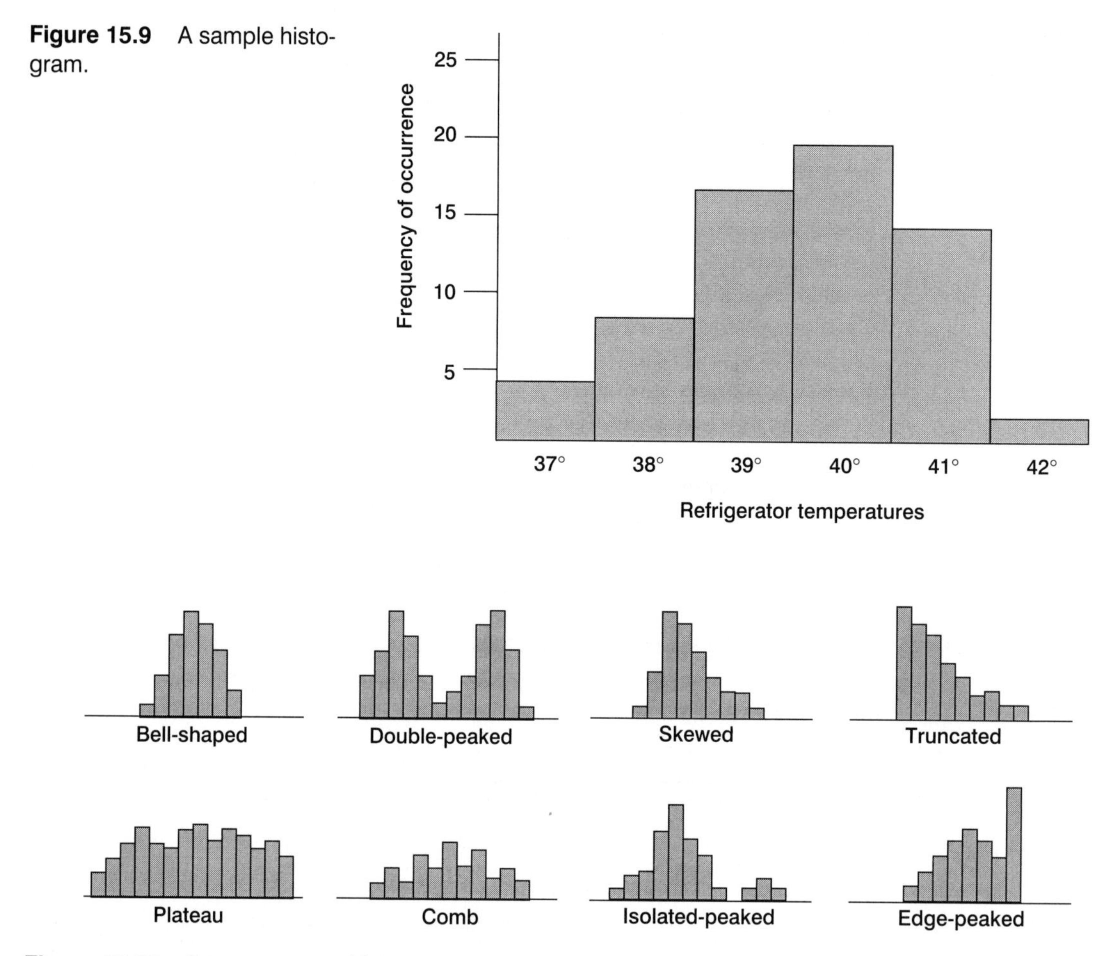

Figure 15.9 A sample histogram.

Figure 15.10 Some common histogram patterns.

Drucker (1992) contends that partnership with the responsible worker is the only way that will work; nothing else will work at all.

Quality standards can be made more quantifiable by establishing **key result areas** (KRAs) such as food and service quality, financial management, human resource management, productivity, planning and marketing, and facilities management. Within each KRA, measurable and quantifiable critical factors are listed.

SUMMARY

The need for raising productivity is urgent; so urgent, in fact, that it has been called the first social responsibility of management in a knowledge society. Research conducted during the past 100 years confirms that productivity can be raised and provides the information on how this can be accomplished. Simply stated, it is a matter of defining the task; concentrating work on it; defining the performance; continuously working on improvement and learning; and making the employee a complete partner in the productivity improvement program.

REVIEW QUESTIONS

1. Define productivity as it relates specifically to foodservice management.
2. Give some examples of useful profitability ratios that would be useful in the management of a foodservice.
3. Describe the quality of work life (QWL) concept. What are some key words that embody the QWL approach?
4. What are the overall objectives of work design?
5. List and briefly discuss some environmental factors that can minimize worker fatigue.
6. Give the formula for determining a worker's efficiency.
7. Describe what is meant by *work simplification.*
8. List and briefly explain the principles of motion economy.
9. List the steps required for the analysis and revision of a job method.
10. Explain the use of work sampling, pathway or flow diagrams, operation and flowcharts, and micromotion studies as tools for a work improvement study.
11. Describe the concept of total quality management (TQM).
12. Why are issues such as work improvement, productivity, and TQM so important in today's work environment?

SELECTED REFERENCES

Aguayo, R.: *Dr. Deming: The American Who Taught the Japanese About Quality.* New York: Lyle Stuart, 1990.

Armstrong, L., and Symonds, W. C.: Beyond 'may I help you?' Bus. Week. 1991; 25:100.

Avery, A. C.: *A Modern Guide to Foodservice Equipment,* rev. ed. Boston: CBI Publishing Company, 1985.

Axline, L. L.: TQM: A look in the mirror. Mgmt. Rev. 1991; 80(7):64.

Baille, A. S.: The Deming approach: Being better than the best. SAM Advanced Mgmt. J. 1986; 51:15.

Bain, D.: *The Productivity Prescription.* New York: McGraw-Hill, 1982.

Barnes, R. M.: *Motion and Time Study—Design and Measurement of Work,* 7th ed. New York: John Wiley & Sons, 1980.

Behnam, N., and Neves, J. S.: The Deming, Baldrige, and European Quality awards. Quality Progress. 1994; 27(4):33.

Bennett, A. C.: *Productivity and the Quality of Work Life in Hospitals.* Los Angeles: American Hospital Publishing, 1983.

Bloch-Flynn, P., and Vlach, K.: Employee awareness paves the way for quality. Human Resource Mag. 1994; 39(7):78.

Bowman, J., and Gift, B.: Continuing to do well. Food Mgmt. 1989; 24:138.

Bowman, J. S.: At last, an alternative to performance appraisal: Total quality management. Public Admin. Rev. 1994; 54(2):129–136.

———: The rising sun in America: Japanese management in the United States—Part one. Pers. Administrator. September 1986; 31: 62—67, 114–119.

———: The rising sun in America: Japanese management in the United States—Part two. Pers. Administrator. October 1986; 31: 81–91.

Brayton, G. N.: Simplified method of measuring productivity identifies opportunities for increasing it. Industrial Eng. 1983; 15:49.

Capezio, P., and Morehouse, D.: *Taking the Mystery Out of TQM: A Practical Guide to Total Quality Management.* Hawthorne, N.J.: Career Press, 1993.

Costin, H. I.: *Readings in Total Quality Management.* Fort Worth, Tex.: The Dryden Press, 1994.

Daily Labor Report: Report on high performance work practices and firm performance issued by Labor Department. Washington, D.C.: Bureau of National Affairs, July 26, 1993.

Deming, W. E.: *Out of the Crisis.* Cambridge, Mass.: The MIT Press, 1986.

———: *Quality, Productivity, and Competitive Position.* Cambridge, Mass.: The MIT Press, 1982.

Drucker, P. F.: *Managing for the Future.* New York: Truman Talley Books/E. P. Dutton, 1992.

Flamholtz, E. E., and Das, T. A. K., eds.: *Human Resource Management and Productivity.* Monograph and Research Series No. 39. Los Angeles: Institute of Industrial Relations, University of California, 1984.

Gift, B.: On the road to TQM. Food Mgmt. 1992; 27:88.

Gitlow, H. S.: A comparison of Japanese total quality control and Deming's theory of management. Am. Statistician. 1994; 48(3):197.

Hudson, B. T.: Industrial cuisine: New approaches to food production in the restaurant industry. Restaurant Mgmt. 1993; 34(6):73.

Imai, M.: *Kaizen.* New York: Random House Business Division, 1986.

Ivancevich, D. M., and Ivancevich, S. H.: TQM in the classroom. Mgmt. Accounting. 1992; 74(4):14.

Judson, A. S.: The awkward truth about productivity. Harvard Bus. Rev. 1982; 60:93.

Kahn, R. L., and Katz, D.: Leadership practices in relation to productivity and morale. In D. Cartwright and A. Zander, eds., *Group Dynamics Research and Theory,* 2nd ed. New York: Harper & Row, 1960.

Katzell, R. A., and Yankelovich, D.: *Work, Productivity, and Job Satisfaction.* New York: The Psychological Corporation, 1975.

Kaud, F. A.: Productivity: Measurements and improvement approaches. In J. C. Rose, ed., *Handbook for Health Care Food Service Management.* Rockville, Md.: Aspen Systems Corporation, 1984.

Kazarian, E. A.: *Foodservice Facilities Planning,* 2nd ed. Westport, Conn.: AVI Publishing Company, 1983.

King, P.: A total quality makeover. Food Mgmt. 1992; 27:96.

Konz, S.: *Work Design: Industrial Ergonomics,* 2nd ed. Columbus, Ohio: Grid Publishing, 1983.

Lawler, E. E., III: *Employee Involvement and Total Quality Management.* San Francisco: Jossey-Bass, 1992.

Lockwood, D. L., and Luthans, F.: Contingent time off: A nonfinancial incentive for improving productivity. Mgmt. Rev. 1984; 73:48.

Matthews, M. E.: Productivity studies reviewed, trends analyzed. Hospitals. 1975; 49:81.

Maun, C.: A commitment to CQI benefits both customers and facility, Dietary Manager. 1994; 3(4):12–13.

Michaelson, G. A.: Deming in his own words. Across the Board. December 1990; 27:44.

Milakovich, M. E.: Total quality management in the public sector. Natl. Productivity Rev. 1991; 10(Spring):195–213.

Miller, L.: Tearing down the barriers between management and labor leads to increased productivity and greater profits. Mgmt. Rev. 1984; 73:8.

Myers, M. S.: *Every Employee a Manager,* New York: McGraw-Hill, 1981.

Parry, S.: The missing "M" in TQM. Training. 1993; 30(9):29.

Pedderson, R.: Motivation and worker productivity. In J. Wilkinson, ed., *Increasing Productivity in Foodservice.* Boston: Cahners Books, 1973.

Rinke, W. J.: Total quality management: Just another management fad? Dietary Manager. 1995;4(1):10-13.

Schecter, M.: The quest for quality. Food Mgmt. 1992; 27:82.

———: Restaurant makes the CQE commitment. Food Mgmt. 1992; 27(4):112.

Schmidt, W., and Finnigan, J. P.: *TQManager: A Practical Guide for Managing in a Total Quality Organization.* San Francisco: Jossey-Bass, 1993.

The roots of quality control in Japan: An interview with W. Edwards Deming. Pacific Basin Quarterly 1985; Spring/Summer:1.

VanEgmond-Pannell, D.: *School Foodservice,* 4th ed. New York: Van Nostrand Reinhold, 1990.

Willig, J. T.: *Environmental TQM,* 2nd ed. New York: McGraw-Hill, 1994.

16

Financial Management

OUTLINE

Financial planning and accountability for the foodservice organization are major responsibilities for the manager. Cost effectiveness is essential if operations are to be successful, especially with today's economy and competitive market.

Every person responsible for the financial management must know, day by day, what transactions have taken place and how they compare with established goals. Otherwise, downward trends may not be detected in time to take corrective action before financial disaster occurs. The primary purpose of this chapter is to provide basic information necessary for managers to be able to (1) plan a budget, (2) set up a system of records that provides financial operating data, (3) use records to prepare financial reports, (4) review and evaluate data from the records and reports by comparing them with budgeted figures, and (5) know what kinds of corrective action can and should be taken to keep financial operations in line with anticipated and desired goals. Basic information required for this kind of cost control and good financial management is the same regardless of whether a computerized or a manual system is used. Decision making regarding objectives and goals, cost justification, control measures, and the kind of corrective action to take when needed remains a personal activity and responsibility of the foodservice manager.

KEY CONCEPTS

1. Financial management is an increasingly important management function and requires a knowledge of managerial accounting techniques.
2. A budget is a financial plan used to forecast the results of future actions such as sales and food expenditures.
3. Although time consuming to prepare, several specific advantages are gained from planning and using a budget.
4. Several different types of budgets are available, including operating, flexible, fixed, and zero-based budgets.
5. The budget planning process has several distinct phases: evaluation, preparation, justification, implementation, and control.
6. The success of a financial plan is monitored through a system of records and reports.
7. Cost control is the cornerstone of financial success and is influenced by a number of factors internal and external to the foodservice operation.
8. Break-even analysis and calculation of the break-even point are important aspects of recipe costing and menu item pricing.

FINANCIAL PLANNING

Good financial planning and management are essential for the success of every foodservice operation. Management must have knowledge of the techniques used to control costs and to provide data for sound financial decision making. The following four actions are basic to achieving financial credibility on the part of the manager and, when followed, should provide a guide to achieving the desired success:

1. Setting financial goals and objectives to be attained, usually through a planned budget;
2. Knowing what is being accomplished through a system of records that provides pertinent data on current operations;
3. Using the data from the records to evaluate progress that has been made toward reaching the set goals, best done through daily, monthly, and yearly reports that compare actual achievement with the preestablished objectives; and
4. Taking corrective measures as necessary to bring operations in line with financial objectives.

The financial goals of various foodservices differ. Some operations are in business to make as large a profit as possible; others are nonprofit but seek to provide the best possible food and service that their available resources allow. In all situations, however, having some type of financial plan in place is one of the keys toward achieving desired departmental goals. Without such a guide, problems may arise before management is aware of them and could lead to financial downfall.

The financial plan most commonly used is a budget. All foodservice managers should prepare and use a budget as a guide for maintaining a sound, healthy organization. Such planning is, of course, done within the policies of the organization.

Budgets

A **budget** is a forecast of future needs. It covers the planned activities of the department for a given period of time, usually the fiscal or calendar year. Budgets are usually thought of as a *financial* plan, but foodservice budgets also include data on volume indicators, such as the number of meals to be served and labor time requirements.

Budgets are based on factual data from past records of income, expenditures, census and labor hours used, and a consideration for any anticipated changes that can affect future operations. This method alone, however, can focus too much on desired additions to the budget unless a cost evaluation is made to justify both past and newly planned expenditures.

For any foodservice that is just beginning operation, a detailed budget based on historical data will not be possible. Since the service has not operated previously, no records of past operations are available on which to base a plan. In these situations management cannot use past trends as the basis for the budget, but must rely instead on a combination of known facts. For example, the manager of a restaurant may rely on industry-wide data for similar operations. There are also standard financial formulas for the food and hospitality industry that may be used to determine anticipated revenue and costs (see "For More Information" at the end of the chapter).

With the current emphasis on cost containment and better management of all resources, it is important that foodservice managers understand the budget planning process and the techniques used to prepare a realistic budget. Too many managers operate without a budget as a guide because it is time consuming to prepare, predictions are difficult, and incomplete records fail to provide the necessary data. The value of a budget and the budgeting process must be clearly understood and accepted by management; otherwise, if one is planned, it may be an empty gesture, not used for its intended purpose.

Value of a Budget. To some, the word *budget* has the connotation of curtailed spending and inflexibility and, therefore, is undesirable. In reality, a budget is a valuable management tool that should serve as a guide for allocation of resources and for comparison with actual operations—the basis for financial control.

The advantages of budget planning and development are many, far outweighing any potential disadvantages. Consider these advantages, for example:

1. Budget planning forces those in management positions to seriously consider the future directions and development of their department, and to reaffirm old or establish new financial goals. All those with decision-making authority in the department should be involved in budget planning and contribute to final decisions.
2. The budget planning committee's review of previous expenditures provides an evaluation process and a base for justifying future requests for funds.

3. A budget is an important control device because it documents goals and objectives in a quantified manner. It gives a standard for comparison against actual transactions. Deviations from the anticipated (budgeted) income and expenditures are evident and can be corrected or justified as they occur.
4. Since persons involved in planning establish the priorities of need, they are more likely to be committed to staying within the limits they were responsible for setting.
5. A budget establishes goals for profits and revenues. Employees made aware of those goals are apt to feel responsibility toward meeting them.
6. A budget provides for continuity in the event of management turnover.
7. A budget documents plans for anticipated changes due to inflation, increases in the cost of living, and other economic factors.
8. The budget serves as a communication tool for management.

Disadvantages of the budget process are minor compared with the positive results of having a budget for a guide. Such disadvantages include the following:

1. A budget that is rigid is often ignored as unworkable. Budgets should be flexible and adjustable according to changing situations.
2. Budget preparation is tedious, requires time, and takes personnel away from other management activities.
3. Unless the entire managerial staff is in support of budgeting and cooperates in the preparation and use of the budget, the process may become merely a gesture and have limited value.
4. Departments within the organization may vie with each other for funds, which could cause undesirable competition and problems.
5. For implementation purposes, budgets must be planned far in advance of actual activities. Unanticipated changes in the economy or the organization itself may alter all budget predictions.

Types of Budgets. Although there are many types of budgets, all have the same intent: to determine how much money (or other resources) is available and at what rate it is to be spent. It is then the manager's responsibility to allocate that amount fairly to cover various expenses plus allow for a margin of profit.

The following are descriptions of some types of budgets used by foodservice managers.

Master Budgets. The overall financial or *master budget* coordinates every aspect of the operation and consists of the operating budget, the cash budget, and the capital budget. These budgets are used to prepare key financial statements including the income statement and the balance sheet.

Operating Budget. The *operating budget* is a plan that includes revenues, expenses, and changes in inventory and other working capital items. It is a forecast of revenue (sales), expenses, and profit for a specified period of time or fiscal year. The operating budget serves as the guide for day-to-day operations of the department and is an important component of the control process as it is used for financial decision making. The operating budget includes the *statistics budget,* which is an estimate of the volume of sales in commercial operations or services in noncommercial operations.

Budget figures are based on historical data from records of past performance. Categories of performance differ between commercial and noncommercial operations. For example, commercial operations use number of sales and covers as the primary indicators of performance, whereas noncommercial operations use a number of other indicators such as resident census in long-term care facilities, meal participation in schools, and meal equivalents per patient day in hospitals. These groups of performance data are sometimes referred to as *volume indicators* and provide the basis for future projections of activity and costs. All operations monitor historical data on expenses including food, supplies, labor, energy, and overhead expenses. These expense budgets are included in the operating budget along with the *revenue budget,* which is a projection of the expected or anticipated income for the financial time period.

Some different types of operating budgets are explained in the following list:

- *Fixed budgets:* A fixed budget is a set dollar amount based on a set, predetermined level of activity or transactions. It is generally based on past activity indicators such as sales and costs with some consideration for future change. Fixed budgets are rigid and are more commonly used for a specific purpose such as a program funded by a grant.
- *Flexible budgets:* A flexible budget, as opposed to a fixed budget, gives a dollar range for low to high levels of predicted activity. This type of operating budget is developed to reflect the variability in performance activities and therefore fluctuations in volume indicators. It is more flexible compared to the fixed budget but more difficult to use as a control tool.
- *Zero-based budgets:* The zero-based budget (ZBB) is newly developed each fiscal year starting with a "blank piece of paper." This approach requires that the manager prepare a budget for each activity of the department, thereby forcing the manager to evaluate all activities each year and justify every request for funds. Zero-based budgets are frequently used for large capital requests such as dishmachines.
- *Cash budget:* A cash budget is developed to project the receipt of revenue and the expenditure of funds. The purpose of this budget is to determine if funds will be available when needed in order to meet the financial obligations or demands of the operation. The cash budget is an illustration of the inflow and outflow of cash, thereby identifying the amount of cash on hand at any given time.

Capital Budgets. A capital budget is a long-term plan prepared to estimate or predict the costs of capital outlays or expenditures and their financing. Examples of capital expenditures include equipment replacement, renovation projects, and facility expansion.

Steps in Budget Planning

As mentioned, budgets are time consuming to prepare and require thought-provoking effort on the part of all who are involved. A timetable should be set up for various phases of data collection and planning. The schedule must allow ample time for careful completion of the project so it can be evaluated and approved well in advance of the beginning of the next fiscal period. Budget planning as a joint endeavor represents many points of view. Final agreement by participants should ensure a fair allocation of funds for the goals to be met.

The budget planning process includes several distinct phases; each one deriving information from the one before:

- The *evaluation phase* dissects an operation's past performance and identifies those factors that are likely to influence future activities.
- The *preparation* or *planning phase* uses information from the evaluation phase to forecast and prepare the first draft of the budget.
- The *justification phase* is a time of review, revision, and final approval. This process involves persons with the authority to grant funding.
- The *implementation* or *execution phase* translates the budget expenditures into the operation functions
- The *control phase* is an ongoing monitoring process to ensure that operations stay in line with budgeted predictions

More specifically, the following steps are usually used in planning the foodservice budget:

1. Collect operating data from records and reports. (See later section in this chapter, "Financial Operations: A System of Records and Reports.")

2. Study these data and evaluate them against departmental goals. Information reviewed should cover actual operating and budget variance figures from the previous 3 or 4 years with justification or explanation of variances; income and expense trends; sales reports and statistics; menus, prices, customer selection, portion sizes, and food cost per portion; such labor statistics as the number of employees, their duties, schedules, and wage rates.

3. Analyze and discuss any factors that can affect future operations. These include both external (outside your control) and internal (within your control) influences that can be identified as having a possible effect on foodservice costs or activities in the future. Examples of *external* factors: the local economy, actions by government (change in taxes or laws), changes in utility costs, new construction that might bring in new business or divert traffic and so reduce patronage, or increases in competition. Examples of *internal* factors: a planned addition to the facility that will change the number of persons to be served; a change in type of foodservice system such as going from conventional to assembly/serve; converting to computerized record keeping; change in hours of service to better accommodate patrons.

4. Discuss and plan for new goals or activities desired, such as a remodeling project, purchasing new equipment, or a new service (e.g., catering) to be offered. These are all considered in light of anticipated changes and current prices.

5. Set priorities and make decisions as to what can be included in the budget for next year. In establishing budgetary priorities, the manager must weigh the relative desires and needs of each unit, search out nonessentials, decide if any items could be provided at a lower cost, plan for upkeep and expansion of the plant, and provide for the development of personnel. Even though the income remains approximately

the same year after year, careful appraisal of past expenses should be made whenever a new budget is prepared in order to prevent overspending for some items to the neglect of others.

The budget must reflect the objectives and policies of the organization for which it is planned. For example, the school foodservice has as its objective the serving of good food at a minimum cost. The foodservice may pay for food, labor, and laundry only, while the school subsidizes the overhead and operating costs. In such a case, the percentage of income spent for food is high. The foodservice budget of a college residence hall that is helping to pay for retirement of bonds is quite different from one whose physical plant and equipment are not encumbered. In each situation, the proportion of the income spent for various expense items varies with the size and type, objectives, and policies of the individual foodservice.

The budget planning committee has responsibility for making decisions as to what expense items can be included in the budget with a fixed income, or how the income can be expanded to cover the cost of the expense items deemed essential. Forward-looking managers establish priorities and plan an orderly way of achieving these goals, which means looking beyond next year's budget.

6. Write the budget for presentation. Although there is no established format for the formal write-up of the budget, it contains an organized listing of expected income, classified as to sources, and the classified list of all expense items. Usually a form similar to the one shown in Figure 16.1 is used as a work sheet to organize data in the budget. Writing the budget follows these steps:

a. List all sources of expected income. Figure 16.2 illustrates those for a college residence hall where income comes from student board fees. Commercial operations derive income from cash sales. Record this year's dollar figures for each of the sources of income and then the anticipated changes for the next budget period. Calculate and record the total expected income.
b. Classify and list the items of expense with the cost calculated for each. Basically, these are food, labor, overhead (e.g., fixed costs such as amortization or rent, taxes, and insurance), and operating costs (e.g., utilities, telephone, supplies). Figure 16.1 shows one such *chart of accounts,* as the listing is called. The National Restaurant Association (NRA) has provided comparable information for commercial operations in its Uniform System of Accounts for Restaurants.
c. Add other pertinent data, such as number of meals served, labor hours for both total meals and per meal served, last period, expected changes, and new totals.
d. Prepare a justification for requests of new funds.
e. Review and make any changes necessary.
f. Write the budget in final form with any cost-benefit statements attached. If the budget is to be forwarded to a higher administrator for approval, an explanation of certain items can prove helpful for better understanding of requests.
g. Once it has been approved, use the budget. The yearly budget figures are divided into appropriate time periods for comparative purposes. For example, a calendar month may be used for those with year-round operations. Schools may wish to use a 20-school-day period, because holidays and vacations result in more or fewer operating days in some months, making comparisons unrealistic.

Budget Work Sheet—Income and Expense	*Estimated Last Year*	*Actual Last Year*	*Anticipated Changes*	*Estimated Next Year*
	Total %	*Total %*	+ −	*Total %*
Income				
Regular Sales				
Special Meals				
Other (Itemize by Foodservice)				
TOTAL				
Expenses				
1. Food				
2. Salaries and wages:				
Regular employees				
Student or part-time employees				
Social security tax				
Other taxes				
Fringe benefits				
TOTAL				
3. Services:				
Laundry				
Utilities				
Telephone				
Exterminator				
Garbage and trash disposal				
TOTAL				
4. Supplies, repairs, and maintenance:				
Cleaning supplies				
Paper supplies				
Office supplies				
Equipment repairs				
Miscellaneous supplies				
Physical plant				
TOTAL				
5. Housing:				
Amortization or rent				
Taxes				
Interest				
Depreciation				
Insurance				
Repairs				
TOTAL				
GRAND TOTALS				
Excess of income over expenses				

Figure 16.1 Budget planning work sheet, adaptable for use in any foodservice.

Sources of Income	*Income Past Period*	*Anticipated Increase*	*Anticipated Decrease*	*Anticipated Income*
Board fees				
Cafeteria receipts				
Guest meals				
Special meals				
Catering				
Special food orders				
Miscellaneous				
Total				

Figure 16.2 Income sources and detailed information for use in budget planning.

A form similar to the one illustrated in Figure 16.3 should be used for recording the budgeted figures. At the close of the month or other specified time period, the actual operating figures are recorded from the income, expense, and census records for easy comparison. A quick glance at this comparative report will show any discrepancies or variances between anticipated and actual business. Any deviation, positive or negative, should be analyzed to determine the reasons for the deviation and to plan correct action. The data on the comparative form should be used in connection with information from the daily food cost report and profit and loss statements to evaluate activities and decide what actions to take. This is discussed in more detail later in the chapter.

FINANCIAL OPERATIONS: A SYSTEM OF RECORDS AND REPORTS

Knowledge of the day-to-day financial transactions and an awareness of "where the money is going before it is gone" are ongoing responsibilities of the foodservice manager. The use of records is essential for providing readily available operating data.

Records, like all forms of control, vary with the type, size, and policies of the institution, so that management must ascertain what information is needed and how it may be obtained with the least expenditure of effort, time, and money.

Most foodservice organizations today utilize the computer for much of their record keeping and reporting. Data required for computer input is essentially the same as that for noncomputerized record keeping; forms and procedures vary. Good manual control and decisions on what information should be provided by computer are prerequisites to a good computerized control system. Designing appropriate forms for data organization is the first step in setting up either a manual or computerized system.

Records for Control

Records deemed essential for a foodservice operation include those that control the major phases of the operations. These essential records may be classified as those

Name of Food Service

Month ________________, 19 ____

	Number of Meals	Food Cost		Payroll Expense		Other Costs and Expenses	
	Total for month	*Total for month*	*Per meal served*	*Total for month*	*Per meal served*	*Total for month*	*Per meal served*
Budgeted							
Actual							
Over + – Under							

Cumulative for Year to Date

	Jan.	Feb.	Mar.	April	May	June	etc.
Budgeted							
Actual							
Over + – Under							

Figure 16.3 Budgeted figures are compared with actual operating figures for management control of finances.

for purchasing and receiving, storage and storeroom control, production and service of food, cash transactions, operating and maintenance, and personnel. More specifically, these include the following:

Purchasing and Receiving Records

Purchase orders
Invoices
Receiving records
Purchase records
Summary of purchases records

Storage and Storeroom Records

Requisition or storeroom issue records
Perpetual inventory
Physical inventory

Food Production
Menu
Standardized recipes
Portion control standards
Production schedule and leftovers report
Menu tally

Service Records
Census reports: number of meals served
Regular and special meals' count
Catering and vending

Cash Transactions Records
Cash sales register reports
Guest checks
Cash disbursement records for all expense items

Operating and Maintenance Records
Maintenance records
Repair invoices

Personnel Records (related to cost control)
Time card/payroll records
Work schedules
Fringe benefits record

Illustrations of record forms are given in the chapters appropriate for each topic. Therefore, only census and income/expense records are presented in this chapter. Since these are given as examples only, each foodservice director should design forms for his or her own department according to the specific situation.

Service Records. A record of the number of people served is vital for forecasting, purchasing, and production needs. The data from a census report are necessary for determining per-meal costs, average sales per person, and the distribution of meals served to various categories of consumers for each meal of the day.

A form designed for recording the census should be large enough to include figures for the entire month (or other accounting period) and for each category of consumers. Figures 16.4 and 16.5 give examples of typical census report forms for two types of foodservices. If more detailed information is desired, patient meals could be divided into "regular" and "therapeutic" or "modified" diets, and personnel meals into "staff" and "employees."

Information to be recorded on the census form comes from cash register reports for personnel and guest meals; dietary count of patient meals and of meals served at any special functions.

Catering Records. A record of catering functions is necessary for foodservices that provide this service in addition to their own regular meal service. Each event has a different menu and demands specialized service, for which customers pay accordingly. In all

Monthly Census Report

Name of Organization ______________________ Date ______________ 19 ______
Month Year

	Regular Guest Count					Employee Meals					Special (Catering) Functions					Grand Totals	
				Totals					Totals					Totals			
Day/ Date	Break-fast	Lunch	Dinner	Today	To Date	Break-fast	Lunch	Dinner	Today	To Date	Break-fast	Lunch	Dinner	Today	To Date	Today	To Date
Su 1																	
Mo 2																	
Tu 3																	
—																	
Tu 31																	

Figure 16.4 Census record-keeping form, which is adaptable to commercial foodservice. The "To Date" figures are cumulative for the month.

MONTHLY MEAL CENSUS

Hospital Date ________________ 19 ____
mo.

	Patient Meals					Personnel Meals					Patient and Personnel	
				Total Meals					Total Meals		Total Meals	
Date	Break-fast	Lunch	Dinner	Today	To Date	Break-fast	Lunch	Dinner	Today	To Date	Today	To Date
1												
2												
3												
—												
30												
31												

Figure 16.5 Meal census summary sheet suitable for health care facilities.

cases, it is necessary to have definite policies and procedures outlined that can be followed through the planning and billing stages of the charges in order to ensure successful and satisfactory operation. Figure 16.6 is a typical agreement form for special group meetings and meals with a record of price and total charges and number of people served.

Income and Expense Records. A record of daily transactions is essential for preparation of monthly financial statements. Managers must know the sources and amounts of income, and where that income goes. Several records are needed to provide that information in a simplified way.

Sales and Cash Receipts. Even small operations such as school lunchrooms handle some cash, and business-like procedures are needed for accounting of the money. Cash registers provide a relatively safe place for money during serving hours and also provide accurate data on number of sales made and total cash received. Larger organizations may use cash registers of varying degrees of sophistication including electronic point-of-sale (POS) computer terminals. These produce summary printouts and proofs of cash collected, which can replace the use of the cash register record shown in Figure 16.7. The POS terminal can be programmed to provide as much information and detail as management desires, which can include the following:

- Number of sales;
- Total sales dollars by cash, check, credit card;
- Tax collected;
- A total customer count of those who paid by cash and a count by number of those customers who received meals other than by the cash system;

Name of Foodservice

Organization ______	Function ______
Date ______ Time ______	Arranged by ______
Room ______	Address ______
Number Guaranteed ______ Served ______	Phone No. ______
Price ______	Booked by ______ Date ______
Total Charge ______	Approved by ______

Menu	Details
	Setup Speaker's Table Flowers Music Public Address Tickets Misc.
Guarantees are not subject to change less than 24 hours in advance of party. We are prepared to serve 10% in excess of the number guaranteed. Accepted ______ Union Office ______	Copies: Manager Food Director Catering Maintenance Kitchen Accounting

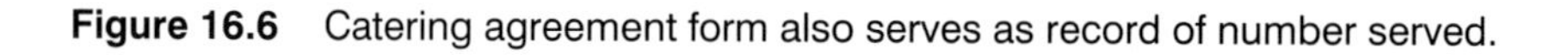

Figure 16.6 Catering agreement form also serves as record of number served.

- The total number of servings for each type of food, such as entrées, vegetables, desserts, salads, and beverages (menu tally); and
- The dollar volume for each type of food sold.

The computer can also perform the following tasks:

- Print an itemized receipt for each customer.
- Calculate automatically the change to be returned to the customer and print the transaction on the receipt.
- Report totals and productivity by hour or shift.

A record of income from sales other than cash sales and of payments made for all expense items is also essential. A *cash receipts and disbursements* book is used by bookkeepers to record these transactions. Also, they can be kept by computer. Sample forms of the two parts of the cash record are shown in Figure 16.8. These should be filled in daily, posting the disbursement amounts from bills received and paid by check; the cash received from the cash register reports; and reports of any other cash payment received. Much of this bookkeeping and record keeping is computerized in today's foodservices. However, the data and information presented here are basic for either manual or computer record keeping.

Form 7158

Report No. ....

FOOD SERVICE

Dep't. No. ....

REPORT OF CASH RECEIVED

Page .... of ....

.... (Unit)

Period from .... to .... Incl.

Present Reading	Less: Previous Reading	Difference	Void	Register Sales	Tax on Register	Over Short	Description	Validated Receipt	Code
							REGISTER 1		
							TOTAL 1		
							REGISTER 1		
							TOTAL 2		
							REGISTER		
							TOTAL 1		
							REGISTER		
							TOTAL 2		
							REGISTER		
							TOTAL 1		
							REGISTER		
							TOTAL 2		
							TOTALS		

Receipt No.	Received from (Name and Description)	Date of Charge or Period Covered	Received on Account (Tax Incl.)	Other Receipts
TOTAL CASH RECEIVED				

Signed .... Supt. or Manager

Total Deposit ....

Figure 16.7 Cash report form to be filled in by cashiers using cash registers.

Cash Receipts Record (a)

Name of Organization ______________________ Month __________ 19 ____

Date	Total Amount Received	Source of Income					
		Food Sales	Beverage Sales	Accounts Paid	Misc. Sales	Other	
						Source	Amount
1							
2							
3							
4							
—							
31							
Totals							

Cash Disbursements Record (b)

Name of Organization ______________________ Month__________ 19 ____

Date	Name of Account	Check No.	Amount Paid	Classification of Expense Accounts					
				Food	Beverages	Supplies	Utilities	Payroll	Rent
1									
2									
3									
4									
—									
31									
Totals									

Figure 16.8 Sample cash receipts and cash disbursements forms for financial control. Income sources and expense accounts vary with type of operation. Totals on each form are posted to the appropriate classification column.

No records, however carefully designed, are of value unless they are kept *daily,* are *accurate,* and are *used by management.*

FINANCIAL ACCOUNTABILITY

Accountability for expenditures made with the company's or institution's money is not ensured by merely keeping records. The data and information provided by the records just described must be used to analyze and improve the financial situation. To do that, certain reports are prepared using data from the records. Essential ones are a daily food cost report; a monthly report, or profit and loss statement; a yearly summary report; and comparative reports of operations.

Reports

Daily Food-Cost Report. A simple daily food-cost report as illustrated in Figure 16.9 is prepared from four records: the cash receipts record of *total income from sales* for the day, the *census record* of number of people served, the *storeroom issues* record, and the *invoices for perishable foods,* issued directly to the kitchen and not kept on inventory (known as direct purchases). A total of storeroom issues and direct purchases gives the cost of food for the day.

This is the most valuable of the three reports as a management tool because it provides up-to-the-minute information about sales, food costs, and number of people served. Expenses other than food usually are not included in the daily report, because they do not fluctuate as greatly as food costs.

The relationship between food cost and income expressed in terms of percent is the most significant single figure to be observed. This percentage figure is found by dividing the cost of food used today by the income for today. For example, if the income from sales (or budgeted allocation) is \$1280 and the food cost is \$530, then

$$\frac{\text{Food Cost: \$530}}{\text{Income: \$1,280}} = .414 \times 100 = 41.4\%$$

41.4% is the food-cost percentage, or
percent of income spent for food

By looking at this figure, the manager can quickly tell whether or not it meets the standard set for his or her operation. Each foodservice sets its own food-cost percentage standard based on the type of organization, its goals, characteristics, and expenses. A luxury restaurant with elaborate service and high overhead may set 20% to 25% of the sales for food as its standard. Other expenses and profit take up the other 75% to 80% of the income.

In contrast, in a school foodservice in which some of the foodservice overhead is paid by the school and in which labor costs are not excessive, 50% to 60% may be the standard for food. Although the percentage is much higher, the actual amount spent for food is considerably lower, since the income per meal is so much greater in the restaurant than in the school.

In addition to the daily food-cost calculation, *cumulative* figures for the month are important (see Figure 16.3). The sales and food-cost figures for the first day of the month are added to those of the second; the total of the first 2 days is added to the third day's figures, and so on throughout the month.

The cumulative (rather than daily) figures should be used to determine the food-cost percentage because they tend to average out the "ups and downs" of a single day's operation and give the picture of operations to date. If the food-cost percent figure varies from day to day within reasonable limits, there is little cause for con-

University Commons
Daily Food-cost Report

Year ________

Day and Date	Tuesday May 1		Wednesday May 2		Thursday May 3		Friday May 4		Saturday May 5		Sunday May 6	
Income and Census:	Census	Sales	Census	Sales	Census	Sales	Census	Sales	Census	Sales	Census	Sales
A. From cafeteria sales (Cash register report)												
B. From parties (Charges)												
C. Total today												
D. Total to date												
E. Cumulative total for month												
Food Cost:												
F. Food cost today												
G. Total food cost to date												
H. Cumulative total for month												
Food-cost percentage												

Figure 16.9 Daily food-cost report form; used continuously for each accounting period, usually a calendar month. Food-cost percentages are calculated on the cumulative totals; the one for the last day is the average for the month.

cern. It is the cumulative or average figure, calculated each day, that should be in line with the desired food-cost percentage. Variations in the figure indicate the need for investigation to determine the reasons and take corrective action.

The *daily* cumulative food-cost report is valuable because it shows deviations from the budget at the time they occur and corrective action can be taken *at once*. An end-of-the-month report (profit and loss statement) shows deviations that occurred sometime during the month, and any action taken does not bring expenditures in line for that month.

Another daily report that can help managers pinpoint and evaluate expenditures is the record of charges for *raw* food sent to various production and/or service units. For example, large foodservice operations distribute food to a pantry, salad unit, and cafeteria among others. On a form, such as that shown in Figure 16.10, the direct and storeroom issue costs are recorded. This information shows trends in unit spending and can also be used for comparison with income generated from production in each unit. These figures also provide a helpful productivity analysis.

A record of cost of *prepared* food sent from a main production center to various service units, such as from a hotel to its coffee shop, main dining room, catering and room service, or in a large medical center to its patient areas, visitor/staff cafeteria, and perhaps separate children's, psychiatric, rehabilitation, and other hospital units, is another necessary cost control record.

The charge for food prepared and sent to these units can be based on the number of portions issued, or on total weight or volume of the food. In either case, precosting with current prices of the various menu items is the preliminary step. A form for recording these daily issues and charges may be developed by the dietitian or foodservice manager so that the appropriate transfer of charges can be made.

The daily food-cost report, although more or less approximate, is usually sufficiently accurate to pinpoint trouble spots before serious financial reverses can occur. If a detailed breakdown of costs each day is important to the manager, a computer

Summary of Daily Food Cost Expense by Production and Service Units

Month ________________, 19 ____

Day and Date		Main Production Unit	Vegetable Preparation	Salad Unit	Bakery	Serving Counter	Totals	
							Today	To Date
Mon. 1	Direct purchases							
	Storeroom issues							
Tues. 2	Direct purchases							
	Storeroom issues							

Figure 16.10 Charges for raw food issued to production and service units can be accounted for by use of this form.

program can be planned to make such information quickly available. A computer printout of essential data from the previous day's transactions can be on the manager's desk each morning for study and evaluation. As with all reports, the daily food-cost report, of whatever form, provides a working tool for foodservice managers, who are expected to know how to use it.

Profit and Loss Statement. The profit and loss report is a summary comparison in dollars and cents of the income, including *all* expenses of the department, to determine the amount of profit or loss for a given period. Usually, it is prepared at the end of each calendar month. In schools, however, the accounting period may be the number of operating days per year divided into equal periods of about 20 or 25 days each. This gives a basis for better comparison of operations from period to period. The profit and loss statement shows the true cost of food used based on purchases, adjusted with inventory numbers, and all other actual expenditures.

The figures for preparing this statement are taken from the cash ledger, income and disbursements, and from the beginning and ending physical inventory figures. A simple summary of the profit and loss statement is shown here:

	Income (sales)
Less:	Cost of food sold
Equals:	Gross profit
Less:	Labor, overhead, and operating costs
Equals:	Net profit or loss

The "cost of food sold" in the profit and loss statement is determined by adjusting the cost of food purchased during the month with the beginning and ending inventories for that month:

Purchases	(Figure obtained from vendors' end-of-month statements verified by manager's check on daily invoices.)
+ Beginning inventory	(Value of goods on hand in the storeroom available for use. Figure obtained from costed inventory taken on *last* day of preceding month, which becomes the beginning inventory on the *first* day of the next month.)
= Cost of goods available to be used	
– Ending inventory	(Value of goods on hand on the last day of the month; goods not used during the month but available for use in the next month.)
= Cost of goods used	

An example of a typical profit and loss statement is given in Figure 16.11. Percentage ratios of the major items of expense and of the profit to the sales are included for better interpretation of operations, because dollar and cents figures in themselves have little meaning here. The percentage figures are determined by dividing each expense item *by* total income; that is, the percentage of income that is spent for food, labor, other expenses, and a profit or, if expenses exceed income, a loss.

As with the daily food-cost report, a cumulative statement of profit and loss, recorded month by month, together with the budgeted figures gives comparative data for the manager's use.

College Commons Profit and Loss Statement
April, 19____
(Operating days, 22)

				Percent of Sales
Sales	$26,476.72			
Less: sales tax	1,040.00			
Net sales		$25,436.72		100.00
Cost of food sold				
Inventory—April 1	1,976.05			
Freight	35.16			
Purchases	10,632.07			
	12,643.28			
Less: inventory April 30	2,258.99			
Net cost of food sold		10,384.29		40.82
Gross profit on food			$15,052.43	59.18
Labor:				
Regular employees	4,651.48			
Student employees	3,081.51			
Supervision	2,479.53			
		10,212.52		40.15
Operating expenses:				
Social security tax	386.96			
Other taxes	291.16			
Maintenance and repair	590.06			
Utilities	757.56			
Supplies (cleaning, paper, office)	766.73			
Laundry	527.39			
Depreciation on equipment	600.00			
		3,919.86		15.41
Total labor and operating expense			14,132.38	
Excess of income over expenses			$920.05	3.62

Figure 16.11 Monthly profit and loss statement for a college foodservice unit.

If this report is to be effective, it must be completed and available as early in the month as possible, and certainly no later than the tenth of the month. Reports coming to the food manager's desk a month or 6 weeks after the end of the operating period have little or no control value at that late date. The amount of profit or loss should be no surprise, however, to the manager who has used the daily reports to "keep a finger on the pulse of operations."

Large organizations where foodservice is but one of several departments, such as hospitals, colleges and universities, and hotels or motels, have their own bookkeeping departments. However, in most situations the foodservice manager generates his or her operating data for closer control and speedier, more complete reports than may come from a central business office. They do provide good checks on each other, however, and managers of both departments are well informed about needs and requests presented in the budgeting process.

Annual Reports. An annual report is usually prepared for higher administrative officials and provides a resumé of the activities and accomplishments of the year just completed, as well as plans for future developments. Although an annual report contains statistical data, it should be much more than a mere listing of facts and figures. It is the interpretation of these data that makes the report significant to those who read it.

The preparation of such a report usually requires staff participation and planning together as a group. It calls for creative thinking and provides opportunity to dream while forecasting ways in which the department can be improved, not just next year but 2, 5, or 10 years hence. If there is a specific message that the managers wish to convey, such as the need for enlarging or remodeling the facility, the report should emphasize that theme for maximum impact.

The cumulative profit and loss statements, census reports, and other records of income and expense are the bases for the financial part of an annual report. Comparisons should be made of the actual and budgeted figures for the year and explanations given for increases or decreases. Often, the presentation of these data in graphic form, such as a pie chart illustration of the expenditure of the income dollar, is more quickly understood than the mere statement of words. The use of bar graphs to show comparisons and the plotting of income or census figures month by month on a graph are other examples of effective reporting.

Information on the personnel situation should be included as well, since the cost of labor is high. Administrators are interested in percentage of turnover, promotions, length of service, and outstanding achievements of the staff and employees. Reasons for terminations of employment are important, too; they may help to point out some weakness in the organization. Interpretations of labor hours worked in terms of ratios to meals served, income received, or other comparisons of special significance are also included.

The physical plant and the equipment changes or repairs that have been made, as well as anticipated needs in all areas of operation, should be reported. Any situation that calls for a major expenditure of money requires explanation and justification.

The annual report also includes a summary of accomplishments or goals reached during the period. Special problems encountered and solved, unusual occurrences or services rendered, and the projected plans for the future complete an annual report.

Factors Affecting Cost Control: Evaluation of Operations

A critical step in achieving a good financial position involves the manager making follow-up decisions and taking actions after review of the records and reports. If operations are in line with the budget plan, no action is indicated. If, however, costs are excessively high and profits not as predicted, or volume indicators are lower than anticipated, a review of the many factors involved in cost control is necessary. There are many alternatives to consider before raising prices or reducing portion size. These actions may be required eventually, but not as the first approach.

Managers with professional preparation and experience search for causes of deviation from the expected by reviewing the many factors that have a bearing on overall expense. These include every activity in the department, from the physical arrangement and layout of equipment to customer satisfaction. A brief review of some of these factors is given here. Table 16.1 summarizes some of the key questions that the manager should consider when analyzing budget deviations.

Food Costs. Food is the most readily controlled item of expenditure and the one subject to greatest fluctuation in the foodservice budget. If control of food costs is to be effective, efficient methods must be employed in planning the menu, purchasing, storing, preparing, and serving. The expenditures for food vary greatly from one type of institution to another and often for institutions of the same type because of the form of food purchased, the amount of on-premise preparation, geographic location, and delivery costs.

In spite of the variation in the amounts spent for food, the underlying bases of food-cost control are the same for all types of foodservices. The effectiveness of control is determined by the menu; menu costing and establishment of the selling price; the purchasing, receiving, storage, and storeroom control procedures; methods used in the production of food, including pre-preparation, cooking, and use of leftovers; portion size and serving waste; and the cost of employees' meals.

Menus. Menu planning is the first and, perhaps, the most important step in the control of food costs. The menu determines what and how many foods must be purchased and prepared. Knowledge of these food costs, as well as labor cost and intensity, and the *precosting* of the menu to determine whether or not it is within budgetary limitations are essential for control procedures. Menu planning is discussed in detail in Chapter 4.

Menus that provide extensive choices require preparation of many kinds of foods, several of which may not be sold in quantities sufficient to pay for their preparation. If a widely diversified selection is offered, the investment of too large a sum in food or labor for its preparation can result. Also, it may result in carrying an extensive inventory of small quantities of food items, or of foods infrequently used.

Foodservice managers should remember also that although menus are made some days or even weeks in advance, they can be adjusted daily to the inventory of food on hand and to local market conditions. Waste can be prevented only by wise utilization of available supplies, which helps to keep food costs under control and adds to variety of the menus.

Table 16.1 Some factors to consider when analyzing budget deviations.

Factor	Key Analytical Questions
Food Costs	
Menus, menu costing, selling price	1. How many choices are offered? 2. What are the food and labor costs of each menu item? 3. Which items are most profitable? 4. How many of each item are sold? 5. Could menu items be better merchandised?
Purchasing	1. Are appropriate specifications used? 2. Has prime vending or group purchasing been considered? 3. Has buyer kept current with market trends and conditions?
Receiving	1. Are deliveries checked against the purchase order? 2. Are qualities and quantities verified with specifications?
Storage	1. How does current turnover rate compare with previous calculations? 2. Is there opportunity for theft? 3. Are storeroom requisitions controlled? 4. Are store areas properly arranged to minimize spoilage? 5. Are proper temperatures maintained in all storage units?
Food production	1. Are standardized recipes available and used? 2. Has establishing central ingredient assembly been considered? 3. Is production equipment adequate and well-maintained? 4. Are forecasts accurate and followed by production staff?
Portion size	1. Have portion sizes been established for each menu item? 2. Are employees aware of the portion sizes, and have they been trained in portion control? 3. Are appropriate and adequate portioning utensils available for employee use?
Employee meals	1. Have the costs of employee meals been calculated? 2. Are proper accounting methods used to justify these costs?
Labor Costs	
Type and extent of service offered	1. Can more self-service opportunities be offered without sacrificing customer satisfaction?

Recipe Costing and Establishment of Selling Price. One important responsibility of foodservice managers is to determine a sound basis for establishing the selling price for food. Haphazard methods may lead to financial disaster or the dissatisfaction of the patron.

Foodservices use a variety of methods for determining a selling price. (Calculations are summarized in Chapter 4.) The most common method is based on the raw-food cost of menu items plus a markup to give a selling price appropriate for the

Table 16.1 *Continued*

Factor	Key Analytical Questions
Hours of service	1. Is there enough customer volume to justify current hours of service? 2. What merchandising techniques can be implemented to stimulate sales during slack times? 3. Are there tasks done at peak times that could be transferred to slack times?
Physical plant	1. Is equipment arranged logically to minimize human energy expenditure? 2. Are carts available and are size and number adequate to minimize trips to storage areas? 3. Does sharing equipment delay production?
Personnel policies and productivity	1. Have production standards been determined and communicated to the employee? 2. Are employees properly trained and supervised to assure productivity standards are met? 3. Can overtime expenditures be justified?
Supervision	1. Are supervisors monitoring departmental activities? 2. Do supervisors recognize key productivity times and take action to assure that standards are met? 3. Does supervisor know how to allocate resources on a day-by-day basis to keep within budgetary guidelines?
Operating and Other Expenses	
Maintenance and repair	1. Is a preventative maintenance system defined and used? 2. Can repair costs be justified compared to equipment replacement? 3. Are employees trained to report broken equipment?
Breakage	1. Are records of china and glass breakage kept and summarized periodically to monitor changes in amounts broken? 2. Are employees aware of costs? 3. Are proper handling techniques used to minimize breakage?
Supplies	1. Do employees have access to supplies? 2. Are procedures for use of chemicals established to minimize waste?
Energy and utility costs	1. Does preventative maintenance include energy efficiency check? 2. Is equipment used properly to minimize energy utilization?

type of organization and the desired food-cost percentage level that the foodservice wishes to maintain.

The *raw-food cost* is found by costing the standardized recipe for each menu item. An example of a costed recipe is shown in Figure 16.12. Storeroom purchase records provide the price of ingredients to use in costing the recipes. Many foodservices have the costed recipes and storeroom records on computer and they use programs to update recipe costs as ingredient prices fluctuate.

Name of product	Quiche Lorraine	Size of Pan	12 X 20 X 2
Yield (total quantity)	2 pans	How Portioned	4 X 6: 24/pan
Size Portion	7 oz	Date Prepared	8/31/97
No. of Portions	48	Prepared by	L.L.

NUMBER OF SERVINGS: 48

INGREDIENTS	EP Weight	Measure	AP Weight	Measure	Unit Price	Cost	
Flour, pastry			3 lbs 2 oz		.523/lb	1	6343
Salt, cooking			2 oz		.058/lb		0072
Shortening			1 lb 12 oz		.73/lb	1	277
Water		2½ c					–
Onion, chopped	4.2 qt		4.5 oz		.238/lb		0069
Milk		1 gal			2.44/gal	2	44
Swiss cheese, grated			2 lb		2.95/lb	5	90
Eggs, fresh, whole		3 doz			.68/doz	2	04
Mustard, dry			½ oz		.92/lb		0287
Ham, ground (optional)	1 lb 4 oz				2.72/lb	3	40

Procedure:

Total Cost $16.7941

Portion Cost .3498

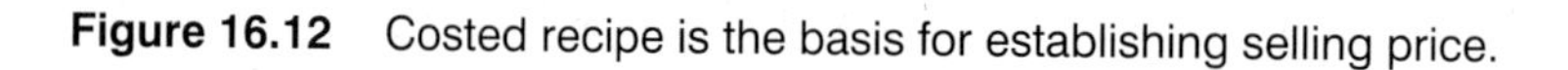

Figure 16.12 Costed recipe is the basis for establishing selling price.

The *markup* is determined by dividing the desired food-cost percentage that the foodservice wishes to maintain into 100 (representing total sales or 100%). The resulting figure is called the *markup factor.* This is the figure by which the raw-food cost is multiplied to obtain a selling price. To illustrate, assume the foodservice wishes to maintain a 40% (of income) food cost:

$$\frac{100 \text{ (represents total sales)}}{40 \text{ (percent of income for food)}} = 2.5$$

2.5 is the markup factor.

Cost of a portion in Fig. 16.12 is .3498 × 2.5 = $.874, or .87, the suggested selling price.

The markup factor cannot be used alone, however, to calculate selling price. There are many "free" items given with a meal that must be added in—salt and pepper, condiments, sugar and cream, and jams, jellies and sauces. These do not show up in recipe costing, but they must be accounted for.

Also, it is imperative that the food manager know not only the raw-food cost of menu items but also the cost of the many hidden losses in preparation, cooking, and serving, which if not controlled add appreciably to the total food cost. Overproduction and unavoidable waste likewise add to the costs, and the wise manager analyzes these, controls what is possible, and considers the other when establishing selling prices. To compensate for these "unproductive and hidden costs," many foodservice managers add 10% (or some such standardized amount) to the recipe cost before markup. Thus, in the illustration given, the suggested selling price of $.87 would be changed:

Raw-food cost of $.3498 × .10 = .03498
\+ .3498 = $.3848
× 2.5 = .961, or $.96, a more realistic selling price.

Pricing of table d'hôte, selective, elective, or other combination menus usually found on printed menus in commercial foodservices follows the same procedure as illustrated. However, all items that are served together at one price (such as meat, potato, vegetable, salad, and beverage) are costed out and the total raw-food cost is obtained before the markup factor is used to calculate the selling price.

Obviously, the exact markup price cannot be used when a fraction or an "awkward" number results. Such numbers are rounded up to the nearest reasonable figure; $1.87 might become $1.90 or even $2, for example.

Another method, called *demand-oriented pricing,* is based on what the customers perceive the value/cost should be and what they are willing to pay for what they receive. This is also known as "what the traffic will bear." Prices are set as high above raw-food cost as the customers will pay, without lowering the number of sales on the item. Close control and watch on sales with good records to support pricing decisions are necessary.

Sometimes managers set their selling prices to be comparable with their competitors. This is unrealistic if used as the only basis, because there are too many variables in each operation. This method may be used for a single food item, perhaps for a cup of coffee, but it is not a "scientific" way of pricing.

These two methods, demand-oriented and competitive pricing for establishing a selling price, have been questioned by some foodservice managers, especially in these times of soaring labor and operating costs. Overhead costs remain about the same regardless of the menu and do not vary from day to day. Labor costs are affected by the menu, however, because preparation time is different for each item of food. Roasts and steaks, for example, usually are high-cost foods but require relatively little labor time to prepare. Stew and quiche, traditionally lower cost items, require much more time to prepare. It seems logical to some operators, therefore, to base selling prices on labor cost plus raw-food cost. In this way, all customers share more equally in the expenses of the foodservice. This method is called *prime-cost pricing* because food and labor are the primary expense items in the budget.

Together, these two items of expense usually make up 70% to 85% of the total of all expenses, depending on the type of organization. The markup in this case is figured on the prime costs:

$$\text{Food } + \frac{\text{Sales: } 100}{\text{Labor: } 70} = 1.43 \text{ or } \frac{100}{85} = 1.18$$

instead of the 2.5 used in the example of the traditional method given previously.

Utilization of the prime-cost system requires accurate records of labor time spent in preparing the various foods. The amount of labor time used multiplied by the employee's wage rate gives this cost of labor.

A comparison of the prime-cost pricing method and the conventional method for steak and for quiche follows:

Conventional Method at 33⅓% food-cost markup = 3x plus 10% of raw-food cost		*Prime-Cost Method* at 33⅓% food cost + 46⅔% labor cost = 80% prime cost; Markup = 1.25 x	
STEAK			
Raw-Food Cost	1.80	Raw-Food Cost	1.80
+ 10% Hidden Cost	+ .18	+ Labor Cost of 5 min. @ 12.00/hr.	+ 1.00
	1.98		2.80
× Markup	× 3	× Markup	× 1.25
	5.94		
	or		
Selling Price =	6.00	Selling Price =	3.50
QUICHE			
Raw-Food Cost	.850	Raw-Food Cost	.85
+ 10% Hidden Cost	+ .085	+ Labor Cost of 20 min. @ 12.00/hr.	+4.00
	.935		4.85
× Markup	× 3	× Markup	× 1.25
	2.805		6.06
	or		or
Selling Price =	2.80	Selling Price =	6.00

Due to customer perception of value/cost, it is doubtful that many portions of quiche would be sold for $6 in the same foodservice in which steak is sold for $3.50. At the same time, management could apply the demand-oriented pricing method to the steak item in the prime-cost method, because customers would be willing, and expect, to pay more for a steak than for quiche.

Regardless of which calculations are used to determine selling price, it is imperative that at some point the manager evaluate the cost-volume-profit relationships to ensure that selling prices are based on sound information specific to the operation's financial objectives. *Break-even analysis* is a cost-volume-profit relationship used by managers of commercial operations who constantly face decisions about selling prices. The manager must make reasonably accurate predictions about costs and revenues or face potentially disastrous results. The **break-even point** (BEP) is a commonly used calculation to initiate break-even analysis. The BEP is the point at which an operation is at $0 net income. In other words, the level of sales volume is such that all costs are covered but no profit is realized. The operation is neither making nor losing money. This concept is important to understand to appreciate how and why all costs must be covered in menu pricing.

The objective of the BEP is to generate a level of sales volume to cover both fixed and variable costs. The formula to calculate the BEP is:

$$\text{BEP} = \frac{\text{Fixed Costs}}{1 - \dfrac{\text{Variable Costs}}{\text{Sales}}}$$

The denominator is referred to as the *contribution margin* and represents the proportion of sales that can be contributed to the fixed costs and profit after variable costs have been covered. *Fixed costs* are those that are incurred regardless of level of sales. Utilities and rent are examples of fixed costs. *Variable costs* are those that fluctuate in relationship to changes in sales volume. Food costs, for example, should decrease during periods of reduced sales.

A simplified BEP for foodservice operations is:

$$\text{BEP} = \frac{\text{Fixed Costs}}{(\text{Selling Price} - \text{Variable Costs})}$$

Purchasing. Food purchasing is fully described in Chapter 5. However, when managers are reviewing and evaluating overall expenses, purchasing methods should be included.

The market is ever changing and the buyer must keep abreast of new developments and learn what best suits the needs of the foodservice and at the most advantageous price. Specifications may need to be changed from time to time according to market trends. Certain costs are controlled through wise purchasing by an informed, capable buyer who is alert to ever-changing market conditions and has a knowledge of new products as they become available.

Receiving. Losses may easily occur at the point of receiving goods if management is negligent about checking in orders as they are received. This task may be entrusted to an assistant, but should be someone with managerial authority.

Storage and Storeroom Control. Protection of the company's large investment of money in the food after it is purchased and received contributes greatly to overall cost control. It has been said that one should buy only the amount that can be used at once or stored adequately. Furthermore, one should store only what is essential for limited periods of time because unnecessarily large inventories tend to increase the possibility of loss through spoilage, waste, or theft.

The value of the inventory can be monitored by calculating the turnover rate, which is a measure of how many times storeroom goods are used and replenished during a specified time period. The rate is determined by dividing the cost of goods sold by the value of the ending inventory (from the profit and loss statement). A turnover of three to five times a month is fairly average for many foodservices, although this varies considerably. A small fast-food restaurant in a large city may have a high turnover of inventory, because foods are used quickly and deliveries can be made frequently. In contrast, a large university that is located in a somewhat geographically isolated area may keep a large inventory of staple items to carry it through a school year, and so the inventory turnover would be very low.

If the turnover rate is excessively high, it may indicate a shortage of funds to purchase in sufficient quantities, and purchases are made in small amounts that are used almost at once. This is an expensive method. It may also limit the foodservice's credit rating and ability to buy competitively.

If the turnover rate is low, too much stock may be remaining on the storeroom shelves too long, or many items may be left unused. Managers should check the inventory from time to time and include on the menu those items that need to be "moved" before they deteriorate and cause "waste" cost.

Food Production: Preparation, Cooking, and Leftovers Control. Foodservice managers are well aware of the many costs and potential losses that can occur in the production of food for service. Foodservice workers, although well trained, need continuous supervision to ensure that standardized recipes are used properly, and equipment is operated appropriately to minimize preparation losses. See Chapter 6 and Appendix A for details.

Reducing the amount of leftover prepared foods is another step that can be taken toward cost containment. The manager's ability to forecast accurately the number of portions that are used or sold is critical and should not be based on guesswork or left to the cooks to decide how much to prepare. Rather, the use of records to show *amounts prepared, amounts sold or used,* and *amounts left* gives a realistic basis for estimating quantities required the next time an item is served.

Portion Size and Serving Wastes. An established portion size is part of the standardized recipe and is one basis for costing and setting the selling price. The size of a portion or serving to be offered to the consumer is a management decision and must be communicated in writing to the employees.

One means of assuring standardized portions is to know size and yield of all pans, measures, ladles, and other small equipment used in the serving. For example, if 1 gallon of soup is to yield 16 1-cup servings, accurate measurements must be made of both the original quantity and the amount taken up in the ladle. A 1-cup ladle should be provided for the server's use, not a three-quarter cup or some other size, to obtain the standard portion. Other appropriate-size serving equipment should be used for other food items.

Employees' Meals. Sometimes employees are given a discount on their meals or are charged at-cost prices rather than the usual marked-up selling price. The value of food consumed by employees should be of real concern to management in attempting to better control both food and labor costs. The philosophy of management regarding employees' meals—to charge for them or not—varies with the individual institution. Managers should question the present policy in view of overall cost control. If meals are provided, their value must be determined for use in the financial statement.

Meals provided as part of employees' compensation should be handled as a labor cost, not food cost, in the profit and loss statement. A cost determination of the value of the meals is made by management, and the total of all employees' meals is deducted from the "cost of food sold" and added to "labor" as an employee benefit to reflect their true place in financial accounting.

Labor Cost Control. Labor costs represent a major component of the total foodservice expense in most organizations today. Until recent years, food was first in importance and labor was second. Together, food and labor made up around 75% of the total expense. With ever-increasing wage rates and employee benefits, it now is estimated that labor constitutes 50% to 70% of the total, as an overall average.

However, there are so many variables in each situation that even "averages" have little meaning. Restaurants with full table service in luxurious surroundings and French-style service have quite a different labor cost than a serve-yourself buffet operation. In the first case, the income per meal may be $20; in the second, $2.95. The labor cost may be 50% in each case—$10 for labor in the luxury restaurant and $1.48 in the self-service establishment—but each shows the same *percentage* of income spent for labor. Dollar figures need to be closely monitored in any evaluation, and managers should not rely on percentages only.

Production employees can prepare more servings of most menu items with little extra time expenditure; supervisors can handle a somewhat larger volume of trade during their time on duty; and probably no additional office help would be required as the volume of business increases. Good managers should be able to determine when additional labor is required to handle increased volume. Generally speaking, the greater the volume of business, the greater the returns on labor dollars spent. Labor costs are less controllable than food costs, and their percentage of payroll costs to sales fluctuates with sales. It is impractical if not impossible to change the number of employees day by day in proportion to the number of customers, patients, or students, as one might change the menu to meet fluctuating needs. Therefore, it is necessary to consider ways to get full returns from the payroll dollar.

The following ratios are commonly used in the foodservice industry to analyze labor productivity and costs:

$$\text{Meals per labor hours} = \frac{\text{Total number of meals served}}{\text{Total hours of labor to produce the meals}}$$

$$\text{Meals per full-time equivalent (FTE)} = \frac{\text{Total number of meals served}}{\text{Total FTEs to produce the meals}}$$

$$\text{Labor minutes per meal} = \frac{\text{Total minutes of labor time to produce the meals}}{\text{Total number of meals served}}$$

Type of Foodservice System. The various types of systems and the labor required in each are described in Chapter 2. Foodservice organizations faced with excessively high labor costs might investigate the feasibility of converting to another system that requires less labor. Or if a complete conversion is not possible, consideration could be given to the use of more pre-prepared frozen food items, thus reducing labor time and cost for food production.

Type and Extent of Services Offered. The extent of service offered within the organization affects total labor costs. In cafeterias, for example, the patrons may carry their own trays and bus their own soiled dishes, or if table service is used, the ratio of servers to guests varies as does the cost of labor. If the menu and service are simple, one waiter is able to serve many guests. When the formality of dining calls for personalized service and several echelons of dining room employees from the maître d' to the head waiter or waitress, server, wine steward, coffee server, and bus person, we can easily understand the high cost of labor in such establishments.

Hours of Service. The hours of service determine the number of "shifts" of personnel as well as the total number of labor hours required to accomplish the work. The hospital cafeteria that is open 7 days a week and serves four or five meals daily—breakfast, lunch, dinner, night supper, and 3 A.M. lunch for the night workers—demands a larger complement of employees than does the school lunchroom that serves only one meal per day for 5 days. The restaurant that is open 24 hours a day, 7 days a week, uses a different labor schedule than the one that is open for business 10 to 12 hours a day, 6 days a week. Each situation has a different labor-cost expense.

Records of patronage and sales by 15-minute time segments throughout the serving period provide valuable data for management when evaluating labor needs and scheduling. Electronic or computer point-of-sale (POS) cash registers provide such data; small foodservices use a simple form such as that shown in Figure 16.13 for posting such data. Use of a graph as in Figure 16.14 helps managers evaluate labor distribution.

Physical Plant: Size and Equipment Arrangement. An efficient kitchen arrangement and a convenient location are positive factors in labor-cost control. Facility planning and layout are discussed in detail in Chapter 10. The foodservice manager may not be the one responsible for the kitchen plan, but if he or she "inherited" one that is poorly arranged, some changes may be necessary. Work flow and productivity analysis provide information to determine whether changes are needed (see Chapter 15).

Menu and Form of Food Purchased. Many questions can be asked about the menu and form of food purchased as they affect overall labor costs. (Various types of menus are presented in Chapter 4.) The number of menu choices offered, the complexity of preparation involved, the labor time and cost required, and the number of dishes to be washed, resulting from the menu items served, are only a few costs. Studies to determine the exact labor time and cost involved in preparing foods from the raw state versus the food and labor costs of using convenience foods give a preliminary basis for decisions on which form of food to buy.

Name of Foodservice					
Date ____ AM	Number of Customers	Amount of Sales	Average Sales	Labor Hours	Labor Cost
7:00–7:15					
7:16–7:30					
7:31–7:45					
(etc.)					

Figure 16.13 Records of census and sales by time segments give managers data helpful for evaluating and scheduling personnel.

Personnel Policies and Productivity. Labor is a commodity that cannot be purchased on short notice. The careful selection and placement of workers are basic to reducing turnover and its inherent costs. A study of overall personnel policies, including wage scales and employee benefits, can give managers possible clues to remedy excessive labor costs. However, the most important aspect of labor cost is probably employee productivity. As noted in Chapter 15, ineffective management is the single greatest cause of declining, or poor, productivity. Good supervision is vital to holding labor costs in line.

Supervision. Supervision is a major factor in the labor-cost picture. The effects of good or bad supervision cannot be underestimated when evaluating total labor cost. Good supervision ensures adherence to established policies and rigid control of work schedules and standards, and influences employee morale; productivity is high and management receives fair returns on the labor dollar invested.

Too often, however, administration views supervision costs as excessive and attempts to cut labor costs by replacing competent, well-trained supervisors with inexperienced, immature ones. Sometimes, an experienced but unqualified person is promoted from the ranks to assume supervisory duties at a relatively low cost. Rarely does such replacement prove satisfactory. Neither the inexperienced nor the experienced untrained worker is able to see the full view of the foodservice operation. Usually the costs begin to rise until any slight saving entailed in the employment of an untrained director is absorbed many times over. Money spent for efficient supervision brings high returns in economic value to the organization. There is no substitute for good supervision.

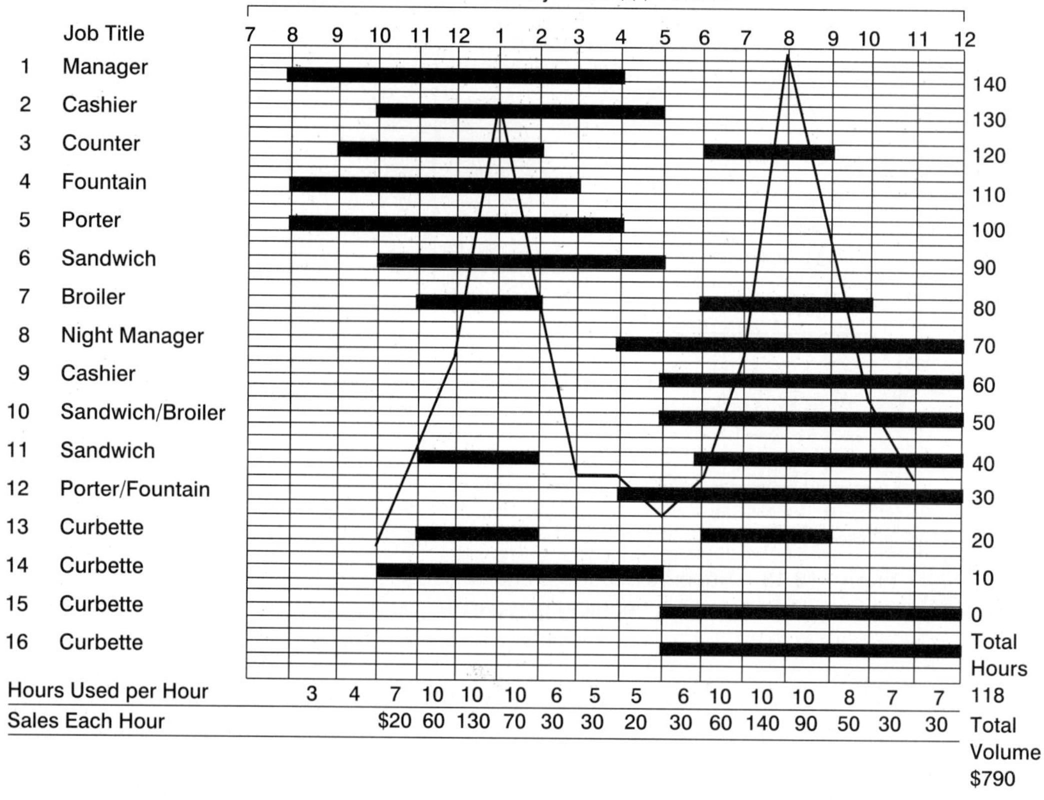

Figure 16.14 Graphic presentation of employees' time schedules in relation to volume of sales in one restaurant. (Courtesy of Ohio State Restaurant Association.)

Operating and Other Expenses. Control should not end with consideration of food and labor costs only; 12% to 18% of the departmental budget will probably be used for other items classified as overhead and operating expenses. These include utilities, laundry and linen supplies, repairs, replacement and maintenance, telephone, printing, paper goods, office supplies and cleaning materials, depreciation, rent or amortization, and insurance and taxes.

In addition, there is a real concern for conservation of energy resources within all foodservice establishments. Not only does conservation meet a national need, but it also helps to reduce departmental operating costs.

Maintenance and Repair. A planned maintenance program with the services of a maintenance engineer helps prevent breakdowns, extends the life of the equipment, and is usually a cost-effective procedure.

Breakage. Excessive breakage quickly adds to operating costs. Procedures should be in place to document breakage and employees should be well trained on materials handling to prevent excess breakage.

Supplies. Supplies include linens, paper goods, cleaning compounds, dishwasher detergents, office supplies, and similar items. Although these items may be considered small in relation to other costs, any waste is costly. Accurate accounting of these items helps control costs.

Energy and Utility Costs. The energy "squeeze" of the 1970s made most foodservice managers and equipment manufacturers well aware of the need for conservation. New equipment designs are energy efficient, and mangers continue to seek ways to conserve and to reduce utility bills. Energy-efficient equipment is discussed in Chapter 11 and built-in energy efficient layout and design in Chapter 10.

SUMMARY

Financial planning and accountability are the responsibility of the foodservice manager. A basic understanding of accounting and financial management concepts is necessary to analyze the financial performance of the foodservice department and to make appropriate decisions for allocation of available resources.

The foodservice manager must recognize that the budget is the key management tool for financial planning and that he or she must take an active role in budget development and implementation. Understanding and getting involved in the budget planning process ensure a managerial commitment to the financial goals of the department.

The foodservice manager must know the day-to-day transactions and operational activities that take place in order to compare them with established goals. A sound system of records and reports provides the database for making these comparisons. Without this close monitoring, downward trends are not detected and financial disaster can occur.

REVIEW QUESTIONS

1. What is the purpose of the budget?
2. Describe the three phases of budget planning.
3. What are the major requirements for any recording-keeping system?
4. What is the formula for determining food-cost percentage?
5. How are standards for food-cost percentages set?
6. Define markup. How is it calculated?
7. What three ratios are commonly used to analyze labor productivity and costs?

SELECTED REFERENCES

Controlling food production and inventory. School Food Service J. September 1993; 45–48.

Controlling food purchasing, receiving, and storage. School Food Service J. October 1993; 37–38, 44.

Controlling labor costs. School Food Service J. November 1993; 51–52.

Daniels, R. D., and Gregorie, M.B.: Use of capital budgeting techniques by foodservice directors in for-profit and not-for-profit hospitals. J. Am. Diet. Assoc. 1993; 93:67–69.

Getting control. School Food Service J. August 1993; 28–29.

Kaud, F. A.: Budgeting: A comparative analysis of techniques and systems. In Rose J. C., ed., *Handbook for Health Care Food Service Management.* Rockville, Md.: Aspen Systems Corporation, 1984.

Keiser, J., and DeMicco, F. J.: *Controlling and Analyzing Costs in Foodservice Operations,* 3rd ed. New York: Macmillan Publishing Company, 1993.

Sneed, J., and Kresse, K. H.: *Understanding Foodservice Financial Management.* Rockville, Md.: Aspen Systems Corporation, 1989.

FOR MORE INFORMATION

The National Restaurant Association
Educational Foundation
250 S. Wacker Drive, Suite 1400
Chicago, IL 60606
Phone: (312)715–1010

17

Marketing and Promotions in Foodservice Organizations

OUTLINE

Marketing has been an essential function of commercial foodservice operations for a long time. In recent years, managers of noncommercial operations (i.e., hospitals, long-term care facilities, schools, and universities) have recognized the value of marketing principles as a means to survive in a highly competitive industry in which resources are increasingly scarce and costly. Serving good food is not enough. Today's customer is more sophisticated and has higher expectations of food and service than ever before. Good food must be accompanied by good service, a clean and pleasant atmosphere, an appropriate price, and convenience. These attributes can create a competitive edge or *point-of-difference*—the point at which a customer chooses one foodservice over another.

An individual does not need a business degree in marketing to develop a successful marketing program for a foodservice operation. This chapter introduces the reader to the basic principles of marketing and offers suggestions on how to develop and implement a successful program. It begins by defining some key marketing terms and concepts. Subsequent sections describe the unique aspects of the marketing process in foodservice, and the final section describes merchandising and promotion techniques appropriate for a foodservice operation. Guidelines are offered on how to plan and implement a specific promotion.

KEY CONCEPTS

1. Marketing is a cyclical function that requires planning, implementation, and evaluation.
2. A target market is a group of people who, as individuals or organizations, have needs for products and possess the ability, willingness, and authority to make a purchase.
3. Market segmentation is related to the demographic, geographic, and psychographic profile of potential customers.
4. The marketing mix includes four elements: product, price, place, and promotion.
5. Staff training on customer relations is essential to the success of a marketing program.
6. Marketing in foodservice operations encompasses several unique aspects, including the tangible and intangible components of the product.
7. Promotions are specifically designed to influence customer behavior and are based on clearly defined objectives.

THE DEFINITION OF MARKETING

Today, the word **marketing** is used in many contexts. Earlier in this book, the marketing system is discussed in terms of agricultural products. In this chapter, however, marketing is examined as a specific managerial function. One definition of marketing is provided by the American Marketing Association, which defines marketing as the performance of business activities that direct the flow of goods and services from the producer to the consumer. Others have defined marketing in more human terms. For example, Kotler and Armstrong (1987) define marketing as a human activity directed at satisfying the needs and wants of customers through an exchange process.

In today's environment of limited resources and increased competition, it is easy to lose sight of the marketing concept in favor of financial stability. However, it is important for the manager to adopt the marketing concept and empower the employees to put customer needs first; without the customer, there would be no foodservice organization. The marketing concept implies having the flexibility to change with customers' evolving needs, wants, and demands.

Many marketing terms are used interchangeably. For the purposes of this chapter, the following terminology, as defined by Underwood (1992), is used:

Marketing is the act or process of selling or buying in a marketplace.
Merchandising is a global term that describes sales promotion as a comprehensive function, including market research, development of new products, manufacturing, marketing, advertising, and selling.
Promotion is a process of action meant to influence, increase, or improve the acceptance and sale of merchandise or services through advertising and publicity.
Advertising results when a promotion has been planned and presented to the public. It is the action of calling something to the attention of the public, making something public or generally known, and emphasizing desirable qualities to arouse a desire to buy a product or patronize a facility.
Publicity consists of acts or devices designed to attract public interest. It is the dissemination of information or promotional material, such as advertising.

Marketing activities identify and attract customers to an organization and its products and services. Marketing also serves an important function in the commercial foodservice industry. Marketing in the noncommercial foodservice setting, however, is a relatively new managerial concern. During the past decade, the health care and education industries have faced numerous challenges, including increased costs, inflation, increased government regulation, and competition for customers. As a result, many hospitals and long-term care facilities are struggling to maintain or increase their patient numbers. Schools, colleges, and universities are challenged with declining enrollment and competition from nearby restaurants.

The foodservices of these organizations can play an important role in attracting and keeping customers by creating good quality food and service and by increasing customer awareness of their availability. For example, good food and service can be a key factor in whether a family is willing to admit a family member to a long-term care facility. Variety and excitement in the school cafeteria can prompt a student to choose the cafeteria over a nearby fast-food restaurant for lunch. Furthermore, patients satisfied with their food will most likely comment favorably on their entire hospital stay.

Marketing is indeed a customer-driven process that includes specific activities. Figure 17.1 illustrates a successful marketing cycle. Steps include (1) identification of customers; (2) development of products, pricing, and distribution; (3) customer purchases; (4) generation of profits; and (5) appropriate action based on profits and customer feedback.

Marketing operates in a dynamic environment influenced by outside and inside forces. Outside forces include politics, the economy, government regulation, laws, social pressures, technology, local competition, industry trends, and, of course, cus-

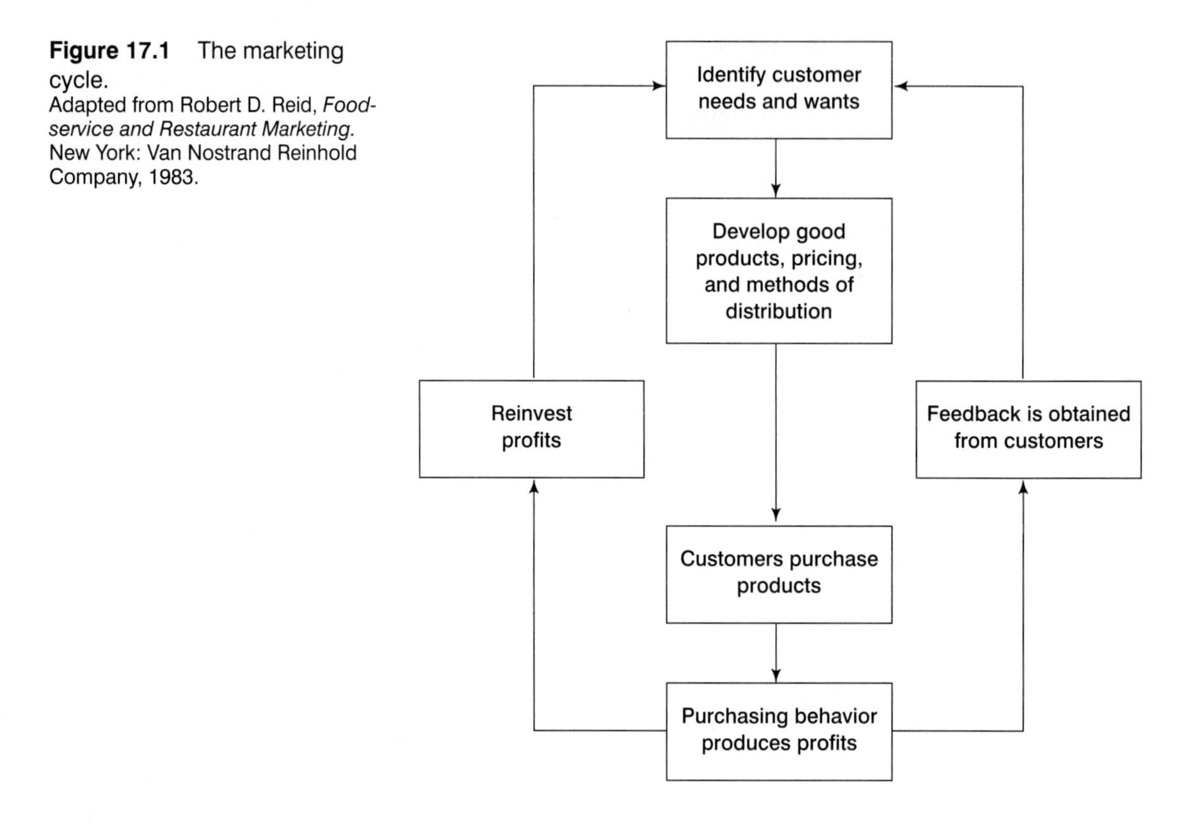

Figure 17.1 The marketing cycle.
Adapted from Robert D. Reid, *Foodservice and Restaurant Marketing.* New York: Van Nostrand Reinhold Company, 1983.

tomer attitudes and behavior. Inside influences can include organizational goals, budgetary constraints, and departmental policy.

THE MARKETING CYCLE

The **marketing cycle** begins by identifying the customers who make up the potential market. The market, or **target market,** is defined as a group of people who, as individuals or an organization, have needs for products and possess the ability, willingness, and authority to make a purchase. The manager should ask the following questions:

- Who are the current customers?
- How many customers are in the potential market?
- Can we attract additional customers?
- What are the unique needs of these customers?

For example, the director of a school lunch program may find that potential customers include students that currently bring their lunch or go home, as well as teachers, staff, and visitors that have not previously participated in the lunch program. Numbers can be estimated by calculating total student enrollment, total teaching and support staff, and the estimated number of visitors to the school each day.

Additional customers often include the elderly population of the community, who may participate in a congregate meal program at the school. Finally, the unique needs of each group must be identified. These unique needs may include fast-food-type preferences for the student, foods of appropriate texture for the elderly, and speed and convenience for the teachers and staff.

Answering these questions helps the manager begin to define the market segments. **Market segmentation** divides the total market into smaller groups of people with similar product needs. Categories of customers emerge based on demographics, geographic, psychographics, and product preference.

Demographic segmentation refers to the statistical data on customer profile characteristics such as age, sex, income, and educational background. **Geographic segmentations** categorize customers according to where they live. **Psychographic segmentation** refers to customers of similar lifestyles, attitudes, and personalities. Finally, *product preference* is segmentation by customer behavior exhibited in a foodservice operation.

Once the needs and wants of potential customers have been identified, the manager can proceed with the marketing process by developing the products and services necessary to satisfy if not exceed these needs and wants. A product may be an object, service, or idea. It must also be available at the right place at the right time and be priced appropriately so that the customer identifies it as meeting a need, is willing to pay for the product, and prefers the product over that offered by the competition. To this end, the marketing process must be customer oriented and customer driven.

THE MARKETING MIX

A well-defined marketing program includes four elements: product, place, price, and promotion. These elements, often called the four P's of marketing, and their unique combinations in any marketing program are called the marketing mix.

Product is the unique combination of goods and services that satisfies a want or need. The product can be objects, services, ideas, places, or an organization. It is what is produced based on knowledge of the market and what is ultimately sold. Foodservice products include all items on the menu, as well as the many types of service options, such as cafeterias, vending machines, and catering, and other desirable attributes, such as pleasant atmosphere.

Place includes distribution and how products are sold. Products must be available at the right time and place, convenient for customers. Many foodservices today are accommodating customer desires for speed and convenience. For example, large-scale hospitals often set up mobile cafeterias or kiosks on nursing units during busy lunch periods as a convenience for medical staff with limited break time. Many organizations are opening food courts similar to those seen in shopping malls.

Price is the amount of money charged for a product or the sum of the values customers exchange for the benefit of the product. Strategic pricing encourages the customer to make a purchase, contributes to product image, and allows products to

Figure 17.2 An example of a "preferred customer" coupon to increase cafeteria sales. Copyright 1992 Board of Regents, University of Wisconsin Systems, University Hospital and Clinics, all rights reserved.

compete in the market. For example, a cafeteria manager may offer coffee at a reduced price during slow periods knowing that a customer is likely to make additional purchases such as sweet rolls, pie, or popcorn (see Figure 17.2).

Promotion involves all communication with the customer. It introduces the customer to, or increases customer awareness of, the available product. (Promotion is discussed in more detail later in the chapter.)

TRAINING STAFF ABOUT THEIR ROLE IN MARKETING

Employees are a very important part of a successful marketing program. They must be recruited and hired with great care and trained immediately on proper customer relations and professional behavior.

Training time and costs should be valued as an investment in a successful marketing program. Reid (1983) suggests the following steps for employee training:

1. The training should be carefully planned and kept simple. An outline should be prepared to detail the subject matter covered in the training session.
2. The trainer should plan to cover only what the trainee can learn in a training session. Specific times should be set aside for training rather than trying to "squeeze it in." The marketing concept and all of its implications can be covered during several short sessions rather than during one long one.
3. The trainer should explain for the trainees specifically what it is they should be able to do; they should then be allowed to practice. For example, a cafeteria manager may wish to teach *suggestive selling* by training each cafeteria line employee to ask "Would you like fries with your hamburger?" or to offer another appropriate suggestion.
4. All employees should be provided with feedback. Employees want to know how they are doing and follow-up is essential to evaluate the effectiveness of the training.

Managers, too, must take responsibility for self-development in marketing concepts, functions, and activities. Participation in professional associations, such as the National Restaurant Association (NRA), is a good way to stay current on a number of industry topics including marketing, planning, promotional ideas, and pricing strategies. Other self-development options include attending professional seminars and reading some of the numerous professional publications specific to the foodservice industry.

MARKETING FOR FOODSERVICE OPERATIONS

Unique Aspects of Foodservice Marketing

Marketing in foodservice requires a unique approach, because unlike many industries, foodservice includes a service component. Service is the application of human or mechanical efforts to people or objects. Service industries such as foodservices differ from most manufacturing industries in product, customer contact, perishability of inventory, and distribution.

Product

Food provided by a foodservice operation is consumed but not possessed, which distinguishes it from other consumer goods such as appliances. Food as a product is unique in that it has both a tangible and an intangible component. The food itself is the *tangible* component, meaning that it is capable of being perceived by the buyer through smell, touch, taste. Service is *intangible* in that it cannot be seen, touched, tasted, or possessed but yet the consumer is very much aware of its presence and certainly aware of its absence. For example, customers are quick to notice lack of friendliness on the part of wait staff.

Customer Contact

The customer takes a more active role in the marketing function in a service industry. For example, in many cafeterias, patrons help themselves to displayed foods or, in the case of table service, there is direct and frequent customer-employee contact. Each contact represents an opportunity for the foodservice to market not only the food product but the organizational image as well.

In the cafeteria self-serve situation, the foodservice operation has the opportunity to entice the customer with attractive well-designed displays. Table service offers more direct and personal contact. For example, wait staff can be trained to anticipate customer wants and needs such as a beverage refill or readiness for the check. One negative customer-employee interaction can generate lasting dissatisfaction that may result in the loss of business not only from the unhappy customer, but from all potential customers that the dissatisfied customer influences through word-of-mouth.

Perishability

Food is unique in that it is highly perishable and difficult to store in inventory. Unlike a tangible product, such as a television set, which can be stored in a warehouse during low-demand periods, food is highly perishable and, if unsold or simply not used, it is lost income, or waste. For example, if the customer count is lower than expected in a cafeteria or school, these potential sales are lost forever and food prepared ahead of time is wasted.

Distribution

In many types of foodservice operations, food must be prepared in advance, held either hot or cold, and transported for distribution. For example, many elementary and high schools receive their food from commissary foodservice operations. Without careful consideration of the conditions during holding and transport, food quality can deteriorate significantly and, thus, it can be rejected by the paying customer.

MARKETING AS A MANAGERIAL FUNCTION

Management must recognize marketing as an essential function, similar to traditional management functions of organizing, leading, and controlling. The marketing function isn't a management fad that, if ignored, will go away. It is the customer who will not return if the marketing function is not understood or properly implemented.

Common marketing mistakes include the following:

- Lack of planning,
- Improper budgeting,
- Poorly defined goals and objectives,
- Lack of product development, and
- Inadequate program evaluation.

With a clear vision of and a commitment to the organization's mission, the wise manager can develop a marketing program that includes planning, implementation, and evaluation.

Planning

Planning begins with a clear understanding of and a commitment to the goals and objectives of the marketing plan. In other words, the manager should ask: "What are we as an organization or a department trying to do?" "What do we hope to accomplish?" For example, is the purpose of the marketing program to attract new customers, retain current customers, or influence specific purchasing behaviors?

General goals need to be defined as part of the organization's long-term or strategic planning. For a goal to become reality, specific objectives must be established. Objectives must be clear and, for purposes of program evaluation, they must be measurable. Responsibility for achieving the objective through specific activities should be assigned and a timetable established for meeting the objective. For example, the goal for a hospital cafeteria may be to increase usage of the cafeteria by employees and visitors. A specific objective would be:

to increase by 10% the number of customers using the cafeteria between 11:00 A.M. and 1:00 P.M.

Personal interviews with potential customers and surveys such as the one shown in Figure 17.3 are appropriate methods to determine the food and service preferences of the target market. Using the results of the interview and surveys, the man-

THE BAYSIDE CAFETERIA

We invite you to rate our food and service to help us improve the quality of our cafeteria. Please circle one choice in each category; (5=very satisfied, 3=meets expectations, 1–dissatisfied). Please comment on ratings of 3 or less.

1. Food: *Comments:*	5	4	3	2	1
2. Price/Value: *Comments:*	5	4	3	2	1
3. Prompt Service: *Comments:*	5	4	3	2	1
4. Courtesy: *Comments:*	5	4	3	2	1
5. Atmosphere: *Comments:*	5	4	3	2	1

When and how often do you use the cafeteria? (Check all that apply and indicate the number of times per week).

		No. of times/week
Breakfast	_____	_____
Morning break	_____	_____
Lunch	_____	_____
Afternoon break	_____	_____
Dinner	_____	_____

What food items would you like to see added to the cafeteria selections?

1. ______________________________
2. ______________________________
3. ______________________________
4. ______________________________
5. ______________________________

Please use the space below for additional comments and suggestions. Thank you.

Figure 17.3 Survey designed to determine preferences of current and potential cafeteria customers.

ager can create specific plans to achieve the objective. Without specific action plans, well-defined objectives may be deserted and desired goals will not be met.

Implementation

Implementation is critical in keeping objectives from becoming good intentions that are left unchallenged. It involves empowering staff to embrace the marketing plan, training employees to successfully execute the plan, defining and developing the promotion plans, effectively communicating marketing messages, and providing support to enable the plan to succeed. Part of this support includes implementing procedures to evaluate the success of the marketing plan.

Evaluation

Evaluation is the process of determining the success of the plan as implemented. The manager must measure the degree to which the previously established objectives were achieved. For example, if the foodservice manager sets an objective of increasing school lunch participation by 5%, then the manager must review daily meal counts to determine if that objective was actually achieved. The knowledge gained from this evaluation can be used to refine objectives and action plans.

Marketing programs, in particular, are costly in terms of time and resources; therefore, management must take steps to ensure that there is an actual, measurable return on this investment. Logically, the evaluation strategies come after implementation of the program and after a specific amount of time has passed. This allows the program ample opportunity to achieve anticipated objectives. Results of the marketing program can then be compared against the projected results. For example, a manager for a long-term care facility may wish to increase customer meal participation on holidays. Figure 17.4 is an example of an invitation used in one long-term care facility to promote such an objective. After the holidays, the manager can simply compare participation following this promotion against guest participation for previous holiday seasons. The change in participation in an indicator of the program's success with this one objective. Obviously, clear and concise record keeping is also critical to ensure that these comparisons are made effectively.

MERCHANDISING AND SALES PROMOTION IN FOODSERVICE OPERATIONS

As described, marketing is a process of identifying customers and their needs, and developing products to satisfy those needs. Merchandising and promotions are distinct functions used to continually pique the interest of customers in an effort to stimulate repeat business, as well as new business.

Merchandising

Merchandising is defined as a comprehensive function that includes market research, product development, manufacturing, marketing, sales promotions, adver-

Figure 17.4 An example of a personalized invitation used to promote increased guest participation at a holiday meal in a long-term care facility.
Courtesy of Maplewood of Sauk-Prairie, Sauk City, Wisconsin.

maplewood
sauk prairie

Skilled Nursing Facility

245 Sycamore Street, Sauk City, Wisconsin 53583
Phone (608) 643-3383

SEASONS GREETINGS!

In celebration of the Holiday Season, you are cordially invited to a complimentary dinner with your spouse at Maplewood on Christmas Day, December 25, 1995.

Dinner will be served to you and your spouse as close to noon as possible.

Please complete and return the attached RSVP by December 10, 1995.

Happy Holidays!

Mary Leonard
Director of Dietary

______ **YES. I will attend Christmas dinner with my spouse.**

______ **NO. I will not attend Christmas dinner with my spouse.**

SIGNED: ____________________

Spouse's Name: ____________________

tising, and selling. Terms such as *marketing, promotions, merchandising,* and *selling* are used interchangeably, and clear, concise definitions are difficult to find.

No amount of merchandising, however, will be effective if the foodservice establishment is not clean, organized, pleasant, and comfortable. With these basic requirements achieved, the manager should consider the numerous opportunities available to tantalize the customer:

- Present all food attractively, regardless of whether it is being served in a dining room, on a patient tray, from a steam table, or from a vending machine. Use garnishes and appropriate plate-fill to visually attract customers to the food.
- Arrange displayed foods in cafeterias or on buffet lines in an eye-appealing manner. Mix colors for preplated salads and arrange in mixed combinations. For

example, alternate crisp, colorful vegetable plates with softer, less colorful salads (i.e., potato salad).
- Use bulk display techniques. Offer fresh fruit from baskets or bowls brimming with polished, clean apples, oranges, plums, and so forth.

Sales Promotion

As discussed earlier in this chapter, promotion is one of the four Ps of a marketing program. Promotion is a distinct function, different from merchandising and advertising. It is the function of influencing the customer's purchase and repurchase behavior, with a primary goal of increasing patronage and, in turn, improving sales and profit. Promotion can also be used for a number of nonprofit goals, such as increasing public awareness of a facility's services.

By planning, implementing, and evaluating promotional strategies, the manager can accomplish several goals. These goals are to (1) present information to the customers, (2) reinforce desired purchasing behavior to stimulate repeat business, (3) stimulate first-time business by arousing curiosity, and (4) enhance the image of an organization. Such strategies are appropriate for all types of institutional foodservices. Obviously, promotions are necessary for cafeterias, vending, and catering, which rely on a profit. There are, however, many other reasons for understanding the basics of promotional marketing.

There are two categories of promotions. First, there are the *share of market* promotions. These are financial, volume-based activities designed to increase patronage, sales, or a combination of the two. Second, there are *share of mind* promotions, which seek to influence the customer's preference or feelings about a particular facility or product. Both of these can be used in institutional foodservices.

Promotion Planning

Promotion planning begins just like any other managerial function—by establishing clear, measurable objectives. The primary objectives of promotions are to increase the frequency of customer visits and the level of customer satisfaction. Table 17.1 suggests specific objectives for various types of foodservice operations.

The manager can generate ideas for specific objective(s) by asking the following questions:

1. What is it that we are trying to accomplish with this promotion?
2. Is it consistent with our mission statement?
3. Is it designed to meet customer needs?
4. How can we evaluate or measure the success of this promotion?

In contrast to marketing and merchandising, promotions are generally designed to run for a short period of time but may be extended or repeated if the objectives remain desirable. For example, a school may try a special promotion during National School Lunch Week to increase participation, offering a free dessert with each meal. If highly successful, the foodservice manager may wish to repeat the promotion at another time during the school year.

Table 17.1 Examples of promotions objectives.

Schools
Increase participation in the school lunch program.
Increase total à la carte sales.
Create an awareness of a new product.
Hospital Cafeteria, Vending
Increase average total profit.
Increase beverage to food ratio.
Increase salad and dessert sales.
Long-Term Care Facilities
Increase awareness of nutritional value of food.
Increase awareness of special services.
Increase family participation in holiday meals.

Some guidelines for the development, implementation, and evaluation of a successful promotion are as follows:

1. Plan well in advance—at least 1 to 3 months before the event is to take place.
2. Establish goals and objectives.
3. Know the current and potential customers.
4. Select a promotional idea consistent with customer need and the organization's mission.
5. Seek advice and ideas from internal and external sources.
6. Verify availability of financial resources and compare costs of promotion to expected benefits.
7. Design a written plan for implementation and review the plan with employers.
8. Execute the plan, paying careful attention to all details.
9. Evaluate the results against the planned objectives and make changes as necessary for future promotions.

There are numerous means by which promotion objectives can be met including coupons, contests, discounts, combination pricing, signs, special events, menu boards, and theme days. The illustrations shown in Figures 17.5 through 17.9 are examples of promotions used in various types of foodservice organizations.

BRANDING: THE HOTTEST MARKETING STRATEGY OF THE 1990S

Since the beginning of the 1990s, there has been an explosion in the use of *branding* and the *branded concept* as a marketing strategy, particularly in noncommercial foodservice operations. The catalyst behind this phenomenon is one of economics. All foodservice operations, and particularly those traditionally recognized as noncommercial, are under increasing pressure to operate "in the black." This pressure has resulted in a paradigm shift referred to as the commercialization of noncommercial operations. For example, many hospital cafeterias are no longer subsidized by their parent organization. Instead, these foodservice operations are expected to at least break even (e.g., generate

Figure 17.5 Examples of promotions to increase customer awareness of and participation in cafeteria options.
Courtesy of St. Joseph's Hospital, Marshfield, Wisconsin.

enough revenue to cover expenses) or actually generate a profit, thereby contributing to the financial well-being of the organization. Traditionally, noncommercial foodservices such as those in hospitals and schools have relied on familiar revenue-generating options such as catering, vending, and take-out foods to boost income. Branding and the branded concept began to emerge in the early 1990s and have continued to gain in popularity ever since. It is important for today's foodservice manager to understand what branding is, why it has the potential to work as a marketing strategy, the types and variations of branding, and management issues such as deciding on which branded items to use and how to avoid potential pitfalls of branding.

As a marketing strategy, **branding** refers to the use of nationally or locally labeled products for sale in an existing foodservice operation. In foodservice operations, branding is used specifically to increase sales through brand promotions that are designed to woo new customers and/or increase the average amount of individual transactions. In practice, branding takes on many forms. The most popular are retail-item, restaurant, and in-house branding.

Retail-item branding, also referred to as manufacturer's branding, has been used for years and simply refers to the sale of nationally recognized items in existing food-service operations. Examples include Skippy peanut butter in school lunch programs and Kraft salad dressings in hospital cafeterias. *Restaurant branding,* on the other hand, is a far more recent approach to branding and refers to the inclusion of a national restaurant chain in an existing operation. This approach may vary from the purchase of a franchise (such as a McDonald's in the lobby of a hospital) to contracting with a chain restaurant for periodic sales (such as offering Pizza Hut pizza in a school cafeteria once a month). Figures 17.10 and 17.11 are examples of how branding is used at a university residence hall. There is at least one variation to branding

PIZZARIFFIC
PIZZA
to Go
"Buy" The Slice
"Buy" The Box
Call 263-1526
To order whole pizza
Ready in 20 minutes
Selections vary daily
Triple Cheeese
Sausage & Mushroom
Garden Veggie
Pepperoni
Hawaiian
Seafood
Deluxe
Pizza served 10:30 a.m. - 6:30 p.m.
University Hospital Cafeteria
(C2)

Figure 17.6 Example of pizza promotion for a hospital cafeteria, designed to be hung on the doorknobs of offices.
Copyright 1992 Board of Regents, University of Wisconsin Systems, University Hospital and Clinics, all rights reserved.

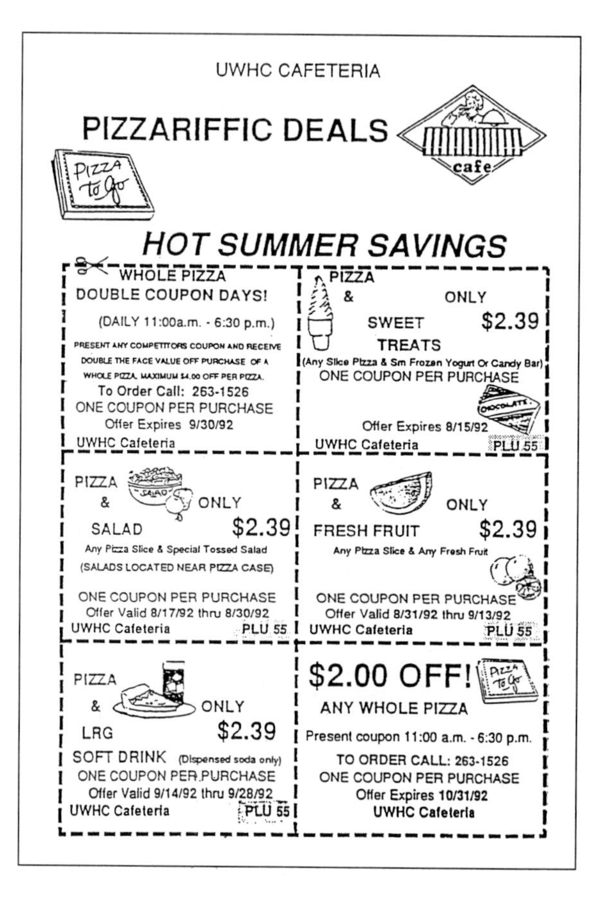

Figure 17.7 Example of pizza promotions for a hospital cafeteria. Copyright 1992 Board of Regents, University of Wisconsin Systems, University Hospital and Clinics, all rights reserved.

with nationally recognized chains and that is to contract with a popular local or regional brand. For example, schools in some Midwest states can contract with Rocky Rococco's Pizza in an effort to increase the variety of branded products and maintain interest in the school lunch program.

Another major type of branding is that of *in-house* or *signature branding.* Signature brands are items prepared within a specific foodservice operation and identified as unique to that operation. For example, a hospital cafeteria may sell a line of sandwiches that customers identify with that cafeteria and recognize for their consistent high quality. The foodservice, in turn, can take that line, develop it as an in-house brand, and promote it using a specially designed logo and other item-specific promotional materials. This approach is sometimes used to offset the potential for "cannibalization" of in-house items when national or local branding is introduced. The in-house brands are designed to compete side by side with the other brands for sales. Figure 17.12 is an example of in-house branding.

A more contemporary term related to branding is that of the **branded concept.** This term refers to a complete marketing package that communicates a recognized and consistent brand identity to the customer. This package is developed and made available by the commercial company and generally consists of two components.

UNIVERSITY OF WISCONSIN HOSPITAL AND CLINICS

Welcome

TO OUR CAFETERIA

Family, friends and guests of our patients
Students, Faculty and Staff

BREAKFAST	***LUNCH***	***DINNER***	***SNACKS***
6:00 - 9:45 A.M.	**11:00 - 2:00**	**4:30 - 7:00**	**ANYTIME**

Day	Meal	Menu
MONDAY 08/03/92	LUNCH:	Grilled Chicken Breast, Orange Roughy, Chicken Fajita, Vegetarian Lasagna with Garlic Bread, Kafta with Tahini Sauce
	DINNER:	Vegetarian Lasagna with Garlic Bread, Sweet & Sour Pork, Roast Whole Turkey, Baked Pollock, Egg Roll
TUESDAY 08/04/92	LUNCH:	Pork Cutlets, Chicken Cordon Bleu', Carved Steamship Round on a Kaiser Roll, Cheese Strata, Lemon Peppered Pollock
	DINNER:	Baked Lasagna, Carved Steamship Round on a Kaiser Roll, Chicken Breast Cordon Bleu', Walleyed Pike, Cheese Strata
WEDNESDAY 08/05/92	LUNCH:	Swiss Steak, Baked Sole, Chicken Breast Hoagie, Baked Sole, Italian Spaghetti, Vegetarian Spaghetti, Baked Potato with BBQ Roast Beef and Cheese
	DINNER:	Vegetarian Spaghetti, Carved Whole Baked Ham, Bacon Burger, Italian Spaghetti & Meat Sauce, Poorman's Lobster
THURSDAY 08/06/92	LUNCH	Baked Meatloaf, Baked Fish with Rice, Falafel, Broccoli Quiche, Beef Liver & Onions
	DINNER	Carved Roast Pork, Baked Chicken, Ranch Steak on Kaiser Roll, Broccoli Quiche
FRIDAY 08/07/92	LUNCH:	Carved Turkey, Pepper Steak Over Rice, Hot Turkey Ham and Cheese Sandwich, Eggroll with Sweet and Sour Sauce, Baked Potato with Broccoli, Cheese and Bacon, Vegetarian Chili, Fish Marinara
	DINNER:	Stir Fried Beef with Vegetables over Rice, Chicken Kiev, Hot Turkey Ham and Cheese Sandwich, Baked Sole, Carrot Mushroom Loaf
SATURDAY 08/08/92	LUNCH:	French Fried Perch, Carved Roast Turkey, Sliced Roast Pork, Manicotti with Garlic Bread, Super Dog
	DINNER:	Carved Roast Turkey, Manicotti, Sliced Roast Pork, Chimichanga
SUNDAY 08/09/92	LUNCH:	Carved Prime Rib of Beef, Baked Filet of Sole, Sliced Baked Ham Baked Potato with Broccoli & Cheese
	DINNER:	Carved Prime Rib of Beef, Baked Filet of Sole, Sliced Baked Ham Baked Potato with Broccoli & Cheese

TRY OUR NEW PIZZARIFFIC PIZZA BY THE SLICE OR WHOLE!!

(2792o)

WE FEATURE LIGHT CHOICES AT EVERY MEAL
GO FOR THE GOLD!
WIN WITH OUR UPCOMING
OLYMPIC SPECIALS

AWARDS

Figure 17.8 Example of a cafeteria menu flyer to increase customer awareness of cafeteria specials.

SHAPE Your Future!

SHAPE YOUR FUTURE with School Lunch
National School Lunch Week – Oct. 12–16, 1992

Skim, 2%, Whole Milk & Chocolate Milk Served Daily

OCT '92 LUNCH MENU

MONDAY	TUESDAY	WEDNESDAY	THURSDAY	FRIDAY
			1 Oven Baked Chicken Fluffy White Rice/Gravy Whole Kernel Corn Choice of Bread/Butter Chilled Peach Slices	2 Cardinal Burger/Bun w/Red & White Toppings Football Fries Victory Green Beans Favorite Fruit Medley Winning Rice Krispie Bar HOMECOMING
5 Chicken Nuggets w/Sauces Macaroni w/ Cheese Sauce Broccoli Cuts Choice of Fresh Fruit	6 Italian Dunkers* (Cheese Bread w/ Meat Sauce) Tossed Salad w/ Choice of Dressing Diced Pears Sugar Cookie	7 French Toast w/ Warm Syrup Little Smokies Choice of Juice Fruit Cup w/Melon Iced Cinnamon Roll YOM KIPPUR	8 Roast Turkey/Gravy Snowy Mashed Potatoes Wisconsin Blend Veggies Dinner Roll/Butter Fruited Jello Square	9 Pepperoni Pizza or Cheese Pizza Fresh Vegetable Sticks w/ Reduced Calorie Dip Pineapple Tidbits
12 DISCOVER ETHNIC FOODS Burrito* Ole w/Queso Lechuga, Tomate Frijoles (Refried Beans) Centennial Bread Applesauce for 500 Italian Ice Discovery COLUMBUS DAY (Observed)	13 Build-Your-Own Burger Cheese Slice, Lett., Tom. Straight Cut Fries Fruit Cocktail Shapes Future Fortune Cookie	14 NATIONAL SCHOOL LUNCH WEEK (October 12-16) Grilled Unbreaded Chicken Breast Patty/Bun Lettuce Leaf & Tomato Sl. Seasoned Green Beans Baked Fruit Bar	15 Curley Noodles w/ Italian Meat* Sauce Bread Stick Confetti Salad/Dressings Peach Smiles PENCIL DAY – Elem. Schools	16 WORLD SERIES TAILGATE Home Run Turkey or Beef Stacker/Bun Cheese Slice/Lettuce Leaf Curve Ball Fries Catch of Fresh Fruit BASEBALL CARD DAY
19 Cheese & Sausage Pizza Fresh Veggies w/ Ranch or Yogurt Dip Mixed Fruit	20 Chicken Nuggets/Sauces Toasty Tritator Green Beans Bread/Butter Strawberry Shortcake	21 Meat & Cheese Lasagna* Garlic Bread Tossed Salad w/ Choice of Dressings Pineapple Bites	22 Sloppy Joe* /Bun Tator Tots Tiny Green Peas Pear Halves TELL THE COOKS YOUR NAME TREAT: Ice Cream	23 Ham & Cheese Melt Waffle Cut Fries Golden Cut Corn Choice of Fresh Fruit
26 Hotdog or Brat/Bun Crinkle Cut Fries Bubbly Baked Beans Cinnamon Apple Crisp	27 Breaded Chicken Patty on Bun w/Lettuce Leaf Tator Tots Broccoli & Cauliflower Choice of Fruit	28 NO SCHOOL PROFESSIONAL DEV./ WORK DAY	29 NO SCHOOL TEACHER'S CONVENTION	30 NO SCHOOL TEACHER'S CONVENTION HALLOWEEN (Saturday, 31st)

* Items may contain ground turkey or pork as well as ground beef

All menus subject to change

Figure 17.9 Example of a school lunch menu designed to promote National School Lunch Week. Copies of the menu are sent to student homes.
Courtesy of Sun Prairie Area School District, Sun Prairie, Wisconsin.

Figure 17.10 Burger King Express™ is one of the national brands featured in this student center's food court.
Courtesy of University of Missouri-Columbia.

The first is the entire point-of-sale environment, which includes all of the materials with logos that are used to promote a specific product or line of products. Examples include product packaging, signage (including menu boards), staff uniforms, table tents, flyers, and so on. The second component of the branded concept refers to the management resources made available through the commercial company. Resources include purchasing assistance, production tools, such as recipes, and service suggestions. Commercial companies currently participating in this practice include Pizza Hut, Burger King, Taco Bell, Chic-fil-a, and McDonald's.

The success of banding and the banded concept is based on the premise that customers are willing to pay more for a branded product. Part of this success can be attributed to the fact that brands are recognized and trusted. More important, however, relative to the growth of branding in today's operations, is the psychological phenomenon that customers are willing to pay for *perceived* quality and value. So even though a product prepared by a foodservice may be bigger, better, and less expensive, consumers will still prefer and pay more for a brand they recognize and trust.

It follows then, that with careful planning and implementation, a foodservice manager can anticipate the following advantages and outcomes from branding:

- Increased volume of business,
- Increased per-capita spending resulting in higher average receipts;
- Increased revenue; and
- Increased cross-over traffic to in-house brands (i.e., customers are attracted to the branded items but will also purchase in-house items).

Each of these advantages contributes to the ultimate goal of improved customer satisfaction.

Branding, with all of its attractive advantages, does not, however, come without its potential pitfalls. Contracts and agreements must be carefully negotiated to ensure

Figure 17.11 Franchised kiosks for Dunkin Donuts™ and Otis Spunkmeyer™ cookies.
Courtesy of University of Missouri-Columbia.

Figure 17.12 The Caffé Fresco coffee shop is an example of an in-house brand.
Courtesy of University of Missouri-Columbia.

that responsibilities and obligations are clearly understood by all parties involved. Pricing of branded items must be carefully considered because customers are well aware of the street prices for popular, national brands. Managers may also have to respond to employee fear of contracting with commercial companies. Depending on the degree to which branding is incorporated into an existing foodservice, it may mean a reduction of in-house staff.

To offset these risks, foodservice managers should carefully study branding and the branded concept before signing a contract or agreement with a commercial company. The decision-making process of brand selection should begin with careful evaluation of the target market of existing and potential customers. Actual and perceived value as defined by the customer must be clearly understood. From there, the selection process should focus on products with the potential for the greatest gross profit. The overall financial investment must be carefully considered. For example, implementing branding and the branded concept in an existing cafeteria may require additional operations or facility renovation. These and other potential investments must be carefully weighed against the potential for increased revenue and other less tangible advantages such as improved customer satisfaction.

SUMMARY

Marketing has been an essential function of commercial foodservice operations for a long time. In recent years, managers of noncommercial operations (e.g., hospitals, long-term care facilities, schools, and universities) have recognized the value of marketing principles as a means to survive in a highly competitive industry with limited resources and ever-increasing costs.

Foodservice managers today must have a sound knowledge base of marketing terminology, the marketing cycle, the marketing concept, and the unique aspects of marketing in a foodservice in order to successfully implement a program. Implementation includes carefully planned training for foodservice employees and an evaluation strategy to assess the degree to which the program objectives were achieved.

Merchandising and promotions are very important activities related to a marketing program. The foodservice manager must become familiar with promotion design, planning procedure, implementation, and evaluation strategies as a means to gain the competitive edge and retain or attract new customers to the foodservice operation.

REVIEW QUESTIONS

1. What is marketing and why is it referred to as a cyclical management function?
2. What is a target market? How can it be identified and segmented?
3. What are the elements of a marketing mix and what factors influence this mix in a given foodservice operation?
4. Differentiate between tangible and intangible components of a foodservice product.

5. What is the difference between a share-of-mind and a share-of-market promotion?
6. Define branding and the branded concept.
7. Why is branding so popular today, particularly in noncommercial foodservices?
8. Define and give examples of the major types of branding.

SELECTED REFERENCES

Boss, D., and Schechter, M.: The brand explosion, part I. Food Mgmt. 1991; 26:90–93.

Boss, D., Schechter, M., and Cohen, M.: The brand explosion, part II. The boom goes on. Food Mgmt. 1991; 26:134–154.

Boss, D., and White, J.: Targeting the trends: Menus, nutrition, and branding. Food Mgmt. October 1993; 28:96–99, 102–104.

Carangle, C.: Branding for profitability. Food Mgmt. 1995; 30:78.

DeBurgh, R.: Firsthand strategies for branding success. School Foodservice Nutr. 1996; 30:34–36, 38, 40–42.

DiDomenico, P.: Promoting Columbus Day. Restaurants USA. 1992; 12:10–13.

Dollar, F.W.: Marketing: A constant commitment. Food Mgmt. 1985; 20:51–52.

Erffmeyer, R.C.: Incorporation of the marketing concept. Application in university contract food service. J. Am. Diet. Assoc. 1987; 87:1215–1216.

Helm, K. K.: *The Competitive Edge: Advanced Marketing for Dietetic Professionals,* 2nd ed. Chicago, Ill.: The American Dietetic Association, 1995.

King, P.: The first foodservice branding conference. Food Mgmt. 1995; 30:68–69, 72, 74, 76, 78, 82–83, 86, 88–89, 92, 94–95.

Kotler, P., and Armstrong, G.: *Marketing: An Introduction.* Englewood Cliffs, N.J.: Prentice Hall, 1987.

Lovelock, C. H.: *Services Marketing,* 2nd ed. Englewood Cliffs, N.J.: Prentice Hall, 1991.

McCabe, G.: Anatomy of a promotion. Food Manager. 1991; 26:74.

Morrison, S.: A trip down logo lane. School Foodservice Nutr. 1996; 50:22–32.

Pickens, C. W., and Shanklin, C. W.: State of the art in marketing hospital foodservice departments. J. Am. Diet. Assoc. 1985; 85:1474–1478.

Reid, R.D.: *Foodservice and Restaurant Marketing.* New York: Van Nostrand Reinhold Company, 1983.

Sagowitz, S.: Anticipating customer demands. Food Mgmt. 1985; 21:67–68.

———: Reach out and get to know your customers. Food Mgmt. 1985; 2:33–34.

Smith, D. I.: Merchandising: Influencing the customer purchase decision. In R. A. Brymer, ed., *Hospitality Management* (pp. 255–268). Dubuque, Iowa: Kendall/Hunt Publishing Company, 1991.

Spertzel, J. K.: Defusing branding bombs. School Foodservice Nutr. 1996; 50:44–46, 50–51.

Underwood, R. F.: *52 Cafeteria Promotions That Really Work.* Gaithersburg, Md.: Aspen Publishers, 1988.

———: Marketing and revenue-producing projects. In K. E. Lawrence and S. N. DiLima, eds., *Hospital Nutrition and Food Service: Forms Checklists and Guidelines* (pp. 4:1–4:2). Gaithersburg, Md.: Aspen Publishers, 1992.

Appendix A: Principles of Basic Cooking

OBJECTIVES OF COOKING

The basic objectives of cooking are to enhance the flavor of food and the attractiveness of the original color, form, and texture; to destroy harmful organisms and substances to ensure that food is safe for human consumption; and to improve digestibility.

HEAT TRANSFER

Cooking is accomplished by the transfer of heat from an energy source to and through the food by one of three methods: conduction, convection, or radiation.

Conduction is the transfer of heat through direct contact from one item or substance to another. The heat is first transferred from a heat source, usually gas or electricity, through a cooking vessel to the food. Conduction is the principal means of heat transfer in grilling, frying, boiling, and, to some extent, in roasting and baking. In pan broiling or grilling, for example, the heat is transferred from the source to the pan or grill and then directly to the food. In panfrying, fat is the transfer agent between the pan and the food.

Convection, in which heat is spread by the movement of air, steam, or liquid, or *radiation,* in which energy is transferred by waves from the source to the food, are additional means of heat transfer. These methods usually involve some intermediary, such as the air in the oven; the water, steam, or fat that may surround the food and/or container; or the metal, glass, or ceramic of which the food container is composed. In convection ovens, for example, heat is circulated by fans, thus transferring the heat more quickly to the food. In a microwave, radiation is used in food production. The radiation waves do not possess energy, but upon entering the food they induce heat by molecular movement.

BASIC COOKING METHODS

Cooking methods are classified as dry heat or moist heat. *Dry heat* methods are those in which the heat is conducted to the food by dry air, hot metal, radiation, or in a minimum amount of fat. Roasting, baking, broiling, grilling, griddling, and frying are examples of dry heat methods. *Moist heat* methods are those in which the heat is conducted to the food product by water or steam. Examples are boiling or simmering, stewing, blanching, poaching, braising, and steaming.

The method used depends on the type and quality of the food and availability of equipment. Different cooking methods are suitable for different kinds of food. For example, tender cuts of meat usually are prepared using a dry heat method, whereas a less tender cut should be cooked using moist heat. The following is a summary of common cooking methods.

Baking. Cooking by dry heat, usually in an oven.

Time and temperature: Baking temperature is determined by type of food and equipment. Oven load and size of containers must be considered when figuring required cooking time.

Equipment: Oven (deck, revolving, conveyor, range, or convection). A convection oven reduces cooking time by 10% to 15% and cooks at a temperature 25°F to 50°F lower than a traditional or conventional oven.

Barbecuing. Cooking on a grill or spit over hot coals, or in an oven, basting intermittently with a highly seasoned sauce.

Typical products: Meat, poultry.

Blanching. Cooking a food item partially and briefly. Food usually is blanched in water, although some foods, such as french fries, are blanched in hot fat. To blanch in water, the food is placed in rapidly boiling water and held until the water returns to a boil, then quickly cooled in cold water.

Equipment: Steam-jacketed or other type of kettle.

Typical products: Vegetables or fruits; to set the color and destroy enzymes, or to loosen skins for easier peeling.

Boiling. Cooking food in a liquid that is bubbling rapidly.

Time and temperature: The temperature of boiling water is 212°F at sea level. This point is raised by the presence of solids in the water and lowered by higher altitudes. At 5000 feet, water boils at about 203°F; therefore, it takes *longer* to boil foods at high altitudes because of the lower boiling temperature.

Equipment: Steam-jacketed kettle, stock pot, or other kettle on top of the range.

Typical products: Vegetables, pasta, cereals, rice. Not generally used for high protein foods (meat, fish, or eggs) because heat toughens the protein, and the rapid movement of boiling breaks delicate foods. This type of food is usually brought to a boil, then the heat is reduced to simmering temperature for the rest of the cooking period.

Braising. A method of cooking that combines cooking in fat with the addition of moisture. Food is browned in a small quantity of fat, then cooked slowly in liquid in a covered utensil.

Equipment: Tilting frypan or steam-jacketed kettle. For smaller amounts, a skillet or wok may be used. After the moisture is added to the browned food, the product may also be finished in the oven at a low temperature.

Typical products: Meats, poultry.

Broiling. Cooking by radiant heat. Food is placed on a rack either below or between the gas or electric heat source. The rack is positioned 3 to 6 inches from the heat source, depending on the type and intensity of the heat. The temperature required depends on the thickness of the food. Traditional broilers lack precise temperature controls and food must be closely monitored during cooking. In *panbroiling,* food is cooked without fat in a frypan. If fat is not poured off as it accumulates, the process becomes panfrying.

Equipment: Specially designed broilers for institutional use.

Typical products: Tender cuts of meat (steaks, chops), fish, poultry.

Deep-Fat Frying. Cooking by submerging food in hot fat. In this type of cooking, some of the medium becomes part of the food during the cooking process, and it is not unusual for foods to absorb 10% to 20% fat during frying. Foods may be dipped in a breading or batter before frying to form a protective coating between food and fat and to give the product crispness, color, and flavor. A well-prepared deep-fat fried food should have minimum fat absorption, an attractive golden color, a crisp surface or coating, and no off-flavors imparted by the frying fat. For the production of a high-quality product, use a good quality fat with a high smoke point, avoid overloading the baskets, fry at proper temperatures, and avoid frying strongly and mildly flavored foods in the same fat. About 10% to 30% of the fat should be replaced with fresh fat before each daily use. Modern fryers are equipped with automatic basket lifts, fat temperature sensors, and computerized timers to aid in the consistent production of good fried foods. Thermostatic control and fast recovery of fat temperature permits the rapid production of consistent quality fried foods.

Equipment: Deep-fat fryer. Pressure fryers and convection fryers are recent developments that reduce the frying time, thus enabling large-volume foodservices to produce fried foods more rapidly. Pressurized fryers are used most frequently in foodservice operations specializing in fried chicken.

Typical products: Fish, shellfish, chicken, vegetables, meat.

Fricasseeing. Browning in fat, then simmering in gravy. Similar to braising, but moisture is in the form of gravy rather than water or other liquid.

Equipment: Tilting frypan, steam-jacketed kettle.

Typical products: Chicken, some meat cuts.

Frying. Cooking in fat or oil. See Deep-Fat Frying, Ovenfrying, Panfrying, and Sautéing for frying methods.

Griddling. Cooking on a solid cooking surface. Food is placed on a flat surface and cooked with or without a small amount of fat.

Equipment: Griddle.

Temperature: The temperature on a griddle is adjustable and is lower than on a grill (usually around 350°F).

Typical products: Meat, eggs, pancakes, sandwiches.

Grilling. Cooking on an open grid over a heat source.

Equipment: Grill, with heat source of charcoal, an electric element, or gas-heated element.

Typical products: Meat, fish.

Ovenfrying. High-temperature cooking in an oven. Food is placed on greased baking sheets, melted fat brushed or drizzled over it, and baked in a hot oven (usually 400°F to 450°F.) The resulting product resembles fried or sautéed food. This method is used when deep-fat fryers are not available or are inadequate to handle the production demand, usually in large health care facilities or other noncommercial foodservice operations.

Typical product: Chicken pieces, fish fillets.

Panfrying. Cooking in a moderate amount of fat in a pan over moderate heat. The amount of fat depends on the food being cooked. Only a small amount is used for eggs, whereas more may be needed for panfried chicken. Most foods must be turned at least once for even cooking. Some larger foods may be removed from the pan and finished in the oven, to prevent excessive surface browning.

Equipment: Skillet, tilting frypan.

Typical products: Meat, chicken, eggs, potatoes, onions.

Poaching. Cooking by immersing food in hot liquid maintained at simmering temperature. Food is added to hot liquid and simmered, keeping the temperature below boiling.

Equipment: Tilting frypan, steamer, oven, or a shallow pan on the range top.

Typical products: Fish, eggs out of the shell, fruit.

Roasting. Baking of meat or poultry (uncovered) in an oven. Cooking uncovered is essential to roasting because if the meat or poultry is covered with a lid or aluminum foil the steam is held in, changing the process from dry-heat to moist-heat cooking.

Equipment: Oven (range, deck, revolving, range, or convection).

Typical products: Poultry, tender cuts of beef, pork, lamb, or veal.

Sautéing. Cooking quickly in a small amount of fat. Food is placed in a preheated skillet with a small amount of fat and cooked quickly. Food should be cut or pounded to an even thickness and not overcrowded in the pan. After a food is sautéed, wine or stock is frequently added to dissolve brown bits of food sticking to the sides or bottom of the pan, a process called deglazing. Generally, this liquid is served with the sautéed item.

Equipment: Skillet, tilting frypan.

Typical products: Poultry, fish fillets, tender cuts of meat.

Searing. Browning of food in fat over high heat before finishing by another method.

Equipment: Skillet, tilting frypan.

Typical product: Meat.

Simmering. Cooking of food in liquid below the boiling point. Liquid should be kept at a temperature ranging from 185°F to just below the boiling point. Most foods cooked in a liquid are simmered because the higher temperatures and intense bubbling of boiling may be detrimental to the texture and appearance of the food. Food may be brought to the boiling point first, then the heat reduced for the rest of the cooking period.

Equipment: Steam-jacketed kettle, stock pot, or other kettle on top of the range.

Typical products: Soups, sauces, meat, poultry.

Steaming. Cooking by exposing foods directly to steam. Steam cookers provide for controlled cooking in constant temperatures. Food properly cooked in a pressure steamer is evenly cooked, retains a high vitamin content as well as its natural color and flavor, and suffers less of the usual cooking losses, such as shrinkage caused by prolonged cooking, boiling over, or burning. Pressure steaming is an extremely rapid method of cooking and must be carefully controlled and timed.

Equipment: Steamers may be low pressure, high pressure, or zero pressure. High-pressure steam cookers are used primarily for fast cooking and small batch cookery of vegetables. In a zero convection, a fan circulates the steam throughout the steamer cavity. Most steamers are designed to accommodate standard-size pans that can be used directly on the serving counter. Steam is the source of heat in jacketed kettles, but is not the cooking medium. The heat is transferred through the walls of the inner lining of the kettle by conduction, but no contact between food and steam is possible. The temperature is higher than in a double boiler because the steam is under pressure. The temperature increases with increase in pressure.

Typical products: Vegetables, fruits, poultry, dumplings, pasta, rice, cereals.

Stewing. Cooking in a small amount of water, which may be either boiling or simmering. Whether a food is to be simmered, stewed, or boiled, the liquid usually is brought to a full boil first. This compensates for the lowering of the temperature when the food is added. The heat is then adjusted to maintain a steady temperature.

Equipment: Steam-jacketed kettle, tilting frypan, kettle on top of the range.

Typical products: Meat, poultry, fruit.

Stir-Frying. Cooking quickly in a small amount of oil over high heat. Food is cut into uniform strips or small pieces and cooked quickly in a small amount of oil. A light tossing and stirring motion is used to preserve the shape of the food.

Equipment: Tilting frypan, skillet, wok.

Typical products: Vegetables, chicken, pork, tender beef, or shrimp.

COOKING METHODS FOR SPECIFIC FOODS

Most foods are cooked by one of the basic methods described in the previous section. Broiling, frying, baking, and simmering are standard processes in food preparation, but there may be variations in exact procedures because of the type of product, available equipment and personnel, and the size and character of the operation. The following information is given to augment the basic definitions previously provided.

Meat

Cooking Methods. Depends on the quality and cut of meat and the quantity that must be prepared at one time. For *beef,* dry heat (broiling, roasting) is generally used for tender cuts, and moist heat cookery (braising, stewing, simmering) for the less tender cuts. Table A.1 suggests methods of cooking appropriate for various cuts and grades of beef. Lower grades of meat and the less tender cuts of higher quality beef may be tenderized by scoring, cubing, grinding, or by the addition of enzymes. Adding tomatoes or vinegar to a meat mixture also has a tenderizing effect. For *veal, pork,* and *lamb,* practically any cut but the shank may be cooked by dry heat, although broiling is not as desirable for pork or veal as it is for lamb or beef. Veal, because of its delicate flavor and lack of fat in the tissues, combines well with sauces and other foods.

Roasting. The time-weight relationship expressed in minutes per pound can be used as a guide, but the most accurate way to determine the doneness of a roast is with a meat thermometer that registers the internal temperature. For ease in roasting and handling, the roasts to be cooked together should be uniform in size. Frozen meat generally is thawed in the refrigerator before cooking to reduce both time and heavy drip losses during preparation. Roasts will continue cooking for a period of time after removal from the oven, and the internal temperature of the roast may rise as much as 5°. The roast should be allowed to sit or rest in a warm place for 15 to 20 minutes before it is carved. The roast becomes more firm, retains more of its juices, and is easier to carve. When cooking a number of roasts, it is possible to offer meat at different stages of doneness by staggering the times that roasts are placed in the oven. The well-done roasts are started first and, when done, are removed from the oven, allowed to stand 20 minutes, sliced, and placed in pans in the warmer or in the oven at low heat. The rare meat is put in the oven last and, when the thermometer reaches 125°F, is removed from the oven, sliced, and sent directly to the serving area. For optimum quality, roasts are cooked and sliced just prior to serving. However, this may not be possible in some foodservice operations. If the meat must be cooked the day before or several hours prior to serving, the quality is better if the cooked roasts are stored in the refrigerator, then sliced and reheated before serving rather than refrigerating and reheating the sliced meat.

Time/Temperature. Yield is an important factor in the cooking of meat. Reduction in the yield of cooked meat may occur as cooking losses or as carving or serving losses. Shrinkage during cooking usually is the major loss involved and it may range from 15% to 30%. Some shrinkage

Table A.1 Suggested cooking guide for beef.

Cut	Prime	Choice	Select
Top round and sirloin tip (steaks and roasts)	Braise, broil, panfry, pot roast, or roast	Braise, broil, panfry, pot roast, or roast	Braise, panfry, pot roast, or roast
Bottom round (steaks and roasts)	Braise, panfry, pot roast, or roast	Braise, panfry, pot roast, or roast	Braise, pot roast, or roast
Rump roast	Roast or pot roast	Roast or pot roast	Roast or pot roast
Sirloin (steaks and roasts)	Broil, panfry, or roast	Broil, panfry, or roast	Broil, panfry, or roast
Porterhouse, T-bone, club, and rib steaks	Broil	Broil	Broil
Rib roast	Roast	Roast	Roast
Chuck, round bone, and blade (roasts and steaks)	Roast, pot roast, or braise	Roast, pot roast, or braise	Roast, pot roast, or braise

This guide lists the most generally accepted method of cooking retail beef cuts of each grade. Flank, plate, brisket, foreshank, and the heel of the round should be prepared in the same manner for all grades of beef. These less tender cuts are used for stewing, braising, pot roasting, or boiling, or are ground for use in meat loaves and similar dishes.

Source: "U.S. Grades of Beef," Marketing Bulletin No. 15, U.S. Department of Agriculture, Washington, D.C.

occurs regardless of the cooking method, but the cooking temperature and the cooking time have a direct bearing on the amount of shrinkage. Low temperatures generally result in fewer cooking losses and the most palatable product. If cooked too long, meat dries out and tends to be less tender. Even meat that requires moist heat and a comparatively long cooking time to become tender will be less tender when overcooked.

Poultry

Cooking Methods. Broiling (if not too large), panfrying, deep-fat frying, ovenfrying, roasting, barbecuing, fricasseeing, stewing, broasting.

Time/Temperature. Moderate heat for tender, juicy, and evenly done meat. High temperatures result in stringy, tough, and unappetizing meat. When roasting, use a thermometer placed in the thickest part of the breast or inner part of the thigh muscle. Make sure the bulb does not rest against a bone. Temperature should reach 170°F to 180°F. Poultry usually is cooked well done, but overcooking results in loss of juiciness. Many foodservices prefer to purchase boneless turkey roasts or rolls for convenience in roasting, slicing, and portion control. Cooking time for ready-to-cook rolls is longer in minutes per pound than for whole turkeys, but the total cooking time is less. (See Table A.2 for a timetable for roasting turkey.) Stuffing or dressing should be baked separately and served with the roasted sliced meat.

Fish

Thawing. Frozen fish steaks or fillets need not be thawed prior to cooking unless they are to be breaded, but any defrosting should be at refrigerator temperature and only long enough to permit ease in preparation. Thawing at room temperature is not recommended. Once thawed, fish should be cooked immediately and never refrozen. Frozen breaded fish portions should not be thawed before cooking.

Cooking Methods. Deep-fat frying, panfrying, ovenfrying, broiling, baking, poaching. The best cooking method is determined by size, fat content, and flavor. Baking and broiling are suitable for fat

Table A.2 Timetable for roasting turkey.

Form of Turkey	Weight of Turkey, Whole or Piece (lb)	Oven Temperature (°F)	Cooking[a] Time (hr)
Turkey, whole, ready-to-cook	12 to 16	325	3½ to 4½
	16 to 21	325	4½ to 6
	21 to 26	325	6 to 7½
Turkey parts			
Breast	8 to 12	325	3 to 4
Leg (drumstick and thigh)	3 to 8	325	1¾ to 3
Halves	8 to 12	325	2¼ to 4
Quarters	3 to 8	325	1½ to 3½
Turkey roasts, boneless			
Frozen	9½ to 11½	350	4 to 4½
Thawed	9½ to 11½	350	2¾ to 3½

[a]Turkey is done when thermometer registers 180° to 185°F in inner thigh or 170° to 180°F in breast of whole turkeys; or 170°F in turkey roasts

Source: Adapted from "Quantity Recipes for Type A School Lunches," PA-631, U.S. Department of Agriculture, Washington, D.C., revised 1971.

fish, such as salmon, trout, and whitefish. If lean fish is baked or broiled, fat is added to prevent dryness, and it often is baked in a sauce. Lean fish, such as haddock, halibut, and sea bass, are often poached, simmered, or steamed, although they may be broiled or baked if basted frequently. Fish cooked in moist heat (poaching) requires very little cooking time and usually is served with a sauce. Frying is suitable for all types, but those with firm flesh that will not break apart easily are best for deep-fat frying. Whatever the method, fish should be served as quickly as possible after cooking for optimum quality. See Table A.3 for suggested cooking methods.

Equipment. Deep-fat frying, skillet, tilting frypan, oven, steamer.

Time-Temperature. Low to moderate heat. Allow extra time for frozen fish that is not defrosted. Fish should be cooked only until the flesh is easily separated from the bones.

Eggs

Cooking Methods. Poaching, frying, scrambling, and cooking in the shell.

Poaching. Cooked to order in a shallow pan of hot water on top of the range, or in quantity, a counter pan deep enough to permit 2 to 2½ inches of water to cover the eggs is used. Eggs should be broken onto saucers and slid into the water toward the side of the pan. The water should be simmering when the eggs are added. The addition of 2 tablespoons of vinegar and 1 tablespoon of salt to 1 gallon of water prevents whites from spreading. Poaching may be done in a shallow pan on the range top or in the oven, tilting frypan, or steamer.

Frying. Usually cooked to order in a frypan or on a griddle. Eggs cooked on a griddle are more apt to spread than those cooked in a small frypan, and so are less attractive. To prevent toughening, eggs should not be fried at a high temperature.

Scrambling. Cooked to order in a frypan or griddle, or in quantity in the steamer or oven. The addition of milk to the eggs keeps them from drying, and medium white sauce added in place of milk prevents the eggs from separating and becoming watery when held on the serving counter.

Cooked in the Shell. In a pan on top of the range, an automatic egg cooker, a wire basket in the steam-jacketed kettle, or in a steamer. If eggs are brought to room temperature before cooking, the shells will not crack when heat is applied.

Time/Temperature. Avoid high temperatures and long cooking times. Eggs should be cooked as close to service as possible or cooked to order.

Pasta and Cereals

Pasta. Pasta is cooked uncovered in a large amount of boiling water in a stock pot or steam-jacketed kettle until tender but firm (*al dente*), then rinsed with cold or hot water and drained. If pasta is to be combined with other ingredients in a casserole, it should be undercooked slightly. If the pasta is not to be served immediately, it may be drained and covered with cold water. When pasta is cold, drain off water and toss lightly with a little salad oil. This will keep pasta from sticking or drying out. Cover tightly and store in the refrigerator. To reheat, place pasta in a colander and immerse in rapidly boiling water just long enough to heat through; or reheat in a microwave oven.

Rice. Rice is cooked in a steamer, the oven, or by boiling. It is cooked until all of the water is absorbed, so the right proportion of rice to water and the correct cooking time are important. Converted or parboiled long-grain white rice requires slightly more water and a longer cooking time than does regular long-grain or medium-grain rice. The cooking time for brown rice is almost double that of white rice.

Cereals. Cereals in quantity are generally cooked in a steam-jacketed kettle or steamer, but they may be prepared in a heavy kettle on top of

Table A.3 Suggested methods of cooking fish and shellfish.

		Baking		Broiling		Boiling, Poaching, or Steaming		Deep-Fat Frying		Panfrying	
Species	**Approximate Weight or Thickness**	**Temperature (°F)**	**Minutes**	**Distance from Heat (in.)**	**Minutes**	**Method**	**Minutes (per lb)**	**Temperature (°F)**	**Minutes**	**Temperature**	**Minutes**
Fish											
Dressed	3 to 4 lb	350	40 to 60			Poach	10	325 to 350	4 to 6		
Pan dressed	½ to 1 lb	350	25 to 30	3	10 to 15	Poach	10	350 to 375	2 to 4	Moderate	10 to 15
Steaks	½ to 1¼ in	350	25 to 35	3	10 to 15	Poach	10	350 to 375	2 to 4	Moderate	10 to 15
Fillets		350	25 to 35	3	8 to 15	Poach	10	350 to 375	2 to 4	Moderate	8 to 10
Portions	1 to 6 oz	350	30 to 40					350	4	Moderate	8 to 10
Sticks	¾ to 1¼ oz	400	15 to 20					350	3	Moderate	8 to 10
Shellfish											
Clams—											
live, shucked		450	12 to 15	4	5 to 8	Steam	5 to 10	350	2 to 3	Moderate	4 to 5
Crabs—											
live, soft-shell				4	8 to 10	Boil	10 to 15	375	2 to 4	Moderate	8 to 10
Lobsters—live	¾ to 1 lb	400	15 to 20	4	12 to 15	Boil	15 to 20	350	2 to 4	Moderate	8 to 10
Spiny lobster	¼ to ½ lb										
tails—frozen		450	20 to 30	4	8 to 12	Boil	10 to 15	350	3 to 5	Moderate	8 to 10
Oysters—											
live, shucked		450	12 to 15	4	5 to 8	Steam	5 to 10	350	2 to 3	Moderate	4 to 5
Scallops—											
shucked		350	25 to 30	3	6 to 8	Boil	3 to 4	350	2 to 3	Moderate	4 to 6
Shrimp											
Headless, raw						Boil	3 to 5				
Headless, raw, peeled		350	20 to 25	3	8 to 10	Boil	3 to 5	350	2 to 3	Moderate	8 to 10

Source: "How to Eye and Buy Seafood," National Marine Fisheries Service, U.S. Department of Commerce, Washington, D.C., 1970.

the range. Add cereal and salt to boiling water, using a wire whip. Stir until some thickening is apparent, then reduce the heat and cook until cereal reaches desired consistency and the raw starch taste has disappeared. Cereal should be thick and creamy but not sticky. Overstirring or overcooking produces a sticky, gummy product.

Fruits and Vegetables

Pre-preparation. Fresh fruits should be washed to remove surface soil, sprays, and preservatives before they are served raw or cooked. Apples, bananas, and peaches discolor rapidly after peeling so they should be immersed in pineapple, orange, or diluted lemon juice. Fruits also may be treated with ascorbic acid or other preparations that prevent oxidation. Berries deteriorate rapidly, so washing should be scheduled as near service as possible. A small amount of sugar sprinkled over the berries after cleaning keeps them fresh-looking.

Fresh vegetables should be washed, trimmed, peeled if necessary, and cut into even-sized pieces for cooking. Preparing fresh vegetables too far in advance causes them to discolor. Covering prepared vegetables with cold water helps retain color but reduces their nutritive value if they are held too long. Many foodservice operations have taken the preliminary preparation of fruits and vegetables out of the individual kitchens; they either centralize this function or buy convenience products that have some or all of the pre-preparation completed. Peeled potatoes and carrots, washed spinach and other leafy vegetables, cut vegetables ready for cooking, and peeled and sectioned citrus fruits and fresh pineapple are examples.

Cooking Methods. Steaming, boiling, baking, frying.

Equipment. Steamer, steam-jacketed kettle, tilting frypan, kettle on top of the range.

Whatever method is used, vegetables should be cooked in as small a quantity at one time as is feasible for the type of service. The needs of most foodservices can be met by the continuous cooking of vegetables in small quantities. Vegetables should be served as soon as possible after cooking for optimum quality and should be handled carefully to prevent breaking or mashing. Appearance is important to customer acceptance of vegetables, as is the seasoning. Frozen vegetables are cooked by the same methods used for fresh vegetables, but because frozen vegetables have been partially cooked, the final cooking time is shorter than for fresh products. Most frozen vegetables do not need to be thawed; they can be cooked from the frozen state and placed directly into steamer pans or boiling salted water. Exceptions are vegetables that are frozen into a solid block, such as spinach and winter squash. Results are more satisfactory if they are thawed in the refrigerator first for more even cooking.

To *steam,* place prepared vegetables not more than 3 to 4 inches deep in stainless steel inset pans. Perforated pans provide the best circulation, but if cooking liquid needs to be retained, use solid pans. To *boil,* add vegetables to boiling salted water in a steam-jacketed kettle or stockpot, in lots no larger than 10 pounds. The amount of water used in cooking all vegetables is important for retention of nutrients. The less water used, the more nutrients retained. Addition of baking soda to the water also causes loss of vitamins. Mature root vegetables that need longer cooking require more water than young, tender vegetables. Spinach and other greens need only the water clinging to their leaves from washing. Cover and bring water quickly back to the boiling point. Green vegetables retain their color better if the lid is removed just before boiling begins; strong-flavored vegetables, such as cabbage, cauliflower, and Brussels sprouts, should be cooked uncovered to prevent development of unpleasant flavors. To *stir-fry,* cut vegetables diagonally or into small uniform pieces. Heat a small amount of oil in a wok, tilting frypan, or steam-jacketed kettle. Cook and stir until vegetables are coated with oil. A small amount of liquid is usually added and the vegetables cooked, covered, until tender but crisp.

One of the main purposes of cooking vegetables is to change the texture. During the cooking process, however, the color and flavor may be altered and some loss of nutrients may occur. The degree to which these characteristics change determines the quality of the cooked vegetables. Many factors affect cooking time, including the type and maturity of the vegetable, the presence of acids, size of the pieces, and the degree of doneness. Vegetables are considered done when they have reached the desired degree of tenderness. The starch in vegetables also affects texture. Dry starchy foods, such as beans or lentils, must be cooked in enough water so that the starch granules can absorb moisture and soften. Most starchy vegetables, such as potatoes and yams, have enough moisture of their own, but they must still be cooked until the starch granules soften. Color is a major factor in consumer acceptance of vegetables, so methods of cookery that retain color, as well as nutritive value, should be selected. The green pigment chlorophyll is the least stable of food pigments, and considerable attention is given to preserving this color in vegetables. Chlorophyll is affected by acid to produce an unattractive olive-gray color. Vegetables are slightly acidic in reaction, and when cooked, the acid is liberated from the cells into the cooking water. Fortunately, much of the acid is volatile and given off in the first few minutes of cooking. If an open kettle is used for cooking green vegetables, the volatile acids may escape easily, aiding in the retention of the green pigment.

Canned vegetables are heated in a steam-jacketed kettle, stockpot, steamer, or oven. Overheating, as with overcooking of fresh and frozen vegetables, results in further loss of nutrients and a soft-textured, unattractive, and poor-flavored product.

Dried vegetables are soaked before cooking to restore the water content and to shorten the cooking time. Legumes will absorb enough water to approximately double their dry weight, with an attendant increase in volume. The length of the soaking period depends on the temperature of the water, with warm water cutting the soaking time to about half. The vegetable may be covered with boiling water, let stand for 1 hour, then cooked until tender; or they may be covered with cold water and soaked overnight, drained, then cooked.

Small batch or continuous cooking of vegetables throughout the meal service is the most satisfactory way to obtain high-quality products. Quantities of not more than 10-pound lots, and preferably 5-pound batches, should be cooked at intervals as needed. High-speed steamers and small tilting steam-jacketed kettles behind the service line are the most useful kinds of equipment for batch cooking of vegetables.

Salads

Preparation of Ingredients. Salad greens should be clean, crisp, chilled, and well drained. Wash in a spray of water or in a large container of water, shake off excess moisture, drain thoroughly, and refrigerate. All ingredients for salads should be chilled thoroughly and drained when necessary.

Arrangement. To make salads efficiently, prepare all ingredients and chill. Arrange salad plates or bowls on trays that have been lined up on a worktable. Place a leafy underliner on each plate, then add the body of the salad to the plates. This may be a mixed salad, measured with a dipper or scoop, or it may be a placed salad in which individual ingredients are arranged on the underliner. Chopped lettuce placed in the lettuce cup gives height to the salad. Top garnishes add the final touch of color and flavor contrast. The trays of salads are then refrigerated until service but should not be held more than a few hours or the salads will wilt. Dressings generally are served separately or added just before serving, except in potato salad and some entrée salads where flavor is improved by standing 2 to 4 hours after mixing.

Salad Bar. The basic salad bar consists of salad greens with a variety of accompaniments and dressings. Lettuce usually is the main ingredient, but other greens and accompaniments are usually added. In addition to vegetables and fruits, chopped hard-cooked eggs, crumbled crisp bacon, croutons, shredded cheese, cottage cheese, cabbage slaw, pasta salads, molded fruit gelatin, and other prepared salads often appear on salad bars. A variety of dressings is offered. The salad bar should be attractively presented and the salad ingredients kept cold. A logical arrangement places chilled plates or bowls on ice first, near the greens, followed by the accompaniments, the prepared salads, and, finally, the dressings. The salad bar offers an opportunity for creativity and can be an effective merchandising tool.

Sandwiches

Sandwiches may be prepared to order in commercial foodservice operations by pantry workers and/or short-order cooks. Fillings are made and refrigerated, margarine or butter softened, lettuce is cleaned and crisped, and other ingredients prepared ready for assembling. Ingredients should be arranged for maximum efficiency, with everything needed within easy reach of both hands.

In large quantity, all sandwiches needed may be made and refrigerated until service or, in cafeteria service, may be assembled a few at a time. Hot sandwiches may be made up and grilled or baked as needed, or for cafeteria service, the fillings may be cooked and the sandwiches assembled on the cafeteria counter. An efficient workstation should be set up for making sandwiches and an assembly-line procedure used. Place bread slices on a baking sheet or tray, and brush with margarine, butter, or mayonnaise, then spread or place the filling, according to the type of sandwich being made, on all slices on the tray. Add the top bread slices to all the sandwiches. For grilled sandwiches, brush the top and bottom slices with melted margarine or butter. Measure fillings with a dipper or scoop and portion sliced meat or cheese according to count or weight. The recipes or instruction sheets should include the directions for portioning.

Sanitation is important in the making of cold sandwiches because of the amount of handling involved and because they are not cooked. Mixed fillings containing meat, poultry, fish, eggs, or mayonnaise should be prepared the day they are to be served, and only in such quantities as will be used during one serving period. Fillings should be refrigerated until needed, and if sandwiches are made ahead they, too, should be refrigerated. Lettuce should be omitted from sandwiches to be stored for some time in the refrigerator because the lettuce will wilt and become unappetizing.

Soups

Most soups can be classified as clear or unthickened, and cream or thick. *Clear* or *unthickened* soups are based on a clear, unthickened broth or stock. Vegetables, pasta, rice, meat, or poultry products may be added. *Bouillon* is a clear soup without solid ingredients. *Consommé* is a concentrated flavorful broth or stock that has been clarified to make it clear and transparent. *Broth* or *stock,* the basic ingredient for all clear soups, is made by simmering meat, poultry, seafood, and/or vegetables in water to extract their flavor. Brown stock, made from beef that has been browned before simmering, and white or light stock, made from veal and/or chicken, are the stocks used most often. Because the making of stock is so time consuming, many foodservice operations use concentrated bases, which are mixed with water to make flavored liquids similar to stocks. Bases vary in quality, with the best products being composed mainly of meat or poultry extracts. These are perishable and must be refrigerated. Many bases have salt as their principal ingredient, so it is important to read the list of ingredients on the label. When using these bases, the amount of salt and other seasonings in the soup recipe may need to be adjusted.

Cream or *thick* soups are made with a thin white sauce combined with mashed, strained, or finely chopped vegetables, chicken, fish, or meat. Chicken stock may be used to replace part of the milk in the sauce to enhance the flavor. *Chowders* are unstrained, thick soups prepared from seafood, poultry, meat, and/or vegetables. *Bisques* are mixtures of chopped shellfish, stock, milk, and seasonings, usually thickened. *Purees* are soups that are naturally thickened by pureeing one or more of their ingredients.

Sauces

Basic to many sauces is a *roux,* which is a cooked mixture of equal parts by weight of fat and flour. A roux may range from white, in which the fat and flour are cooked only for a short time, to brown, cooked until it is light brown in color and has a nutty taste and aroma.

Many meat and vegetable sauces are modifications of basic recipes, such as white sauce, bechamel sauce, and brown sauce. *White sauce* is made with a roux of fat, usually margarine or butter, and flour, with milk as the liquid. White sauce is used as a basis for cream soups, as a sauce for vegetables, and as an ingredient in many casseroles. *Bechamel sauce* and its variations use milk and chicken stock as the liquid and are generally served with seafood, eggs, poultry, or vegetables. *Brown sauce* is made with a well-browned roux and beef stock and is used mainly with meats.

Bakeshop Production

Breads, cakes, cookies, pies, and other desserts may be produced in a separate bakeshop or made in an area of the main kitchen in which ovens, mixers, and other equipment are available. Although some foodservice operations purchase all or some of these items already prepared, others prefer to make these on the premises either from mixes or "from scratch."

The choice of baking mixes influences the finished product and should be made only after testing and comparing more than one brand. Some large foodservices contract with manufacturers to make mixes to their specifications. If mixes are used, the baked product may be individualized by variations in finishing and presentation. For example, a basic white cake may be baked as a sheet cake or made into a layer cake by cutting the sheet cake into two pieces and placing one on top of the other. A variety of icings may be used with a basic plain cake. Many possibilities exist also when making up plain or sweet roll dough.

If mixes are not used for breads and cakes, balanced formulas should be developed and standardized for the pan sizes used in the foodservice. Many variations are possible if good basic recipes are developed for butter, white, and chocolate cakes, and for biscuits, muffins, and rolls. An important factor in successful bakeshop production is the weighing and portioning of batters and doughs. Each recipe should include information on what size of pan to use and the weight of batter for each pan for products such as cakes, coffee cakes, and loaf breads; the weight of each roll or bun; and the size dipper for muffins and drop cookies.

A baking sheet 18 × 26 inches and 1½ inches deep, with straight sides is used for cakes, cookies, and some quick breads. A half-size baking sheet 12 × 18 inches is often used and is especially good for layer cakes. If a baking pan 12 × 18 inches, or 12 × 20 inches, with 2½- to 3-inch sides is used, special attention must be given to the amount of batter in the pans. Loaf pans for pound cakes and quick breads, and tube pans for foam cakes, vary in size and should be selected according to the size serving desired. Pans should be prepared before mixing of the products begins. Pans for angel food and sponge cakes are not greased; baking powder biscuits and cookies with high fat content are usually placed on ungreased baking sheets. Most other pans are either lightly greased, greased and floured, or covered with a parchment paper liner.

Table A.4 Approximate yields expressed in weight of selected ready-to-cook or ready-to-serve raw foods from one pound as purchased.

Food Item	Yield	Food Item	Yield
Meat and meat alternates		Apple, fresh	.91
Ground beef (≤26% fat)	.72	Asparagus	.53
Ground beef (≤20% fat)	.74	Bananas	.65
Ground beef (≤15% fat)	.75	Beans, green or wax	.88
Ground beef (≤10% fat)	.76	Beets	.77
Roast, boneless chuck	.63	Blueberries	.87
Roast, boneless rump	.68	Broccoli	.81
Steak, round, boneless	.63	Cantaloupe	.52
Pork chops, with bone	.45	Carrots	.70
Stew meat	.58	Celery	.83
Sausage	.62	Corn on the cob	.33
Chicken, fryer, with skin	.66	Grapes, seedless	.97
Chicken breast, with skin	.64	Lettuce	.76
Drumsticks	.49	Mushrooms	.98
Thigh	.52	Peaches	.76
Whole chicken	.41	Potatoes, white	.81
Ham, without bone	.63	Squash, acorn	.87
Whole turkey	.48	Tomatoes	.99

Source: Adapted from "Food Buying Guide for Child Nutrition Programs," PA-1331, U.S. Department of Agriculture, Washington, D.C., 1984; supplements added in 1993.

Table A.5 Common abbreviations used in food production.

AP	As purchased	lb	Pound
AS	As served	pkg	Package
C	Cup	psi	Pounds per square inch
EP	Edible portion		
°F	Degrees Fahrenheit	pt	Pint
		qt	Quart
fl oz	Fluid ounce	tsp	Teaspoon
gal	Gallon	Tbsp	Tablespoon

Appendix B: Foodservice Equipment

COOKING EQUIPMENT

Ranges

Simple design, easily cleanable; heavy, well-braced angle iron frame; sturdy riveted or welded construction; body—sheet steel with baked-on black Japan or porcelain enamel smooth finish, or stainless steel; with or without ovens and high backs; heating elements of burners with individual controls; automatic pilot; removable drip trays to prevent spillage under elements or burners; may be mounted on casters or flush-to-wall. A new innovation is an automatic flame regulator that triggers a flame when a pan is on the burner and shuts it off when the pan is removed.

Types: The types of ranges include the following:

1. *Heavy-duty ranges:* Durable and well-suited for large-volume foodservice operations with constant usage, as in hotels, large restaurants, colleges, hospitals. Approximate sizes of sections: electric—36 inches wide, 36 inches deep, 32 inches high; gas—31 to 34 inches wide, 34 to 42 inches deep, 33 to 34 inches high.
2. *Medium weight or restaurant type:* Lighter in construction than heavy-duty ranges and used where demands are less constant such as short-order cooking or where use is intermittent such as in churches and clubs. Complete units, 6, 8, or 10 burners, or combination with fry-top and/or even-heat top; 1 or 2 ovens. Approximate size 35 to 64 inches wide, 27 to 32 inches deep, 34 inches high; ovens, 26 inches wide, 22 inches deep, 15 inches high.

Tops: Tops are usually polished chrome-nickel-iron alloy, with high strength and heat-absorption qualities; resistant to warping, chipping, and corrosion; accurate thermostats provide controlled heat surfaces as desired. The types of tops include the following:

1. *Open or hot plate top:* Usually associated with short-order preparation. Heat concentrated under kettles; heating elements and grates simple in design, easily removable for cleaning; gas cones elevated so combustion and ventilation can be complete; burners can be turned on and off as needed; instant heat available; high Btu output by means of small blower to force air into burner (Figure B.1).
2. *Closed top:* Styled for heavy-duty continuous cooking as entire surface area is heated; various burner arrangements. Gas ranges have these types of tops: *Uniform hot tops,* which provide even heat distribution from rows of bar burners set in fire brick under a smooth top; a depression in the brick around the edge acts as a duct to the flue in gas range; *graduated heat,* which heats by means of concentric ring burners with separate controls; intense heat in the center (approximately 1100°F) to low heat at the edge (450°F); projections on the underneath side of the top help direct heat to edges; and *front-fired tops,* which have a row of burners under the front of range top; heat is concentrated at front with gradation in degrees of heat intensity toward back (Figure B.2).

Figure B.1 Open-top range with oven base.
Courtesy of The Garland Group, 185 East South Street, Freeland, Pennsylvania 18224.

3. *Fry- or griddle-top:* Even heat; solid top with edges raised to prevent overflow of grease; fitted with grease trough and drain to receptacle (Figure B.3).

Range Ovens: Even heat distribution, automatic pilot, and heat control; high-quality insulation; walls, top, bottom, and removable racks or shelves of smooth, durable, cleanable material or finish; sturdy counterbalanced door with nonbreakable hinges, cool handle; door to support at least 200 pounds. Designed so spillage will drain to front for easier cleaning. Approximate size: 26 inches wide, 28 inches deep, 15 inches high (inside measurements).

Legs and Feet: Simple design; rigid support; adjustable legs; shaped at floor contact to prevent accumulation of dirt or harborage of vermin; sealed hollow sections; minimum clearance of 6 inches between floor and lowest horizontal parts unless mounted on raised masonry island at least 2 inches high and sealed to floor.

Installation: Heavy-duty range sections are often joined to other modular units such as broilers and fryers to make a complete cooking unit.

Griddles

Separate griddle units to supplement or substitute for range sections; mobile griddles give use where needed, such as kitchen or counter; extra-heavy, highly polished plates to hold heat and recover rapidly; even heat distribution; chromium or stainless front and ends; oversized, cool valve handles; sloped to grease drain-off. Sizes from 7 by 14 inches to 36 by 72 inches. Capacity expressed in terms of food that can be cooked at one time.

Two-Sided Cookers

A relative newcomer to the cooking equipment line-up, this type of cooker offers the benefits of both a griddle and a grill (see Figure B.4). Some

Figure B.2 Closed-top range with oven base.
Courtesy of U.S. Range and The Garland Group, 185 East South Street, Freeland, Penn., 18224.

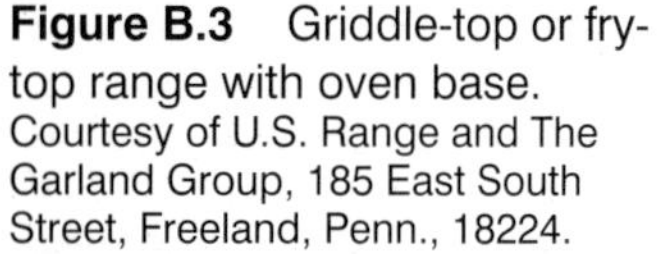

Figure B.3 Griddle-top or fry-top range with oven base.
Courtesy of U.S. Range and The Garland Group, 185 East South Street, Freeland, Penn., 18224.

models sandwich food between a heated top and bottom plate, or platen. This eliminates the need for turning. Other models utilize a bottom plate and an infrared, noncontact broiler on top. Both gas and electric cookers are available and are constructed of stainless steel, varying in widths from 2 to 6 feet. The platens are made of cast iron, highly polished steel, cold-rolled steel, or a chrome finish.

Broilers

Sheet steel of 16 gauge or better, with smooth, baked-on black Japan finish, or stainless steel, rigidly reinforced with angle support; warp-resistant heating units with radiant ceramic or alloy materials to give even heat distribution, lining of long-wearing reflective materials; spring-balanced raising or lowering device; right- or left-hand operation; safety stop locks; close-fitting cast-iron grids, removable for easy cleaning, adjustable over distances of 1½ to 8 inches from heat source; removable drip tray; drain to receptacle; size of grid determines capacity.

Types: Broilers come in these types:

1. *Unit of heavy-duty broilers:* Designed for large-volume production and fast continuous broiling; grid area varies from 3.3 to 5.0 square feet; may be same height as range section and with high shelf above or integral with an overhead oven, heated by burners in broiler below, or mounted on a conventional range-type oven and with or without overhead warming oven.

Figure B.4 A two-sided cooker, with a griddle on the bottom and a broiler on the top, cuts broiling time in half.
Courtesy of Lang Manufacturing Company.

2. *Combination broiler and griddle units:* Suitable for small kitchen where space is limited. Fry griddle forms top of broiler and both are heated by the same set of burners but simultaneous use is not recommended.
3. *Salamander or elevated miniature broiler:* Mounted above the top of a heavy-duty closed range or over a spreader plate between units of cooking equipment. Features are similar to heavy-duty broiler except for a smaller grid area of only 1.6 to 2.8 square feet. Advantage is that it requires no floor space but may be mounted on separate legs or stand if desired. Used where small amounts of broiled foods are served.
4. *Hearth-type or open-top broilers:* Utilize a heavy cast iron grate horizontally above the heat source. Charcoal or chunks of irregular size ceramic or other refractory material above gas or electric burners form the radiant bed of heat. Juice and fat drippings cause smoking and flaming that necessitate an efficient exhaust fan over the broiler. Available in multiple sections of any desired length (Figure B.5).

Figure B.5 This counter top, underfired charbroiler has a tilting grate to provide variable heat zones on the cooking surface. Courtesy of U.S. Range and The Garland Group, 185 East South Street, Freeland, Penn., 18224.

Fryers

Chromium-plated steel, stainless steel; automatic temperature control with signal light and timer; quick heat recovery; cool sediment zone; self-draining device; easy removal of sediment and filtering of fat; capacity expressed in pounds of fat or pounds cooked per hour; fuel input used to determine production capacity also; should fry from 1½ to 2 times the weight of fat per hour.

Types: Fryers are available in these types:

1. *Conventional instant fat fryers:* Sizes from 11 by 11 to 24 by 24 inches with fat capacities of 13 to 130 pounds. Models are available as free-standing, counter, or built-in, single, or multiple units. The bank of deep-fat fryers shown in Figure B.6 is computer-controlled to automatically control temperature and cooking time by lowering and raising the frying baskets at precisely timed intervals.
2. *Pressure fryers:* Equipped with tightly sealed cover, allowing moisture given off during cooking to build up steam pressure within kettle; cooking accomplished in approximately ⅓ normal time.
3. *Semiautomatic—speed production model:* Equipped with conveyer to permit continuous batch cooking and automatic discharge of product as completed.
4. *Convection fryers:* Combine convection cooking, continuous fat filtration, and a heat exchanger to produce an energy-efficient, highly productive piece of equipment (see Figure B.7).

Installation: Adequate ventilation necessary, venting into hood recommended; flue venting from fryer to general vent flue not desirable; table or work space adjacent to fryer is necessary.

Tilting Frypan

Versatile piece of equipment; can be used as a frypan, braising pan, griddle, kettle, steamer, thawer, proofer, bagel maker, oven, food warmer-server; eliminates most top-of-stove cooking, provides for one-step preparation of many menu items, and can double as a sink to assist with

Figure B.6 A system of computer-controlled fryers.
Courtesy of Frymaster Corp.

cleanup chores if necessary. All surfaces, interior and exterior heavy-duty stainless steel; contoured pouring spout; one-piece counterbalanced hinged cover; self-locking worm-and-gear tilt mechanism; even-heat smooth flat bottom (either gas or electric); automatic thermostat heat controls for wide range of temperatures. Available in several sizes and capacities as floor models mounted on tubular legs with or without casters, wall-mounted, or small electric table-mounted (see Figure B.8); conserves fuel and labor; quick-connect installation conducive to rapid rearrangement, easy maintenance, and good sanitation. Easy to clean and reduces use of pots and pans and their washing.

Ovens

The two basic oven designs are *still-air radiation,* in which heated air circulates around outside of the heating chamber and radiates through a lining, and *convection,* where heated air from a heat source is forced over and around food racks by fans located on the rear wall of the oven.

All welded construction of structural steel for durable rigid frames; inner lining of 18-gauge rust-proof sheet metal reinforced to prevent buckling; minimum of 4 inches on nonsagging insulation on all sides, up to 10 inches in large bakery ovens; thermostatic heat control precise between 150°F and 550°F; signal lights and timer; level oven floor or deck of steel, tile, or transite (concrete and asbestos combination); well-insulated, counter-balanced doors that open level with bottom of oven to support a minimum 150-pound weight; nonbreakable hinges; concealed manifolds and wiring; cool handles; system designed to eject vapors and prevent flowback of condensate; light operated from outside oven; steam injector for baking of hard rolls; thermocouple attachments

Figure B.7 Convection fryer.
Courtesy of Hobart Corporation.

for internal food-temperature record, and glass windows in doors available on request.

Types: Ovens come in these types:

1. *Deck:* Units stacked to save space; separate heating elements and controls for each unit and good insulation between decks; decks at good working heights; 7 or 8 inches clearance for baking, 12 to 16 inches high for roasting; capacity expressed in number of 18- × 26-inch bun pans per deck; pie, cake, or baking pans should be sized to fit multiples of that dimension; floor space requirements and inside dimensions vary with types; example of a typical one-section oven of compact design on 23-inch legs:
   - Floor space requirements: 60½ inches wide, 39½ inches deep without flue deflector
   - Inside dimensions: 42 inches wide, 32 inches deep, 7 inches high
   - Capacity: Two 18- × 26-inch bun pans; 24 one-pound loaves of bread; 12 ten-inch pies
   - Btu hour: 50,000.
2. *Convection oven:* Forced air circulation cabinet, which employs high-speed centrifugal fan to force air circulation and guarantee even-heat distribution by an airflow pattern over and around product in a minimum of time or from one-third to three-quarters of time required in a conventional oven. More cooking is accomplished in smaller space because food is placed on multiple racks instead of on a single deck (see Figure B.9). Sizes vary with the manufacturers, but a typical convection oven measures 36 inches wide by 33 inches deep or larger models 45 inches wide by 24½ inches deep. Removable rack

Figure B.8 A tilting frypan.
Courtesy of Groen—A Dover Industries Co.

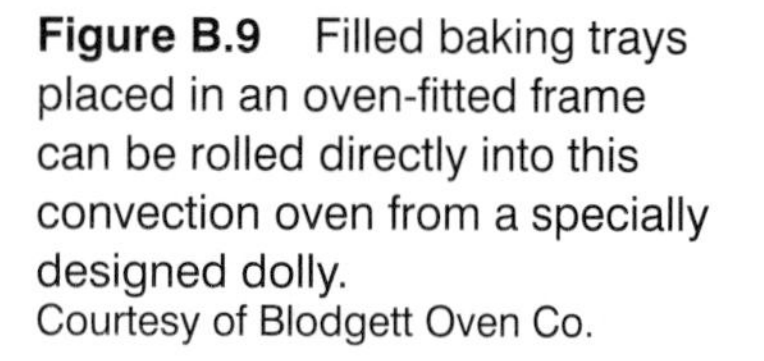

Figure B.9 Filled baking trays placed in an oven-fitted frame can be rolled directly into this convection oven from a specially designed dolly.
Courtesy of Blodgett Oven Co.

glides designed to accommodate 8 or 9 trays or baking sheets, 2 inches apart, thus holding more than other ovens that require greater floor space. Units may be stacked to double the output in the relatively small floor space. Convection ovens must be well insulated; may have interiors of stainless steel or vitreous enameled steel. Shelves and shelf supports lift out for easy cleaning; fitted with inside lights, timer, thermostatic heat control, glass doors or window in doors, removable spillage pan. Quick-connect installation and addition of casters made for flexibility in arrangement. Muffle-type seal on doors for roasting and baking reduce shrinkage because of moisture retention and reduced time for cooking.

3. *Revolving tray or reel ovens:* Flat tray decks suspended between two revolving spiders in a ferris-wheel type of rotation; compact, space saving; welded steel, heat-tight construction; all parts highly resistant to heat and corrosion; main bearings and entire tray load supported independently of side walls; trays stabilized to keep level and sway proof; each tray equipped with individual emergency release; heavy-duty motor; smooth roller-chain drive, self-adjusting, automatic controls; example of relative dimensions; four trays, each 96 inches long × 26 inches wide; capacity, twenty 18- × 26-inch bun pans; outside, 10 feet 2 inches wide, 7 feet 4 inches deep, 6 feet 7 inches high. Small units 3½ feet deep and six-pan units available for small foodservice operations.
4. *Rotary ovens:* Similar to revolving tray ovens except rotation is on a vertical axis instead of a horizontal one. Both revolving tray and rotary type ovens are most suitable for large-volume baking.
5. *Microwave ovens:* Electromagnetic energy directed into heating cavity by magnetrons producing microwaves that penetrate food, create a magnetic field, and set up friction, causing almost instantaneous cooking of the

(a)

Figure B.10 (a) This combination steamer oven is well suited for a small operation. (b) This conveyor oven uses air impingement technology.

Figure B.10 *Continued.*
(a) Courtesy of Groen—A Dover Industries Co. (b) Courtesy of Lincoln Food Service Products, Inc.

(b)

food; energy produced at given rate is not stored nor does it heat the air surrounding or the dish containing the food (glass, china, plastic, paper); components include heating cavity of stainless steel, radio-frequency generator, power supply, usually 220 volts, between 30 and 50 amperes; must pass close inspection to ensure safety during use; automatic shut-off before door can be opened. Can be stacked; used extensively for fast reheating of prepared bulk or plated foods, but items can also be cooked quickly and served immediately on the same dish.

6. *Combination ovens:* (a) an oven that combines a convection oven, pressureless steamer, proof cabinet, and cook-and-hold oven in a single compact unit (Figure B.10a); (b) an ultra-high-speed oven that uses a combination of microwaves and high-velocity convection heat to cook food at speeds that surpass a microwave; (c) an oven that uses a combination of intense light and infrared energy to cook foods quickly.
7. *Conveyor ovens:* Programmable for temperature/speed/heat zones, typically uses one of three technologies: (1) infrared—a radiant heating process that doesn't heat the air surrounding the food but transfers heat directly to the surfaces it contacts; (2) jet sweep—sometimes called air impingement, hundreds of air ducts under and over the food sweep away cold air, cooking the food uniformly (see Figure B.10b); (c) convection—hot air is circulated by fans in the oven cavity.
8. *Cook-and-hold ovens:* Food temperature rises until nearly done, then burner turns off and a fan continues to circulate stored heat. Once the hold temperature is reached, the burner and fan cycle to maintain heat.
9. Other specialty ovens such as wood-burning ovens and gas-fired brick ovens.

Steam Equipment

Steam may be supplied from a central heating plant, directly connected to the equipment; or steam may be generated at point of use, which requires water connection and means of heating it to form the steam; pressures vary according to needs, with automatic pressure control and safety valve if supply is above 5 to 8 pounds per square inch (psi); equipment of stainless steel or aluminum for rust resistance; smooth exterior and interior surfaces for easy cleaning and sanitation; timing and automatic shut-off devices; concealed control valves; steam cookers offer fast cooking in two general types: cabinet steamers and steam-jacketed kettles.

Cabinet Cookers. Steam injected into cooking chamber comes in direct contact with food—to ensure that steam is clean, the supply may need to be generated on-premise from a tap-water source instead of from the steam system for a group of buildings; door gaskets to seal; doors of full-floating type, with automatic bar-type slide-out shelves linked to doors; timers and automatic shut-off, and safety throttle valve for each compartment so doors cannot be opened until steam pressure is reduced; perforated or solid baskets for food; capacity in terms of number of 12- × 20-inch counter pans side by side each shelf or 10- × 23-inch bulk pans. Counter pans are used both for cooking and serving.

Types: The following types of cabinet cookers are available:

1. *Heavy-duty, direct connected steamers:* Compartments fabricated to form one-piece body and entire interior of stainless steel; 5 to 8 pounds per square inch with continuous steam inflow and drain-off of condensate; 1 to 4 compartments with adjustable shelves; inside dimensions 28 × 21 inches desirable to accommodate two 12- × 20-inch counter pans on each shelf, and 10 to 16 inches high.
2. *Pressure cookers:* Operate at 15 pounds steam pressure for small-batch speed cooking; reheating frozen meals of thawing and cooking frozen foods; smaller than free-venting cabinets; self-sealing inside door cannot be opened under pressure; 15-pound safety valve and 30-pound gauge; automatic timers and cutoffs. Inside capacities, from 12 to 40 inches wide, 14 to 28 inches high, 18 to 31½ inches deep; 1 to 3 cooking compartments.
3. *Self-steam-generating (nonpressure):* Intended for installations without direct steam supply; requires water (hot preferred) connection and adequate source of heat supply to produce the steam; steam generators fit below cookers; designs and capacities similar to heavy-duty steamers.
4. *Pressureless forced convection steamers:* High-speed steam cookers with convection generators producing turbulent steam, without pressure, in the cooking compartment. Doors may be safely opened at any time during cooking cycle, and cooking is faster than in the conventional pressure cooker.

Installation: Heavy-duty steamers of cabinet type may have pedestal support or be equipped with feet and have at least 6-inch clearance from floor, or be wall mounted to save space; install in drip pan or floor depression with drain; modular units available in many combinations with other steam equipment (Figure B.11).

Steam-Jacketed Kettles. Two bowl-like sections of drawn, shaped, welded aluminum or stainless steel with air space between for circulation of steam to heat inner shell (Figure B.12); food does not come in contact with steam; steam outlet safety valve and pressure gauge; steam pressure inside the jacket determines the kettle's operating temperature (e.g., 50 psi = 298°F); direct-connected or self-generated steam supply; full or two-thirds jacketed; stationary or tilting; open or fitted with no-drip, hinged and balanced cover; mounted on tubular legs, pedestal, or wall brackets, or set on table. Power twin-shaft agitator mixer attachment for stirring heavy mixtures while cooking (Figure B.13), and electrically operated device to automatically meter water into kettle are available; may have cold water connection to jacket to cool products quickly after cooking; modular design (square jacket) for easy combining with other modular equipment to save space. Basket insets available for removing and draining vegetables easily.

Types: The following types of steam-jacketed kettles are available:

1. *Deep kettles, fully or two-thirds jacketed:* Best for soups, puddings, pie fillings.
2. *Shallow kettles, always full-jacketed:* Suitable for braising and browning meats, stews;

Figure B.11 Compartment pressureless steamers and jacketed kettle powered by a pressure boiler-in-base.
Groen—A Dover Industries Co.

prevents crushing of underlayers of food as in deep type.

3. *Trunnion or tilting kettles:* Mounted on trunnions with tilting device and pouring lip for easy unloading; either power-driven or manual mechanism; self-locking devices to secure kettle in any position; large floor models, or small units mounted to table to form battery; used on deep or shallow-type kettles. Capacities: from 1 quart to 80 gallons; up to 12-gallon size suitable for table mounting and rotation vegetable cookery.
4. *Stationary types for liquids or thin mixtures:* Tangent outlet for straight-flow drain-off; capacities from 10 to 500 gallons.

Installation: Kettle set for easy draw-off of food; and drip into grated drain in floor or table; mixing swivel faucet over kettle to fill or clean; table models at height convenient for workers; adequate voltage or gas supply for self-generating models.

Figures B.14 and B.15 show equipment designed to be used in the cook/chill or cook/freeze foodservice systems. As the diagram in Figure B.14 shows, food is cooked, packaged in a special airtight and watertight plastic casing at or above pasteurization temperature, chilled rapidly in ice water, and stored up to 45 days in 28°F to 32°F storage. Pumpable foods are cooked in mixer kettles, pumped into the casings, sealed, and chilled in the tumble chiller. Solid foods, such as roasts, are browned, encased, and vacuum-sealed

Figure B.12 Floor-mounted, stationary steam-jacketed kettle.
Courtesy of Groen—A Dover Industries Co.

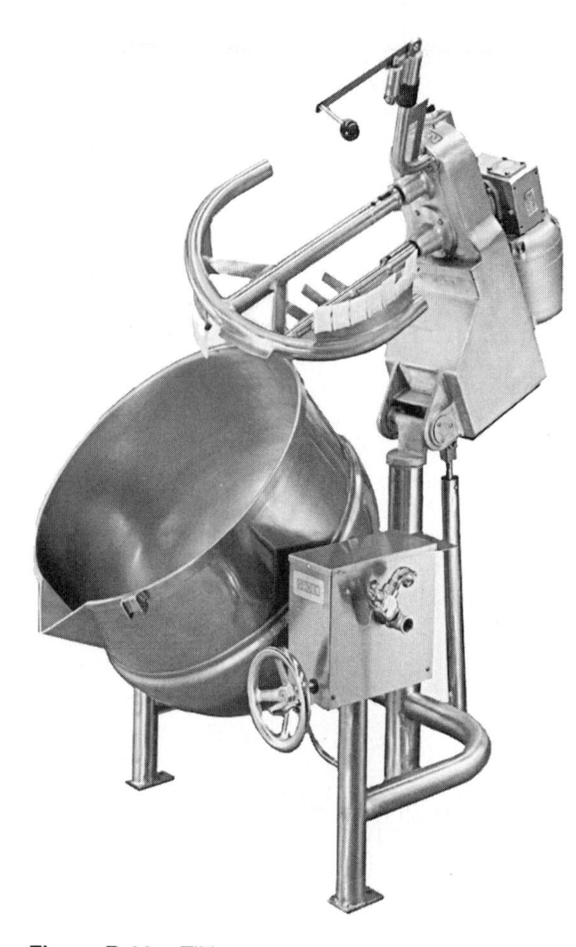

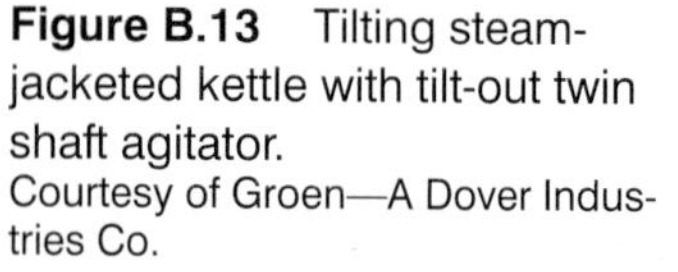

Figure B.13 Tilting steam-jacketed kettle with tilt-out twin shaft agitator.
Courtesy of Groen—A Dover Industries Co.

in casings, and then cooked in the cook/chill tank. In this tank, the product is water bath cooked and then rapidly cooled by circulation of ice water in the same tank. **Rethermalization** may be accomplished by convection oven, combination oven, a convection steamer, pressure steamer, or steam kettles. Figure B.15 shows a cook/chill system designed for a small foodservice operation.

Mixers

Bench models (Figure B.16) for use on tables, counters, and back bars; floor models; three- or four-speed transmission, ball-bearing action; timed mixing control with automatic shut-off; action designed for thorough blending, mixing, and aerating of all ingredients in bowl; electrically controlled brake; possible to change speeds while in action on some machines; durable washable finish as stainless steel or anodized aluminum. Bowls: heavily tinned steel or stainless steel. On some models, a safety ring prevents operation of the mixer unless the ring is locked in place.

Standard equipment includes one bowl, one flat beater, one wire whip; other attachments are available such as dough hook, pastry knife (see Figure B.17), chopper, slicer, dicer, oil dripper, bowl splash cover, dolly, purchased separately; most models have one or two adapters with smaller bowls, beaters, and whips that may be used on same machine. Capacities of 5 to 200 quarts.

Choppers, Cutters, Slicers

Some foodservices meet their needs for chopping, slicing, and shredding through the use of mixer attachments; others need specialized pieces of equipment in certain work areas. Various sizes and capacities of such machines are available in pedestal or bench models or mounted on portable stands. A typical slicer is shown in Figure B.18. All should be made of smooth, noncorrosive metals, have encased motors, safety protectors over blades, and parts removable for cleaning and should slice in horizontal or angle-fed troughs. Figure B.19 illustrates how a slicer and portion control scale may be combined for greater efficiency.

Vertical Speed Cutter Mixer

High-speed vertical cutter-mixer (Figure B.20), gray enamel cast iron base, stainless steel or aluminum bowl; blades move at 1140 rpm. Mixes, cuts, blends, whips, creams, grates, kneads, chops, emulsifies, and homogenizes. Counterbalanced see-through bowl cover interlocks with motor; easy tilt design for emptying. Mounted on tubular steel frame; variety of cutting blades,

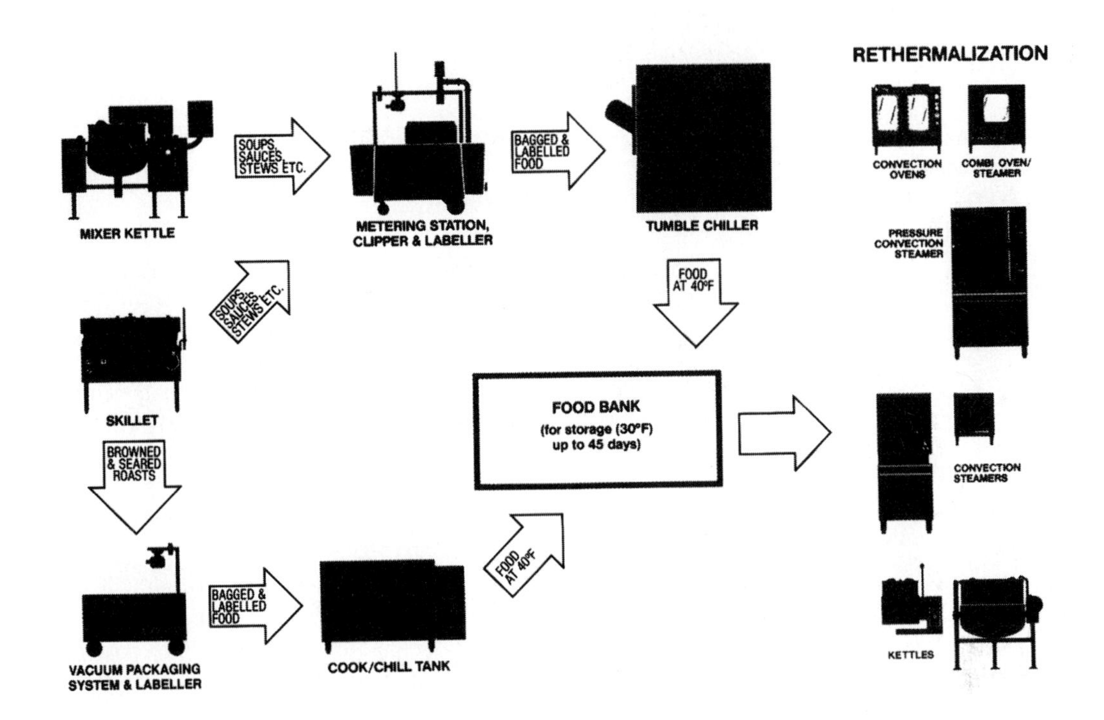

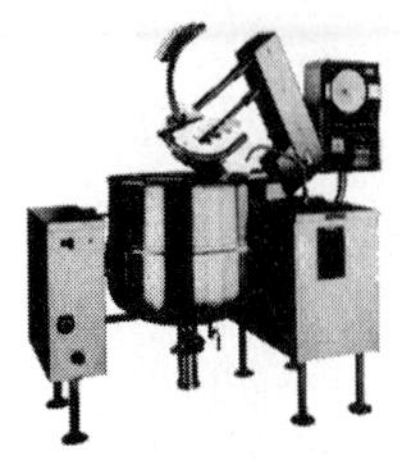

Vertical; Sweep & Fold™ Tilting Mixer Kettles

COOK/CHILL MIXER KETTLES are especially designed and equipped for cooking pumpable food products. A sequential microprocessor and time/temperature recorder combined with automatic water cooling and water metering assures that the product is always cooked to your specifications. The infinitely variable scraper speed and adjustable scraper/agitator ratio assures the best possible mixing conditions for any product and proper suspension during pumping.

Horizontal; Stationary, Mixer Kettles

Horizontal agitator, MIXER KETTLES have processed food for years in industrial applications. This system is now also available for the food service industry. The patented, easily removable agitator is ideal for blending thick and heavy products at a gentle speed. The removable, sanitary temperature sensor is connected to the temperature set point controls and chart recorder.

Metering Filling Stations

METERING FILLING STATIONS are designed to pump cooked foods from MIXER KETTLES and TILT SKILLETS and accurately meter into special Cryovac® casings or other containers. A clipper and labeler are mounted conveniently on the cross bar for one or two person operation. The air operated piston pump has infinitely variable volume adjustment and is designed for easy cleaning and long life.

Tumbler Chillers

TUMBLE CHILLERS cool casing packaged products from 180°F to 40°F in 25 to 65 minutes depending on the product. The gentle tumble action in recirculating ice water assures maximum storage life without product damage. Casings are loaded through the entry chute by hand or conveyor. Fully automated controls assure correct chilling prior to removal and 30°F refrigerated storage.

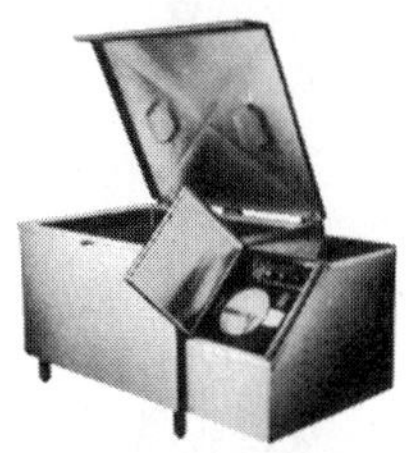

Cook/Chill Tanks

COOK/CHILL TANKS are designed for water bath cooking of packaged meats and other prepared items. Low temperature cooking reduces shrinkage, ensures tenderness while retaining natural juices and flavor. When the desired cooking process is complete, hot water is automatically exchanged for ice water, efficiently reducing food temperature to below 40°F in minimum time.

Ice Builders

ICE BUILDERS are designed to provide a continuous supply of ice water to TUMBLE CHILLERS, COOK/CHILL TANKS and COOK/CHILL MIXER KETTLES. The system economically builds ice up to 24 hours a day and circulates the ice water when required through heat exchangers to maintain sanitary water at the chilling source.

Figure B.14 A cook/chill system and its components.
Courtesy of Cleveland Range, Inc.

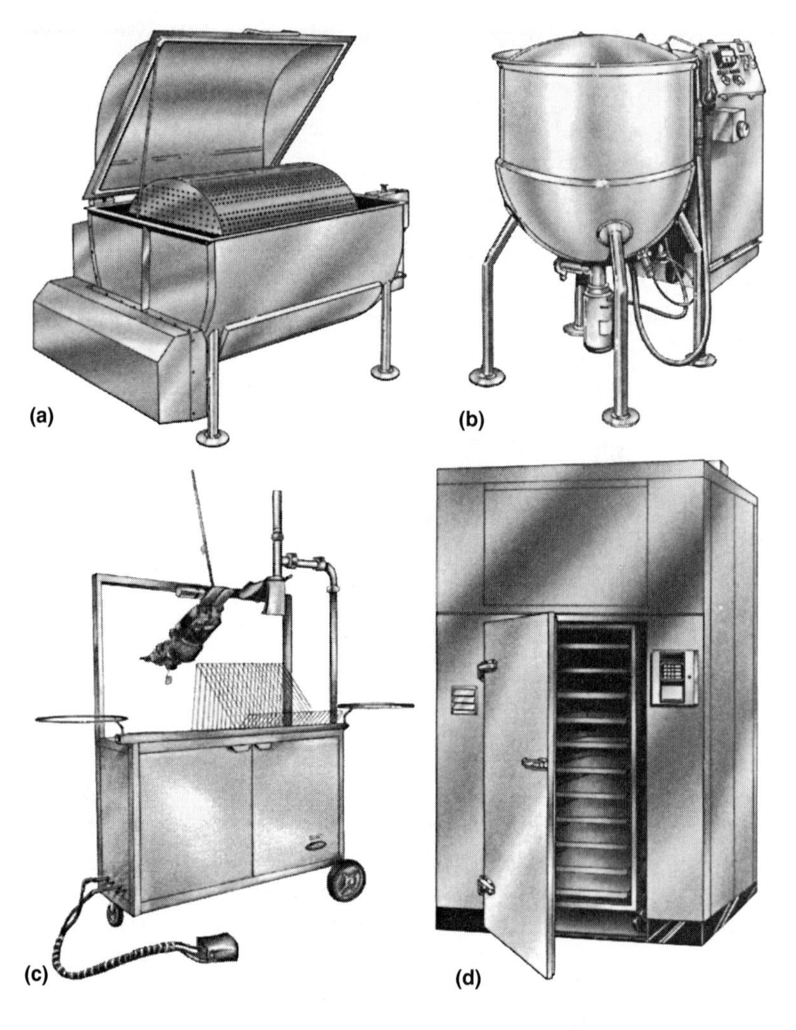

Figure B.15 A cook/chill system with (a) a combination tumble chiller/cook tank, (b) a mixer kettle, (c) a metering/filling station, and (d) a rapid-chill refrigerator. Courtesy of Aladdin Synergetics, Inc.

shafts, and baffles for specific uses. Capacities: 30 and 45 quarts; chops 10 heads lettuce in 3 seconds, makes 32 quarts of salad dressing in 60 seconds, grates 20 pounds of cheese in 30 seconds, makes 40 pounds of ham salad in 90 seconds.

Refrigerators

Detailed information in Chapter 8. There are three basic categories of refrigerators: reach-ins, walk-ins, and blast chiller/freezers. Central or self-contained units; water- or air-cooled compressors; pass-through, cabinet convertible temperatures; efficient nonabsorbent insulation; tight-fitting doors, strong no-sag hinges, strong catches; all surfaces and parts cleanable. *Reach-in:* fitted with tray glides to accommodate standard tray sizes, or removable wire or slatted stainless steel shelves (note Figure B.21). *Walk-in:* portable, sectional, slatted metal shelving. Some reach-in models can be detached from motor unit to provide portable, temporary refrigerated storage. *Counter units:* individual compressors for salad, frozen dessert, milk storage areas; self-leveling dispensers for cold or freezer storage and service. *Ice makers:* central and self-contained units; cubes, tubes,

Figure B.16 Bench-model mixer.
Courtesy of Blakeslee Co.

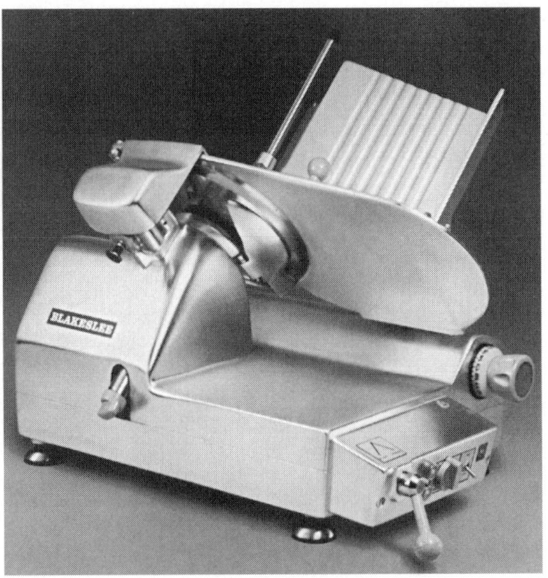

Figure B.18 A slicer with an angle-fed trough.
Courtesy of Blakeslee Co.

flakes, capacity, measured in output per hour; many models and sizes are available. *Water coolers:* glass filler or bubbler faucet; capacity; depends on cooling volume per hour and size of storage tank; designed for convenient storage of clean glasses. *Bottle chillers:* top opening cabinet most usually found in bar operations. *Wine refrigerators:* reach-in units designed to hold red and white wines at optimal serving temperatures. *Display refrigerators:* reach-ins designed to merchandise products often with well-lit interiors, revolving shelves, multiple doors for self-service. *Dough retarders:* upright reach-ins or undercounter units designed to hold unbaked dough at a consistent temperature and high humidity. *Undercounter drawers:* holds foods at refrigerator or freezer temperatures under a countertop griddle/grill or fryer. *Wall and overcounter:* wall-mounted refrigeration to provide extra storage over a work or service station.

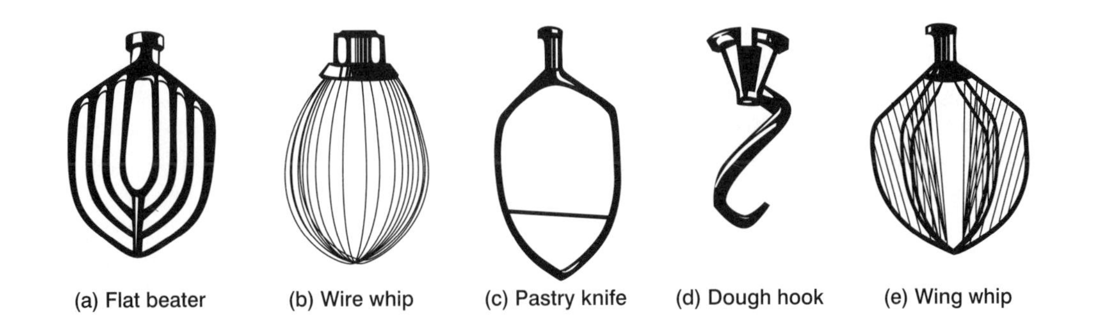

Figure B.17 Mixer attachments.

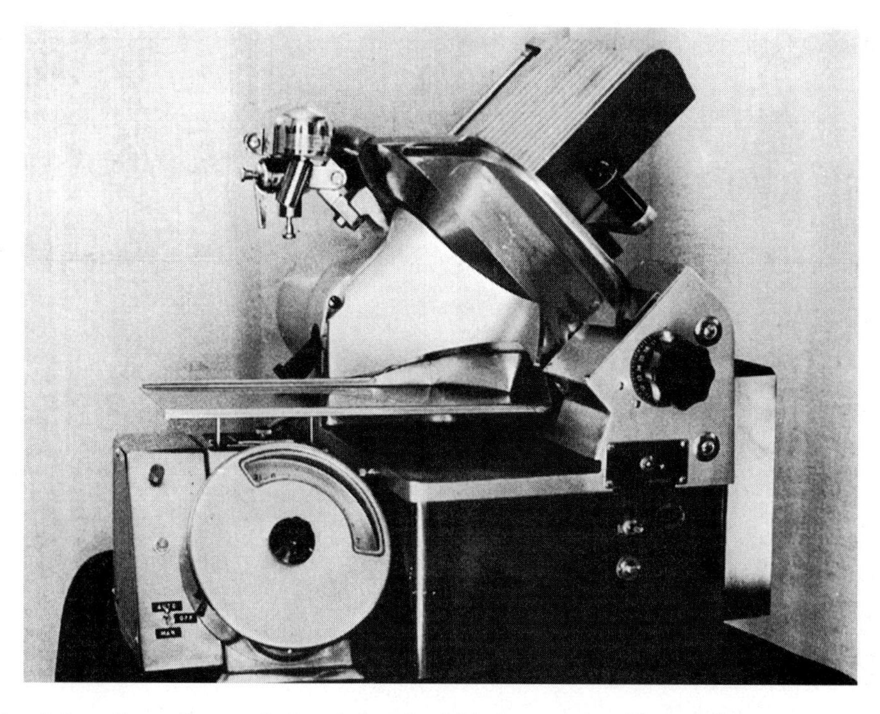

Figure B.19 Automatic portion-control scale is adaptable to operate with most late model gravity-feed automatic slicers. It is designed to shut off the slicer when a preset weight of thinly sliced meat drops onto the scale platter.
Courtesy of E. L. Sly Co.

DISH AND UTENSIL CLEANING EQUIPMENT

Stationary Warewashers

Undercounter or *upright, door-type warewashers* that may be operated by one person usually used in small-volume operations. Undercounter models are similar to home-style dishwashers in that they may fit under a counter or be freestanding. Foodservice models are designed to withstand heavier and more frequent use and to clean faster and with more power, often completing an entire cycle in as little as 90 seconds. Models may include a booster heater, low-detergent alert signal, detergent pump, and pump drains.

Door-type or single-rack warewashers have a 35 to 55 rack per hour capacity and may be designed as a corner or straight-through model.

Moving Warewashers

Rack conveyors are designed to transport racks of ware between wash and rinse arms. They range from single-tank machines capable of washing approximately 125 to 200 racks per hour to multiple tank machines that wash between 250 to more than 300 racks per hour. Optional features may include recirculating prewash and power prewash cycles, corner scraper units, gas-heated and/or low-water models, automatic activators that run the machine only when racks are in it, and automatic loaders and unloaders.

The largest operators require the highest speed warewashers, the *flight-type* or *circular conveyer machines* that are capable of handling between 8000 and 24,000 dishes per hour. Because of the design of the conveyor on a flight-type machine, dishes and trays do not require a rack to move through the prewash, wash, rinse, and final rinse

Figure B.20 High-speed cutter-mixer prepares foods in seconds.
Courtesy of Hobart Corporation.

tanks of the machine. Options on this type of warewasher include straight-line or circular configuration, high- or low-temperature operation, customized length (minimum of 13 feet) and width, left-to-right or right-to-left operation, extra water- and energy-saving capability, noise-reducing insulation, custom conveyors, variable speeds, automatic "eyes" to shut off the cycles when no ware is present, dryer attachments, special designs for insulated trays and silverware troughs, and theft-proof/tamper-resistant designs for correctional facilities. Examples of single-tank, double-tank, and flight-type dishwashers are shown in Figure B.22.

Specialty Warewashers

Pot and pan/utensil warewashers feature high-pressure water scraping capabilities with a longer wash cycle than standard warewashers. *Flatware washers* eliminate the problems of nesting spoons and dried-on, difficult-to-clean foods such as eggs that often adhere to flatware. *Tray washers* are designed to hold all sizes of trays, full sheet pans, and other large, flat-surface items (Figure B.23).

Support Systems

Items that may increase efficiency and lower costs in this area are garbage troughs, food waste pulpers, tray accumulators, tray conveyors, blowers and tray dryers, scrape and sort tables, soak sinks, exhaust condensers, automated dispensers, and hot water boosters.

WASTE DISPOSERS

One system for the disposal of waste may solve the problem in a given situation but, in many cases, it may be feasible to combine two or more of the following methods.

Unit disposers for food waste at vegetable and salad preparation sinks and dish scrapping areas eliminate the need for garbage can collections, storage, and outside pickup unless their installation and use are prohibited by environmental controls. All waste paper, cardboard cartons, wood crates, plastics, tin cans, broken china, and glassware (and garbage) might need to be discharged into trash bins for pickup if incineration of burnable waste is restricted by antipollution regulations in the community.

Can and bottle crushers are capable of reducing this type of disposable bulk up to 90% and cut labor costs, refuse space, and cost of pickup. Capacities of models vary from 50 cans and bottles per minute to 7500 per hour. In the crushing mechanism shown in Figure B.24, rollers set in a "V" design prevent clogging and progressively reduce cans to smallest bulk possible.

The use of *compactors* to reduce the volume is a convenient and economical aid to the disposal of all waste in many foodservices. One model with 13,000 pounds of force can compact paper, milk cartons, cans, bottles, and food scraps to a mini-

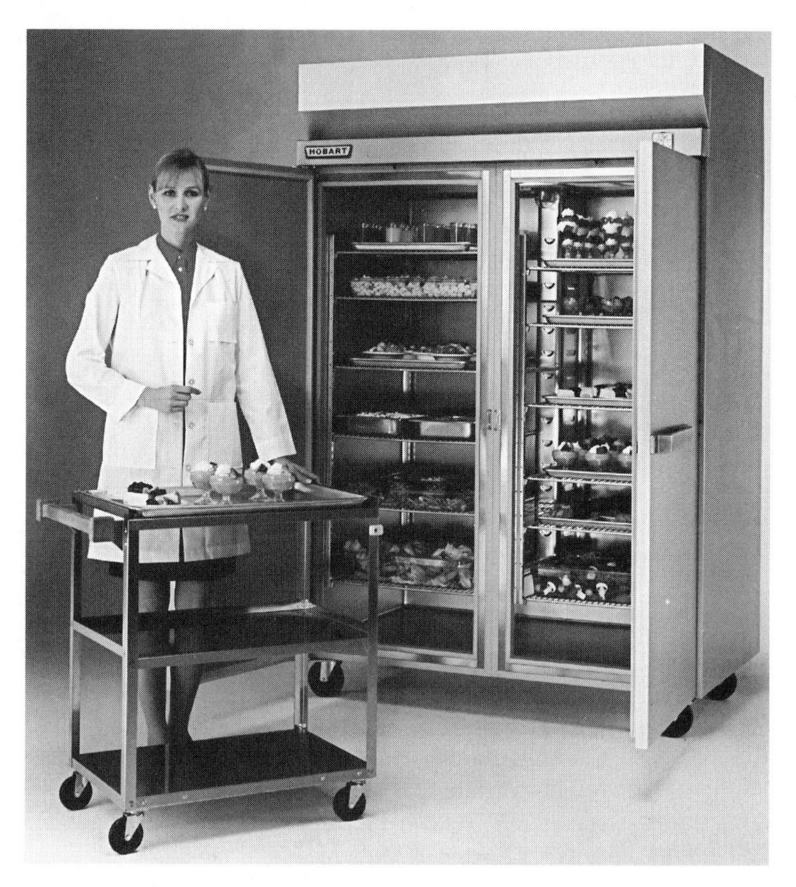

Figure B.21 Convenient refrigerator unit may be fitted with pan slides or pull-out shelves; doors glass or same as outside thermal-bonded vinyl finish.
Courtesy of Hobart Corporation.

mum 5-to-1 ratio or as high as 200-to-1, depending on the combination of materials. Discharge of the compacted material, up to 50 pounds, into a polybag or carton on a dolly makes it ready for short-time storage and haul-away. Most machines operate on a 120-volt, 20-ampere outlet, have safety interlocks throughout for operating protection, and a sanitizing-deodorizing spray that may be released at each return stroke of the compaction ram to avoid any objectionable odor from the compacted mass (Figure B.25).

The *pulping system* reduces the volume of disposable materials such as food scraps, paper, plastic, and cooked bones up to 85%, depending on the mix. Cans, silverware, and some glass are tolerated, but are automatically ejected from the pulping tank into a trash box. Durable teeth on a rotating disk and cutters pulp the material in the tank. It is then circulated to a powerful waterpress above, reducing the pulp to a semidry form that is forced into a discharge chute to containers for removal as low-volume waste. The water from the press recirculates to the pulping tank. This equipment is available in several sizes. It may be incorporated into the dishwashing system or other area where pulpable waste originates (Figure B.26).

TRANSPORT EQUIPMENT

Usually powered equipment for transport of food and supplies within a foodservice is kept to minimum distances by careful planning of area relationships. A thorough study of the advantages, capabilities, and maintenance factors should precede the selection of a system for a particular situation. Also, automatic and emergency shutoffs, enclosed but easy access to working parts, safety, and cleanability are important features to consider.

Conveyors and Subveyors

Reverse for two-way service; emergency brakes, safety guards; automatic stop and start with removal of tray, or continuous flow. *Conveyors:* horizontal transportation; stationary or mobile units for flexibility of tray or food assembly. *Subveyors:* vertical conveying, used where space may be limited on a single floor and work or serving units are on different floors.

Monorail and Driverless Vehicles

These require special equipment and installation; reduce labor and hand-pushing of carts; speedy;

Figure B.22 Examples of (a) single-tank, (b) double-tank, and (c) flight-type dishwashers.
Courtesy of Hobart Corporation.

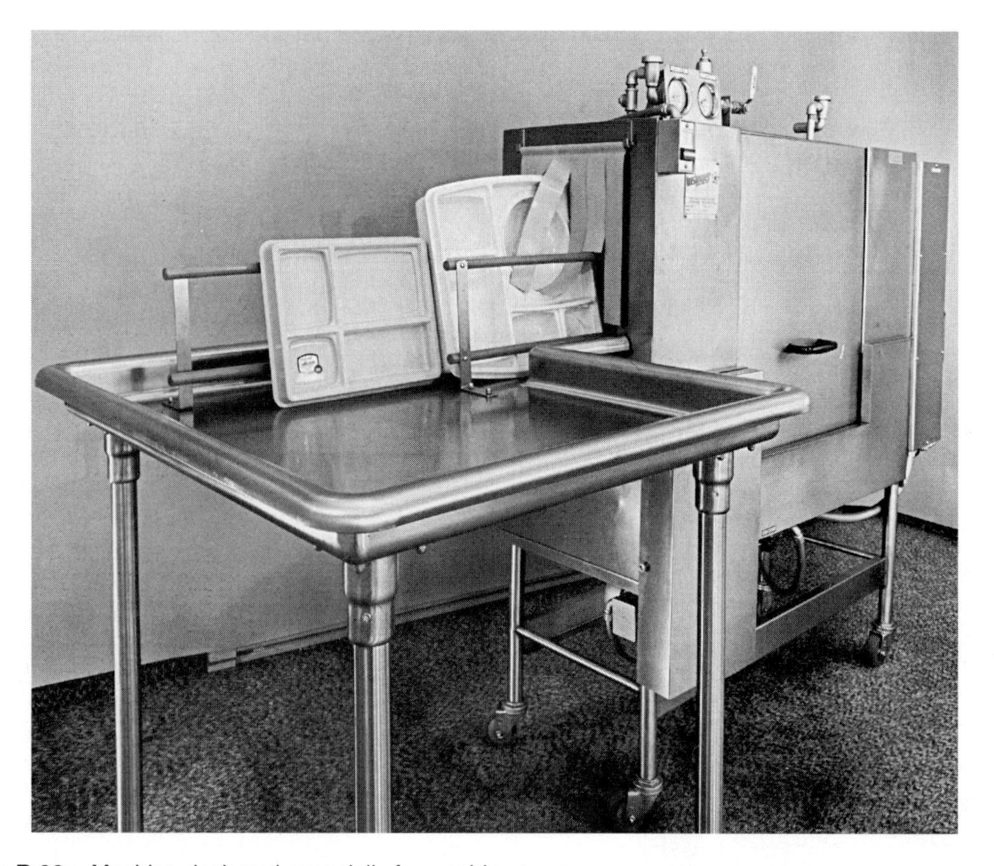

Figure B.23 Machine designed especially for washing trays.
Courtesy of Insinger Machine Company.

relatively expensive to install. Monorail requires overhead rail and "Amsco" system, a special electronic track under the floor. Cars of the latter are monitored from a control panel; powered by batteries; directed over the track to locations on the same floor or to a bank of special elevators that automatically open and close on signal and exit on the assigned floor.

NONMECHANICAL KITCHEN EQUIPMENT

Tables and Sinks

Often fabricated by specification order to fit space and need; stainless steel, No. 12 or 14 gauge. No. 4 grind; welded and polished joinings; rounded corner construction; seamless stainless steel tubular supports with welded cross rails and braces of same material; adjustable inside threaded stainless steel rounded or pear-shaped feet; worktables may be fitted with ball-bearing rubber-tired casters, 2 swivel and 2 stationary, brakes on 2 casters.

Tables. Top of one sheet without seams; edges integrally finished, rolled edge, raised rolled edge where liquids are used, turned up as flange or splashback with rolled edge. Legs and feet: tubular, welded, or seamless metal; adjustable; simple design; provide a minimum of 6 inches of space between bottom of unit and floor; drawers: operate on ball bearings, equipped with stop, removable.

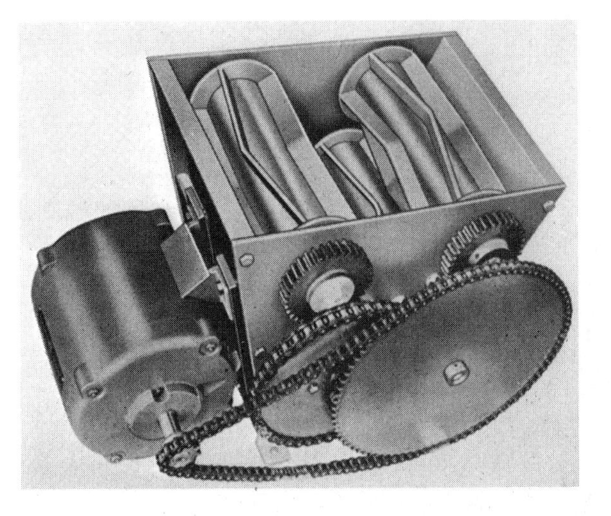

Figure B.24 Mechanism of can and bottle crusher.
Courtesy of Qualheim, Inc.

Undershelves: stationary bar, slatted, solid, removable sections; sink or bain-marie (Figure B.27).

Dimensions: Standard, length 48, 60, 72, 84, 96, 108, 120 inches; width 24, 30, 36 inches; height 34 inches; other dimensions by individual specification.

Types: *Baker's tables*—fitted with drawers; separate storage bins as specified. *Salad tables*—with or without refrigerated work space and storage, sinks. *Sandwich tables*—refrigerated storage for fillings; removable cutting boards. *Dish tables*—well-braced sturdy understructure; 3-inch upturned and rolled edges, higher if joined to wall; scrap block, waste drain, sinks for soaking, over-and-under shelves, rack return, tray rest; adequate space for receiving, soiled dishes, clean dishes, preflush.

Sinks. One, two, three compartments; all-welded seamless construction, drainboard and splashback integral from one sheet of metal, rolled edges; corners fully rounded with 1-inch radius, coves spherical in shape at intersection of corners; bottom of each compartment scored and pitched to outlet; outlet recessed 5 inches in diameter, ½-inch deep, fitted with nonclog waste outlet; partitions: two thicknesses formed of one sheet of metal, folded and welded to bottom and sides of sink; provision for overflow; drainboards pitched to drain into sink; drainboards supported by channel braces to sink legs or wall-bracketed, if longer than 42 inches usually supported by two pipe legs at end away from sink; removable strainer at waste outlet; external level control for outlet valve; stationary or swing faucets.

Dimensions: Standard single compartment, 20, 24, 30, 36, 48, 60, 72 inches long, 20, 24, 30 inches wide, 14, 16 inches deep; two-compartment, 36, 42, 48, 54, 60, 72 inches long, 18, 22, 24, 30 inches wide, 14, 16 inches deep. Others by individual specification; 38-inch height convenient for sinks.

Storage Cabinets, Racks, Carts

Cabinets and racks stationary or portable. Open or closed; shelves: attached, removable, cantilevered, adjustable, tray slides; sturdy construction for use; metal or polymer; solid floor; bolted or welded; doors, hinged, or side sliding, side sliding removable, suspension hung. Both stationary or portable types can be heated or refrigerated. Size determined by needs and space.

Wall-mounted storage and/or workstations can be designed with combinations of grids, shelving, and accessories to store supplies or to transform a traffic aisle by folding down to form temporary workstations.

Scales

Heavy-duty *platform scale* built into floor of receiving room area for weighing in supplies and

Figure B.25 Mobile, hydraulic trash compactor reduces trash to $1/10$ of its original size.
Courtesy of Precision Metal Products, Inc.

food; weight indicator should be plainly visible from both front and back. *Exact-weight* floor or table models in storeroom, ingredient room, bakeshop, and where recipes are made up. *Portion scales* for weighing individual servings where needed. All types of scales are now available with lighted electronic display (LED) readouts, locking in accurate weight almost instantly on an easy-to-read screen, as in Figure B.28.

Cooking Utensils

Strong and durable to withstand heavy wear; nontoxic material; resistant to chipping, dents, cracks, acids, alkalis; cleanable; even heat spread; highly polished metal reflects heat, dull metal absorbs and browns food more readily in baking; variety of sizes of sauce pans, sauce pots, stock pots, frypans, roast and bake pans; *aluminum heavy-duty*—double-thick bottoms, extra-thick edge; *semiheavy*—lighter weight, uniform thickness, rolled edge; *stainless steel*—uniform thickness, spot heats over direct fire. Small equipment as pudding pans, pie and cake pans, quart and gallon measures, mixing bowls of lighter weight metals. Pudding and counter pans selected to fit serving table, refrigerator, and mobile racks for flexibility of use: 12- × 20-inch size recommended. Clamped-on lids cut spillage losses in transporting prepared foods.

Cutlery

High-carbon tolled steel or high-carbon chrome-vanadium steel; full tang construction, compression-type nickel-silver rivets; shapes of handles and sizes of items varied to meet needs; handle and blade weight balanced for easy handling.

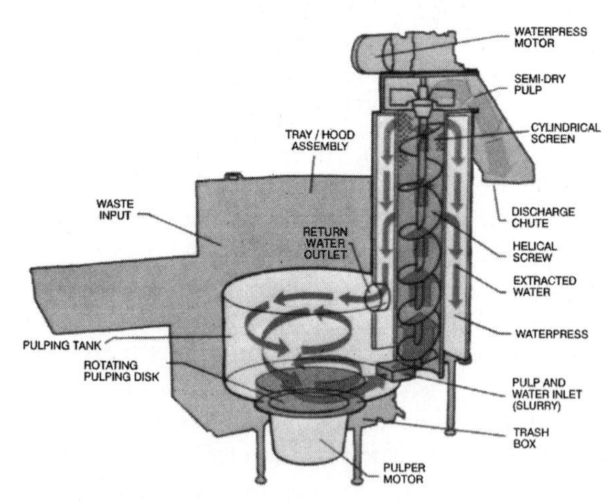

Figure B.26 Diagram showing how pulpable waste can be reduced in volume and form.
Courtesy of Hobart Corporation.

Figure B.27 Kitchen worktable.
Courtesy of Restaurant Equippers.

SERVING EQUIPMENT

Counters

Attractive, compact, efficient arrangement designed for specific foodservice; welded and polished in one piece; hot and cold units well insulated; easily cleaned; separate temperature controls for each unit of heated section; counter guard shields for open food display sections; portable or built-in self-leveling dish and tray storage may be desired; adequate tray slide to prevent accidents.

Serving Utensils

Variety of sizes of ladles, long-handled spoons, perforated, slotted, and solid; spatulas; and ice cream dippers; selected to give predetermined portions size. Capacity or size marked on ladle handles and on dippers.

Special Counter Equipment

Convenient arrangement: easily operated automatic heat controls. Coffee maker—urn or battery vacuum makers with cup storage near; toasters, egg cookers, grills with hoods; temporary storage cabinets for hot cooked foods, rolls—controls for temperature and moisture content;

Figure B.28 An electronic portion-control food scale gives accurate weight almost immediately on a large LED readout. This model is shown with an AC adaptor and optional foot pedal control.
Courtesy of Pelouze Scale Company.

freezer cabinet unit for ice creams; bread dispensers; milk-dispensing machines.

Self-Leveling Dispensers

Counterweighted springs bring platform to uniform level on removal of item; for foods, dishes (Figure B.29), containers; heated, refrigerated, or freezer storage; mobile, stationary, or built-in; open or closed frames of stainless, galvanized, carbon steel or aluminum; noncorrosive springs. Tube type: for plates, saucers, bowls; chassis type; accommodates square or rectangular trays, or racks, empty or filled; adjustable to vary dispensing height.

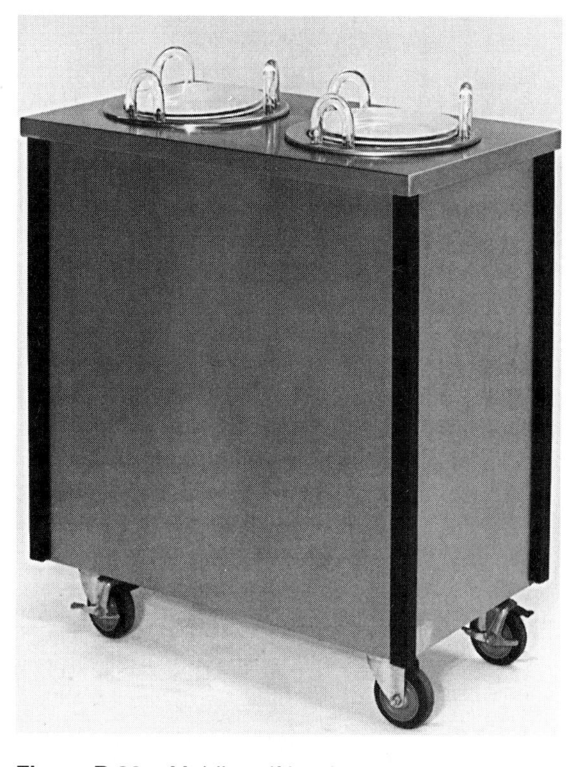

Figure B.29 Mobile self-leveling china dispenser.
Courtesy of Precision Metal Products, Inc.

Coffee Makers

Coffee-making equipment falls into two general types: (1) urns for making large quantities of coffee when many people are served in a short period of time and (2) small electronic automatic brewing units for a continuous fresh supply of the beverage. Requirements are fairly simple in either case, but important to the making of an acceptable product: glass or stainless steel liners for urns, glass or stainless steel decanters for the automatic brewing machines; fluted paper filters; controlled hot water temperature, coffee and water measurement, infusion time, brewing speed and holding temperatures; easily cleaned. Installation with quick-disconnect outlets provides for easy relocation of equipment (Figure B.30). The use of freeze-dried coffee simplifies the process, reduces time and labor, and eliminates the necessity of discarding coffee grounds.

Mobile Food Serving Carts

Specialized equipment for transporting bulk or served food some distance to the consumer; well-insulated, automatic temperature controls; engineered for ease in moving and turning; circumference bumper guards; designed for easy cleaning; may require high-voltage outlets; combination heated, nonheated, low-temperature, and refrigerated sections; beverage dispensers and other accessories as found on serving counters (see Chapter 6 for details).

The selection of foodservice equipment by any arbitrary rule would be unwise and ill advised. Each operation must be studied to determine the real needs and purchases made accordingly. *The foregoing statements are suggested as a guide only in helping to recall basic considerations regarding various items of equipment.*

China

The three types of china are vitrified, semivitrified, and pottery. Of these, only vitrified is considered durable enough for use in most foodservices.

Vitrified china, also known as porcelain, is made of excellent quality clay free from iron, with

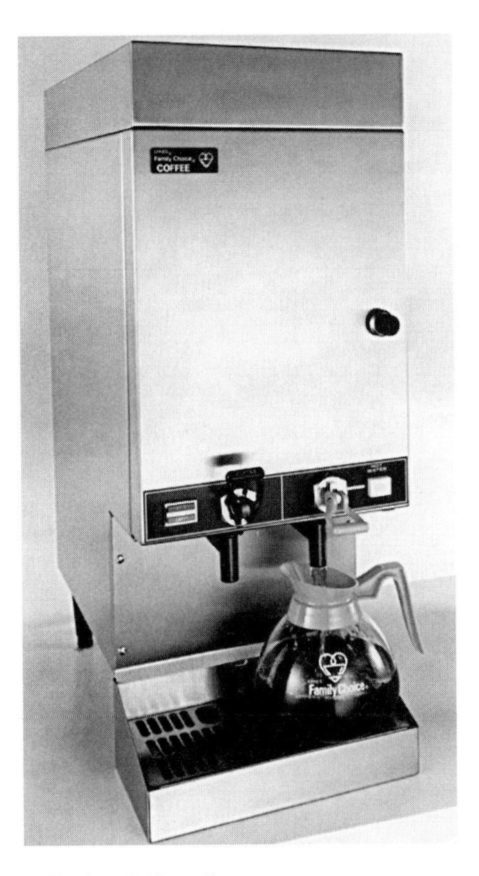

Figure B.30 Coffee dispenser uses brewed liquid coffee concentrate.
Courtesy of Pasco Packing Company.

flint and feldspar added. These materials plus water can be shaped and fired to a high temperature for at least 60 hours, which fuses the mixture into a homogeneous body that is durable and virtually nonabsorbent. At this point, the shaped piece is known as "bisque." A notable improvement in making vitrified china was effected by the introduction of a metallic ion "alumina" into the body of the materials. This enabled the industry to make a thinner, whiter, stronger piece of china with greater edge chip resistance, greater impact strength, and smoother body that has faster surface cleanability than china made without alumina.

The U.S. Bureau of Standards has established three standards for vitrified china for institutional use: (1) *thick:* 5/16 to 3/8 inches thick (which is quite heavy); (2) *hotel:* 5/23 to 1/4 inches thick, with a rolled underedge; and (3) *medium-weight:* sold on the market as "banquet" weight, thinner than hotel weight, with straight edges.

The bureau also tests and sets limits for moisture absorption for each size piece, and tests for durability by use of chipping, impact, and breakage tests under stated conditions. Figure B.31 shows a resistance-to-breakage test being conducted by one china manufacturer to ensure that government standards are met.

It is essential that the buyer recognize that weight does not mean strength and long life for china. Durability and strength are far more directly related to the quality of materials used and the methods of manufacture employed than they are to weight.

The thick china is commonly used for lunch counter service or other situations where extra-heavy service is demanded of the table appoint-

Figure B.31 Measuring china's resistance to breakage.
Courtesy of Syracuse China Company.

ments. It is clumsy to handle and apt to be unattractive in appearance. All hotel-weight china except cups have a roll under the outer edge that gives the effect of weight and also lessens chipping on the upper side of the plate. This type of china is well adapted for use in institutions such as hospitals, residence halls, and restaurants. It is highly resistant to shock, easy to handle, and available in many designs and colors. Banquet-weight china is used extensively in exclusive restaurants, clubs, and the private room service of hospitals. It resembles more closely household dinnerware.

Vitrified heat-resistant ware of good quality is nonabsorbent, stainproof, and withstands high temperatures without crazing or breaking. Items are available in a variety of attractive colors and designs and include coffee pots, teapots, casseroles, ramekins, and individual pudding or pie dishes.

Semivitrified china is a good quality earthenware that has been fired insufficiently to obtain vitrification. This treatment results in a soft body, which is, therefore, porous and absorbent. Semivitrified china has been given a glaze that seals and finishes the dish, but the glaze may be sensitive to heat shocks and cracks easily. The design may not be permanent as it is applied after the china is glazed and fired in making semivitrified and semiporcelain china.

Decoration. Three methods are used to put color, designs, and decorations on china: lining, printing, and decalcomania (decals). In *lining,* a line design is applied to the edge or rim of the dish by machine; only one color can be used. In *printing,* any type of design may be applied by stamping or printing-on. In *decalcomania* the design is transferred from a specially prepared paper; any number of colors can be used. After the colored design has been applied on vitrified china by whatever method, the item is dipped into a glaze and fired at a high temperature. The glaze is a molten glass that is applied as a coating to the shaped, fired, and decorated dish and is fused to it. This process seals the surface of the bisque, covers and protects the design, further strengthens the body, makes the surface smooth, and is then highly resistant to chemicals and to cracking, crazing, or marring by physical shock.

Certain colors such as some blue pigments and the application of gold trim are affected by the high temperatures used to fire the glaze, so may be applied after the glazing process. This design over the glaze is not as satisfactory as an underglaze for institutional foodservice use in most situations. The colors and gold that are put on top of the glaze are less durable and wear away faster than those that are put on under the glaze and protected by it.

Factors in Selection of China. The things to consider when selecting china are its weight, the color and design, budget, availability of replacement, shapes, sizes, and capacities. In addition, in purchasing china, "firsts," the most perfect pieces that can be selected from each run of the kiln after the firing process, are the most desirable. They are free from warping, chips, faults in the glaze, thin or uneven glaze, large scars on the underside from the pins on which the china was held during firing, and uneven or poorly applied designs. Other pieces are graded as "seconds" or "thirds," depending on the degree of imperfection. Warped plates are detected by rolling several plates on edge simultaneously. The warped ones show up plainly in the rolling in contrast to the first selection. Close inspection of each piece by experienced workers completes the grading process.

The *color* and *design* of the china selected should be in harmony with the overall motifs and general atmosphere desired in the dining area. Pigments and processes have been so perfected that at the present time there is practically no limit to the color and design possibilities of china. Conservative but attractive designs enhance the beauty of any dining room and ordinarily do not detract from interest in the food. Colors primarily used for the body of china are white, off-white, or ivory, and buff or tan. These complement natural food colors and serve as a good background for them. Design colors should harmonize. Gaudy and naturalistic designs in the center of plates seem to leave little room for food. Also, such designs may add 5% to 25% to the cost of each plate. In contrast, an inexpensive design may be created with a colored edge for the dominant note.

That the choice of china is influenced by the size of the budget is evident. Not infrequently, the budget limits the choice to china that has a simple border pattern, which may or may not have artistic appeal. Managers need to weigh values of beauty and durability along with cost in the selection of china. Interest in and demand for good design in less expensive china has influenced manufacturers to produce such items.

Another factor that may influence choice is the available designs for which *replacements* can be obtained within a reasonable time period. Stock types of patterns are usually available for immediate shipment. Specially made china, such as that having a monogram or crest, must be ordered weeks in advance. This fact must be considered along with the relatively higher cost of such special china in selecting a pattern for any specific service. Even open-stock types of patterns may be discontinued with limited notice; so when the initial selection is made, the possibility of replacements with identical china or that similar in type should be considered.

Another consideration in selecting china is the *shape* of the pieces because there are many different ones on the market. Plates are available with a wide flat 1 to 1½-inch rim; Econo-Rim, which is ¼ to ⅜-inch rolled under edge, designed to save space on trays and in storage and give extra strength to the edge; and coupe-shape, which is a no-rim design with the body of the plate scooped or slightly concave. Cups are made in a low, wide shape and in a taller, slender shape with a much narrower opening, which holds heat in the beverage longer and stores more easily than the more open shape. In addition, mugs of all sizes and shapes, some footed and some not, have become popular in many foodservices and eliminate the need for saucers.

A wide range of *sizes* and *capacities* of china is available and can vary somewhat from manufacturer to manufacturer. The present trend in purchasing is to limit the number of different-size dishes, supplying one size for several uses. For example, instead of buying both 4-inch bread/butter plates and 6-inch salad/dessert plates, a 5-inch plate to satisfy both uses can be purchased. Or, instead of ordering different size bowls for soup, cereals, and similar items, one size is purchased for all. This simplification is advantageous from the standpoint of service, inventory, dishwashing, replacement, and storage.

The size of plates is the measurement from outer rim to outer rim, and that is the size specified when ordering. Cups, bowls, sugar bowls, creamers, and pitchers are specified by capacity in terms of ounces.

Amounts to Purchase. The quantity of dinnerware to be purchased for equipping a foodservice depends on many factors: the seating capacity and total number of people to be served, the length of the serving period, the type of menu and the price of the meal, the kind of service, the dishwashing facilities and whether they are used intermittently or continuously, and the caliber and speed of the employees. Other factors not to be overlooked are the variety of sizes of each item to be stocked and the frequency of use of the piece. For example, if only one size plate is purchased for multiple use as a bread and butter, salad, and underliner plate, fewer total pieces would be required than if three different-size plates had been selected. Also, a larger number of coffee cups used many times a day must be purchased to provide a margin of safety than would be necessary for bouillon cups that may be used only one or twice a week.

Any listing of quantities must be determined by the needs of a particular institution and not by a set formula. Table B.1 suggests the number of each item of dinnerware needed per customer in a foodservice using an intermittent dishwashing cycle and might be helpful as a basis for initial planning.

Care. China has a much longer life when handled carefully and cleaned properly. It is believed that most breakage is caused by china hitting against china, and that 75% to 80% of all breakage occurs in the soiled dish and washing area.

Table B.1 Dinnerware needs per customer.

Item	Number Per Customer
Cups	1.25
Saucers	1.25
9-inch plates	1.0
5-inch plates	0.66
Bowls, cereal	0.5
Sauce dishes	0.5

Careful training and supervision of the personnel can do much to prevent breakage and keep dishes looking bright and clean. Procedures to reduce the number of times a piece of china is handled will also assist in this. Examples are to separate and stack soiled dishes into like kinds before taking them to the dishroom for washing, and to store clean cups and glasses in the racks in which they are washed. Rubber plate scrapers and collars on openings in scrapping tables not only decrease noise but help to reduce breakage of dishes. Also helpful are the use of plastic and other synthetic-coated metal dishracks and plastic or nylon pegs on dishmachine conveyors.

Suitable washing compound, proper temperature of wash and rinse waters, in addition to careful attention throughout the scrapping, washing, and stacking and storing of clean dishes contribute to the life span of china.

The soiled dishes, scraped and ready for the machine, are placed on belts or in racks, so that all surfaces are exposed. Sorting and stacking dishes into piles of dishes in the racks or on the conveyor belt speed washing at both the loading and unloading ends of the machine and ensure better wash action, since there is no overlapping of larger dishes to block the spray. After the washing and rinsing, the china is air dried. Plates of like size are stacked carefully so that the bottom rim of one does not mar the surface of the plate beneath it. Cups and glasses are stored in the wash racks, stacked on dollies, and wheeled to the unit where they will be used next. Plates and bowls likewise may be placed directly from the dishmachine into self-leveling mobile units or onto dish storage trucks where they remain until needed. Thus, breakage is lowered through reduced handling, and fewer labor hours are required for this one-time handling as opposed to storing dishes in a cupboard and then having to remove them when needed.

Generally, breakage is highest on small plates, saucers, and fruit dishes; they are often stacked too high and slide off the trays or carts; handles are broken from cups; and the edges of large heavy plates may be chipped if stacked carelessly. As a means of reducing breakage through carelessness, it is advisable to make a frequent inventory of stock in circulation. Thus, the workers become aware that a constant check of breakage is being made. Often a price list of china is posted, so the total loss to the foodservice through breakage is made known to the workers. Supervisors, too, should try to determine how and where breakage is occurring. If the breakage seems unreasonably high, they should include corrective procedures in training sessions held for those workers involved in dish handling.

Replacement of china may be made as breakage demands, or provision may be made for a stockroom supply ample for the probable yearly need. In this case, replacements of the storeroom stock can be made following the annual inventory. Managers should be aware of the supply of dishes in circulation and an ample quantity made available so service is not slowed because of a lack of clean dishes when needed.

Glass Dinnerware

A popular dinnerware made by the Corning Glass Company is a basic glass in which a percentage of the sand used in its manufacture is replaced with aluminum powder. The resulting dinnerware is strong, thin, well-tempered, has a smooth surface, and is highly resistant to stains, heat, scratching, and breakage. It is available in a variety of sizes, shapes, and decorative designs on a white background. The cost is less than that of some high-quality vitrified china dinnerware. The amounts to order and the care of Corning dinnerware are comparable to china.

Plastic Dinnerware

The introduction and availability of synthetic compounds for molded dinnerware has provided competition for china and glass for use in some types of institution foodservices. The history of the development of a suitable and highly acceptable product has been a long one.

Celluloid (1868), an early synthetic thermoplastic compound and a forerunner of modern plastics, was made of cellulose nitrate and camphor. Its nonresistance to heat, high inflammability, and camphor odor and flavor made it unsuitable for dishware. In the next period of development (1908), phenol and formaldehyde were incorporated into a thermosetting compound that was capable of being molded under pressure and heat into forms that would retain their shapes under mechanical strains, at well above the temperature of boiling water. This type of compound has had wide and varied usage, but because of its odor and unattractive brownish color, its use in the foodservice industry was limited mainly to counter and serving trays. The substitution of urea for phenol made it possible to produce a white compound of great strength that would take colors well. The basic cost of this material was high; therefore, it was often made into thin dishes suitable only for picnic or limited use and that could be sold for a reasonably low price.

During World War II, it was found that melamine could be combined with formaldehyde to give a tough resin that could withstand the demands on it in high-altitude flying equipment. This type of melamine plastic compound is now used in the production of dinnerware, often called melamine ware.

The first heavy-duty dinnerware made of melamine-formaldehyde compounds contained a chopped cotton cloth filler. The products had a high tensile and impact strength but were unattractive and limited to a low color range. Compounds made by blending long-fiber, high-grade paper stock with melamine resin and colorfast pigments are used in the production of dinnerware at the present time. This material is known as alpha-cellulose-filled, melamine-formaldehyde, thermosetting molding compound, and the products made from it are available in a wide range of colors and designs.

The melamine compound undergoes chemical change in the molding process under pressure of some 3000 to 3500 pounds per square inch at 335°F, which gives the dinnerware pieces a smooth lustrous surface, resistant to scratching, chippage, breakage, detergents, and grease. Also, it is not affected by the hot water used in dishwashing. Because the color pigment is thoroughly blended with the compound before molding, there is no fading of the finished product.

The permanent decoration of melamine dinnerware is made possible by opening the press when the material has just been shaped and adding a melamine-impregnated overlay, with the lithographed side placed down onto the dish. The mold is closed and, during the cure, the overlay becomes an integral part of the base material, and the resulting product has a smooth wear-resistant and protective glaze over the design.

Factors in Selecting Plasticware. The U.S. Department of Commerce has established standards for heavy-duty type of melamine dishes. Foodservices should specify that the ware being purchased complies with Commercial Standards (CS) 173–50 that relate to thickness, resistance to acids, boiling water, dry heat, and the finished product.

Sample pieces of plasticware may be purchased or requested for testing before an order is placed. Special attention should be given to balance and to any marring of the surface by normal cutting and use.

The original cost of plastic dinnerware may be somewhat higher than for medium-weight china, but the replacement costs are estimated to be only about one-tenth of that for china. Differences in shape, density, and balance in design account in some measure for the price range in melamine ware. Competition is keen between the molding companies to produce items from this common basic material that are attractive in color and design and that meet the needs of the food industry. A price quotation from several companies should precede purchase.

The choice between melamine or the long-accepted china dinnerware may pose problems for the prospective buyer. The light weight of melamine, which is about one-third that of ordinary dinnerware, its low breakage, minimum handling noise, and attractive colors make it especially acceptable in many types of foodservice operations, especially in hospitals and other health care facilities, and in school foodservices.

One disadvantage of melamine may be staining and difficulty in cleaning, although improvements in manufacture have reduced this as a major problem. Although melamine products possess low thermal conductivity, thus eliminating the need to preheat them for service, they may present some dishwashing problems. This ware does not air dry quickly and may remain damp for storage. Bacteriological tests on such dishes, however, indicate no cause for concern over this condition.

Care. The same care in dishwashing as described for china should be followed for plasticware dishes. However, the staining of cups may require the extra step of soaking them to remove the stain. China cups may also require this step.

Many manufacturers of melamine dinnerware have successfully incorporated stain-resistant compounds into the thermosetting resin, which prevents much of the objectionable staining and adds to the life of this type of dinnerware. The development of new washing compounds and closer attention to washing techniques have eliminated the problem somewhat. Alkaline detergents are recommended for washing. Abrasives cannot be used successfully on plastic surfaces; therefore, chemical rinses must be depended on, preferably those without chlorine. Some users believe that frequent cup replacement is the answer and is justly compensated by the high resistance to breaking, chipping, cracking, and crazing under ordinary conditions, the lightness of weight, the low noise level in handling, the attractive coloring and luster, and the relatively low upkeep and replacement costs.

Amounts to Purchase. The initial stock of plastic dinnerware is comparable to that given for china. Sizes of dishes are also comparable. Another guide for amounts of dinnerware to select for the average foodservice would be an allowance of three times the number of dining room seats for items such as bread and butter plates, salad dishes, dinner plates, saucers, fruit and/or cereal dishes, and four times for cups. The amounts of these and other items would depend much on the menu pattern and other conditions mentioned earlier.

Disposable Dinnerware

One-time use items for table service are available in many different materials, including paper, plasticized paper, clear or colored thin plastics, styrofoam, and aluminum foil. They are available in a wide range of sizes, shapes, colors, and quality by weight. Some are made for use with cold foods only; others withstand considerable heat, making them suitable for oven or microwave use, and for serving hot foods.

The selection of disposable dishware over other types may well be justified, especially for any foodservice using the assembly/serve system, and for fast-food, carry-out businesses.

Factors in Selection. Consideration should be given to initial and replacement *costs* of conventional dinnerware, space and equipment for dishwashing, and labor for handling in comparison to the initial and repeat cost of paper or plastic; and its *disposal* and *acceptability* by the persons to be served. In any case, all foodservices should have ready access to some disposable dishware for times of emergency.

Disposal of large quantities of "disposables" poses problems in some situations. The availability on the premises of a large trash compactor is a necessity to handle this waste without undue bulk.

Quantities to purchase are determined by the amount of space to store the large cartons of paper or plastic goods, the relative closeness to a marked supply, and, of course, the number of persons to be served in a given period of time and the menu items offered.

Silver Tableware

Silverware. Quality silverware has been used in discriminating foodservices because of the demands for and interest of the residents or clientele in attractive service. It lends dignity and charm to dining tables, perhaps because of the association of the idea that silver, a precious metal, is found where people know and appreciate gracious, comfortable living. Some knowledge of the manufacture of silverware will help the foodservice manager make a wise decision in the selection of this item.

"Blanks" serve as the basic forms for flatware as well as hollow ware. They are made of 18% nickel-silver, a metal that gives the utensil the needed strength and resistance to bending or twisting to which institutional silverware is often subjected. The design, shape, and thickness or weight of the blanks should be conducive to heavy wear and beauty.

Nine pounds per gross is the standard weight of blanks of ordinary teaspoons sold for public service. The principal weights of blanks used are heavy, 10½ pounds; regular, 9 pounds; medium, 7½ to 8 pounds. The 9-pound blanks are desirable for hospital tray service, whereas the 10½-pound patterns may be advisable for heavy-duty silverware for certain commercial restaurants and cafeterias. The weight of the blank used influences the price of the silverware.

Flatware blanks are stamped, graded, and rolled until they are the corresponding size of forks or spoons. They are then placed in various presses, and the fork tines or spoon bowls are shaped. In the next step, they are struck with the pattern die, after which the edges are trimmed and smoothed down so that the articles resemble the finished products. Forks should have well-designed tines, durable and heat treated, to give maximum strength, and both forks and spoons should have heavy reinforced shanks to give the best wearing qualities. After being cleaned and polished, the articles are ready for plating with the silver.

The steps in the manufacture of knives prior to plating differ from those in the making of blanks for forks and spoons. The 18% nickel-silver base was found to produce a blade that bent easily and refused to take an edge sharp enough for practical use in cutlery. Stainless steel has become widely used for knife blades, and noncorrosive alloys have been made that prove satisfactory as the base for solid-handle knives that are to be plated. The popular hollow-handle knife, made with the 18% nickel-silver as the base of the handle, has been largely replaced by the one-piece stainless steel knife with the plate handle. An improvement in the design of knives was the change in style from the long blade, short handle to the short blade, long handle type that permits the user to press down with the forefinger on the back of the handle instead of on the narrow edge of the steel blade.

For better qualities of flatware there is an intervening step between the making of the blank and its plating. Reinforcements of an extra disk of silver are made on blanks at the point of greatest wear: the heel of the bowls of spoons and the base of fork tines (Figure B.32). Such treatment is referred to as *overlay, sectional plate,* or *reinforced plate,* and increases many times the length of wear.

The plating of silverware is accomplished by electrolysis. Pure silver bars or ingots are placed around the side of a plating tank, and the articles to be plated are hung in the solution in the tank. By means of an electric current, the silver passes

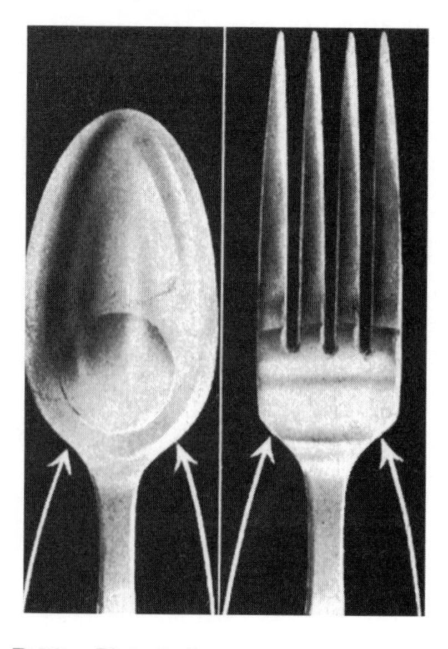

Figure B.32 Plated silver should be reinforced on the points of greatest wear.
Courtesy of Oneida, Ltd.

from the bars and is deposited on the blanks, the length of time and the strength of the current determining the amount of silver deposited.

After the articles are removed from the plating tank, they are sent to the finishing rooms. Better grades are burnished, or rubbed under pressure with a round pointed steel tool, to harden and smooth out the plate. It is then polished and colored. The better qualities of silverware are given extra burnishing. The various finishes are butler or dull finish, hotel finish or medium bright, and bright. The finishes are obtained by using different types of buffs and polishing compounds and by carrying the polishing process to different degrees.

The plating of institutional silverware is heavier than for the silverware generally used in the home, the most common institution ware being known as *triple plate,* or three times full standard. In triple plate, 6 ounces of pure silver have been applied to 1 gross of teaspoons, with other items in proportion—for example, tablespoons with 12 ounces of silver to the gross. A much lighter plating known as *full standard* carries a deposit of 5 ounces of pure silver to the gross of tablespoons and only 2½ ounces to the gross of teaspoons. *Half standard,* as its name implies, carries half the amount of silver deposit of full standard. Full standard plate quality is the lightest grade recommended for use in institutions.

The leading manufacturers of silver plate generally make, under their own trade name, a better quality of silver plate than those noted earlier. An example of such silverware is heavy hotel teaspoons, which weigh 11½ to 12 pounds per gross. An extra-heavy plate deposit is used on 10½-pound blanks in their production. Usually the silver overlay on the tips and backs of bowls and tines is invisible on any 10½- and 11-pound qualities. The heavy finely finished metal blanks, the heavy plating standard, and the fine finish of this quality of silverware make the initial cost greater than the ordinary commercial grades of plate, but the cost is offset by the long-wearing qualities and satisfactory service of the various items.

Hollow Ware. Silver hollow ware items such as serving bowls, platters, sugar bowls, creamers, pitchers, teapots, and coffee pots are made from the same materials as flatware and are plated in a similar manner to varying qualities. The bodies of the various items are die shaped, and the several pieces for each are assembled and hand soldered by expert craftspeople before plating. The quality of materials and workmanship and the design determine cost.

Features to consider in selecting hollow ware include the following: Sharp corners are to be avoided; short spouts are easier to clean than long ones; simple designs are usually more pleasing than ornate ones and are easier to clean. However, plain silverware can become badly scratched with ordinary handling; hence, a pleasing, simple design that breaks the smooth surface may be more practical than plain silver. Simplicity is always the keynote of good taste.

Standard designs and patterns are often made individual by stamping or engraving the name, crest of the organization, or a special decorative motif on the otherwise plain surface. If silverware is to be stamped, the stamping should be done on the back of the item before it is plated. The name of the manufacturer is stamped on the bottom.

Silver hollow ware may seem an extravagance, but when the cost is considered over a period of years, it may be more economical than china or glassware. Furthermore, the satisfaction and prestige gained through its use are not to be discounted.

Care. The care of silverware has much to do with its appearance and wearing qualities. Careful handling prevents many scratches. The following procedure is suggested for cleaning silverware and keeping it in good condition: sort, then wash in a machine to which has been added the proper cleaning compound, at a temperature of 140°F to 150°F, and rinse thoroughly. A final dip in a solution with high-wetting properties prevents spotting of air-dried silver. It is advisable to presoak flatware or wash immediately after use. If washed in flat-bottom racks, the silver should be scattered loosely over the rack surface, sorted after washing, placed into perforated dispensing cylinders, and rewashed to ensure sanitization. If silver is sorted into cylinders before washing, the handles of the utensils are down so that all surfaces of knife blades, fork tines, and bowls of spoons are subjected to the wash and rise processes. Care must be taken not to overcrowd the containers. The washed silver is left in the cylinders to dry. This system is convenient, especially in self-service units, because clean dispensing cylinders may be inverted over those used in washing, turning the silver upside down so handles are up, and placed on the counter without handling of the clean silver.

Tarnishing of silver occurs readily when exposed to smoke or natural gas, or when it comes in contact with rubber, certain fibers, or sulphur-containing foods. Detarnishing is accomplished quickly and easily by immersing the silver, placed in a wire basket, into a solution of water plus a cleaning compound containing trisodium phosphate in an aluminum kettle reserved for this purpose.

The tarnish (oxide) forms a salt with the aluminum, and can be removed through a mild electrolytic action. The cleaning compound also cuts and dissolves any grease or dirt on the silver. The silver is left in the solution *only long enough to remove the tarnish.* It is then rinsed in boiling water and dipped in a solution of high wetting qualities. Burnishing machines are used for silver polishing in large foodservices. Care must be taken to see that the machine is not overloaded and that there are enough steel shots of various sizes in the barrel of the machine to be effective in contacting all surfaces to be polished. Also, the right amount of proper detergent must be added to the water in the burnishing machine to produce the required concentration of the solution. There can be no set rule about the frequency of detarnishing and polishing; each foodservice must set up its own standards.

Stainless Steel Tableware

Stainless steel tableware has gained wide acceptance for heavy-duty tableware in many foodservices. The flatware is fairly inexpensive, is highly resistant to heat, scratches, and wear, and will not rust, stain, peel, chip, or tarnish. It stays bright indefinitely with ordinary washing and offers a wide selection of attractive designs from which to choose. Flatware in stainless steel is available in light, standard, and heavy weights. Cheap quality stainless steel utensils have appeared on the market, but they are not really suitable for most establishments. These are made from rolled sheets of the metal and are die-stamped into desired shapes. The resulting pieces are the same thickness throughout and have poor balance, and fork tines and bowls of spoons are too thick to pick up food easily. This quality should be avoided.

Good quality ware is rolled and tapered as needed to give good balance and to be comfortable in the hand. A test for good balance is to place a fork or a spoon at the base of its bowl or tines on an index finger: the utensil should balance equally between handle and bowl or tines. In poor quality, the bowl or tines will overbalance the handle and the utensil will fall off the finger.

Another consideration in selecting stainless steel tableware is the size and shape of the handles. Older persons particularly find it difficult to hold a slim handle and much prefer a wider, easier-to-grasp shape and size.

Water pitchers and individual teapots of stainless steel are considered a lifetime investment, although the initial cost is high in comparison to these items in ordinary glass or pottery. The same methods of sorting, washing, and drying are recommended for stainless steel tableware as for silver.

Amounts of Tableware to Purchase

The menu to be offered determines what items of tableware the foodservice must supply. It is more difficult to calculate the quantities of each piece to stock. As with dinnerware, the trend in use of flatware is toward as few different pieces as possible. For instance, knives and forks of dessert size can be used for many purposes and are usually preferred to knives and forks of dinner size. Dessert spoons can be used for soup and serving spoons as well as for certain desserts.

A good quantity estimate of flatware for cafeterias is twice the seating capacity for all the flatware items required. Should the dishwashing facilities be limited and the turnover of patrons rapid, this quantity might need to be increased to three or even four times the seating capacity. For table or tray service, 3 teaspoons per cover, 3 forks, using a dessert fork for all purposes, and 2 knives per cover usually are sufficient. All other items are estimated on the basis of 1½ per cover, or according to the needs as for banquet or special party service. This may call for limited quantities of specialized items such as oyster or fish forks, bouillon spoons, butter spreaders, or iced tea spoons.

On the basis of total investment in tableware, an estimate found to be about average is that 2.5% of the budget for all foodservice equipment is required.

Glassware

Glassware is classed as lead or lime glass, depending on the use of lead or lime oxide in the manufacturing process. Lead glass is of better quality, is clearer, and has more brilliance than lime glass, which is less expensive. Articles of glass are made from a molten compound, formed by blowing them into shape by machine or hand processes or by pressing molten glass into molds by means of a machine. The blowing method is the more expensive and produces a thinner glass of finer texture, higher luster, and clearer ring. Hand-blown lead glass is superior to all other glassware because of its brilliance, light weight, and variety of styles. Lime-blown glassware possesses these characteristics in a lesser degree, is less expensive, and is used extensively in institutions. It is usually machine blown. The style, color, and decoration determine the cost of manufacture of blown-glass articles.

Pressed lime glass is used in many institutions. It is serviceable, and better qualities of it are comparatively free from bubbles and cloudiness. Moreover, it is relatively inexpensive and can be obtained in many styles. A good quality of glassware should be selected for the institution, regardless of whether pressed glass or blown glass is to be used. Desirable characteristics are clearness, luster, medium weight, freedom from such defects as marks and bubbles, and a clear ring. Also, it should be designed so that it is not easily tipped over.

Glassware must pass boiling and shock tests without showing signs of corrosion, chipping, scumming, or cracking in order to meet federal specifications. In the boiling test, articles are suspended for 6 hours in boiling water in a closed container with vent. The shock test is made by immersing articles in tap water at 18.5°C ± 2½° (65°F ± 5°) for a 10-minute period, then suddenly transferring them into boiling water. This procedure is repeated five times. Not all the glassware sold meets federal specifications, and there is no labeling to indicate which, if any, is of that quality.

The sizes of glassware used most commonly in institutions are glasses of 5- or 6-ounce capacity for fruit juice, glasses of 9- or 10-ounce capacity for milk and water, and glasses of 12- or 14-ounce capacity for iced tea. A wide range of sizes and shapes must be stocked for bar service.

Goblets and footed dessert dishes are other items of glass selected for some foodservices. The portion size of specific menu items will determine the capacity size required for these items.

Care. Glassware to be washed is sorted and often washed in a separate dishwashing machine from that used for other dishes, or in a glass-washer built for that purpose. If glassware must be washed in the same machine used for dishes, it should be segregated and washed first while the water is entirely free from grease and food particles, or left until after the dishes are finished and the soiled water is replaced by clean. In either case, with a rack machine, all items to be washed are placed upside down in racks after they are transported from the dining room and remain in the same racks to wash and drain. They are then loaded onto carts for transport to the point of storage or use without rehandling.

Glassware should be under constant scrutiny to maintain in service only those pieces that are not chipped, cracked, clouded, or scratched in appearance. Filmed glasses may be caused by low rinse pressure and volume, a rinse cycle that is too short, nonaligned spray jets, and a hard-water precipitate. Tea stains may be removed by using a chlorinated detergent in the glass washing machine. Water spots may be caused by slow drying or the need for softened rinse water. The effect of an otherwise attractive dining service may be spoiled by damaged or poorly washed glassware on the tables.

The breakage of glassware in institutions is often high and results from careless handling and storage, choice of improper designs for heavy service wear, use of poor quality of glassware, and subjection to high temperature during washing. The shape of the glass has much to do with the breakage anticipated. Straight side tumblers that can be stacked are a decided breakage hazard, and there are patented shapes available that make it impossible to stack tumblers. Many styles of glassware curve in slightly at the top so that the edges do not touch when they are set down together, the contact coming at a reinforced part away from the edge of the glass. Other styles have reinforced edges at the top, advertised as making them more highly resistant to chipping. This feature is also found around the foot of some stemmed ware.

Amounts to Purchase. The amount and kind of glassware to supply vary as for tableware and dinnerware. Choice is based primarily on the menu, type of service, seating capacity and rate of turnover, dishwashing facilities, skill of persons employed in dish handling, and whether scheduling is continuous or intermittent. However, since glassware is more fragile than other tableware, foodservices should have an ample stock on hand of the most frequently used items: tumblers, fruit juice glasses, and sherbet or dessert dishes (if glass ones are used). A suggested rule for quantities to purchase is two pieces for each person to be served; one piece in use and one-half in the dishroom, plus 50% of that total in reserve in the stock room.

Cloth Table Covers

In some localities, tablecloths may be rented from local laundries, thus relieving the foodservice of purchasing and storing this item. However, if cloths are to be purchased, there are many materials from which to choose: pure linen, union, rayon, cotton, mercerized cotton, linenized damasks, and polyester-cotton blends.

Cotton fibers may be used in combination with linen or rayon in the union damasks to produce durable and satisfactory table coverings and napkins. Rayon and cotton blend table napkins are highly resistant to wear and often superior to all-cotton or all-linen napkins in appearance and breaking strength. Cotton is used alone in plain cotton, mercerized, or linenized fabrics. After being woven, the last two fabrics are treated such that a permanent finish is produced to give the cloth characteristics similar to linen. The wearing quality of cotton fabrics is better than that of linen, less loss in strength occurs through laundering, and cotton does not lint. Linen gives satisfaction in use, is attractive, and lintless.

Because of the high initial and maintenance costs of both linen and cotton cloths, those of a 50–50 blend of polyester and cotton with a no-iron finish are rapidly replacing the former in other than the most sophisticated foodservices. Polyester yarns are used in the making of lace cloths as well as the plain woven ones.

Tablecloths may be purchased in white, in colors, or in white with colored borders or designs. Colored linen is popular as place mats and luncheon cloths for breakfast and luncheon services, as well as to help create "atmosphere" in many dining rooms. Fabrics may be purchased by the yard and made up for the specific size tables of the individual establishment, or the cloths may be purchased ready-made.

The size and shape of the tables determine the sizes of cloths needed. The cloth should be large enough to hang 7 to 12 inches below the table top at both sides and ends, with allowances made for shrinkage according to the material selected. The usual sizes of tablecloths are 52 by 52 square, 60 by 80 inches, 67 by 90 to 102 inches or longer, depending on the length of banquet tables. Some places use a table-size top over the regular cloth, which allows for frequent changes and reduced laundry costs.

Common sizes for cloth dinner napkins in institutions are 18, 20, or 22 inches square.

Paper Place Mats and Napkins

The range of colors and designs available in paper products is so large that selection of appropriate covers should be relatively easy. Size is dependent on the size of tray for tray service. For use on table tops, 12- × 18-inch mats provide generously for each cover, although the 11- × 14-inch size is frequently used. Often the name, logo, or design of the foodservice is imprinted on the mat and/or napkin, which serves as good advertising for the establishment.

Glossary

Accident: An unintentional incident that results in injury, loss, or damage of property.

Aesthetics: Having an appreciation for beauty. In foodservice applies to food presentation.

A la carte menu: Menu where food items are listed and priced separately.

Americans with Disabilities Act (ADA): Prohibits discrimination against qualified persons with disabilities in all aspects of employment.

Artificial intelligence (AI): Information technology that includes robotics, expert systems, and neural networks.

As purchased (AP): Refers to weight before trimming or removal of undesired parts.

Assembly/serve system: Also known as the "kitchenless kitchen," fully prepared foods are purchased, stored, assembled, heated, and served.

Banquet: An elaborate, intensive feast where the service and menu are preset for a given number of people for a specific time of day.

Baseline measurements: In total quality management, data against which progress toward goals may be assessed.

Behaviorally anchored rating scales (BARS): Performance appraisal scales that contain specific behaviors identified for each performance level for each job category.

Benchmarking: The total quality management measurement tool that provides an opportunity for a company to set attainable goals based on what other companies are achieving.

Biological hazard: The threat to food safety caused by contamination of food with pathogenic microorganisms.

Brainstorming: A technique for generating ideas about problems and opportunities for improvement.

Brand: A particular make of a good or product usually identified by a trademark or label.

Branded concept: A complete marketing package that communicates a recognized and consistent brand identity to the customer.

Branding: The use of nationally or locally labeled products for sale in an existing foodservice operation.

Break-even point: Level of activity such as sales volume where total revenues and total expenses are equal.

Brix: The percent of sugar by weight in a sugar solution. Expressed as degrees brix and usually applied to canned fruits packed in syrup.

Broker: A wholesaler who does not assume ownership of goods, but whose responsibility is to bring buyers and sellers together.

Budget: A prediction of the amount of money needed for a specific period of time.

Budget variance: The difference between actual expenditures and budgeted amounts.

Budgeting: Fiscal planning, accounting, and controlling.

Case: An individual that experiences illness after eating an incriminated food.

Centers for Disease Control and Prevention: Agency of the U.S. Department of Health and Human Services charged with protecting the nation's public health.

Chef de rang: The principal food server in French-style table service. Responsible for all table-side preparation.

Chemical hazard: The threat to food safety caused by contamination of food with chemical substances such as cleaning compounds and pesticides.

Chronocyclegraph: A photographic technique to show motion patterns of hands in performing rapid repetitive operations.

Classical theory: A historical theory of management that focuses on tasks, structure, and authority.

Closed shops: Illegal shops that obligate an employer to hire only union members.

Code: A collection of regulations.

Collective bargaining: An obligation to meet and discuss terms with an open mind.

Commis de rang: An assistant in French-style table service. Carries the food to the table and removes dishes as guests complete the courses.

Commissary system: A central production kitchen or food factory with centralized food purchasing and delivery to off-site facilities for final preparation.

Communicable disease: An illness that is transmitted from one person to another through direct or indirect means.

Computer-aided design (CAD): Software programs used to assist in the design and layout of a facility.

Conceptual skills: Understanding and integrating all the activities and interests of the organization toward a common objective.

Consideration: Behavior that indicates friendship, mutual trust, respect, and warmth between the leader and the work group.

Contamination: The unintended presence of harmful substances such as microorganisms in food and water.

Contingency approach: Holds that managerial activities should be adjusted to fit the situation.

Conventional system: Raw foods are purchased, prepared on site, and served soon after preparation.

Cook/chill method: Food production method in which food is prepared and cooked by conventional or other methods, then chilled and refrigerated for use at a later time.

Cook/freeze method: Food production method in which food is prepared and cooked by conventional or other methods, then frozen for use at a later time.

Coordinating: The functional activity of interrelating the various parts of a process to create a smooth work flow.

Cost: The resources expended or sacrificed in order to achieve an objective.

Coved: A curved rather than an angled joining, such as at a floor and wall joining.

Critical control point: A food safety and self-inspection system that identifies potentially hazardous foods and proper handling procedures.

Critical incident: Records are kept of critical incidents for each employee to be used for performance appraisal purposes.

Critical limit: A specific criterion that must be met for each preventive measure identified for a critical control point.

Cross-contamination: The transfer of harmful microorganisms from one item of food to another via a nonfood surface such as human hands, equipments, or utensils. May also be a direct transfer from a raw to a cooked food item.

Cryogenic: Very low temperature.

Cycle menu: A carefully planned set of menus that is rotated at definite time intervals.

Danger zone: The temperature range between 45°F and 140°F in which most bacteria grow and multiply.

Demographic segmentation: Dividing or segmenting a market into groups of people based on variables such as age, sex, income, education, religion, and race.

Detergent: Cleaning agents, solvents, or any substances that will remove foreign or soiling material from surface.

Directing: The continuous process of making decisions, conveying them to subordinates, and assuring appropriate actions.

Disseminator role: Transmitting information gathered outside the organization to members inside.

Disturbance handler: One who handles unexpected change.

Du jour menu: Menu of the day.

Edible portion (EP): Weight after trimming, preparation, and cooking.

Equal employment opportunity (EEO): The umbrella term that encompasses all laws and regulations prohibiting discrimination and/or requiring affirmative action.

Equal Employment Opportunity Commission (EEOC): Provides guidance to management for compliance with EEO statutes.

Emerging pathogen: A pathogen that is increasingly recognized as causing foodborne illness.

Entrepreneurial role: Initiating controlled change in the organization to adapt and keep pace with changing conditions in the environment.

Ethics: The science of morals in human behavior.

Expectancy theory: Theory that states motivation is a function of the person's ability to accomplish the task and his or her desire to do so.

Expert systems: Software products designed to replace specific tasks performed by a specific person.

Extended care facility: An organization concerned with the provision of suitable care for nursing the sick or elderly for long periods of time.

Feedback: Information on how operations worked or failed, or how they should be changed or modified.

Figurehead role: Performing duties of a symbolic, legal, or social nature because of one's position in the organization.

First-in/first-out (FIFO): An inventory method in which stock is rotated to assure that items in storage are used in the order in which they are delivered.

Fish diagram: A cause-and-effect diagram used to focus on the different causes of a problem that allows for grouping and organizing efforts to improve a process.

Flowchart or flow diagram: A written sketch of the movement of people and/or materials from one step or process to the next, such as on a floor plan of a kitchen with arrows drawn to show routes of workers and/or supplies in logical sequence through the facility.

Flow of food: The route or path food follows through a foodservice or food processing operation.

Foodborne illness: Illness that results from ingesting foods containing live microorganisms.

Foodborne infection: Illness that results from ingesting foods containing live microorganisms.

Foodborne intoxication: Illness that results from eating food containing toxins produced by bacteria or molds present in the food.

Foodborne toxin-medicated infection: Illness that results from eating a food containing pathogenic organisms. Once ingested, the microorganisms produced toxins.

Forecasting: A prediction of food needs for a day or other specific period of time.

Garnish: To decorate food, or a decoration used to decorate food.

Geographic segmentation: Dividing a market into different units based on variables such as nations, states, regions, cities, or neighborhoods.

Grade: The rating of quality of meats, poultry, and eggs. Grading is a voluntary service.

Gray water: A biological, chemical, or physical property of a food that threatens the safety of that food.

Hazard: A biological, chemical, or physical property that may cause an unacceptable consumer health risk.

Hazard Analysis and Critical Control Points (HACCP): A food safety and self-inspection system that identifies potentially hazardous foods and proper handling procedures.

Human relations theory: A historical theory of management that views the organization as a social system and recognizes the existence of the informal organization.

Human skills: Understanding and motivating individuals and groups.

Infection control: Specific procedures to prevent the entrance of pathogenic organisms into the body.

Initiating structure: The relationship between the leader and the members of a work group.

Integrated solid waste management system: The complementary use of a variety of waste management practices to safely and effectively handle the municipal solid waste stream with the least adverse impact on human health and the environment.

Integrated staffing: An orderly plan for moving people into, through, and eventually out of the organization.

Interdependency: A key concept in systems theory that the elements of a system interact with one another.

International Organization for Standardization standards (ISO 9000): A series of five international standards that describe elements of an effective quality system.

Interstate commerce: Financial transactions (buying and selling of goods) carried on between states.

Intrastate commerce: Financial transactions (buying and selling of goods) carried on within state boundaries.

Inventory: A detailed and complete list of goods in stock.

Invoice: A list of goods shipped or delivered. Includes prices and service charges.

Job description: An organized list of duties, skills, and responsibilities required in a specific position.

Job specification: A written statement of the minimum standards that must be met by an applicant for a particular job.

Just-in-time (JIT) inventory control: An inventory management system that links suppliers and customers to minimize total inventory-related costs.

Key result areas: Areas of operation used to quantify quality standards.

Labor-Management Relations Act: Prevents unions from coercing employees to join, outlaws union and closed shops, and provides for collective bargaining. Also known as the Taft-Hartley Act.

Labor-Management Reporting and Disclosure Act: Contains a bill of rights for union members that requires financial disclosures, prescribes union officer election procedures, and provides civil and criminal remedies for financial abuses. Also known as the Landrum-Griffin Act.

Layout: The arrangement of equipment on a floor plan.

Leader role: Establishing the work atmosphere within the organization and activating subordinates to achieve organizational goals.

Liaison role: Establishing and maintaining contacts outside the organization to obtain information and cooperation.

Line and staff: Support and advisory activities are provided for the main functions of the organization.

Linking processes: Methods used to unify a system.

Management information system (MIS): Computerized data processing to facilitate management functions.

Managerial grid: A graphical representation of management styles based on the relationship between concern for people and concern for production.

Market: The set of actual or potential customers for a product or service.

Market, local: Suppliers within close proximity to the buyer; includes farmer's markets, nearby supermarkets, and local fish and produce growers.

Market segment: A group of customers, actual or potential, who behave or react similarly in the marketplace.

Market segmentation: The process of placing customers into groups of like characteristics such as by demographics or geographic location.

Marketing: An activity directed at satisfying the needs and wants of customers.

Marketing channel: The food processing and distribution system, beginning with the grower of raw food products and ending at the final customer or point of consumption.

Marketing concept: The management philosophy that holds that determining and satisfying the needs and wants of customers are the primary objectives of the organization.

Marketing cycle: A recurrent series of activities designed to meet the wants and needs of customers. The cycle is driven by customer feedback.

Menu: A detailed list of foods to be served at a meal or, in a broader sense, a total list of items offered by a foodservice.

Menu pattern: An outline of food to be included in each meal, and the extent of choice at each meal.

Merchandising: A marketing term that describes sales promotions. It is a comprehensive managerial function that may include market research, product development, advertising, and selling.

Micromotion study: A technique whereby movements of the worker are photographed and recorded permanently on film.

Misbranded: A food product whose label either does not include the information mandated by law or the label information is misleading.

Mission statement: A summary of an organization's purpose, customers, products, and services.

Modular: A module is a standard or unit of measure. Modular is that size to which all units, such as pieces of equipment, are proportioned; compatible in size to fit together.

Monitor role: Collecting all information relevant to the organization.

Municipal solid waste: The solid waste produced at residences, commercial, and industrial establishments.

National Labor Relations Act: Established the National Labor Relations Board (NLRB), regulates the right of employees to join a union, and provides for collective bargaining. Also known as the Wagner Act.

Negotiator role: Assigned to a person who deals with individuals and other organizations.

Neural networks: Advanced computer technology used predominantly in medical research.

Nonselective menu: A menu that offers no choice of food items.

Occupational Safety and Health Act (OSHA): Requires employers to furnish employment free from safety hazards.

Open system: A system that interacts with external forces in its environment.

Operations: The work performed to transform inputs.

Organization chart: A graphic representation of the basic groupings and relationships of positions and functions.

Organizing: The function of management that involves the development of the formal structure through which work is divided, defined, and coordinated.

Outbreak: An incidence of foodborne illness that involves two or more people who ate a common food, which has been confirmed through laboratory analysis as the source of the outbreak.

Pareto charts: Bar charts in which the strategy is to work on the tallest bar or problem that occurs most frequently.

Path-goal theory: Functions of a leader should consist of increasing personal rewards and clarifying pathways for goal attainment for subordinates.

Pathogen: A disease-causing microorganism.

Pathway chart: A scale drawing of an area on which the path of the worker or movement of material during a given process may be indicated and measured. Also called a flow diagram.

Pellet: A preheated metal disk used to maintain the temperature of an individual portion of plated hot food.

Perpetual inventory: A running record of the balance on hand for each item of goods in a storeroom.

Physical hazard: A threat to food safety by the presence of any particle not typically part of that food.

Physical inventory: An actual account of items in all storage areas.

Pilfer: To steal in small quantities at a time.

Planning: The function of management that involves developing, in broad outline, the activities required to accomplish organizational objectives and the most effective ways of doing so.

Point-of-sale (POS) terminals: Cash registers that interact with computers to provide data on items sold.

Potentially hazardous foods: Foods that are more likely than others to be implicated in an outbreak of foodborne illness.

Primary market: The basic source of food supply including growing regions and processing plants.

Procedures: Detailed guidelines for planned activities that occur regularly.

Process chart: A form on which to record and analyze the breakdown of a job.

Procurement: The process of obtaining. Used synonymously with the term *purchasing*.

Production schedule: A detailed list of food items to be produced for the current day's menu plus any advance preparation needed.

Productivity: A measure of the output of goods or services in relation to the input of resources.

Program evaluation and review technique (PERT): Used to plan and control the functions of management.

Prospectus: A written plan for a building/designing project that details all elements of the situation being planned; used as a guide and communication tool to aid clear understanding by all who are involved in the planning.

Psychographic segmentation: Dividing a market into groups based on variables such as social class, lifestyle, or personality traits.

Punch list: A detailed checklist that would reveal any defective, substitute, or inferior equipment so that corrections could be made prior to an opening or training date for a new or renovated facility.

Purchase order: Written requests to a vendor to sell goods or services to a facility.

Purchasing: The act of buying.

Quality: In its broadest sense, anything that can be improved.

Quality assurance: Assuring that the quality of the product is satisfactory, reliable, and yet economical for the customer.

Quality control: A system of means to produce goods or services economically that satisfy customer requirements.

Quality control circle: A small group that voluntarily performs quality control activities in the workplace.

Quality of work life (QWL): An approach to management that takes into consideration the quality of human experiences in the workplace.

Quota: An almost always illegal fixed, inflexible percentage or number of positions that an employer decides can be filled only by members of a certain minority group.

Quotation: An amount stated in current price for a desired product or service.

Ready-prepared system: Also known as cook/chill or cook/freeze systems, foods are prepared on site, then chilled or frozen, and stored for reheating at a later time.

Receiving: In foodservice, it is the point at which a foodservice operation inspects product and takes legal possession of the product ordered.

Rechaud: A small heater placed on a small table. Used for table-side temperature maintenance of hot foods.

Recommended Daily Allowance (RDA): Levels of intake of essential nutrients considered to be adequate to meet known nutritional needs of practically all healthy persons.

Regulation: A written government control that has the power of force or law.

Reporting: Keeping supervisors, managers, and subordinates informed concerning responsibility through records, research, inspection, and other methods.

Requisition: A list of desired products. Generally originates in a department and is submitted to the purchasing department.

Resource allocation: Making decisions concerning priorities for utilization of organizational resources.

Rethermalization: Returning to eating temperature.

Rinsing agent: A compound designed to remove and flush away soils and cleaners so they are not redeposited on surfaces being washed.

Risk: An estimate of the likelihood or probability of occurrence of a hazard.

Robotics: Computer-driven devices that are reprogrammable and multifunctional, and are designed to perform specific tasks.

Rules: Written statements of what must be done.

Saponify: To turn fats into soap by reaction with an alkali.

Scientific management: Popular theory in the early 1900s, concerned primarily with the "best" method and "right" wage for a job.

Secondary market: The physical, functional unit of the marketing system in which products are accepted from the primary markets and distributed to buyers.

Selective menu: A menu that includes two or more food choices in each menu category such as appetizers, entrees, vegetables, salads, and desserts.

Semiselective menu: A menu that includes one or more food choices in at least one menu category.

Sequestering: The isolating of substances such as a chemical ion so it cannot react. In foodservice this is a desired characteristic of polyphosphate detergents to bind lime and magnesium of hard water. Results in clear solution with insoluble precipitates.

Service: An intangible good or product that is offered for sale. A service is intangible and does not result in the ownership of anything.

Sexual harassment: Unwanted sexual advances, or visual, verbal, or physical conduct of a sexual nature.

Single-use menu: A menu specially planned and used only once, usually for a holiday or catered event.

Single-use plans: Plans that are to be used only once or infrequently.

Situational management: Effectiveness of a leader is a function of the individual leader, the subordinate, socioeconomic interests, and situational variables.

Skills matrix system: An organized plan that allows employees to plan their own professional growth within the organization.

Social Security: A federal program of insurance to protect wage earners and their families against loss of income due to old age, disability, and death.

Sociotechnical system: A program to improve work processes that begins with an analysis of the existing flow diagram focusing on improvements in technical systems.

Source reduction: The design and manufacture of products and packaging with minimum toxic content and minimum volume of material and/or a longer life.

Specification: A detailed description of a product, stated in terms that are clearly understood by both buyer and seller.

Spokesperson role: Transmitting information from inside the organization to outsiders.

Staffing: The personnel function of employing, training, and maintaining favorable work conditions.

Standard: A measure used to define and evaluate compliance with a regulation.

Standard of identity: Defines what a food product must contain to be called a certain name.

Standard of quality: Describes the ingredients that go into a product; applies mainly to canned fruits and vegetables.

Standard operating procedures: See *Procedures.*

Standardized recipe: A recipe that has been carefully tested under controlled conditions. A recipe is considered standardized only when it has been tried and adapted for use by a specific foodservice.

Standards of fill: Regulate the quantity of food in a container.

Standing plans: Policies and procedures that do not change over time.

Static menu: A menu that is used each day such as a restaurant-style menu.

Statistical process control (SPC): A program to improve work processes that uses statistics to establish control limits for a process.

Strategic planning: Decision making based on environmental conditions, competition, forecasts, and resources available.

Subsystems: The interdependent parts of a system.

Subsystems analysis: A method of problem solving or decision making.

Suspension: The action of a cleaning agent required to hold the loosened soil in the washing solution so it can be flushed away and not redeposited.

Synergism: The combined effects of individual units exceeds the sum of the individual effects. This principle is applied to certain types of insulated patient meal trays.

System: A set of interdependent parts that work together to achieve a common goal.

System inputs: The necessary resources for a system to operate.

System outputs: The finished product or services.

Systems management: The application of systems theory to managing.

Systems theory: Viewing the organization as a whole made up of interdependent parts.

Table d'hôte menu: Menu that offers a complete meal at a fixed price.

Target market: A market segment identified by the seller as having specific wants or needs. Once the segment is identified, the seller develops products to satisfy their wants and needs.

Technical skills: Skills that allow one to perform specialized activities.

Theory X: The traditional set of managerial assumptions that employees have an inherent dislike of work and will avoid it if possible.

Theory Y: The attitude held by the emerging manager of the 1960s and 1970s that employees, under the proper conditions, will seek and accept responsibility, be motivated to achieve organizational objectives, and will exercise creativity and imagination in solving organization problems.

Therbligs: The 17 subdivisions of basic hand movements employed in job performance.

Total quality management (TQM): A management philosophy combining principles of quality assurance, participative management, and QWL.

Toxin: A poison specifically produced by a microorganism.

Union shop: An illegal shop that requires an employee to become a member of a union in order to keep a job.

Unity of command: Each person should be responsible to only one superior.

Vendor: A seller. A source of supply.

Vermin: Pests that infect foodservice operations. Includes rodents, insects, and birds.

Vision: The organization's view of the future.

Wetting: The action of a cleaning agent to penetrate between particles of soil and between the layers of soil and a surface to which the soil adheres. This action reduces surface tension and makes penetration possible.

Wholesale: The sales of products in large quantities to be retailed or resold by others.

Wholism: The systems theory doctrine that the whole of an entity is more than the sum of its parts.

Work sampling: Random sampling of work to measure the activities and delays of people or machines to determine percentage of productive time.

Work schedule: An outline of work to be performed, procedures to be used, and time schedule for a particular position.

Workmen's compensation insurance: A program administered by the state in which premiums are paid by the employer to cover employee accidents.

Index